Lecture Notes in Electrical Engineering

Volume 1572

The book series *Lecture Notes in Electrical Engineering* (LNEE) publishes the latest developments in Electrical Engineering—quickly, informally and in high quality. While original research reported in proceedings and monographs has traditionally formed the core of LNEE, we also encourage authors to submit books devoted to supporting student education and professional training in the various fields and applications areas of electrical engineering. The series cover classical and emerging topics concerning:

- Communication Engineering, Information Theory and Networks
- Electronics Engineering and Microelectronics
- Signal, Image and Speech Processing
- Wireless and Mobile Communication
- Circuits and Systems
- Energy Systems, Power Electronics and Electrical Machines
- Electro-optical Engineering
- Instrumentation Engineering
- Avionics Engineering
- Control Systems
- Internet-of-Things and Cybersecurity
- Biomedical Devices, MEMS and NEMS

For general information about this book series, comments or suggestions, please contact leontina.dicecco@springer.com.

To submit a proposal or request further information, please contact the Publishing Editor in your country:

China

Jasmine Dou, Editor (jasmine.dou@springer.com)

India, Japan, Rest of Asia

Swati Meherishi, Editorial Director (Swati.Meherishi@springer.com)

Southeast Asia, Australia, New Zealand

Ramesh Nath Premnath, Editor (ramesh.premnath@springernature.com)

USA, Canada

Michael Luby, Senior Editor (michael.luby@springer.com)

All other Countries

Leontina Di Cecco, Senior Editor (leontina.dicecco@springer.com)

**** This series is indexed by EI Compendex and Scopus databases. ****

H. C. Nagaraj · Ajit Danti · Sreekanth Rallapalli
Editors

Advancements in Artificial Intelligence and Machine Learning

Proceedings of ERCICAM 2025, Volume 2

Springer

Editors
H. C. Nagaraj
Nitte Meenakshi Institute of Technology
Nitte (Deemed to be University)
Bengaluru, Karnataka, India

Ajit Danti
Department of AI
City University Ajman
Ajman, United Arab Emirates

Sreekanth Rallapalli
Department of MCA
Nitte Meenakshi Institute of Technology
Bengaluru, Karnataka, India

ISSN 1876-1100 ISSN 1876-1119 (electronic)
Lecture Notes in Electrical Engineering
ISBN 978-981-95-7448-3 ISBN 978-981-95-7449-0 (eBook)
https://doi.org/10.1007/978-981-95-7449-0

This Springer imprint is published by the registered company Springer Nature Singapore Pte Ltd.
The registered company address is: 152 Beach Road, #21-01/04 Gateway East, Singapore 189721, Singapore

Preface

The Second International Conference on "Emerging Research in Computing, Information, Communication, Artificial Intelligence and Machine Learning", ERCICAM 2025 is an annual event organized at Nitte Meenakshi Institute of Technology (NMIT), Yelahanka, Bengaluru, India. ERCICAM aims to provide an interdisciplinary forum for discussion among researchers, engineers, and scientists to promote research and exchange of knowledge in computing, information, communication, artificial intelligence, and machine learning. This conference will provide a platform for networking of academicians, engineers, and scientists and will encourage the participants to undertake high-end research in the above thrust areas. ERCICAM 2025 received more than 300 papers from all over the world, viz. from Nepal, Bangladesh, Malaysia, Bahrain, China, and India. The ERCICAM Technical Review Committee has followed all necessary steps to screen more than 300 papers by going through various rounds of quality checks on each paper before selection for presentation/publication in Springer's LNEE (Lecture Notes in Electrical Engineering) proceedings, which is SCOPUS indexed. The acceptance ratio is 1:3.

Bengaluru, India
March 2025

N. Nalini
Sreekanth Rallapalli

Acknowledgments

First, we would like to thank Prof. N. R. Shetty, Advisor, Nitte Education Trust, who has always been the guiding force behind this event's success. It was his dream that we have striven to make a reality. Our sincere thanks to Mr. Rohit Punja, Administrator, Nitte Education Trust, Dr. Sandeep Shastri, Director-Academic, Nitte Education Trust, and Dr. Sudheer Reddy, Dean-Academic, NMIT, for their encouragement in organizing this conference. Our thanks to Dr. H. C. Nagaraj, Principal, NMIT, who has monitored the whole activity of the conference from the beginning till its successful end. Our special thanks to Springer for agreeing to publish the quality papers in their LNEE Series. We would like to express our gratitude to all the review committee members of all the themes of computing, information, communication, artificial intelligence, and machine learning and the best paper award review committee members. Finally, we would like to express our heartfelt gratitude and warmest thanks to the ERCICAM 2025 organizing committee members for their hard work and outstanding efforts. We know how much time and energy this assignment demanded, and we deeply appreciate all the efforts to make it a grand success. Our special thanks to all the authors who have contributed to publish their research work in this conference and participated to make this conference a grand success. Thanks to everyone who have directly or indirectly contributed to the success of this conference ERCICAM 2025.

Program Chairs
ERCICAM 2025

About the Conference

ERCICAM 2025

The Second International Conference on "Emerging Research in Computing, Information, Communication, Artificial Intelligence and Machine Learning", ERCICAM 2025 is an annual event held during March 7th and 8th, 2025, at Nitte Meenakshi Institute of Technology (NMIT), Yelahanka, Bengaluru, India. ERCICAM 2025 is organized under the patronage of Prof. N. R. Shetty, Advisor, Nitte Education Trust. Mr. Rohit Punja, Administrator, Nitte Education Trust. Dr. Sandeep Shastri, Director-Academic, Nitte Education Trust, and Dr. Sudheer Reddy, Dean-Academic, NMIT. Dr. H. C. Nagaraj who is the principal of NMIT acted as a conference chair and the Program Chairs of the conference were Dr. N. Nalini, Professor, CSE and Dr. Sreekanth Rallapalli, Professor and Head, MCA, NMIT, Bengaluru, Karnataka, India. ERCICAM aims to provide an interdisciplinary forum for discussion among researchers, engineers, and scientists to promote research and exchange of knowledge in computing, information, communication, artificial intelligence, and machine learning. This conference will provide a platform for networking of academicians, engineers, and scientists and will encourage the participants to undertake high-end research in the above thrust areas. For ERCICAM-2026, authors are invited to submit the manuscripts of their original and unpublished research contributions to ercicam2026@nmit.ac.in (ERCICAM website: https://www.nmit.ac.in/ercicam). All the submitted papers will go through a peer review process, and the corresponding authors will be notified about the outcome of the review process. There will be many rounds of quality checks on each paper before selection for Presentation/Publication. Authors of the selected papers to present their papers during the conference.

ERCICAM 2025

The Second International Conference on "Emerging Research in Computing, Information, Communication, Artificial Intelligence and Machine Learning" (ERCICAM 2024) was held during March 07–08, 2025, at Nitte Meenakshi Institute of Technology (NMIT), Bengaluru.

ERCICAM 2025—Committee

Chief Patrons

Dr. N. V. Hegde, President, Nitte Education Trust, Mangaluru, India
Dr. N. R. Shetty, Chancellor-Central University of Karnataka, Kalaburagi and Advisor, Nitte Education Trust, Mangaluru, India
Mr. Rohit Punja, Administrator, NET, Mangaluru, India
Dr. S. Vidyashankar, Hon'ble Vice Chancellor, Visvesvaraya Technological University, Belagavi, Karnataka, India

Patrons

Dr. Sandeep Shastri, Director (Academics), Nitte Education Trust, Mangaluru, Karnataka, India
Dr. V. Sridhar, Dean (Academics), NMIT, Bengaluru, India

Conference Chair

Dr. H. C. Nagaraj, Principal, NMIT, Bengaluru, India

Program Chairs

Dr. N. Nalini, Professor (CSE), NMIT, Bengaluru, Karnataka, India
Dr. Sreekanth Rallapalli, Professor and Head, Department of MCA, NMIT, Bengaluru, Karnataka, India

General Chairs

Dr. S. Meenakshi Sundaram, Professor and HOD, Department of CSE, NMIT, Bengaluru, Karnataka, India
Dr. Piyush Kumar Parikh, Professor and HOD, Department of AIML, NMIT, Bengaluru, Karnataka, India

Advisory Committee

Dr. L. M. Patnaik, INSA Senior Scientist, NIAS, Bengaluru, India
Dr. K. D. Nayak, Former OS and CC, R&D (MED and MIST), DRDO, India
Dr. Kalidas Shetty, Founding Director of Global Institute of Food Security and International Agriculture (GIFSIA), North Dakota State University, Fargo, USA
Dr. Varun M. Deshpande, Edison Digital Expert, Schneider Electric, Franklin, Tennessee, USA
Dr. K. R. Venugopal, Former Vice-Chancellor, Bangalore University, Bengaluru, Karnataka, India
Mr. Neal Wagner, Complex Systems Scientist, Systems and Technology Research, Massachusetts, USA
Dr. Vincenzo Piuri, Professor, University of Milan, Italy
Mr. Victor Méndez Munoz, Head of Software Development/Consultant Professor IUL, S.A/Universität Oberta de Catalunya, Spain
Dr. Ho Chin Kuan, Vice Chancellor, Asia Pacific University of Technology and Innovation Technology, Malaysia
Dr. Sohail S. Chaudhry, Professor, Villanova University, USA
Mr. Arvind Kumar Bhardwaj, Senior Software Architect, Capgemini, Houston, Texas, USA
Dr. Wei-Chiang Hong, Professor, Oriental Institute of Technology, Taiwan
Mr. B. L. Narasimharaju, Associate Professor, National Institute of Technology, Warangal, India
Dr. Saurabh Kumar Pandey, Assistant Professor, Indian Institute of Technology, Patna, India
Dr. P. Deepa Shenoy, 2022 Chair, IEEE Bangalore Section, University Visvesvaraya College of Engineering, Bengaluru, Karnataka, India

Program Committee

Dr. Ayush Goyal, Assistant Professor, Texas A&M University—Kingsville, USA
Dr. Jordan Pascual Espada, Deputy Director of the School of Computer Engineering, University of Oviedo, USA
Dr. Bui Thanh Khoa, Industrial University of Ho Chi Minh City, Vietnam
Dr. Michael E. Auer, Prof. Dr. (mult.), Carinthia University of Applied Sciences, Austria
Dr. Man Qi, Senior Lecturer Canterbury Christ Church University, UK
Dr. Y-h. Taguchi, Professor, Chuo University, Japan
John Acken, Professor, Department of Electrical and Computer Engineering, Portland State University, Portland, Oregon, USA
Dr. S. G. Mohan, Professor and Head, Department of ISE, NMIT, Bengaluru, Karnataka, India

Dr. B. D. Parameshachari, Professor and Head, Department of ECE, NMIT, Bengaluru, Karnataka, India
Dr. N. Samanvita, Professor and Head, Department of EEE, NMIT, Bengaluru, Karnataka, India

Organizing Committee

Dr. Krishna Rao Venkatesh, Professor, Department of CSE, NMIT, Karnataka, India
Dr. H. Sarojadevi, Professor, Department of CSE, NMIT, Bengaluru, India
Dr. S. Vijaya Shetty, Professor, Department of CSE, NMIT, Karnataka, India
Dr. V. Vani, Professor, Department of CSE, NMIT, Karnataka, India
Dr. Jyothi Neeli, Professor, Department of CSE, NMIT, Karnataka, India
Mr. A. Balachandra, Professor, Department of ISE, NMIT, Karnataka, India
Dr. D. B. Srinivas, Professor, Department of ISE, NMIT, Karnataka, India
Dr. K. Aditya Shastri, Professor, Department of ISE, NMIT, Karnataka, India
Dr. Vidyadevi G. Biradar, Professor, Department of ISE, NMIT, Karnataka, India
Dr. A. C. Ramachandra, Professor, Department of ECE, NMIT, Karnataka, India
Prof. V. Sitaram Yaji, Professor, Department of ECE, NMIT, Karnataka, India
Dr. N. Rajesh, Professor, Department of ECE, NMIT, Karnataka, India
Dr. B. M. Manjula, Professor, Department of ECE, NMIT, Karnataka, India
Dr. B. Karunakara Rai, Professor, Department of ECE, NMIT, Karnataka, India
Dr. Prashanth V., Professor (EEE) and Controller of Examinations, NMIT, Karnataka, India
Dr. M. R. Dileep, Associate Professor, Department of MCA, NMIT, Karnataka, India

Web Management Committee

Mr. Navaneeth A. V., Assistant Professor, Department of MCA, NMIT, Karnataka, India
Mr. Vidya Sagar S. D., Assistant Professor, Department of MCA, NMIT, Karnataka, India

Contents

Leveraging YOLO for Accurate Fruits and Vegetables Calorie Prediction with Machine Learning

S. R. Nandini and M. K. Likhithashree

Abstract The prevalence of obesity has increased over the past ten years worldwide, which has led to a greater emphasis on creative approaches to address this problem. One well-known field of study is how to regulate weight with meal diaries and caloric analysis. As the importance of leading a healthy lifestyle increases, many apps have surfaced that use image recognition technology to automatically register meals. Maintaining a healthy lifestyle requires careful monitoring of daily caloric consumption. The purpose of this work is to improve the accuracy of calorie calculation from food photos by combining image analysis techniques with sophisticated item detection. The suggested method is using picture segmentation to extract specific information from food photos, which is then analyzed mathematically to accurately calculate the calorie content. We provide a smartphone program that uses photos to compute meal calories. To accurately calculate the calorie value of food products, the application employs You Only Look Once (YOLO) for object identification and photo segmentation algorithms by leveraging datasets that are specific to Indian cuisine.

Keywords Obesity management · Food calorie estimation · Image recognition in food logging · Advanced object detection · Caloric analysis from images

1 Introduction

In nutrition and health studies, estimating food products' calorie content is essential. Accurate calorie estimation is becoming more and more necessary for successful dietary management and health monitoring as diet-related illnesses including obesity,

S. R. Nandini
P Department. of CSE, BGS Institute of Technology, Adichunchanagiri University, Mandya, India
e-mail: nandinisr@bgsit.ac.in

M. K. Likhithashree (✉)
BGS Institute of Technology, Adichunchanagiri University, Mandya, India
e-mail: likhithashreemk@gmail.com

H. Nagaraj et al. (eds.), *Advancements in Artificial Intelligence and Machine Learning*,
Lecture Notes in Electrical Engineering 1572,
https://doi.org/10.1007/978-981-95-7449-0_1

diabetes, and heart disease continue to rise. Because traditional calorie calculation methods often entail physically tracking food and consulting nutritional databases, they are often laborious and prone to human error. This has led to an increasing interest in using developments in computer vision and deep learning to create automated and more accurate calorie calculation methods.

Food is the primary source of necessary nutrients and energy for human life. But during the past few decades, dietary preferences have changed dramatically, with a decline in vegetable intake and an increase in trans and saturated fats, salt, and other lipids. An increase in medical conditions, such as cancer and other ailments, has been.

Connected to these changes. Many people are increasingly concentrating on adopting healthier eating habits as a result. People frequently maintain thorough meal logs to monitor their dietary consumption, but manual tracking is difficult and time-consum-ing, and it can be difficult to interpret nutritional data. An entirely new project has been created to simplify this procedure.

Applications for computerized food classification span widely, from novel ideas like waiter-less restaurants to the control of personal health. For example, mobile food classification systems can tell users about their daily caloric intake and the types of nutrients they need. The goal of this project is to develop an app that can take pictures of meals in real time and calculate the number of calories they contain, assisting users in making better dietary decisions and living healthier lives.

Considerable progress in image processing methods during the last couple decades have produced impressive tools for image analysis and interpretation. With the development of deep learning, convolutional neural networks (CNNs) have significantly changed tasks like object detection, segmentation, and classification. You Only Look Once (YOLO), one of these models, has become well-known for its remarkable accuracy and real-time object identification performance. For applications requiring quick and dependable detection, an updated version of YOLO with increased speed and precision is a great option.

YOLOv5 is designed to recognize objects in photos with remarkable speed and accuracy. YOLOv5 employs a single neural network to reliably forecast class probabilities and bounding boxes from whole images in a single assessment, as opposed to earlier techniques that need several runs based on region suggestions. Because of this, it works especially effectively in real-time applications where making decisions quickly is crucial. When it comes to food item calorie calculation, YOLOv5 can be used as a starting point since it is good at recognizing and categorizing different foods in a picture.

The automation and precision that come with employing YOLOv5 and image processing techniques for calorie estimate are the main advantages. Large amounts of data may be processed quickly by automated systems, and consumers can receive feedback in real time. Furthermore, after training, these models show resistance to changes in backdrops, lighting, and angles, which enables them to successfully adjust to new images.

However, there are other challenges. The disparity in food presentation is one of the main challenges. Food identification and classification can be made more difficult by the substantial changes in food look that might result from varying cooking techniques, portion sizes, and presentation styles. Furthermore, precisely estimating the volume of food items from 2D photos is a challenging issue that frequently calls for advanced methods and extra information.

An encouraging step toward automated food tracking is the integration of YOLOv5 with image processing for calorie estimate. We can create systems that can provide precise, real-time calorie estimations by utilizing deep learning and sophisticated image processing. This would help people make better dietary decisions and enhance their general health. Subsequent investigations will concentrate on surmounting present obstacles, improving the precision of volume estimation, and expanding the system's scope to accommodate a broader variety of food items and situations.

1.1 Importance of Calorie Prediction

In the health-conscious world of today, eating a balanced diet is essential to general wellbeing. Fruits are an essential part of a balanced diet since they are a great source of fiber, vitamins, minerals, and antioxidants. But knowing how many calories fruits contain is just as vital for anyone watching their calorie consumption. Automated fruit calorie prediction systems can provide accurate and timely information, enabling people to make well-informed dietary decisions.

1.2 Health and Nutrition

Fruits are an important component of a diet that is well-rounded since they are a great source of vital nutrients. They supply vital nutrients that are required to maintain optimum health, such as potassium, vitamin C, and dietary fiber. But fruits also naturally contain calories and sugars that can add up, especially for people following a low-calorie diet. Accurate calorie counting helps people enjoy the health advantages of fruits while controlling their daily caloric intake.

1.3 Weight Management

Calorie counting's main goal is to effectively manage weight. Finding and keeping a healthy weight requires a balance between calories burned and calories consumed. Ow-ing to their inherent sweetness, fruits are frequently mistaken for low-calorie snacks, which increases the risk of overindulging. Inadvertent overindulgence in calories may arise from this. People may precisely track how many fruits they eat thanks to automated devices that calculate their calorie content. This prevents overindulgence and promotes healthy weight management.

1.4 Diabetes Management

Controlling blood sugar levels on a daily basis is crucial for people with diabetes. Fruits are full of health advantages, but their carbohydrate content varies, which may affect blood sugar levels. Precise evaluations of the energy and carbohydrate composition of fruits can aid diabetics in meal planning, enabling them to take use of the fruits' nutritional advantages without running the danger of blood sugar rises. Real-time data is provided by automated devices, enabling diabetics to swiftly and confidently make decisions about how much fruit they eat.

1.5 Dietary Planning and Monitoring

Dietitians and other medical professionals frequently support individualized diet programs that are designed to fit the needs of each patient. Precise computation of fruit calories has the potential to improve these programs' efficacy. It is very important to know the precise calorie and carbohydrate content of fruits for people following particular diets, such as low-carb or ketogenic regimens. Automated solutions can make diet planning and monitoring easier, which will help people stick to their dietary goals.

1.6 Technological Advancements in Calorie Prediction

Calorie prediction has undergone a revolution thanks to the introduction of cutting-edge technology like computer vision, deep learning, and image processing. Calorie content of fruits can be determined by using tools such as YOLOv4, which is capable of reliably identifying and classifying different fruits in photos. These devices are extremely useful for managing diets and tracking health since they can provide accurate calorie estimates by interacting with nutritional databases.

Precise calculation of fruit calories is essential for anyone trying to control their weight, keep a balanced diet, or control their blood sugar levels. Fruit calorie estimation is made accurate and efficient by automated processes that use deep learning and advanced image processing techniques. With the advancement of technology, these techniques are becoming increasingly precise and intuitive, enabling people to make more confident decisions about their diets.

2 Related Works

To estimate calories for nutritional evaluation, Tatsuya Miyazaki, Gamhewage C de Silva, and Kiyoharu Aizawa [2] proposed an image analysis method. They created an album with 6000 photos and dietitian-calculated meal log calorie counts. SURF choices, color histograms, and correlograms were among the many criteria used to compare the food image with the food log dataset. Patients and dietitians were helped in managing Parisa Pouladzadeh, Shervin Shirmohammadi, and Rana Al-Maghrabi's method for tracking daily food intake in terms of calories and nutrition [3]. The article suggested creating a smartphone application that would let users capture images of themselves before and after meals in order to assess their calorie intake and nutritional value. Ross Girshick, Ali Farhadi, Santosh Divvala, and Joseph Redmon presented A unified paradigm for object detection is called YOLO [6]. This model may be trained directly on full-length images and is simple to construct.

Unlike methods that rely on classifiers, YOLO trains the entire model at once as it is based on a loss function that is directly related to detection performance. Manal Chokr and Shady Elbassuoni [9] proposed that the simplest food items to estimate the number of calories were pizzas, doughnuts, chicken, and sandwiches since the simplest technique entailed evaluating the item's size by running compressed picture through a regressor. Real-time detection systems, YOLOv2 and YOLO9000, were introduced by Joseph Redmon and Ali Farhadi [8]. YOLOv2 has substantially faster detection than earlier methods. A real-time system called YOLO9000 maximizes both detection and classification for more than 9000 distinct item kinds. Sujata Chaudhari, Nisha Malkan, Ayesha Momin, and Mohan Bonde [12] used Yolo to instantly recognize things. This article employs YOLO as a modern way to attain high accuracy in object recognition. The rise in accuracy throughout the training period is explained by this study. In this paper, they have detailed the creation of CNN and illustrated how object detection is done out using CNN and YOLO. The main objective of the research conducted by Alexey Bochkovskiy, Chien-Yao Wang, and Hong-Yuan Mark Liao [13] is to create an object detector that can operate quickly enough to be used in production systems. They have developed a robust and successful object detection model. This makes the state-of-the-art approaches more efficient for training on a single GPU. Several features are shown to improve the detector and classifier's accuracy (Fig. 1).

3 Implementation

We extract food item volumes from photos using a variety of image processing methods. A variety of segmentation techniques are used to distinguish between the contours of the meal and a calibration thumb that is placed next to it, including Otsu's approach, watershed segmentation, morphological processes, and Canny edge

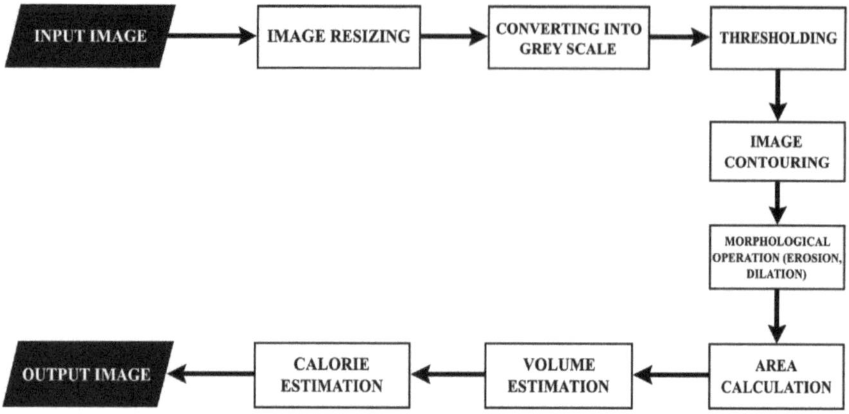

Fig. 1 Proposed model

detection. Using this thumb as a point of reference throughout the photo shoot enables an accurate calculation of the actual size and volume of the dish.

After the food item has been identified, its volume is computed by simulating it as a cylinder or sphere, among other geometric shapes. The mass of the food item is calculated by applying standard density values after the volume has been determined. Lastly, the food's estimated calorie content is determined using the mass and volume data.

There are multiple steps involved in determining calories:

(a) Image Segmentation: By dividing an image into discrete areas or segments according to pixel data, this approach improves the recognition of edges and objects. Included in the segmentation process are the following elements:

  (i) Image Resizing: To guarantee uniformity in analysis and to standardize the pixel count, the original image is resized to 400x400 pixels.
  (ii) BGR to Grayscale Conversion: There are two methods for turning a BGR (Blue, Green, and Red) picture into grayscale:

    Average Method: The RGB components of each pixel are averaged to determine the grayscale value. This method's formula is as follows: Three B + G + R is grayscale.

    Weighted Method: The red, green, and blue channels are given varying weights ac-cording to how much brightness they are thought to provide. This conversion's formula is: Grayscale: R + G + B = 0.299G + 0.114B
  (iii) Image Limitation: This is a simple, yet powerful, image segmentation method. When converting a grayscale image to a binary format, it is especially helpful. In order to facilitate additional image processing, this method helps separate the pixels of the object or foreground from those in the background.
  (iv) Determining Curvature: A curve that joins every continuous point along a border with the same hue or intensity is called a contour. For greater

accuracy, contours are usually recognized using binary pictures. Prior to the contour identification procedure, measures like thresholding or sophisticated edge detection algorithms are used to in-crease accuracy.

(v) Segmentation Based on HSV Color Space: Image thresholding is a basic tech-nique for segmentation; nevertheless, HSV color space segmentation works very well. Hue, Saturation, and Value, or HSV for short, stand for color, brightness, and satura-tion, respectively. Value specifies brightness or intensity, Saturation describes the amount of white light blended with the hue, and Hue is the predominant color as seen by humans. Similar to RGB, the HSV color space is useful for object identification based on specific colors while mitigating the impact of light intensity. The distance from the HSV cylinder's center axis is used to assess saturation, which helps with pre-cise object isolation.

(vi) Erosion and Dilation: Resizing an image modifies the number of pixels in it, readying it for morphological processes such as erosion and dilatation. In an image, dilation adds pixels to the edges of objects, whereas erosion removes pixels from those edges. The size and form of the structural element utilized in these processes determines how much pixel is added or removed.

(b) Calculating Area: Three important metrics are obtained following image segmen-tation: the food's pixel area, the skin area, and the approximate size of the thumb. In order to determine the pixel areas of the skin and food, the picture is converted to the HSV color space. With a skin multiplier of 5 × 2.3 and an assumed thumb size of roughly 5 cm, the actual skin area can be determined using the formula below: pix_to_cm_multiplier = pix_height5.0.

Using the three metrics found, the predicted meal area may be computed using the formula below:

Estimated Food Area = Skin Pixel AreaFood Pixel Area x Actual Skin Area

(c) Volume Estimation: This technique helps with food portion mass estimation, which is necessary for nutritional computations based on food mass. It is possible to calculate the calorie content once the mass has been determined. The finger, which serves as a calibration reference, is compared to the entire area of the meal in the picture to determine the volume. The recognized food item is modeled as a geometric shape, such as a sphere, cylinder, or rectangle, in order to determine its volume. Next, based on this volume, the mass is calculated using the following formula:

$$\mathbf{M} = \rho \times \mathbf{V}$$

where \(M \) is the food item's mass, and \(\rho \) is its density, both of which can be found in the tables that already exist.

(d) Estimating Calories: The calorie content of the food item is estimated using the mass and volume values. The following formulas can be used to calculate calories:

Estimated Weight = Actual Density×Estimated Volume Estimated Calories = Esti-mated Weight×Calories per 100 g100\text{Estimated Calories} = \frac{\text{Estimated Weight} \times \text{Calories per 100 g}}{100}Estimated Calories = 100Estimated Weight×Calories per 100 g.

4 Results and Discussions

The Tkinter package makes it easier to create windows, dialog boxes, and other visual elements by acting as Python's default Graphical User Interface (GUI) toolkit. It provides an extensive range of tools for creating interactive applications, including a selection of widgets like text boxes, buttons, and labels. Tkinter's cross-platform interoperability, which lets programs run on Linux, macOS, and Windows without any changes, is one of its advantages. Its popularity among novices and developers looking for effective GUI solutions is partly due to its easy-to-use interface and smooth inter-action with Python. We'll look at a calorie detection model's user interface in Fig. 2, where we list the main elements and features that are usually present in an interface of this kind.

The design of this login page places a high priority on a safe and simple authentication process. The user login interface is shown in Figs. 3, 4, and 5. Before moving on to the next page for the calorie calculation using an input image, users must enter their username and password.

In-depth User Communication:

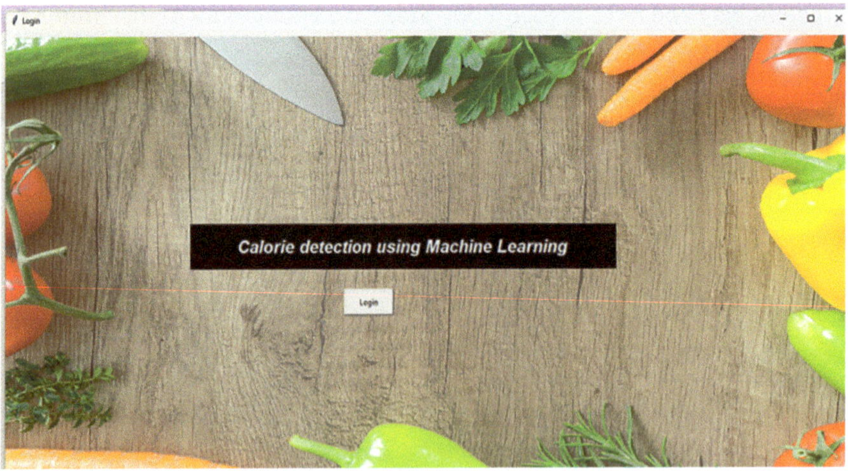

Fig. 2 User interface for calories detection model

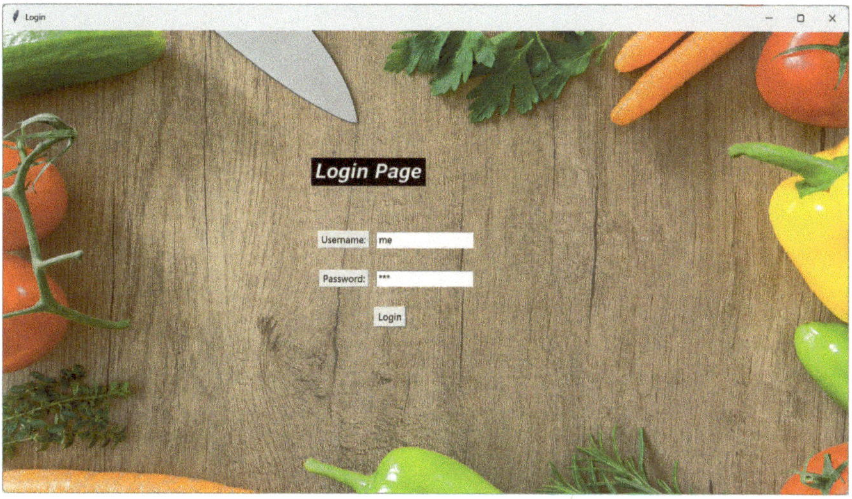

Fig. 3 User login page

Fig. 4 Load the fruit image for calories calculation

- Entering Credentials: The user fills in the "Password" and "Username" fields with their password.
- Log in: The user submits the data by clicking the "Login" button in order to authenticate their credentials.

Fig. 5 Load the fruit image for calories calculation

5 Conclusion and Future Scope

Using image processing techniques for accurate volume estimation and YOLOv4 for real-time object recognition, the suggested system for calorie calculation of food items represents a considerable improvement over conventional method. The technology en-sures precise calorie predictions through the integration of large nutritional databases, and its everyday usability is further enhanced by an intuitive mobile application. Through the use of continuous learning, fast feedback, and user input, this technology is an effective tool for controlling food intake and encouraging healthier eating habits. The system's usefulness is proven by extensive validation and testing in a variety of real-world settings, which also solve many of the shortcomings of previous methods.

Upgrades in the future might improve the system's performance even more. A more comprehensive understanding of health could be provided by combining calorie pre-diction with extra dietary and activity variables via wearable device compatibility. By adding augmented reality (AR) capabilities, the procedure might become more partici-patory and offer instantaneous visual feedback regarding the caloric content. Accuracy would be increased by developments in 3D reconstruction and volume estimate ap-proaches, especially for complicated dishes. More accurate and individualized guidance may result from personalizing dietary recommendations and growing the nutritional database. Improved community features and multilingual assistance could potentially increase the system's attractiveness. In the end, creating a complete health management system that works with various platforms and applications for health would make it easy for users to track and meet their wellness and nutrition objectives.

References

1. Bell Pepper Leaf Disease Classification Using CNN (2021). Ieeexplore.ieee.org. Ieeexplore (https://ieeexplore.ieee.org/document/9170728). Accessed 11 Apr 2021
2. J. Trivedi, Y. Shamnani and R. Gajjar, (2020) Plant leaf disease detection using machine learning In: Communications in Computer and Information Science, pp 267–276. https://doi.org/10.1007/978-981-15-7219-7_23 Accessed 31 March 2021
3. J. Amara, B. Bouaziz and A. Algergawy, (2017) A deep learning-based approach for Banana leaf diseases classification. Accessed 16 Feb 2021
4. W. Jeon and S. Rhee, (2017) Plant leaf recognition using a convolution neural network, Intenational J Fuzzy Log Intell Syst 17(1):26–34. https://doi.org/10.5391/ijfis.2017.17.1.26 Accessed 16 Feb 2021
5. Bhadur, Rani R (2020) Agricultural crops disease identification and classification through leaf images using machine learning and deep learning technique: a review. SSRN Electron J. https://doi.org/10.2139/ssrn.3564973
6. Ma B, Li X, Xia Y, Zhang Y (2020) Autonomous deep learning: a genetic DCNN designer for image classification. Neurocomputing 379:152–161. https://doi.org/10.1016/j.neucom.2019.10.007
7. K. Jagan, M. Balasubramanian and S. Palanivel, (2016) Detection and recognition of Diseases from Paddy Plant leaf images, Int J Comput Appl 144(12):34–41, https://doi.org/10.5120/ijca2016910505 Accessed 16 Feb 2021
8. Suyash S, Patil, Sandeep A, Thorat Early detection of grapes diseases using machine learning and IOT. In: 2016 second international conference on cognitive computing and information processing (CCIP)
9. Mohsin Kabir, M., Quwsar Ohi, A., Mridha, M.F. A multi-plant disease diagnosis method using convolutional neural network. arXiv 2020, https://arxiv.org/abs/2011.05151
10. Prodeep, A.R., Hoque, A.M., Kabir, M.M., Rahman, M.S., Mridha, M.F. Plant disease identification from leaf images using deep CNN'S EfficientNet. In: Proceedings of the 2022 international conference on decision aid sciences and applications (DASA), Chiangrai, Thailand, (2022) pp 523–527

Detecting Signs of Depression Using AI

Akshitha Rao, Pradeep Kanchan, Amulya Nayak, Aditi Vasanth Kamath, and Pujar Gireesha

Abstract Depression detection is vital for early intervention, which helps improve outcomes for those affected. The paper highlights how multimodal machine learning can improve detection accuracy by analysing text, facial expressions, and audio cues together. By comparing how each type of data performs individually and in combination, the research shows that multimodal approaches significantly boost detection performance. We explore finding indicators of depression from social media posts and their facial microexpression imagery. We present a system that detects signs of depression based on a video of the person. We find microexpressions that could be indicative of depression and analyse what the person says to find if he is not depressed, or if depressed, moderately or severely depressed.

Keywords Depression · Machine learning · Mental health

A. Rao · P. Kanchan (✉) · A. Nayak · A. V. Kamath · P. Gireesha
Department of Computer Science and Engineering, NMAM Institute of Technology, Nitte (Deemed to be University), Nitte Off-Campus Centre, Karkala, India
e-mail: pradeepkanchan@nitte.edu.in

A. Rao
e-mail: 4nm21cs016@nmamit.in

A. Nayak
e-mail: 4nm21cs022@nmamit.in

A. V. Kamath
e-mail: 4nm21cs005@nmamit.in

P. Gireesha
e-mail: 4nm22cs411@nmamit.in

H. Nagaraj et al. (eds.), *Advancements in Artificial Intelligence and Machine Learning*, Lecture Notes in Electrical Engineering 1572,
https://doi.org/10.1007/978-981-95-7449-0_2

1 Introduction

Depression is a prevalent mental health condition impacting millions of people worldwide, with far-reaching effects on personal well-being and societal costs. Traditional detection methods, such as clinical interviews and self-reported questionnaires, often fail due to delays, social stigma, and subjective reporting. Sometimes, moderate depression may be overlooked, since they hesitate to meet mental health professionals. These limitations motivate the use of machine learning to analyse objective data sources—text (e.g. social media posts), facial expressions. One single type of data may not be sufficient to obtain all the context required for the same. The advantage of a multimodal machine learning approach is that it integrates several data types, providing a fuller, more reliable view of a person's mental health. This study's aim is to assess how effective various combinations of text, facial, and audio cues about the patient that are for depression detection. This is because when patients meet a mental health professional, they may miss out on the exact cause of their condition. This could either be because they are not sure of the reason, or that they just choose not to say it because talking certain things makes them uncomfortable. In any case, more information about the cause of depression would give a more holistic view, and would make it easier to help the patient out of depression.

2 Related Work

Text data has been analysed using Natural Language Processing to detect depression by identifying linguistic features such as emotion, sentiment, or the presence of specific words related to sadness or loneliness. Social media is an effective tool to identify such signs, since people post freely there, and are also able to post their problems anonymously. In [1], Orabi et al. used Twitter data for the same and found that CNNs outperformed RNNs for text analysis. They preprocessed the data by removing elements useless to the model, such as links and @mentions [2]. used the CLEF eRisk 2017 dataset. They used hybrid embeddings, which consisted of TrainableEmbed, GloveEmbed, Word2VecEmbed, FastTextEmbed and handcrafted metadata features and LIWC features normalized using min-max scaling. Their model consisted of BiLSTM layer followed by fully connected hidden layers with ReLU and sigmoid activation functions. Word2VecEmbed + Meta performed the best in their case. In [16], Lin et al. made a system that used both the text and the pictures posted along with it. They concatenated both using low-rank tensors and the result of this process was analysed using neural network [19]. used transformer-based models like EmoBERTa and CLIP with temporal embeddings, enhancing sensitivity to the sequence and frequency of posts, crucial for understanding longitudinal changes in mental health indicators.

Research into facial expressions has shown that subtle, unconscious emotional cues like microexpressions or changes in eye movement can indicate depression.

According to [4], depressed individuals often show less frequent smiling, more frequent lip pressing, prolonged corrugator muscle movement, and more negative/ neutral expressions. Techniques like Local Binary Patterns (LBP) and Histograms of Oriented Optical Flow (HOOF) were used to detect these expressions, while machine learning classifiers such as Support Vector Machines (SVMs) and CNNs were applied to categorize emotions by [3]. They also developed a new database for the same, classifying the expressions as sad, neutral and blocked (hidden sadness). In [17], Pampouchidou et al. analysed video frames for microexpressions using local curvelet binary patterns. They proposed two algorithms, frame-based and video-based analysis. The face in the video was detected, and tracked using KLT tracker. Curvelet transforms were applied on each frame, then LBP were computed for the XY plane for each face and classified using the KNN in the frame-based analysis. For video-based analysis, they further computed LBPs for the XZ and YZ planes and concatenated them, and, like the frame-based analysis, classified using KNN.

Speech characteristics like pitch, loudness, and the rate of speech are strong indicators of depression. Depressed people generally speak slowly, and in a less energetic way. Pitch, energy, formants, jitter, shimmer and loudness of the speech are good indicators of depression as well. Tools like openSMILE and COVAREP extract these features, and various machine learning algorithms analyse them for depressive speech patterns [4–6]. Although each modality has shown promise, recent research suggests that combining them could provide a more accurate and comprehensive depression detection tool. Dumpala et al. [18] showed that speaker embeddings trained using LibriSpeech and Voxceleb2, combined with OpenSMILE acoustic features improved the accuracy with which depression can be detected [5]. Combined and analysed audio, video, and textual data. From the video, they extracted the head pose, gaze, and facial landmarks. They considered acoustic features such as pitch, quality, and spectral properties extracted using COVAREP. They analysed textual input such as word count, sentence structure, and emotional content, including negative words, which are closely related to depression. They trained eight models, each giving the intensity for each question in the PHQ-8 scale. They used a combination of SVM and Neural Networks (only for the Fisher vectors), which yielded good results.

We aim not only to detect depression and its intensity, but also to find out why the person is depressed. We aim to find the root cause of the person being depressed, so that it can be easier to help the patient recover, with more information. This was not done in most of the literature that we referred to. In fact, one of them, [16] has done it to some extent. They have made functions to analyse why the patient is classified as depressed. However, they have only used social media pictures and texts for the same.

2.1 Datasets

Text data comes from publicly available datasets, such as the CLEF eRisk dataset [7, 8] and depression-labelled Reddit data. These datasets contain user-generated content annotated with depression symptoms. The CLEF eRisk 2024 dataset has labelled texts, each individual's texts in separate files [9, 10]. have created a text dataset with labels 'not depressed', 'moderately depressed', 'severely depressed'. This data was further preprocessed and more Reddit texts were added by [11].

One source of facial expressions data is the DAIC-WOZ dataset. It contains the OpenFace facial landmarks of the participants faces captured during an interview. The Extended DAIC dataset includes conversational recordings from clinical interviews. These recordings are annotated for prosodic features that can be linked to depression, like tone, speech rhythm, and pitch. Another source of data that people have used are the CASME and SMIC datasets.

3 Methodology

3.1 Data Collection

A good and well-trained model is a result of a good dataset. Hence, data collection was a vital step to begin with our project.

For the text classifier, we used the dataset by [9–11]. We chose this since it helped us classify not only severe, but also moderate depression, hence considering various levels of depression.

To train our facial microexpressions classifier, we used a dataset from Kaggle [12], which contains images cropped to the face for the microexpressions: angry, disgust, fear, neutral, sadness, happiness and surprise.

3.2 Preparing the Text Classifier

Data preprocessing. The training text data was first converted to a form that is suitable to train the model on. First, the words were stemmed to convert them into their simpler forms. Then, the stopwords were removed using the nltk library in Python. Negative words like "not" or "isn't", however, were retained. This was done to make it less computationally expensive and also to remove words that do not add meaning to the sentence.

Feature extraction. The text was converted into a bag of words using the CountVectorizer, which the model was trained on. Scikit-learn [13, 14] package was used for this purpose. We also trained another set of models on the data encoded using empath [15], which is an open-source tool similar to LIWC [2]. Final results were noted for both encoding methods.

Following figure, Fig. 1 shows the algorithm for text data preprocessing and feature extraction:

Fig. 1 Algorithm for text data preprocessing and feature extraction

```
initialise corpus array

for each text in dataset do:
    remove punctuation marks
    for each word in text do:
        if not stopword:
            stem the word
        join the words to make it a single text again
    append the text to the corpus array

create a bag of words with max_features 1500
```

Training the model. Models were trained using various machine learning algorithms and the results were noted. The models were trained using Gaussian Naive Bayes, Support Vector Machine and Random Forest Classifiers. The best performing algorithm was downloaded using joblib and used for all further analyses.

Evaluating the model. The test data was first preprocessed and features were extracted in the same way as done with the train data. The test data was classified using the trained models and the performance of each model was evaluated using the following metrics:

$$accuracy = \frac{TP + TN}{TP + FP + TN + FN} \tag{1}$$

$$precision = \frac{TP}{TP + FP} \tag{2}$$

$$recall = \frac{TP + TN}{TP + FN} \tag{3}$$

$$F1 = \frac{2 \times precision \times recall}{precision + recall} \tag{4}$$

3.3 Preparing the Facial Expressions Classifier

Data preprocessing and feature extraction. Since the images were already resized into 80*80 pixel images by the dataset creators, we just proceeded with feature extraction. We read each image using the cv2 library in Python. Since the important features were only the facial expressions and the wrinkles on the face resulting from it, the position of the eyebrows, etc., the colour of the images would not have been of any use to the prediction process, so the original images were read in grayscale. The features were extracted from the images by applying wavelet transforms on the images. We tried various wavelets, and the results were documented. Based on the results, the Biorthogonal 2.2 wavelet was chosen, with 4 levels. Wavelet transform was performed using the PyWavelets Python package. The original (grayscale) images were concatenated with the wavelet transforms of the images and passed on for model training. We referred to [20] for the function to perform wavelet transform and reconstruct the transformed image, except that we loaded the image in black-and-white and performed used these images for the analysis without converting colour, since skin colour and such details are not necessary for our use case.

Model training. The model was trained using Support Vector Machine, with RBF kernel, for each feature representation method that we tried, and results were noted. This was done using the Scikit-learn package. Figure 2 shows the steps we followed and steps for preprocessing and feature extraction from the facial images.

Evaluation. These models were evaluated using accuracy score, precision, recall and F1 scores. We also plotted confusion matrices to see the number of samples classified correctly.

3.4 Integrating Both to Analyse Video Data

The video was first transcribed using the Whisper Base model available on HuggingFace. This transcript was then analysed using the text model that we had trained. The output of this stage would be that the person is not depressed/moderately/severely depressed.

To analyse a video, we first read all the frames of the video in grayscale, and performed wavelet transforms on each frame and analysed the frames for their microexpressions using the best working facial expressions classifier, that we had downloaded as a joblib file. The respective segments of the transcript corresponding to the timing of that particular frame in the video were then mapped with the video frames, thus giving an idea of what makes the person scared, sad, etc. Figure 3 shows the flowchart of the system.

Fig. 2 Preprocessing and feature extraction from image data

```
def w2d(img, wavelet, level):
    imArray = img
    imArray = np.float32(imArray)
    imArray /= 255  # scale the pixel values to [0, 1]

    # Perform 2D wavelet decomposition
    coeffs = pywt.wavedec2(imArray, wavelet, level=level)
    coeffs_H = list(coeffs)

    # Remove low-frequency details
    coeffs_H[0] *= 0
    arrH = pywt.waverec2(coeffs_H, wavelet)

    # Add 255 to restore the intensity range
    arrH += 255
    arrH = np.uint8(arrH)
    return arrH
```

4 Results and Discussions

4.1 Results

Following results were the results of evaluation of the text and microexpression classifier models trained using various ML algorithms. The results of each classifier are shown in Table 1.

Based on these results, we decided to go with Random Forest Classifier with empath for feature representation. Although Multinomial Naive Bayes gave a higher accuracy, we decided not to choose that as the F1 score was too low, indicating that there could be a higher false positive or false negative rate if that was used. For the image classifier, we used SVM with bior2.2 wavelet.

Table 2 shows the results of the microexpressions classifier with different wavelets used for feature extraction.

The two systems combined, generated an analysis based on video data. The following figure (Fig 4) shows the reconstructed wavelet transform of a frame of a sample video, the microexpression shown in each frame and the verdict of the model about whether the person is depressed or not.

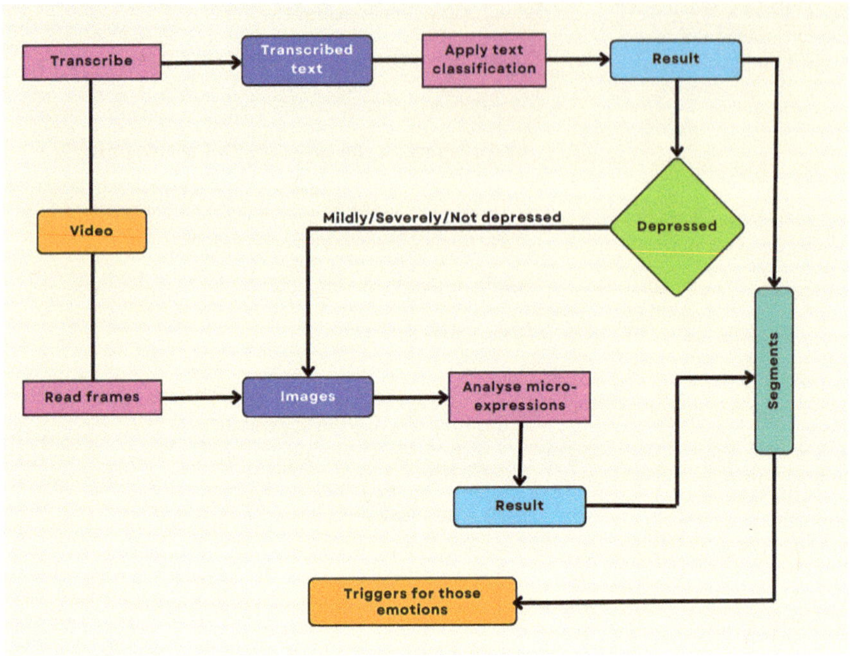

Fig. 3 Flowchart

Table 1 Results of evaluation of the text classifier

Algorithm	Feature representation	Accuracy %	Precision %	Recall %	F1 %
Gaussian naive bayes	CountVectorizer	33.77	38.385	41.765	31.256
Support vector machine	CountVectorizer	46.50	33.617	37.315	31.749
Random forest	CountVectorizer	61.910	40.548	36.534	35.734
Random forest	empath	63.482	50.418	40.172	40.357
Multinomial naïve bayes	empath	66.841	22.280	33.333	26.708

Table 2 Results of evaluation of the microexpressions classifier

Algorithm	Wavelet	Accuracy score %	Precision score %	Recall score %	F1 score %
Support vector machine	db1	60.16	57.14	59.45	57.72
Support vector machine	bior2.2	61.02	58.065	60.10	58.53
Support vector machine	bior2.4	60.053	56.955	58.682	57.302

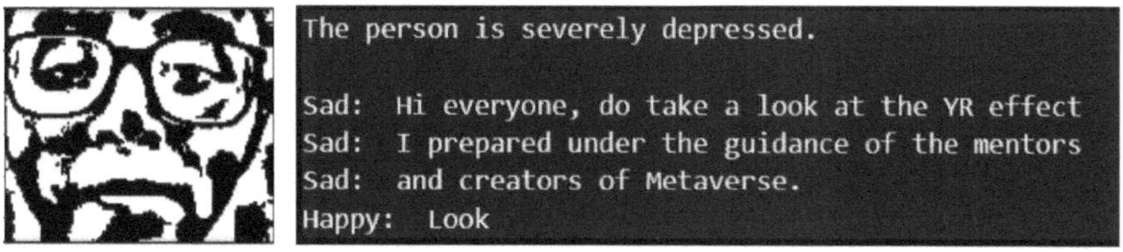

Fig. 4 The first image shows the reconstructed wavelet transform of one of the frames in the video. The second one shows the analysis of the video file

4.2 Discussions and Future Work

One limitation of the study is that the image dataset that we used was not very clear and was a result of web-scraping, as per the creator of the dataset. Also, the text data that was used had been labelled based on individual texts using a set of rules, without much details about the person [9]. Hence, future scope of work includes integrating more sources of data, to enable a more holistic analysis. One such dataset could be the iSAFE dataset [21], which has clearer images, hence more prominent features. Also we have decomposed the image signals into 4 levels while performing wavelet transforms on it, since it worked well and extracted only the relevant features from the training data. But its performance may not always be the same on test data; if the test images are very clear, 4 levels of decomposition may extract more than just the required features. One way of solving this problem would be to check how sparse the reconstructed wavelet transform is, and iteratively increasing or decreasing the number of levels until it is within a required range, as observed on the train data.

Additionally, we have used traditional machine learning algorithms here to build our prototype. We plan to try out deep learning models, to increase the effectiveness of the analysis. We aim to reduce the false positive rate on test data. Using deep learning models may help solve this issue.

Acknowledgements The authors acknowledge their college and their project guide for his guidance. They also acknowledge the makers of the datasets who made their datasets available for use by researchers.

References

1. Husseini Orabi A, Buddhitha P, Husseini Orabi M, Inkpen D (2018) Deep Learning for Depression Detection of Twitter Users. In: Proceedings of the Fifth Workshop on Computational Linguistics and Clinical Psychology: From Keyboard to Clinic. Association for Computational Linguistics, Stroudsburg, PA, USA, pp 88–97. https://doi.org/10.18653/v1/W18-0609
2. Shah FM, Ahmed F, Saha Joy SK, Ahmed S, Sadek S, Shil R, Md. H K (2020) Early depression detection from social network using deep learning techniques. In: 2020 IEEE region 10 symposium (TENSYMP). IEEE, pp 823–826. https://doi.org/10.1109/TENSYMP50017.2020.9231008

3. Gorbova J, Colovic M, Marjanovic M, Njegus A, Anbarjafari G (2019) Going deeper in hidden sadness recognition using spontaneous micro expressions database. Multimed Tools Appl 78:23161–23178. https://doi.org/10.1007/s11042-019-7658-5
4. Mamidisetti S, M. R A (2022) Multimodal depression detection using audio, visual and textual cues: a survey. NeuroQuantology 20:325–336. https://doi.org/10.14704/nq.2022.20.4.NQ22127
5. Dham, S, Sharma, A, Dhall, A (2017) Depression scale recognition from audio, visual and text analysis
6. Yadav U, Sharma AK (2021) Review on automated depression detection from audio visual clue using sentiment analysis. In: 2021 second international conference on electronics and sustainable communication systems (ICESC). IEEE, pp 1462–1467. https://doi.org/10.1109/ICESC51422.2021.9532751
7. Crestani F, Losada DE, Parapar J (2022) Early detection of mental health disorders by social media monitoring. Springer International Publishing, Cham. https://doi.org/10.1007/978-3-031-04431-1
8. Parapar, J, Martín, P, Losada DE and Crestani, F (2024) Overview of eRisk 2024: early risk prediction on the internet. In: Experimental IR meets multilinguality, multimodality, and interaction proceedings of the fifteenth international conference of the clef association (CLEF 2024). Springer International Publishing
9. Sampath K, Durairaj T (2022) Data set creation and empirical analysis for detecting signs of depression from social media postings. Presented at the. https://doi.org/10.1007/978-3-031-16364-7_11
10. S Kayalvizhi, Durairaj, T., Chakravarthi, BR, C, JM (2022) Findings of the Shared Task on Detecting Signs of Depression from Social Media. In: Proceedings of the second workshop on language technology for equality, diversity and inclusion. pp 331–338. Association for Computational Linguistics, Stroudsburg, PA, USA. https://doi.org/10.18653/v1/2022.ltedi-1.51
11. Poświata R, Perełkiewicz M (2022) OPI@LT-EDI-ACL2022: detecting signs of depression from Social media text using RoBERTa Pre-trained language models. In: Proceedings of the second workshop on language technology for equality, diversity and inclusion. Association for Computational Linguistics, Stroudsburg, PA, USA, pp 276–282. https://doi.org/10.18653/v1/2022.ltedi-1.40
12. Irfan, KM Micro-expression dataset. https://www.kaggle.com/datasets/kmirfan/micro-expressions. Last accessed 30 Oct 2024
13. Pedregosa F, Varoquaux G, Gramfort A, Michel V, Thirion B, Grisel O, Blondel M, Prettenhofer P, Weiss R, Dubourg V, Vanderplas J, Passos A, Cournapeau D, Brucher M, Perrot M, Duchesnay É (2011) Scikit-learn: machine learning in python. J Mach Learn Res 12:2825–2830
14. Buitinck, L, Louppe, G, Blondel, M, Pedregosa, F, Mueller, A, Grisel, O, Niculae, V, Prettenhofer, P, Gramfort, A, Grobler, J, Layton, R, Vanderplas, J, Joly, A, Holt, B, Varoquaux, G (2013) API design for machine learning software: experiences from the scikit-learn project
15. Fast, E, Chen, B, Bernstein, M.S. (2016) Empath. In: Proceedings of the 2016 CHI conference on human factors in computing systems. pp 4647–4657. ACM, New York, NY, USA. https://doi.org/10.1145/2858036.2858535
16. Lin, C., Hu, P., Su, H., Li, S., Mei, J., Zhou, J., Leung, H. (2020) SenseMood: depression detection on social media. In: ICMR 2020—proceedings of the 2020 international conference on multimedia retrieval. pp 407–411. Association for Computing Machinery, Inc https://doi.org/10.1145/3372278.3391932
17. Pampouchidou A, Marias K, Tsiknakis M, Simos P, Yang F, Lemaitre G, Meriaudeau F (2016) Video-based depression detection using local Curvelet binary patterns in pairwise orthogonal planes. In: 2016 38th annual international conference of the IEEE engineering in medicine and biology society (EMBC). IEEE, pp 3835–3838. https://doi.org/10.1109/EMBC.2016.7591564
18. Dumpala SH, Dikaios K, Rodriguez S, Langley R, Rempel S, Uher R, Oore S (2023) Manifestation of depression in speech overlaps with characteristics used to represent and recognize speaker identity. Sci Rep 13:11155. https://doi.org/10.1038/s41598-023-35184-7

19. Bucur, A-M, Cosma, A, Rosso, P, Dinu, L.P (2023) It's Just a matter of time: detecting depression with time-enriched multimodal transformers. In 200–215. https://doi.org/10.1007/978-3-031-28244-7_13
20. Navaneeth How to combine pyWavelet and openCV for image processing? How to Combine pyWavelet and openCV for image processing Last accessed 23 Jan 2025
21. Singh, S, Benedict, S (2020) Indian semi-acted facial expression (iSAFE) dataset for human emotions recognition. Presented at the https://doi.org/10.1007/978-981-15-4828-4_13

SchemeSmart: AI Assistant for Government Schemes

Pinjari Haneef, Krishna Vamsi Kotha, and K. Venkatraman

Abstract Government schemes empower citizens, but getting information on schemes available across various states in India remains a complex, many-sided affair. This pa per presents an AI-driven platform meant to streamline and make easier access to all of the government schemes across all 29 Indian states, through a conversational chatbot based on Mistral 8 x 7B models. The chatbot offers a totally smooth interface, by which citizens can ask questions and retrieve relevant information so as to better understand various schemes operating for their benefit. The advanced NLP techniques implemented by the chatbot facilitate the handling of large amounts of scheme data to personalize recommendations and summaries for us errors. Our results demonstrate that the AI chatbot significantly improves users' discovery, understanding, and navigation abilities through rigorous user testing and interaction analysis. Integrating the Mistral 8x7B model makes the response from the chatbot faster and more accurate so that citizens may quickly access any information deemed essential. This research proves that AI based solutions can involve more citizens in government activities and makes more informed decisions. The ease of access and comprehension of government services is essential to their efficacy. For information access and engagement, a user-friendly system is consequently crucial. In order to solve this, our AI-powered chatbot uses natural language comprehension to have a conversation.

Keywords Mistral-8x7B technology · Government schemes · Natural language processing · Chatbot

P. Haneef · K. V. Kotha (✉) · K. Venkatraman
Department of Computer Science, Amrita School of Computing, Chennai, India
e-mail: chowdarykrishnavamsi@gmail.com

P. Haneef
e-mail: phaneef8@gmail.com

H. Nagaraj et al. (eds.), *Advancements in Artificial Intelligence and Machine Learning*, Lecture Notes in Electrical Engineering 1572,
https://doi.org/10.1007/978-981-95-7449-0_3

1 Introduction

For many citizens, accessing government services and programs can become a pretty nerve-racking task, mainly because of eligibility and complex application processes. In general, most citizens who have the potential to take advantage of the initiatives are unable to find them or are puzzled in navigating the bureaucratic setting. To address this problem, we proposed the Mistral QA Chatbot, a chatbot powered by Mistral-8x7B Technology. This innovative platform,

therefore, shall be able to act as an easy-to-use interface for getting such vital information about government schemes easily and adequately, hence making it easier to check whether or not one is eligible for access.

The NLP capability used in the chatbot allows for intuitive interactions that empower the user by putting them in control of making well-informed decisions regarding their access to government services. Actually, the Mistral-8x7B model is designed to be as accurate as it responds. The model is focused on understanding and processing natural language inquiries [1]. This means that this chatbot will know the user's intention, retrieve relevant information, and give correct answers by relating them to big datasets of government schemes. Engaging with a virtual assistant, the chatbot sparks meaningful conversations with its audience, uncovering each question's or re quest's context so that appropriate and personalized responses to users are provided and lead people through programs and regulations.

Early tests of the chatbot have shown encouraging outcomes, demonstrating its capacity to furnish users with the proper, updated information. User testing and interaction analysis were carried out, and it was found that participants could go for more refinement in the understanding of content in the program, more confident regarding the request for help, and more on the comprehension of the application procedure. It is here that the interpretation capability of natural language by a chatbot allows increasing its interaction levels with users and, thereby increases the satisfaction level of users pretty sharply. Mistral-8x7B technology blends with providing quick and accurate response while performing other operations. This actually underscores the capacity of new AI tools for bridging the gap between citizens and core functions of the government in general, with a move toward enhancing access and efficacy of community services [2].

2 Background Study

2.1 *Why Mistral-8x7B?*

Mistral-8x7B, a large language model (LLM) created especially for conversational applications, has a number of benefits. It can comprehend and respond to intricate inquiries about government services due to its seven billion parameters. The model's proficiency in natural language processing allows it to evaluate user intent, extract pertinent data, and generate precise answers from the extensive datasets of government programs. By providing human-like interactions and tailored responses that cater to individual user demands, the chatbot's sophisticated contextual awareness boosts user confidence and engagement.

With advanced contextual understanding, a chatbot finds Mistral-8x7B more understandable, which will make it effective in interpreting queries from users and coming up with the correct answers or information fairly relevant to each user's needs. Responsiveness is needed to foster a good experience with the chatbot because it builds trust and encourages interaction with the end product. Mistral-8x7B is reliable and well-informed, thus enhancing the relationship.

between citizens and the chatbot. In this way, dealings with the chatbot can be very smooth, and confidence will increase in the ability of the user to cope with the complexity of government services [3].

However, the strong architecture of the technology also gives natural language processing capabilities to the chatbot, enabling understanding of both linguistic and context nuances. This ensures that the user receives personalized responses which meet their specific query needs, thus further enhancing the effectiveness of the service. Overall, the inclusion of Mistral-8x7B contributes not only toward enhancing the ability to serve users effectively by the chatbot but also toward a larger aim of increasing access to essential government services.

2.2 *Comparative Study*

We compared how effective Mistral-8x7B was for our particular use case, to other well-known LLMs. The reasoning, math/science, coding, and question–answering skills were among the main benchmarks on which we assessed performance. In this comparison investigation, the models LLAMA 2 (with variants 7B and 13B), Mistral (7B), and Gemma (7B) were used. These benchmarks are defined as follows:

- **Mistral 7B versuss GPT-3**: This comparison of Mistral-8x7B and GPT-3 con tains more information about their capabilities and applications in natural language processing (NLP). Mistral-8x7B has 7 billion parameters, making it a powerful model, but it is much smaller than GPT-3, which has 175 billion parameters. This slight difference reflects the complexity of each model's performance. Mistral-8x7B's user manual, like GPT-3's, contains many articles to assist the owner and create a human-like message.
- **Mistral 7B versus BART**: Mistral-8x7B and BART (Bidirectional and Auto regressive Transformer) are two complex languages with different features. Mistral-8x7B has 7 billion features, more than BART's 406 million features. Mistral-8x7B is trained with extensive text and excels at language understanding and data retrieval, relying on powerful context recognition to provide the appropriate response. BART, on the other hand, specializes in activities such as writing articles, translating, creating text, content creation, and other tasks that yield great

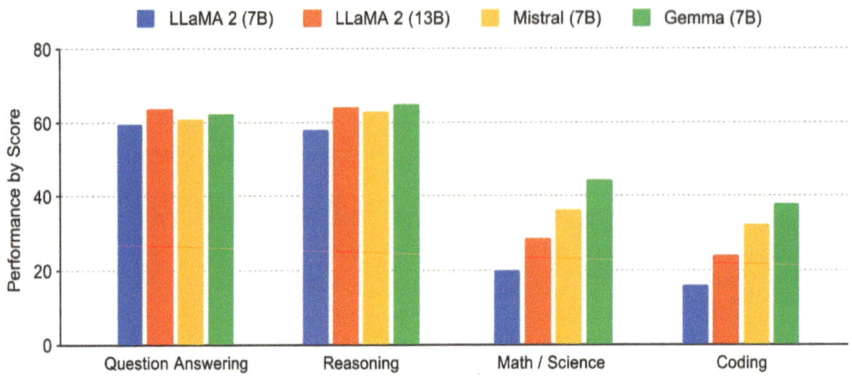

Fig. 1 Flow chart

results. While Mistral-8x7B has many applications due to its larger size, BART's intelligence makes it suitable for certain NLP tasks. From Fig. 1, the LLMs compared are LLAMA 2 (with parameter variants 7B and 13B), Mistral (7B), and Gemma (7B). Model Gemma (7B) outperformed the other three models in all four categories. LLAMA 2 (13B) ranks second in all functions except coding. It is worth noting that LLAMA 2. From Fig. 1, the LLMs compared are LLAMA 2 (with parameter variants 7B and 13B), Mistral (7B), and Gemma (7B). Model Gemma (7B) outperformed the other three models in all four categories. LLAMA 2 (13B) ranks second in all functions except coding. It is worth noting that LLAMA 2.

Figure 1 Flow chart (13B) has approximately twice as many parameters as Gemma (7B), so it should be more energy efficient. The performance of the remaining two models (LLAMA 2 (7B) and Mistral (7B)) was more mixed. LLAMA 2 (7B) out per forms Mistral (7B) in question–answering and reasoning, while Mistral (7B) out per forms LLAMA 2 (7B) in encoding

Fig. 2 Comparative Study Graph: Based on the comparative study of different models shown in Fig. 3, in order to test the model and compare it with other LLMs, we selected a few samples from the dataset and tested them with the samples. We then compared the generated answers with the reference answers, and their performances were evaluated using certain metrics. By calculating the probability of the generated answer relative to the reference answer, we found that, after testing, the Mistral-8x7B outperformed the other models and demonstrated better contextual understanding compared to the other models [4]

- **MMLU**: Evaluates a model's performance across academic disciplines, testing general knowledge.

- **HellaSwag**: Assesses a model's ability to understand context and predict the correct sentence ending.
- **Winograd**: Tests a model's ability to resolve pronoun references.

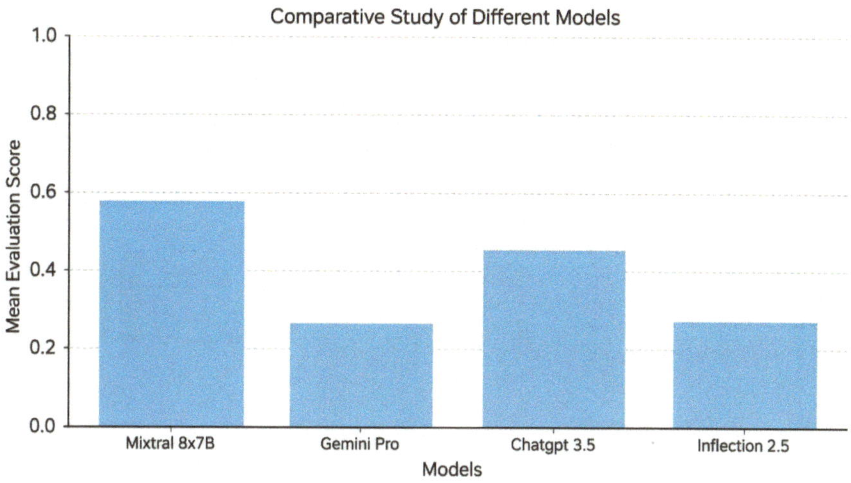

Fig. 2 Comparitive study graph

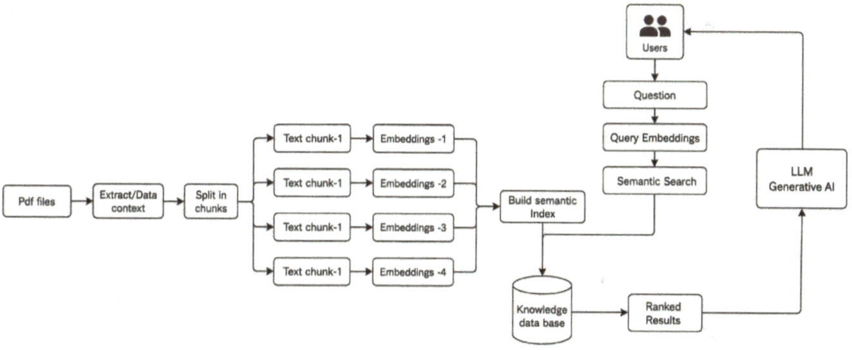

Fig. 3 Block diagram

- **Arc C(5) & Arc C(25)**: Measure reasoning and problem-solving abilities.
- **TriQA**: Evaluates the model's ability to answer multiple choice questions related to text.
- **TruthfulQA**: Tests if a model avoids hallucinating incorrect information.

From the Table 1(Metrices Evaluation) the performance of the Mistral Large, LLaMa 2.70B, GPT-3.5, GPT-4, Claude 2, Gemini Pro 1, and Mistral-8x7B models is evaluated using various benchmarks. MMLU, HellaS, WinoG, Arc C(5), Arc C(25), TriQA, and TruthfulQA are metrics used to measure multitasking, language comprehension, problem-solving, responsiveness, reading comprehension, fairness, and truthfulness. Mistral-8x7B performs well on MMLU (79.6%), HellaS (86.7%), WinoG (81.2%), and Arc C (85.5%), while GPT-4 performs well on HellaS (95.3%)

Table 1 Metrics evaluation

Model	MMLU (%)	Hel (%)	LaSwag (%)	Wingo (%)	Arc (C5) (%)	Arc (25) (%)	TriQA (%)
Mistral 8x7B	81.2	89.2	86.7	94.2	–	–	94.0
LLaMa 2 70B	69.9	87.1	83.2	86.0	85.1	77.6	–
GPT 3.5	70.0	85.5	85.2	85.2	44.7	–	–
GPT 4	–	86.4	95.3	87.5	–	96.3	–
Claude 2	–	78.5	–	–	91.0	–	87.5
Gemini Pro 1	71.8	84.7	–	–	–	–	–
Mistral 8x7B	79.6	86.7	81.2	–	85.8	–	–

and Arc C(25) with excellent performance (96.3%). These evaluations provide valuable data for educators and manufacturers to choose the best model for specific needs.

3 Literature Review

The development of conversational QA chatbots has received a lot of interest in recent years, with numerous techniques and technologies being tested to improve user engagement and access to information. Several research studies have focused on using sophisticated natural language processing (NLP) techniques and models to increase the efficacy of chatbot interaction.

In their separate literature, "Conversational AI: The State of the Art" by Madison Milne et al. and "The Effectiveness of Artificial Intelligence Conversational Agents in Health Care: A Systematic Review" by Sharma et al., Milne et al. [5].Discuss the evolution of automated response systems. From early automated response systems that used basic rule-based algorithms, the development of conversational agents has been an exciting journey. These systems evolved from simple text-based interactions to the sophisticated machine learning models we see today, as technology expanded, including increasingly sophisticated methodologies. Advances in natural language processing have driven this shift by making it possible for chatbots to comprehend and produce responses that are more human-like. Additionally, the application of AI in industries such as healthcare demonstrates its revolutionary potential, since chatbots may expedite patient contacts and deliver real-time information [6].

There are still issues, though, especially with guaranteeing response accuracy and taking into account various linguistic subtleties, both of which are essential for broad acceptance and user confidence. These challenges have advanced our knowledge of interactive AI technologies. Researchers examine the development of chatbots from early systems to current machine learning models, including the problems of natural language interpretation and the emergence of sophisticated standards like GPT-3 [7]. Sharma et al. investigate the impact of AI on public services, particularly healthcare, and emphasize the potential of chatbots to enhance public involvement and access to government services [8]. They also address significant issues such as the limitations of current systems in proper translation and voice recognition.

In their study, "Natural Language Processing in Radiology: A Systematic Review," Ewoud Pons et al. investigate how NLP techniques can be applied to improve media coverage of government officials. Their study provides pipelines that recognize significant entities, answer questions, and provide summaries to give users crucial information and context [9]. The paper "A Survey on Conversational Agents/Chatbots Classification and Design Techniques" emphasizes the need to understand user intent for conversational chatbots. The survey explores rule-based approaches, machine learning models, and hybrids for target classification [10]. This inquiry sheds light on challenges such as regulating curiosity and the potential of deep learning algorithms to improve chatbot accuracy.

Conversational agents, or chatbots, have grown in popularity in public services as a means of improving public relations and engaging with the government. Kumar et al. discuss how AI-powered chatbots can swiftly and efficiently answer user inquiries and facilitate contact with government organizations and services [11]. According to Simone Angioni et al., providing visual information to a chatbot can improve its usefulness. "Integrating interactive humans and knowledge maps into the learning space" explores how chatbots can respond to questions more effectively by using design and event information from a knowledge map [12].

In "Interactive storytelling for children: A case-study of design and development considerations for ethical conversational AI," Jennifer Chubb et al., discuss the use of social media intelligence to enhance access to government services for underprivileged individuals. Donagh Horgan et al.[13].'s essay, "Frameworks for Citizens' Participation in Planning: From Conversational to Smart Tools" addresses how chatbots can promote public engagement and transparency in government processes [13]. They discuss critical issues such as privacy and security, as well as the practical challenges of implementing AI into public opinion polls and civic engagement initiatives.

In "Multimodal Learning for Identifying Opportunities for Empathetic Responses," Leili Tavabi et al.[14] explore the integration of various input and output models within the realm of AI applications in the public sector, aiming to enhance user experiences and accessibility to government services. This study investigates the use of auditory, visual, and other adjustments to increase the overall efficacy of these services. Similarly, the study "The AI-enabled Chatbot Framework for Intelligent Citizen-Government Interaction for Delivery of Services" examines the challenges of implementing and supporting conversational AI systems at scale in government agencies. The authors suggest methods to manage the knowledge base, manage multiple user queries, and ensure security and performance [15].

The literature review for the project "Survey of Text Document Summarization Based on Ensemble Topic Vector Clustering Model" by G Bharathi Mohan and R Prasanna Kumar explores how to use content-specific common vector clustering techniques to realize efficient and effective text and content aggregation. The authors present a two-level approach: textual abstraction and abstract text [16]. The content editor extracts the most important information from the original document, while the content editor gathers the information back to create the relevant content of the query.

Overall, these connected publications highlight the many methodologies and technologies used in the creation of conversational QA chatbots. Researchers aim to use advanced NLP models, objective search, and graph information to create chatbots that can accurately answer user questions based on context [17].

4 Methodology

4.1 Dataset Description

The Government Information Schemes dataset is an all-encompassing collection of data pertaining to different schemes undertaken by the Government of India. There are various issues from health and education to agriculture and welfare that it deals with. The documentation covers 29 states, with each state document having around 10,000 lines. Among these papers, there are 11 important lines that disclose significant names of the program, purpose, right of usage, and beneficiaries of the respective schemes. This will be helpful in understanding and analyzing the endeavor the Indian government has undertaken for welfare and development activities. It also aids researchers and policymakers in making proper assessments in arriving at a conclusion that measures taken by the government were successful or not in influencing other sectors [18].

4.2 Data Preprocessing

- **Start**: This indicates the start of the chatbot programme or a specific function inside the programme. Marking the chatbot application's entrance point or a specific function inside it and indicates where the chatbot's operations begin.
- **Process PDF's**: The langchain library uses the PyPDFLoader to load PDF files from a chosen directory, effectively extracting text content from the files and incorporating this into a coherent collection of document texts. This is basically an important step for ensuring that the chatbot will have information that is derived from government schemes; thus, pertinent and relevant information returning answers to the user inquiries.

4.3 Splitting Documents into Chunks

Immediately after importing the content of the PDF documents, the chatbot splits the content into smaller pieces for effective processing. The splitting here is done by using the Recursive Character Text Splitter, splitting the documents into chunks of 10,000 characters, but this time with a 20-character overlap between consecutive chunks. This continues to help the chatbot manage bigger text sizes while still keeping contextual integrity for better understanding and response generation.

4.4 Creating Embeddings

To understand semantics, text chunks are converted into numerical vector representations using the 'sentence-transformers/all-MiniLM-L6-v2' model from Hugging Face. Trained to capture semantic meaning, this model generates dense vector embeddings. This lets the chatbot understand the context and find information more accurately. The process includes:

- **Word Encoding**: Mapping words in a text chunk to vector embeddings.
- **Contextual Embeddings**: Considering the relationships of words in a text sequence.
- **Average Pooling**: Generating a chunk embedding by averaging all words' contextual embeddings. Mathematically, this is:

$$Average\ Pooling = \frac{1}{n}\sum_{i=1} v_i \tag{1}$$

Where v_i is the vector embedding for word i.

4.5 Creating Vector Store

The vector store uses the FAISS library (Facebook AI Similarity Search), optimized for fast similarity search and retrieval of dense vectors. FAISS efficiently stores and queries text vector representations, enabling the chatbot to quickly retrieve relevant information and ensure real-time interaction for users. FAISS supports operations including insertion, deletion, indexing, and searching. It is ideal for high-dimensional data and fast retrieval, improving our model's performance with large datasets.

4.6 Model Architecture

Mistral-8x7B is an extremely large model built custom to be used for the purpose of producing functional applications of language. It belongs to the Transformerbased family, which is quite famous in its abilities to understand and generate human-like text. However, with a significant improvement in training methodologies and design architecture, it finds its place among the likes of GPT-3 or InstructGPT models.

- **Model Size**: Mistral-8x7B has 7 billion parameters, which enable it to determine intricate patterns and relationships in data sets and thus provides better comprehension and response generation.
- **Transformer Architecture**: Similar to most other Transformer-based mod-.

els, Mistral-8x7B is based on the Transformer architecture, which has now become the golden standard for most modern natural language processing models. It allows text processing and generation to be performed in a uniform manner and makes it a good fit for a variety of applications in NLP [19].

4.7 Model Training

- **Create Chain**: At this stage, you are required to create a chain object, which is a key constituent of the workflow and the entire processing mechanism for your chatbot. The chain object brings together various elements that constitute the processing part of the operation in the chatbot to interact freely and to retrieve information compellingly.

4.8 Update History

- **Botrieve from (using LLM)**: This is the critical information retrieval phase in the chatbot process where information is retrieved with the assistance of a specialized model called the Large Language Model (LLM). This utilizes the power of the LLM in an efficient manner to process user queries correctly and retrieve information from the vector store to create an efficient and responsive chatbot [20].

This phase updates the history of the conversation by the chatbot, adding on new interactions or messages. The functionality allows important details of previous interactions critical to providing accurate and contextually relevant responses for the ongoing conversation to be stored by the chatbot.

4.9 Return Responses

This is the final stage of the information retrieval process and can be used to show the history of a conversation. It may indicate that an interaction of a certain kind has ended or that the entire chatbot session has concluded. The return response functionality therefore allows the user to read back from previous exchanges, which also helps provide better transparency and general user experience.

4.10 Block Diagram

From Fig. 4, The block diagram describes a semantic search and information retrieval system for processing and analyzing PDF files for effective querying.

Data Ingestion:

- **Extract/Data Context:** Extract important information and context from the given PDF archive.
- **Split in Chunks:** Split data into smaller, more manageable files.

Text Embedding:

- **Embeddings-1, Embeddings-2, Embeddings-3, Embeddings-4**: Use methods such as word patterns or word embeddings to create visual representations (embeddings) of text. Embeddings preserve the semantics of the text.

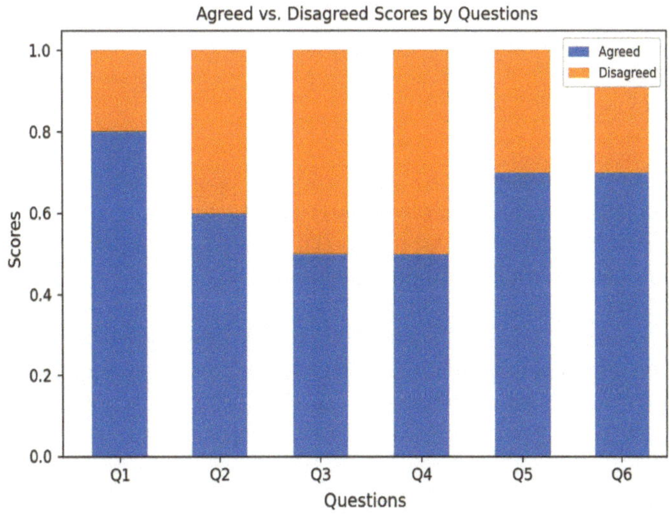

Fig. 4 Analysis graph

Index Building

- The generated embeddings are leveraged to create semantic indexes, enabling fast similarity search and retrieval.

Query Processing:

- **Question**: Accepts user queries or questions as input.
- **Query Embeddings**: Create a block text equivalent embed for the given query.
- **Semantic Search**: Perform a similar search across query embeddings and index embeddings to find relevance.

4.11 Result Retrieval

- **Knowledge Database**: Acts as a knowledge base by storing indexed information and embedded information.
- **LLM (Generative AI)**: A language or artificial intelligence tool designed to help process and generate relevant responses based on collected data.
- **Ranked Results**: The obtained results are ranked according to their relevance or similarity to the query.

4.12 User Interaction

- **Users**: Represents the end-users who submit queries and interact with the system.

5 Results and Discussion

The system's performance was evaluated using user satisfaction and response accuracy. We achieved a 73.02% accuracy, a promising result for a new system. The following section elaborates on the results and areas for improvement. The results of the system call for the efficiency of the Conversational Chatbot System in helping the people with their queries about the multiple schemes managed by the government of India across all the states. The user satis faction is also shown quite satisfactory, and there are many users who are claiming that the information received from the chatbot was much in formative, relevant, and precise as well.

An evaluation process was conducted by administering surveys to the users, as a result of which information was gathered that showed that the chatbot had indeed succeeded in bettering the user experience by streamlining the access route for information and enhancing the confidence of the users in navigating through the

Table 2 Questions to users

Question number	Question description
1	Are you able to follow the chatbot's conversation?
2	How satisfied are you with the performance of the Chatbot?
3	How accurate were the responses provided by the Chatbot?
4	Do you have trouble finding information on government schemes?
5	Would you like to use this chatbot in the future?
6	Have you found the chatbot helpful in understanding government schemes?

complexity of govern ment services. However, limiting factors related to the treatment by the chatbot of the ambiguous queries and the depth of information on the schemes were also observed in the evaluation process. These insights underscore the hopeful role of such AI-driven solutions as Mistral-8x7B in bridging the gap between citizens and government ser vices. There are, however, some areas that require further enhancement to ensure full and effective user engagement.

Feedback Collection Process

Users are asked to use the model to choose the video that suits their mood. After being contacted, they received a feedback form to evaluate their experience. A total of thirty responses were collected for analysis (Table 2).

Calculation of Score

From Eq. (2), the "score" statistic is calculated to measure the chatbot's performance in providing appropriate responses. The score is defined as the probability of a good answer (such as "yes" or "satisfactory") relative to all answers. The formula used to calculate the score (score) is as follows:

$$\text{score} = \frac{\text{Number of good answers}}{\text{Total number of answers}} \tag{2}$$

By analyzing Fig. 5, it is evident that around 80% of users were able to follow the chatbot's conversation.

6 Limitations and Future Work

Although results are promising, the system has limitations that offer opportunities for future enhancement. We will discuss these limitations and proposed solutions.

6.1 Response Accuracy

- **Accuracy of Response**: Complex qualifying requirements and fine linguistic nuances might occasionally be challenging for the chatbot to understand. Incomplete responses to multi-parameter queries could result from it misinterpreting words, acronyms, or double negatives.

6.2 Future Enhancements Will Concentrate on

Using a dataset of questions with intricate eligibility restrictions to finetune the model.

Putting in place an error detection system to check responses before sending them back to the user.

Unanswered Questions

Some user inquiries are ambiguous or subject to multiple interpretations. For instance, asking "What are the benefits?" without mentioning a plan may lead to the return of too much or unrelated information.

- **Future enhancements will concentrate on**:
  - Better user intent identification that integrates contextual awareness and keyword analysis.
  - Asking follow-up questions, such as "Are you referring to scheme A or B?" to elicit the user's intention.

Information Depth

The current chatbot uses only PDF text, which might lack all the details of the scheme and real-world nuances.

- Future improvements will focus on:
  - Integrating with external databases and APIs to provide real-time and comprehensive information.
  - Exploring ways to display complex information clearly.

6.3 User Engagement

While the initial user feedback is positive, there's room to improve user engagement.

- **Future improvements will focus on:**
  - User interface enhancements to be more user-friendly.

- Gamification techniques to make the interactions more engaging.

Privacy

Given the potentially sensitive nature of user queries, ensuring data security and privacy is crucial.

Implementation

- Scaling the chatbot to a national level presents various challenges, including infrastructure requirements.
- **Future improvements will focus on:**
  - Collaborating with government agencies for deployment and adoption.
  - Developing a cross-platform model.
  - Improving data management.

7 Conclusion

In conclusion, our Mistral-8x7B-powered Q&A chatbot significantly enhances access to government schemes nationwide. Achieving 73.02% accuracy, the system highlights the potential of large language models for improving user interaction and public service communication. Identified limitations like response accuracy, ambiguity handling, and information depth, point to specific areas for refinement. Proposed solutions like fine tuning with targeted datasets, better intent recognition, integrating external data, and prioritizing data security, will help improve the chatbot for use by the people.

As we continue to refine and develop this technology, we anticipate a future where the engagement between citizens and government functions becomes increasingly efficient and effective. This initiative not only aims to empower individuals with essential information about various government programs but also fosters greater transparency and accessibility in public service delivery. Through ongoing improvements and user 14 feedback, the chatbot can evolve to meet the diverse needs of the population, ultimately contributing to a more informed and engaged citizenry.

References

1. Jiang, AQ, Sablayrolles, A, Mensch, A, Bamford, C, Chaplot, DS, Casas, DDL, Sayed, WE (2023). Mistral 7B. https://doi.org/10.48550/arXiv.2310.06825
2. Oosterbeek, JW, Stern, M, Braune, H, Ewert, K, Hirsch, M, Hollmann, F, Zubieta-Lupo, RJ (2023) MISTRAL campaign in support of W7-X long pulse operation. In: EPJ web of conferences Vol 277, p 04009. EDP Sciences. https://doi.org/10.1051/epjconf/202327704009
3. Colombo P, Pires TP, Boudiaf M, Culver D, Melo R, Corro C, Desa M (2024a) SaulLM-7B: a pioneering large language model for law. https://doi.org/10.48550/arXiv.2403.03883

4. Allaham M, Diakopoulos N (2024) Supporting anticipatory governance using LLMs: evaluating and aligning large language models with the news media to anticipate the negative impacts of AI. https://doi.org/10.48550/arXiv.2401.18028
5. Schoenegger P, Tuminauskaite I, Park PS, Tetlock PE (2024) Wisdom of the silicon crowd: LLM ensemble prediction capabilities match human crowd accuracy. https://doi.org/10.48550/arXiv.2402.19379
6. Touvron H, Martin L, Stone K, Albert P, Almahairi A, Babaei Y, Scialom T (2023) Llama 2: open foundation and fine-tuned chat models. https://doi.org/10.48550/arXiv.2307.09288
7. Khatri C, Venkatesh A, Hedayatnia B, Gabriel R, Ram A, Prasad R (2018) Alexa prize—state of the art in conversational AI. AI Mag 39(3):40–55. https://doi.org/10.1609/aimag.v39i3.2810
8. Milne-Ives M, de Cock C, Lim E, Shehadeh MH, de Pennington N, Mole G, Meinert E (2020) The effectiveness of artificial intelligence conversational agents in health care: systematic review. J Med Internet Res 22(10):e20346. https://doi.org/10.2196/20346
9. Pons E, Braun LM, Hunink MM, Kors JA (2016) Natural language processing in radiology: A systematic review. Radiology 279(2):329–343. https://doi.org/10.1148/radiol.16142770
10. Hussain S, Ameri Sianaki O, Ababneh N (2019) A survey on conversational agents/chatbots classification and design techniques. In: Web, artificial intelligence and network applications: proceedings of the workshops of the 33rd international conference on advanced information networking and applications (WAINA-2019), vol 33. Publishing, Springer International, pp 946–956. https://doi.org/10.1007/978-3-030-15035-8_93
11. Nirala KK, Singh NK, Purani VS (2022) A survey on providing customer and public administration based services using AI: chatbot. Multimed Tools Appl 81:22215–22246. https://doi.org/10.1007/s11042-021-11458-y
12. Meloni A, Angioni S, Salatino A, Osborne F, Reforgiato Recupero E, Motta E (2023) Integrating conversational agents and knowledge graphs within the scholarly domain. IEEE Access 11:22468–22489. https://doi.org/10.1109/ACCESS.2023.3253388
13. Chubb J, Missaoui S, Concannon S, Maloney L, Walker JA (2022) Interactive storytelling for children: a case-study of design and development considerations for ethical conversational AI. Int J Child Comput Interact 32:100403. https://doi.org/10.1016/j.ijcci.2021.100403
14. Horgan D, Dimitrijevi'c B (2019) Frameworks for citizens participation in planning: from conversational to smart tools. Sustain Cities Soca 48:101550. https://doi.org/10.1016/j.scs.2019.101550
15. Tavabi L, Stefanov K, Nasihati Gilani S, Traum D, Soleymani M (2019) Multimodal learning for identifying opportunities for empathetic responses. In: 2019 International Conference on Multimodal Interaction, pp 95–104. https://doi.org/10.1145/3340555.3353750
16. Devlin J, Chang MW, Lee K, Toutanova K (2018) BERT: pre-training of deep bidirectional transformers for language understanding. https://doi.org/10.48550/arXiv.1810.04805
17. Hasan, I, Rizvi, S, Jain, S, Huria, S. (2021). The AI-enabled Chatbot framework for intelligent citizen-government interaction for delivery of services. In 8th international conference on computing for sustainable global development (INDIACom), New Delhi, India, pp 601–606
18. Bharathi Mohan, G, Prasanna Kumar, R (2022). Survey of text document summarization based on ensemble topic vector clustering model. In IoT based control networks and intelligent systems: proceedings of 3rd ICICNIS 2022 pp 831–847. Springer nature Singapore. https://doi.org/10.1007/978-981-19-5845-8_60
19. Senthilnathan S, Chandrasekar D, K (2023) Government schemes for the success of women entrepreneurs. Int J Sci Res Eng Manag 07:1–11. https://doi.org/10.55041/IJSREM27668
20. Colombo P, Pires TP, Boudiaf M, Culver D, Melo R, Corro C, Desa M (2024b) SaulLM-7B pioneer large lang model law. https://doi.org/10.48550/arXiv.2403.03883

Test Case Generation from User Requirements Using LLMs & gen. AI: A Survey

Sumit Suryawanshi, Awadhoot Khutwad, Soumya Garg, Ayush Bulbule, Madhuri Wakode, Geetanjali Kale, and Nikhil Karkhanis

Abstract Manual test case creation is time-consuming, error-prone, and often misses edge cases, leading to incomplete test coverage and undetected bugs. Existing tools struggle with complex validation logic and lack automated test generation from visual inputs, creating gaps in functional and visual validation (<author>, First International Symposium on Empirical Software Engineering and Measurement (ESEM 2007). IEEE, <publisher-loc>, 2007). To address these issues, we propose an AI-driven approach leveraging multimodal large language models (LLMs) (Fan, Angela, Beliz Gokkaya, Mark Harman, Mitya Lyubarskiy, Shubho Sengupta, Shin Yoo, and Jie M. Zhang, *In 2023 IEEE/ACM International Conference on Software Engineering: Future of Software Engineering (ICSE-FoSE). IEEE*, <publisher-loc>, *2023*). LLMs automate diverse test case generation, including functional, edge, and boundary cases, by analyzing code, recognizing validation logic, and generating test scripts [1, 2]. Additionally, they enable test case generation from images, bridging functional and visual validation gaps. A literature review (2000–2023) explores UML-based and NLP-driven test generation. Future research aims to enhance NLP techniques, non-functional testing, and traceability, improving software quality and reducing testing costs in complex development environments (Siniaalto, Maria, and Pekka Abrahamsson, First International Symposium on Empirical Software Engineering and Measurement (ESEM 2007). IEEE, <publisher-loc>, 2007).

S. Suryawanshi (✉) · A. Khutwad · S. Garg · A. Bulbule · M. Wakode · G. Kale
Department of Computer Engineering, SCTR's Pune Institute of Computer Technology, Pune, India
e-mail: suryawanshisumit8787@gmail.com

M. Wakode
e-mail: mswakode@pict.edu

G. Kale
e-mail: gvkale@pict.edu

N. Karkhanis
Sarvatra Technologies, Pune, India

H. Nagaraj et al. (eds.), *Advancements in Artificial Intelligence and Machine Learning*, Lecture Notes in Electrical Engineering 1572,
https://doi.org/10.1007/978-981-95-7449-0_4

1 Introduction

Software testing is crucial for ensuring software quality, directly impacting cost and delivery timelines [3–5]. It verifies that a system meets specified requirements, with effectiveness often measured by customer satisfaction. As defined in [6], software testing is *"a process of exercising or evaluating a system or system components by manual or automated means to verify that it satisfies specified requirements"*.

Traditional test case generation is manual, labor-intensive, and error-prone. However, Large Language Models (LLMs) offer a promising solution by automating test case generation from user requirements, enhancing efficiency and accuracy [1, 2, 7, 8].

This survey explores LLM-based test case generation, reviewing methodologies, strengths, and limitations [9–13]. We highlight key advancements, challenges, and future directions, aiming to advance automated testing frameworks and improve software quality.

2 Literature Review

- Generating Test Scenarios from NL Requirements using Retrieval-Augmented LLMs: An Industrial Study [9]

  Chetan Arora, Tomas Herda, Verena Homm introduce RAGTAG (Retrieval-Augmented Generation-based Test scenario Generation), a tool that processes NL requirements, an optional test scenario, domain documents, and a glossary. By retrieving context from a vectorized database, it ensures domain-specific relevance. Tested on Austrian Post projects, it showed high accuracy but required expert refinement. Its ability to handle bilingual requirements and improve efficiency makes it valuable for complex domains.
- **A Tool for Test Case Scenarios Generation using Large Language Models** [10] Abdul Malik Sami and Zeeshan Rasheed (Tampere University) introduce a web-based tool using OpenAI's LLMs to generate prioritized test case scenarios from user requirements, streamlining software testing. While reducing manual effort, challenges like "LLM hallucinations" affect reliability. The authors suggest integrating Retrieval-Augmented Generation (RAG) to address this. Content analysis automation is still developing, requiring further refinement for accuracy and robustness.
- **Large Language Models as Test Case Generators: Performance Evaluation and Enhancement** [11] Kefan Li and Yuan Yuan (Beihang University) propose TestChain, a multi-agent framework enhancing LLM test case generation. It includes a Designer Agent for diverse inputs and a Calculator Agent for output mapping via Python, improving accuracy. However, it relies on robust models like GPT-3.5/4, limiting weaker models. Findings are based on HumanEval and LeetCode-hard, which may not generalize to real-world tasks.

- **Reinforcement Learning from Automatic Feedback for High-Quality Unit Test Generation** [14] Benjamin Steenhoek, Michele Tufano, Neel Sundaresan, and Alexey Svyatkovskiy (Microsoft, USA) explore reinforcement learning (RL) to improve LLM-generated test cases. They address issues like test smells, missing assertions, and improper method calls. Using RL with static quality metrics, test quality improved by 21.4% in method invocation and assertion usage, outperforming GPT-4 on four of seven quality metrics. However, challenges remain in generating descriptive test documentation, and RL training instabilities sometimes reduced test case diversity.
- **Generation of Test Cases from Software Requirements Using Natural Language Processing** [12] Ravi Prakash Verma and Dr. (Prof.) Md. Rizwan Beg demonstrate the use of NLP techniques like POS tagging and parsing to generate test cases from natural language requirements, reducing ambiguity early in development. Parsed requirements are transformed into knowledge representation graphs, enabling structured test case extraction. The approach ensures comprehensive coverage using boundary value analysis and graph traversal. However, it struggles with complex or ambiguous requirements and relies on POS tagging accuracy. Manual intervention is needed for tasks like graph merging, and a fully automated tool is lacking.

3 Methodologies

3.1 Ragtag [9]

RAGTAG automates test scenario (TS) generation from natural language (NL) requirements using Retrieval-Augmented Generation (RAG) with Large Language Models (LLMs). It integrates domain-specific knowledge and addresses bilingual and domain-specific challenges.

- *NL Requirements*: Supports bilingual inputs for broader applicability.
- *Few-shot Prompting*: Uses example TSs to improve LLM performance.
- *Structured Prompting*: Ensures format consistency through predefined templates.

To improve TS generation accuracy, RAGTAG retrieves relevant domain-specific context by:

- *Document Indexing*: Splitting documents into 512-token chunks and embedding them with Sentence-BERT.
- *Context Retrieval*: Using cosine similarity to extract top-k relevant passages.
- *LLM-Based Generation*: Producing structured TSs while managing token limits and bilingual adaptation.

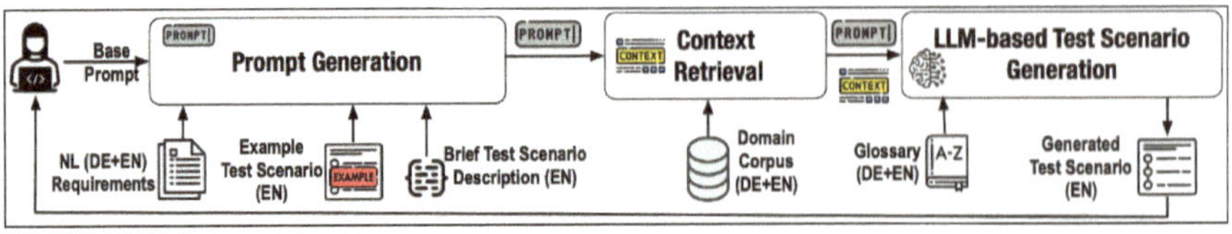

Fig. 1 Retrieval-augmented generation (RAG)-based test scenario generation approach

Evaluation, Findings, and Trends: RAGTAG was tested in industrial projects at Austrian Post Group IT, comparing different configurations (e.g., GPT-3.5 versus GPT-4.0, few-shot versus zero-shot). Results showed:

- *Best Performance*: GPT-3.5 in few-shot mode with one context chunk yielded the highest BLEU, ROUGE, and METEOR scores.
- *Expert Validation*: Four domain experts confirmed strong alignment with real-world testing needs.
- *LLM Behavior*: GPT-4.0 responses were overly verbose, whereas GPT-3.5 produced more concise, practical TSs.

Challenges:

- *Bilingual Mistranslations*: Occasional inaccuracies requiring human correction.
- *Context Sensitivity*: Effective test generation depends on precise context retrieval.
- *Human Intervention*: Manual corrections are needed for edge cases and domain-specific gaps.

Conclusion: RAGTAG enhances test scenario generation from NL requirements, streamlining quality assurance. Despite challenges with bilingual and domain-specific nuances, experts found the approach feasible and coherent. Further refinements can improve scalability, making RAGTAG a valuable tool for large-scale software projects (Fig. 1).

3.2 TestChain Framework [11]

This study evaluates LLMs for test case generation through experiments on four models: two open-source (StarChat, CodeLlama) and two closed-source (GPT3.5, GPT-4). The comparative analysis assesses their effectiveness in generating high-quality test cases (Fig. 2).

Data Collection and Prompt Design:

- **Datasets**: The study used HumanEval (Python benchmark with canonical solutions) and LeetCode-hard (manually verified complex tasks).
- **Prompt Design**: Standardized prompts instructed LLMs to generate basic, edge, and large-scale test cases while avoiding bias by removing example-based conditioning.

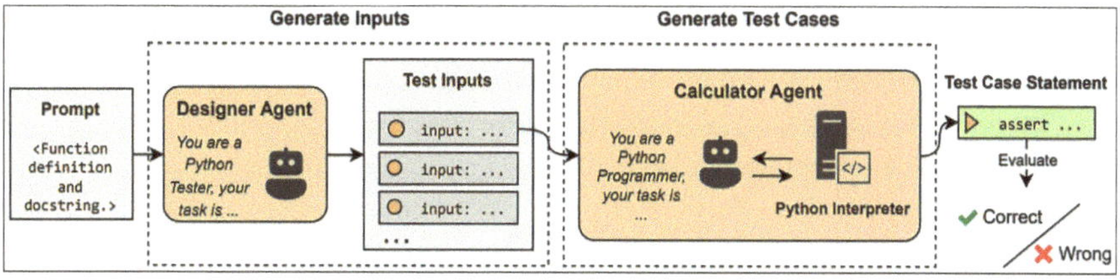

Fig. 2 Illustration of the TestChain framework

Experimental Setup: The results are evaluated using three key metrics: *Correctness*, *Strength*, *Overall Quality* of the output test cases. Two prompting paradigms were explored:

- *0-shot Prompting*: No prior examples given.
- *1-shot Prompting*: Single example guides LLM output.

Design: A multi-agent framework enhancing LLM-based test generation through:

- *Designer Agent*: Generates diverse test inputs.
- *Calculator Agent*: Uses a Python interpreter to ensure accurate computation.

TestChain outperformed baseline methods, improving accuracy significantly.

Findings and Trends:

- LLMs excel in simple tasks but struggle with complexity, leading to high error rates in advanced cases.
- Test case quality remains critical for reliable testing.
- Growing focus on structured, modular frameworks and expanded quality metrics.
- External tool integration enhances LLM-driven test generation, improving accuracy and overcoming limitations.

Challenges and Gaps:

- Complex input-output mappings and difficult problems reduce accuracy and reliability.
- Quality inconsistencies and dependency on high-end models limit accessibility.
- Lack of research on weaker models, improved metrics, and real-world complexity representation.
- Integration challenges remain underexplored.

Conclusion: TestChain enhances LLM-based test generation but faces challenges in model capability, evaluation, and real-world application. Future research should focus on improving weaker models, refining metrics, and strengthening tool integration.

3.3 RLSQM [14]

RLSQM enhances test case generation by integrating Reinforcement Learning (RL) with Proximal Policy Optimization (PPO) to refine Large Language Models (LLMs) like Codex and GPT. It rewards models for assertions, method calls, and syntax correctness, iteratively improving test reliability (Fig. 3).

- *Baseline Test Generation*: LLMs generate initial test cases, often lacking assertions and containing test smells.
- *Reward Model*: Evaluates test quality based on syntax, assertions, and redundancy.
- *PPO Optimization*: RL refines test case generation iteratively.
- *Supervised Fine-Tuning*: LLMs are pretrained on structured, high-quality test datasets before RL.
- *Evaluation Metrics*: Tests are assessed on correctness, strength, and test smells.

Evaluation, Findings, and Trends: RLSQM was tested on open-source unit test datasets, comparing LLM-generated tests with and without RL. Results demonstrated:

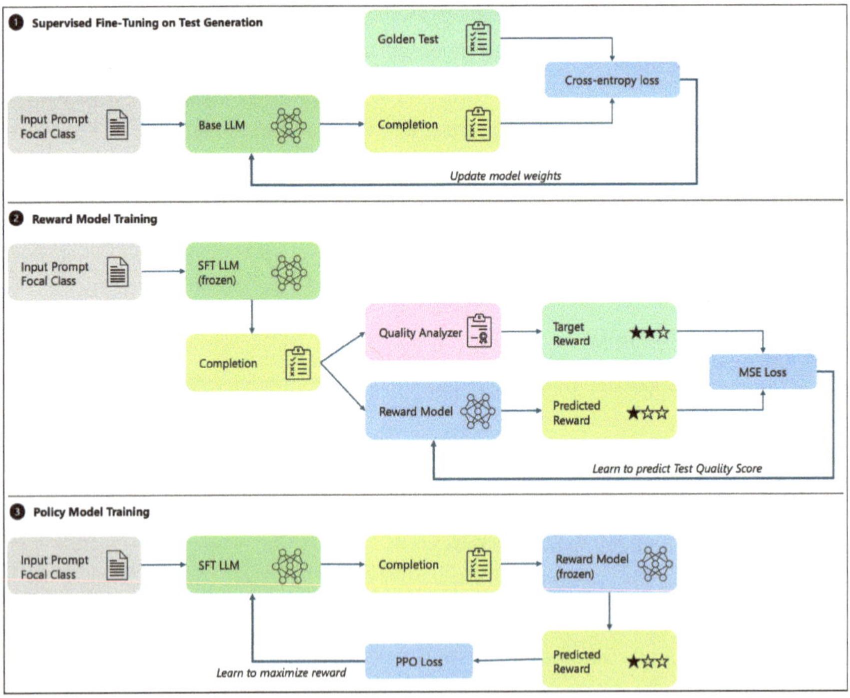

Fig. 3 Overview of RLSQM: [9] Fine-tunes Codex with quality test cases [10]. Uses a reward model to score tests [15]. Optimizes generation with PPO

- *Test Quality Improvements*: RL increased assertion presence and correct method calls by 20%, reducing test smells.
- *Static + Dynamic Metrics*: Combining syntax and test coverage metrics enhances accuracy.
- *Reward-Based Optimization*: Guides LLMs toward best practices, minimizing redundant logic.
- *Human Feedback*: Helps address complex and ambiguous testing scenarios.

Challenges:

- *Test Quality*: LLMs often produce redundant logic, missing assertions, or inefficient tests.
- *Scalability*: RL is computationally expensive, limiting large-scale deployment.
- *Generalization Across Domains*: LLMs struggle to adapt across languages and frameworks.
- *Training Bias*: Publicly available datasets may introduce biases, affecting test quality.

Conclusion: RLSQM enhances LLM-generated test cases by improving assertions and syntax correctness through RL. However, scalability, domain generalization, and training biases remain challenges. Future research should focus on RL efficiency, dynamic metric integration, and human feedback to improve real-world applicability.

3.4 Graph-Based NLP Test Case Generation [12]

Software testing consumes 40–70% of the testing effort, and traditional methods struggle with capturing non-functional requirements. This paper surveys an approach using Natural Language Processing (NLP) for automated test case generation from software requirements written in natural language, reducing ambiguities and improving efficiency.

Approach:

- *Preprocessing the Requirements*: Text normalization, stop-word removal, stemming, and sentence simplification. Domain-specific terms are handled via dictionaries.
- *Part-of-Speech (POS) Tagging*: Assigns part-of-speech labels to words with 96% accuracy, reducing ambiguity.
- *Parsing*: A syntactic tree is generated using a maximum entropy model trained on the Penn Treebank corpus.
- *Knowledge Representation* via *Graphs*: Nouns become nodes, verbs and prepositions form edges, and conditions are labeled as transitions.
- *Merging Graphs*: Subgraphs from individual sentences are merged into a unified representation, removing redundancies.

6 Future Work Direction

Future work should enhance the **quality**, **accuracy**, and **scalability** of LLM-generated test cases while integrating human feedback and dynamic metrics. Key research directions include:

1. Improve bilingual and **domain-specific** requirement handling in RAGTAG with advanced **NLP**.
2. Expand TestChain for weaker LLMs and develop **comprehensive** test quality metrics [18].
3. Scale RLSQM by integrating human feedback for better test generation.
4. Develop NLP and graph models for complex requirements with logic and formulas [10].
5. Integrate LLM-based test generation into **CI/CD** pipelines for regression testing and security detection.

References

1. Bhatia S, Gandhi T, Kumar D, Jalote P (2024) Unit test generation using generative ai: A comparative performance analysis of autogeneration tools. In: Proceedings of the 1st international workshop on large language models for code, pp 54–61
2. Ma, W., S. Liu, W. Wang, Q. Hu, Y. Liu, C. Zhang, L. Nie, and Y. Liu (2023*)* The scope of ChatGPT in software engineering: a thorough investigation
3. Myers GJ, Sandler C, Badgett T (2011) The art of software testing. John Wiley and Sons
4. Pezze M (2008) Software testing and analysis: process, principles, and techniques. John Wiley ans Sons
5. Anwar N, Kar S (2019) Review paper on various software testing techniques and strategies. Global J Comp Sci Technol 19(2):43–49
6. IEEE, (2011) ISO/IEC/IEEE international standard—systems and software engineering, ISO/IEC/IEEE 29148:2011(E), pp 1–94
7. Minaee, S, T Mikolov, N Nikzad, M Chenaghlu, R Socher, X Amatriain, and J Gao (2024) Large language models: a survey. https://arxiv.org/abs/2402.06196
8. Fan A, Gokkaya B, Harman M, Lyubarskiy M, Sengupta S, Yoo S, Zhang JM (2023) Large language models for software engineering: Survey and open problems. In: 2023 IEEE/ACM international conference on software engineering: future of software engineering (ICSE-FoSE). IEEE, pp 31–53
9. C Arora, T Herda, Verena Homm generating test scenarios from NL requirements using retrieval-augmented LLMs: an industrial study, Monash University, Melbourne, Australia
10. Sami AM, Rasheed Z (2024) A tool for test case scenarios generation using large language models. Tampere University
11. Li K, Yuan Y (2024) Large language models as test case generators: performance evaluation and enhancement. Beihang University
12. Ravi Prakash Verma, Dr. (Prof.) Md. Rizwan Beg: generation of test cases from software requirements using natural language processing, Integral University, Lucknow
13. Wang J, Huang Y, Chen C, Liu Z, Wang S, Wang Q (2024) Software testing with large language models: survey, landscape, and vision. IEEE Trans Softw Eng
14. Steenhoek B, Tufano M, Sundaresan N, Svyatkovskiy A (2024) Reinforcement learning from automatic feedback for. High-Quality Unit Test Generation, Microsoft

15. Alshahwan N, Chheda J, Finegenova A, Gokkaya B, Harman M, Harper I, Marginean A, Sengupta S, Wang E (2024) Automated unit test improvement using large language models at meta. Meta
16. Z Xue, L Li, S Tian, X Chen, P Li, L Chen, T Jiang, M Zhang LLM4Fin: fully automating LLM-powered test case generation for FinTech software acceptance testing
17. Siniaalto M, Abrahamsson P (2007) A comparative case study on the impact of test-driven development on program design and test coverage. In: First international symposium on empirical software engineering and measurement (ESEM 2007). IEEE
18. K Li, Y Yuan. Large language models as test case generators: performance evaluation and enhancement. arXiv:2404.13340v1 [cs.SE] 20 Apr 2024
19. Hurani, M, and H Idris (2024) Investigating the use of LLMs for automated test generation: challenges, benefits, and suitability
20. Dakhel AM, Nikanjam A, Majdinasab V, Khomh F, Desmarais MC (2024) Effective test generation using pre-trained large language models and mutation testing. Inf Softw Technol 171:107468

Cervical Cancer Classification Using Convolutional Neural Networks

Bindu Madhavi Tummala, Dodda Preethi, and Muramalla Jay Rajesh

Abstract The preventive and curative management of cervical cancer is mainly through early detection, as this becomes one of the major causes of death among women. Recent works have been able to show great outcomes on cervical cancer feature extraction and feature classification such as ResNet152, DenseNet201, and ResNet50. Although models show considerable results, they often depict poor speed and unreliable precision in cervical cancer classification when the dataset is small and overfits in the end. Thus, we are planning to develop Convolutional Neural Networks within the framework of Deep Learning for classifying cervical cancer from Pap smear images. In the present work, EfficientNetV2 is used with the aim of improving the accuracy and robustness of high-speed and accuracy diagnoses of cervical cancer, particularly for small datasets and poor-quality images. This study introduces an improved CNN model that has been trained on SIPaKMed dataset using a few preprocessing approaches. The model has demonstrated reliable performance on the unknown data. After testing, this model tested yielded a 92.42% accuracy rate and a false-positive rate of less than 7% which highlights the potential of EfficientNetV2 as a robust and scalable tool for early detection of cervical cancer, paving a way for improved screening and diagnostic practices.

Keywords Cervical cancer · EfficientNetV2 · Image classification

1 Introduction

Cervical cancer still remains an important cause of cancer deaths in women worldwide, mainly in low- and middle-income nations with poor access to regular screening and health care services. It has been demonstrated that screening programs can significantly lower mortality rates by identifying precancerous changes in cervical

B. M. Tummala (✉) · D. Preethi · M. J. Rajesh
Siddhartha Academy of Higher Education (DEEMED TO BE UNIVERSITY), Kannur, Andhra Pradesh, India
e-mail: bindumadhavi@vrsiddhartha.ac.in

H. Nagaraj et al. (eds.), *Advancements in Artificial Intelligence and Machine Learning*, Lecture Notes in Electrical Engineering 1572,
https://doi.org/10.1007/978-981-95-7449-0_5

cells early. However, Pap smears and HPV testing are examples of established diagnostic techniques whose accessibility and accuracy are frequently restricted in low-income areas by human error, interpretation variability, and resource limitations. With the advancements of computer vision and machine learning, automated diagnostic techniques present a viable way to improve cervical cancer screening's scalability, speed, and accuracy. Recent innovations in the classification of cervical cancer have improved the possibility of detecting cervical cancer at early stages. However, cervical cancer detection from of smaller datasets and problems in accurate feature extraction along with the overfitting issue. The existing models, such as ResNet-101, DenseNet-201, etc., often offer unreliable outcomes in complex scenarios. Even though with the developed model, the time taken to obtain the results from lab is approximately 1–3 weeks. Meanwhile if the victim is in stage 4, the disease can progress rapidly from stage 4 to stage 5 in that time which becomes untreatable. To address this urgency, we developed a web-based interface which not only creates awareness among women regarding symptoms etc., but also significantly aids laboratory researchers, providing faster and more efficient diagnostic support. The paper comes up with an improved CNN model that has demonstrated outstanding performance in different conditions. Our model has been improved using a variety of preprocessing and augmentation methods that overcome the difficulties in cervical cancer detection. In this work, we use SIPaKMed dataset of cervical cell images to investigate EfficientNetV2's potential for classifying cervical cancer. In order to develop a highly accurate and reliable tool for early diagnosis, we use transfer learning to refine the pretrained EfficientNetV2 model to differentiate between cancerous and non-cancerous cervical cells. Our research mainly focuses to advance automated cervical cancer screening, especially in resource-constrained conditions where early detection can save lives.

1.1 Motivation

The goal of this research project is to increase in survival rates. From a survey, it is the fourth most frequent cancer in women worldwide, with an estimated 660,000 new cases and 350,000 fatalities scheduled for 2022. Human error and the late detection stage are the primary causes of this. Traditional cervical cancer screening methods, such as Pap smears and HPV tests, have limitations in terms of availability, accuracy, and precision. Convolutional neural networks have shown impressive results in the identification and classification of cancer. In past works, models like DenseNet and ResNet have shown strong accuracy in image classification, but they require more parameters and computational resources. So, through this project we aim to develop a model that takes less parameters and make accurate predictions, that would eventually enhance patient outcomes. More extensive cervical cancer screening could be made possible by an accurate, easily accessible, a flexible approach, which could lower the disease's global burden.

1.2 Problem Statement

One of the most important components nowadays would be to detect cervical cancer. Manually analyzing cervical cell images is a time-consuming process, vulnerable to human errors, and needs high specialization and experience which cannot be easily acquired. There are specific issues and difficulties in cervical cancer detection process even though there are some models accessible for the detection by using cervical cancer images. Improvement of methods for cervical cancer detection such as accurate classification, especially concerning the distinction made between healthy cells, precancerous cells and diverse stages of cancerous cells increases generalization about different datasets and promotes early diagnosis of cervical cancer. Also, with the available models the lab researchers may take 1–3 weeks for classifying the obtained microscopic images into five classes of cervical cells, during which the disease can rapidly grow and might become untreatable, if reached fifth stage. So, it is necessary to shorten the classification time which further results in early treatment.

2 Literature Review

An approach using Alexnet, ResNet-101, ResNet-152, and InceptionV3 pretrained deep neural network models to extract features is proposed by Mathivanan et al. [1]. Generally, the more deep models being combined will produce better classification accuracy than a single model. William et al. [2] conducted an elaborate experiment that covered the Herlev dataset in an effort to eliminate the risk of error by automating the process of cervical cancer classification form Pap smear images. They essentially focused on the error of classification that can be achieved through FCM. Tan et al. [3] came up with a new design approach for deep learning models without depending on the segmentation methods or specially designed features for automatic cervical cancer detection. Wang et al. [4] have conducted an experiment getting the regions of interest for the segmentation of cell nuclei using a Mean-Shift clustering algorithm. The work targets high accuracy rates in both cases: segmentation and, of course, classification, thus enhancing the reliability of cervical cancer detection. Kim et al. [5] employed a methodology which would determine that one of the approaches is easy to use, efficient, with improved performance in terms of classification of cervical cancer diagnosis. It was found that deep learning 0.15 above three machine learning methods at an average of 0.82. Alcaraz Chavez et al. [6] presented a new method which utilizes video sequences captured by camera cell phones, SIFT descriptors for detection and tracking precursor cervical cancer cells. Jahan et al. [7] presented a study reporting the performance of the eight most popular classification algorithms where different top feature sets are chosen from the dataset for the detection of cervical cancer. Akhila et al. [8] also reported the usage of Autoencoder (AE) for enhancing the training procedure. Also, to extract features and lower their dimension a cascaded multilayer perceptron (c-MLP) is used to differentiate between normal and abnormal cervical cells. Tummala et al. [9] came up with an approach that highlights the effectiveness of encoder-decoder architectures in medical image segmentation to improve detection accuracy. The study uses multiscale residual blocks and dilated convolutions to capture features at feature resolutions. Tummala et al. [10] introduces Curriculum Learning, where the model is trained progressively from easy to difficult samples for better generalization. The work focuses on using Overcomplete U-Net, which enhances feature extraction by using more convolutional layers and skip connections. Tummala et al. [11] introduces DeepLabV3+ that uses Atrous Spatial Pyramid Pooling (ASPP) to capture features at multiple scales, improving segmentation accuracy in MRI images. The encoder-decoder architecture in DeepLabV3+ improves boundary detection and fine details in medical images. Tummala et al. [12] highlights U-Net's effectiveness in segmenting GI tract images with high precision. The study compares different loss functions (Dice Loss, Cross-Entropy, Focal Loss, etc.) in U-Net for GI tract segmentation. Tummala et al. [13] conducts a study that demonstrates how TransUNet outperforms traditional U-Net by handling complex anatomical variations in GI tract segmentation. TransUNet combines CNNs and Transformers, leveraging CNNs for local feature extraction and Transformers for global context awareness in GI tract segmentation.

2.1 Research Gap

Recent developed methods show promising results in the extraction of feature for Pap Smear images [1]. However, all the models used are very computationally intensive and require a fair amount of computational resources at both the training and inference levels. The recent developments over the last couple of years helped us design a model that will consume less memory and increase the inference speed [3]. Hence, there is a great demand for the application of a model that may hold a better accuracy-speed-computational efficiency-model complexity trade-off. The present models may show high accuracies of classification on experiments with fewer numbers of images and features [4], but the performance is not consistent when low-quality images are presented. Given these challenges, there is a huge need for a reliable model promising accurate results under any circumstances. There aren't many studies that focus on how models like EfficientNetV2 can be used in actual clinical operation [1]. Although cervical cancer classification has advanced with models such as ResNet-152, DenseNet-201, it is still essential to raise awareness among women regarding symptoms to further lower death rates. Therefore, our research focuses on satisfying these demands by developing a better CNN model that has been properly trained using the SipakMed Dataset that holds 4068 pictures, with preprocessing techniques applied on the dataset and create a web interface in which we deploy our trained model that mainly helps lab researchers to ensure accurate and quick results in classifying cervical cancer cells and creates awareness among women.

2.2 *Area of Study and Data Preparation*

Most deep learning applications begin with data preparation. First, the prepared dataset must be split into training, testing, and validating portions before it can be released to improve the performance of the model. The dataset used for this application comes from [1] (Fig. 1)

3 Proposed System

The Enhanced CNN architecture of the model consists of convolutional layers, batch normalization, and excitation layers. It utilizes balanced compound scaling like resolution, width, depth of a network for achieving optimal performance (Fig. 2).

The images of the cervical cells are taken into input layer as an input from the dataset. The prominent layer of the EfficientNetV2 architecture in CNNs is Convolutional Layers, which extract features of cell which are shape, color, cell texture. These layers work especially well for catching minute patterns in medical imaging, such as the texture of the cytoplasm and the shape of the nucleus. It further deepens the training procedure by normalizing inputs, typically by batch normalization with a mean of zero and a variance of one, stabilizing and accelerating convergence. The Swish activation function is included to inject nonlinearity and improve the flow of gradients while backpropagating. Pooling layers prevent overfitting and retain the most relevant features by reducing their computational load through a reduction in spatial dimensions of features. The Fully Connected Layer links up all neurons in the previous layer to the current layer, making it quite possible for the network to classify and determine cervical cancer by combining and classifying retrieved characteristics of cervical cells. SoftMax Activation is used in the final layer to predict the classification of cervical cells into five groups. This can calculate the probability of each image in the SipakMed dataset coming under a particular class because probability range will be nearly 0.1. final layer to predict the classification of cervical cells into five groups. This can calculate the probability of each image in the SipakMed dataset coming under a particular class because probability range will be nearly 0.1. Final layer to predict the classification of cervical cells into five groups. This can calculate the probability of each image in the SipakMed dataset coming under a particular class because probability range will be nearly 0.1 (Table 1).

3.1 *Data Collection*

In this Research, we used the SIPaKMeD Database, containing 4049 pictures of isolated cervical cells. The set consists of images of 966 cluster cells extracted from Pap smear slides using a CCD camera attached to an optical microscope and then

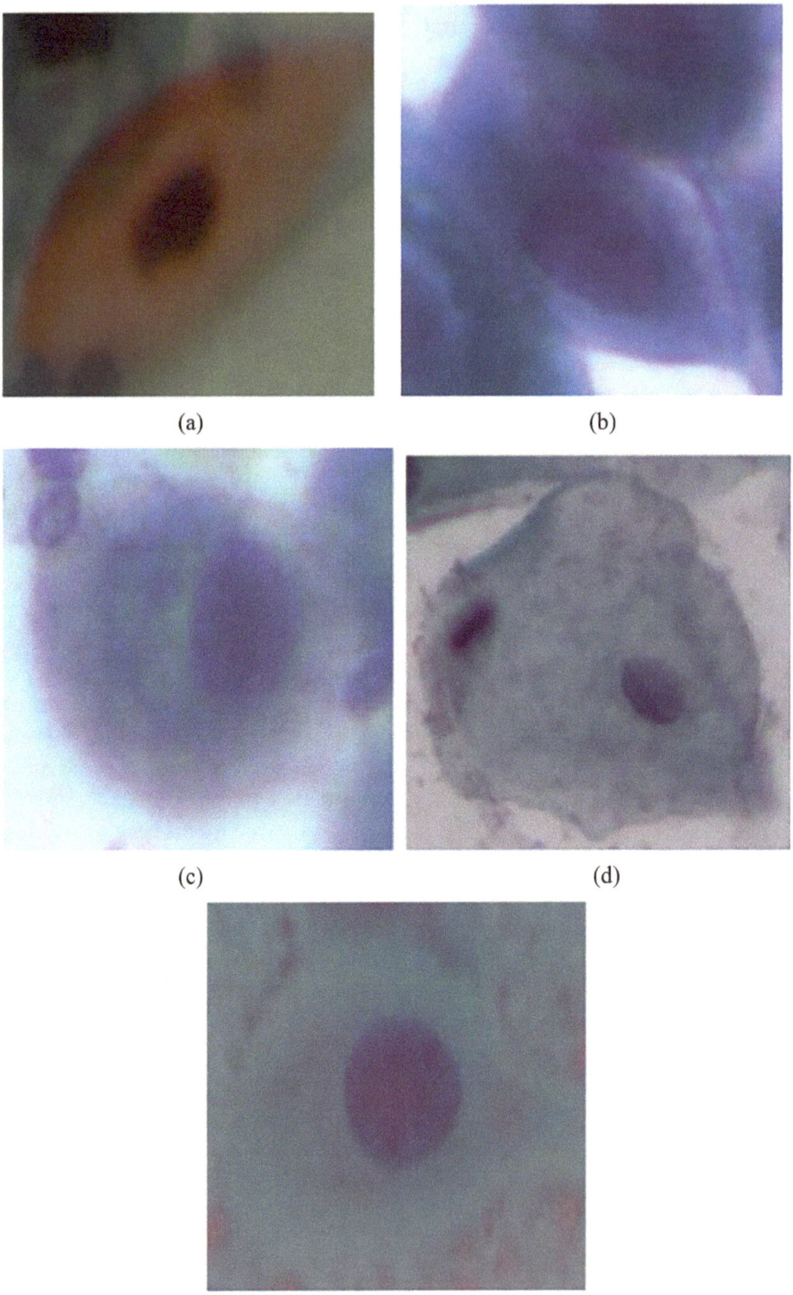

(e)

Fig. 1 Cropped images from each class: (**a**) Dyskeratotic Cells, (**b**) Koilocytotic Cells, (**c**) Metaplastic, (**d**) Parabasal Cells, (**e**) Superficial Cell

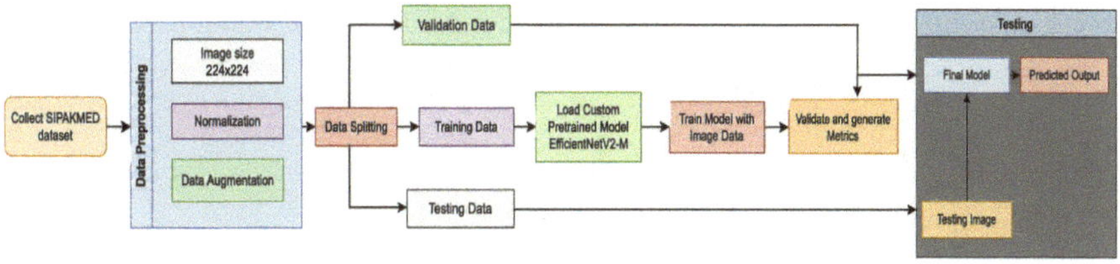

Fig. 2 Sequence followed for proposed system that illustrates the architecture of customized EfficientNetV2-M

Table 1 Model comparision overview

Parameters	Default model EfficientnetV2M	Proposed model
No of input image channels	3	3
Image dimensions	224×224×3	224×224×3
Input kernel size	3×3	3×3
Initial filters number	32	32
Strides	1	1
Variables	8M	~0.2M
Layers number	9	21
Resulting image channel	224×224×3	224×224×3

Table 2 Model comparision overview

Class labels	Image count
1. Superficial intermediate cells (Normal)	831
2. koliocytotic cells (Normal)	825
3.Parabasal cells (Benign cell)	787
4.Dyskeratotic cells (Abnormal)	813
5.Metaplastic cells (Abnormal)	739
TOTAL	3995

cropped individually. Five types of cervical cells are present in the dataset: normal, abnormal, and benign. All images are fed as raw RGB inputs and resized to 224 × 224 pixels for uniform processing (Table 2).

3.2 *Preprocessing*

Images were resized to 224×224 pixels for uniformity throughout the entire database within the preprocessing stage. Apart from this, pixel values were scaled between 0 and 1. To achieve normalization (1), images were rescaled by a factor of 1/255, which allows stability while accelerating the neural network training process.

$$Normalized\ Pixel\ Value = \frac{Original\ Pixel\ Value}{255} \quad (1)$$

Data Augmentation.

The problem chosen was to improve the robustness of the model and reduce overfitting by applying some data augmentation techniques to the dataset. These methods included a 20-degree rotation, 20% width and height shifts at 20%, 20% shear and zoom ranges, and horizontal flipping for widening the training dataset. All these were done to diversify the training environment so that the model could perform well with unseen data.

Shear, Width and Height Shift, Zoom, in object location and scaling. Equations. 2, 3, 4, 5 and 6.

$$\begin{pmatrix} x^1 \\ y^1 \end{pmatrix} = \begin{pmatrix} x+s \cdot y \\ y \end{pmatrix} \quad (2)$$

$$w_{Shift} = width \times 0.2 \quad (3)$$

$$h_{Shift} = height \times 0.2 \quad (4)$$

$$New\ Position = Original\ Position\left(width + w_{Shift}, width + h_{Shift}\right) \quad (5)$$

$$Zoomed\ Image = Original\ Image \times Z \quad (6)$$

Model Creation

The EfficientNetV2-M architecture has been adopted as the backbone of the customized model for the identification of five unique cervical cancer cell types. This model takes photographs of size 224 × 224 as input and extracts useful features from these using pretrained weights from ImageNet. Output from the base model is presented in a successive chain of fully linked layers with ReLU activation functions. The ReLU activation function (Eq. 8) introduces nonlinearity into the model so that it can learn complex patterns. After each dense layer, we have added a batch normalization layer for stabilization and to speed up the training process. Then, we have added some dropout layers with a dropout rate of 30% to avoid overfitting. Finally, in the output layer, Softmax activation function (Eq. 7) was used, which calculates The probabilities for each class to perform multi-class classification. The model is trained with categorical cross entropy loss and tested at both phases of training and testing using appropriate metrics to include accuracy, precision, and recall for its assessment.

$$Soft\ max z_i = \frac{e^{z_i}}{\sum_{j=1}^{c} + e^{z_j}} \quad (7)$$

x denotes the input image. The ReLU layer gives output in between zero and input value. This model computes input in a multi-layered way that contains the EfficientNetV2-M convolutional layer.

$$ReLU(x) = \max(0, x) \quad (8)$$

We can see from Eq. 8 how the ReLU activation function keeps the output in between zero and the input value. The architecture of EfficientNetV2-M is designed so that it would successfully extract features from cervical cell images such that the model is capable of pointing out patterns related to any type of cervical cancer. Finally, the fully connected layer will make predictions based on the extracted features.

Training Model

Following the architecture proposal for CNN, the model continues to process images of isolated cervical cells clipped, with great pain, from 966 images of cluster cells taken from Pap smears, yielding a total of 4049 images. The dataset in question has three categories: normal, abnormal, and benign cells. We have trained the model as in Algorithm 1.

Algorithm 1:

Training the Model.

Input: Pretrained Model (EfficientNetV2-M).

Output: Final Model.

Step 1: Split the dataset into 80% training and 20% validation set.

Step 2: Load the pretrained EfficientNetV2-M model along with ImageNet weights.

Step 3: Dropout layers to avoid overfitting.

Step 4: Compile the model along with rate Categorical cross entropy loss and Adam optimizer having a 0.0001 learning.

Step 5: Train model for 10 epochs with a batch size of 32, using early stopping based on validation accuracy to save the best model.

Testing Model.

After training we tested the model, using 704 images from its test dataset, to predict which cells were malignant or benign.

Algorithm 2:

Testing the Model.

Input: Final model trained.

Output: Predict the Output.

Step 1: Upload the unseen images from the test dataset (704 images).

Step 2: Resize and normalize the unseen images with another round of noise removal.

Step 3: Feed all the preprocessed images to a final model to test them.

Step 4: The classifications into either malignant or benign must be predicted and the output must be provided.

Performance Evaluation of Model

To evaluate the performance of the cervical cancer classification model, the following set of metrics was applied: accuracy, precision, recall, and F1-score. The set of these metrics is provided in Eqs. (9–13).

$$Accuracy = \frac{TP + FP}{TP + TN + FP + FN} \tag{9}$$

$$\Pr ecision = \frac{TP}{TP + FP} \tag{10}$$

$$Recall = \frac{TP}{TP + FN} \tag{11}$$

$$True\,Postive\,Rate = \frac{TP}{TP + FN} \tag{12}$$

$$False\,Positive\,Rate = \frac{FP}{FP + TN} \tag{13}$$

Deployment

For training, we used the EfficientNet V2 M model in machine learning and saved it as 'model.h5' We then used the 'model.h5' file in this embedding. We deployed it using Flask web framework which makes it easier to create an app for online usage (Fig. 3).

Clicking the 'Predict' button reveals the splitting of images into five types of cell classes, color-coded Fig. 5 for easy interpretation, including blue-colored cells as mostly safe, orange for mild danger requiring emergency medical attention, and red for dangerous cancerous cells that need immediate medical treatments. We also included this new feature that enables the submission of images using the QR code, thus making it easy for users to take pictures direct from a microscope's display using cell mobile cameras. This innovation makes it easier for consumers who require urgent predictions of cervical cell images (Table 3).

4 Results and Analysis

The metrics of the model result are graphically shown below. We noticed that as the epochs were increased, so was the model's accuracy as shown in the graph below. From the graph below, we can see that the loss decreases with the increase of the epochs and the epochs begin generating dependable data. Model loss and accuracy are mentioned in below Fig. 4.

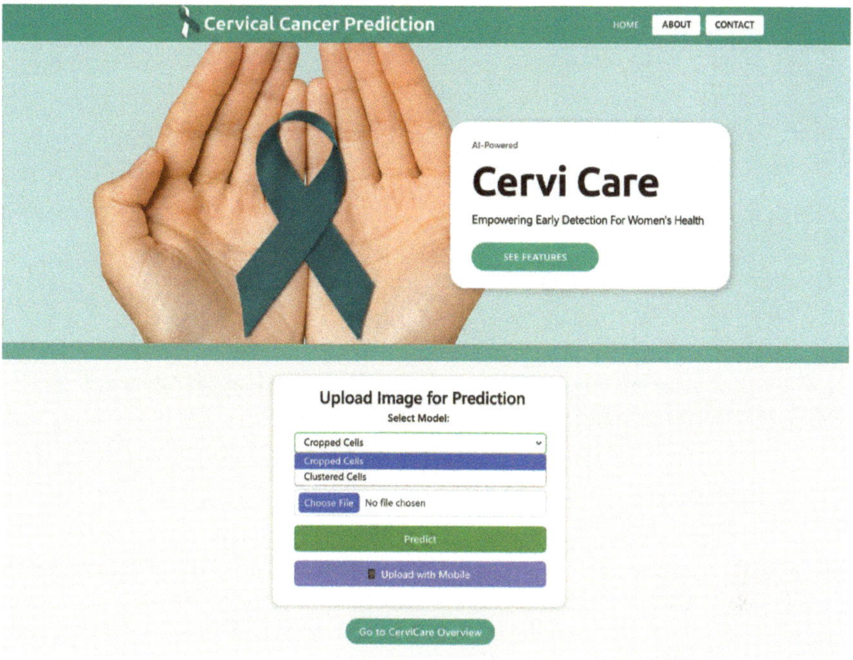

Fig. 3 Deployed webapp interface

Table 3 Performance comparison

Work	Algorithm	Model accuracy	Data source
[14]	BORUTA	Accuracy: 95%	UCI dataset
	ResNet-50	Accuracy: 95%	SipakMed dataset
[16]	Multilayer perceptron and random forest	Accuracy: 95%	HCI dataset
[17]	RESNET-152	Accuracy: 95%	Herlev dataset
Default model (9 layers)	EfficientNetV2-M	Accuracy: 95%	SipakMed dataset
Proposed model (21 layers)	EfficientNetV2-M	Accuracy: 95%	SipakMed dataset

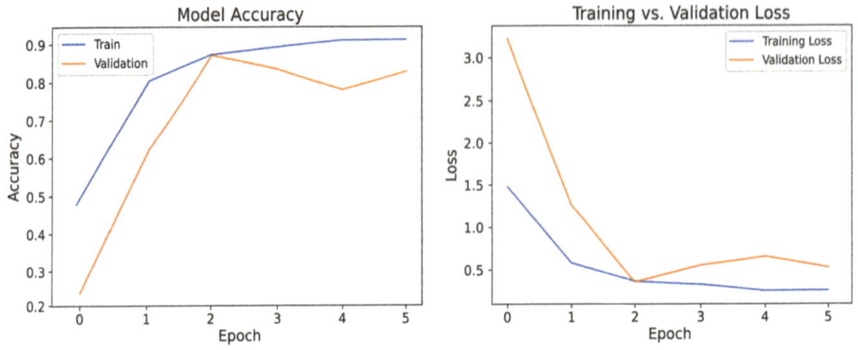

Fig. 4 Image of model loss cervical cancer detection model was applied, and the model was trained through ten epochs at a batch size of 32

5 Limitations

Our model gives best and accurate results for large datasets. But if it is tested on small datasets, then it might result in unreliable outcomes as the smaller dataset. The developed interface is a valuable tool for lab researchers to upload and classify microscopic cervical cells. But nonstandard or low-quality images could result in unreliable classification results affecting the trustworthiness of the web tool. As web interface is engaged with medical images, there is a necessity to provide security and privacy for the data protection, but it can be challenging.

6 Conclusions and Future Work

We present a solution in terms of a convolutional neural network for cervical cancer classification and detection and achieved 92% accuracy which is a good result. The method devised here improves early detection. This model is a member of EfficientNetV2 family. Thus, it allows in achieving high accuracy rates even with less resources on image classification across various datasets, showing promise for real-world applications. We designed a web interface that enhances awareness among women and could, potentially enable researchers to classify cervical cells more rapidly, which is then applied to cut down screening time by a huge margin. As possible extensions of this research, the trained models will be integrated into IoT devices for real-time cervical cancer screening, thus bringing diagnosis closer to home while working in resource-constrained environments.

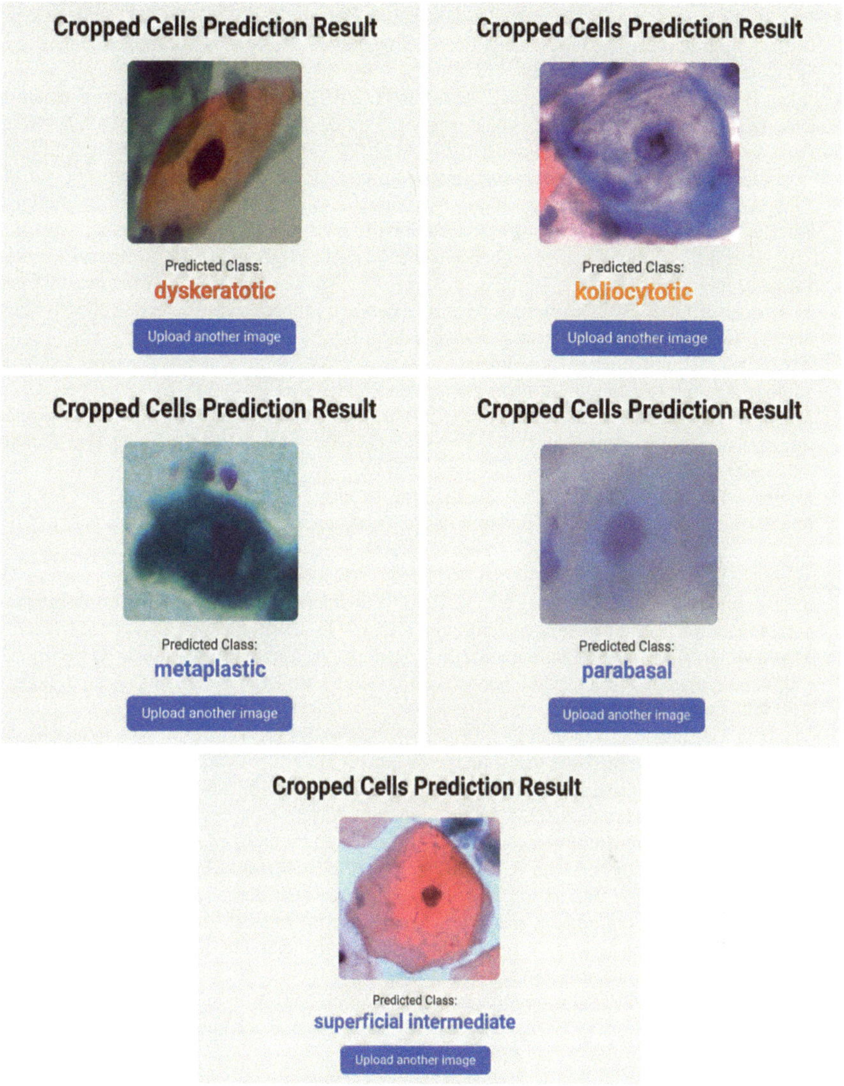

Fig. 5 The output of our custom model

References

1. Mathivanan SK, Francis D, Srinivasan S, Khatavkar V, Karthikeyan P, Shah MA (2024) Enhancing cervical cancer detection and robust classification through a fusion of deep learning models. Sci Rep 14(1):10812
2. William W, Ware A, Basaza-Ejiri AH, Obungoloch J (2019) Cervical cancer classification from pap-smears using an enhanced fuzzy C-means algorithm. Inform Med Unlocked 14:23–33

3. Tan SL, Selvachandran G, Ding W, Paramesran R, Kotecha K (2024) Cervical cancer classification from pap smear images using deep convolutional neural network models. Interdiscip Sci Comput Life Sci 16(1):16–38
4. Wang P, Wang L, Li Y, Song Q, Lv S, Xianling H (2019) Automatic cell nuclei segmentation and classification of cervical pap smear images. Biomed Signal Process Control 48:93–103
5. Park, Yerang, Young Jae Kim, Woong Ju, Kyehyun Nam, Soonyung Kim, and Kwang Gi Kim. "Classification of cervical cancer using deep learning and machine learning approach." (2021)
6. Alcaraz-Chavez JE, del Carmen A, Téllez-Anguiano JC, Olivares-Rojas, and Ricardo Martínez-Parrales. (2024) Real time tracking and detection of cervical cancer precursor cells: leveraging SIFT descriptors in mobile video sequences for enhanced early diagnosis. Algorithms 17(7):309
7. Jahan S, Saimun Islam MD, Islam L, Rashme TY, Prova AA, Paul BK, Manowarul Islam MD, Mosharof MK (2021) Automated invasive cervical cancer disease detection at early stage through suitable machine learning model. SN Appl Sci 3:1–17
8. Akhila KR, Muthukumaran N, Ahilan A (2024) Classification of cervical cancer using an autoencoder and cascaded multilayer perceptron. IETE J Res 70(1):26–36
9. Madhavi TB, Barpanda SS (2022) Liver tumor segmentation from computed tomography images using multiscale residual dilated encoder-decoder network. Int J Imaging Syst Technol 32(2):600–613
10. Madhavi TB, Barpanda SS (2023) Curriculum learning based overcomplete U-net for liver tumor segmentation from computed tomography images. Bull Electr Eng Inform 12(3):1620–1629
11. Veeraiah DC, Tummala BM, Jaladi R, Peruri AK (2024) DeeplabV3+ for precise GI tract segmentation in MRI images. In: 2024 15th international conference on computing communication and networking technologies (ICCCNT). IEEE, pp 1–6
12. Madhavi TB, Chavva SR, Yallamandaiah S, Radhika A, Veeraiah DC, Jaladi R, Peruri AK (2024) Automated GI tract segmentation with U-net: a comparative study of loss functions. J Adv Inf Technol 15(12)
13. Madhavi TB, Jaladi R, Veeraiah DC, Peruri AK (2024) TransUNet for precise and robust GI tract segmentation in MRI images. J Image Graph 12(3)
14. Lilhore UK, Poongodi M, Kaur A, Simaiya S, Algarni AD, Hela Elmannai V, Vijayakumar GB, Tunze, and Mounir Hamdi. (2022) Hybrid model for detection of cervical cancer using causal analysis and machine learning techniques. Comput Math Methods Med 2022(1):4688327
15. Roy SD, Saha P, Nath N, Datta A, Bhowmik MK (2022) AGMC-TU pap-smear cytological image dataset: creation, annotation, and analysis towards early detection of cervical cancer. In: 2022 IEEE 10th international conference on healthcare informatics (ICHI). IEEE, pp 42–47
16. Hou X, Shen G, Zhou L, Li Y, Wang T, Ma X (2022) Artificial intelligence in cervical cancer screening and diagnosis. Front Oncol 12:851367
17. Mahajan P, Kaur P (2024) Improving cervical cancer classification in PAP smear images with enhanced segmentation and deep progressive learning-based techniques. Diagn Cytopathol 52(6):313–324

AI-Driven Multilingual Document Analysis and Interaction System

E. G. Satish, Mohan Bangalore Anjaneyalu, G. N. Basavaraj, P. Ramesh Naidu, and T. R. Vinay

Abstract This research paper introduces a pioneering application that integrates the capabilities of Artificial Intelligence (AI) with document-based interactions. Users can effortlessly upload their preferred documents to the system, enabling AI analysis of the document's content. Following this analysis, users can pose questions related to the document's content. The application harnesses natural language understanding and AI-driven processing to provide comprehensive and insightful answers, effectively transforming static documents into dynamic repositories of interactive knowledge. By seamlessly integrating document processing and natural language comprehension, this paper aims to redefine and elevate the user experience of interacting with information. Whether utilized for research, learning endeavors, or exploration, this application stands out to enhance user engagement and, facilitating smarter and more interactive interactions with documents.

Keywords LLM models · NLP integration · Document analysis · AI-driven interaction · Knowledge extraction

---

E. G. Satish (✉) · P. Ramesh Naidu
Department of Computer Science and Engineering, Nitte Meenakshi Institute of Technology, Bengaluru, India
e-mail: satish.eg@nmit.ac.in

P. Ramesh Naidu
e-mail: ramesh.naidu@nmit.ac.in

M. B. Anjaneyalu · G. N. Basavaraj
Department of Computer Science and Engineering, BMS Institute of Technology and Management, Bengaluru, India
e-mail: ba.mohan@bmsit.in

G. N. Basavaraj
e-mail: drbasavarajgn@bmsit.in

T. R. Vinay
Department of Artificial Intelligence and Data Science, Ramaiah Institute of Technology, Bengaluru, India

H. Nagaraj et al. (eds.), *Advancements in Artificial Intelligence and Machine Learning*, Lecture Notes in Electrical Engineering 1572,
https://doi.org/10.1007/978-981-95-7449-0_6

1 Introduction

Conventional document processing methods face significant challenges in processing tremendous amount of information, hidden in documents, extracting meaningful insights in the documents in an efficient manner. Shortcomings of these traditional methods, mainly due to an inability to manage the enormity and complexity of unstructured data, emphasizing the need for more advanced methodologies that can discover the information that is encapsulated within the documents. AI has caused a great revolution in the paradigms shift in document analysis and comprehension. The convergence of AI technologies like Machine Learning (ML) and Natural Language Processing (NLP) with advanced document analysis techniques creates an opportunity for transformation the existing techniques. By harnessing capabilities of AI, the proposed work taps into ML algorithms and NLP models to help navigate the complex insights in unstructured textual data. The main goal here is to give users a more engaging experience than simply reading documents. The ability of ML-powered document analysis and NLP-enabled language understanding to convert the content of documents into easily grasped insights is sought in the proposed work. The transformation is focused on enabling smooth interactions, where the user can query document repositories in natural language and extract relevant information.

1.1 Comparative Analysis of Large Language Models (LLM) for Structured and Unstructured Data in QA Systems

In smart Question and Answer (QA) systems, various large language models play a critical role in handling structured and unstructured data depending on their strengths and approaches. Table 1 shows different models with the capabilities and applications. For structured data, models like SQLNet and TREQS have shown proficiency. The SQLNet leverages deep learning approach to generate SQL queries directly from natural language, specifically catering to structured databases. It applies a sequence-to-sequence model to interpret user queries and translates them into executable SQL statements. This model is highly effective in environment where it is dealing with structured data. In contrast, for unstructured data, models such as BERT and its derivatives, like RoBERTa and GPT-3, have revolutionized text processing. BERT is trained on a huge amount of unstructured text, and so, by definition, it has a good understanding of the nuances of natural language; it does well in answering questions based on paragraphs, email, or articles.

The strength of the models against each other in this case depends a lot on the data they process. For structured databases and generation of specific queries, SQLNet-like models with their specialized architectures are more suitable. On the other hand, BERT and GPT-3 give more flexibility and wider applicability for handling diversity in the formats of unstructured data at an additional cost in computational resources. In terms of performance metrics, models such as SQLNet have shown high accuracy

Table 1 Comparison of large language models for QA

Model name	Type of data	Key capabilities	Performance metrics	Application examples	Computational resources required
SQLNet	Structured	SQL query generation from natural language	Accuracy in SQL query generation (e.g., >90 on spider dataset)	Database querying, Business data analysis	Moderate
BERT	Unstructured	Contextual understanding of text, Answer extraction	F1 score and exact match in SQuAD 2.0 (e.g., 84.5% F1 score)	Content summarization, Customer service automation	High
GPT-3	Unstructured	Text generation contextual interface	Performance in open-domain question answering	Creative writing, Chatbots, Knowledge discovery	Very high
TREQS	Structured	Natural language interface to databases	Precision and recall in structured data interpretation	Data analytics, Reporting tools	Moderate
RoBERTa	Unstructured	Improved text understanding and predictions	Enhanced results in GLUE and SuperGLUE benchmarks	Sentiment analysis, Language translation	High

in structured query language generation tasks, often surpassing 90% in benchmark datasets such as the Spider dataset. However, in unstructured environments, BERT-based models have proven state-of-the-art results in NLP tasks, with a notable score of 84.5% in the SQuAD 2.0 Question Answering task. Different Large Language Models, each with their unique strengths, highlight the significance of selecting a suitable model depending on the data and precise needs of the QA system being created [1, 2].

1.2 Comparison of Storage Techniques for Structured and Unstructured Data

In data storage domain, the selection of the suitable technology is based on the nature of the data, weather it is structured or unstructured data. Table 2 shows different storage techniques for both structured and unstructured data. The relational databases such as MySQL and PostgreSQL have been the cornerstone of structured data storage, offering robust transactional support and efficient querying through SQL. Their table-based organization and ACID compliance make them ideal for applications that demand high data integrity, such as financial systems and customer record management. However, they often face scalability limitations when handling massive volumes of data or rapidly evolving schemas.

On the other hand, NoSQL databases, such as MongoDB and Cassandra, come in handy when dealing with unstructured or semi-structured data. Their schema-less nature makes them more flexible and capable of handling different data types without requiring a predefined structure for the same. These databases are scalable and hence appropriate for applications involving large-scale data storage, real-time analytics, and big data applications. Though providing flexibility and scalability, they often compromise on transactional consistency offered by relational databases.

Each of these storage methods has its different merits and best-fit scenarios. Choices among them depend on data types, scalability requirements, performance needs, and application requirements. As data grows in volume, variety, and velocity, the choice of appropriate storage solution becomes very critical in the architecture of solid and efficient data management systems.

1.3 Components for Implementing a Multilingual System in QA

Developing a multilingual system in the QA framework is complex, and it involves different components as discussed in Table 3. The starting point of this development process is language detection. The system would be able to quickly and precisely identify the language of the query with libraries such as FastText or Langid. These are followed by the translation service, critical to translating the query into a base language—like English—which the system can understand. These services call to retranslate the system's response back into the user's native language, thereby providing the user with a seamless experience. However, translation quality and latency remain the critical factor to keep effectiveness and efficiency in the system.

NLP is the soul of a multilingual QA system. In most cases, multilingual NLP models like BERT Multilingual and XLM-R are pre-trained in various languages in order to capture better understanding and text processing in multiple languages. These models are very language-agnostic in nature and provide quite a good range of language coverage, capturing contextual tones across different languages easily. Besides, cross-lingual transfer learning techniques can be used with these models to extend their capabilities further if limited amounts of training data are available for some languages. This approach borrows knowledge transferred from data-rich languages to help improve performance on less-represented languages [3–5].

This could also draw upon localized knowledge sources to then provide relevant and accurate answers, especially to region-specific questions. It would make use of

Table 2 Comparison of storage techniques for structured and unstructured data

Storage technique	Data type	Key features	Scalability	Performance	Use cases
Relational databases (e.g., MySql, PostgreSQL)	Structured	Table-based storage, ACID compliance, SQL querying	Moderate	High for structured queries	Financial systems, Customer records, Inventory management
NoSQL databases (e.g., MongoDB, Cassandra)	Both (primarily unstructured)	Schema-less design, Horizontal scaling, Flexible data models	High	Varies; generally high for unstructured data	Big data applications, Real-time analytics, Content management
Data warehouses (e.g., Amazon Redshift, Snowflake)	Structured	Optimized for complex queries, High throughput for read operations	High	High for analytical queries	Business intelligence, Reporting, Data analysis
Object storage (e.g., Amazon S3, Google Cloud Storage)	Unstructured	Highly scalable, Data stored as objects, API access	Very high	Moderate; optimized for durability and availability	Multimedia storage, Backup and archival, Big data storage
File storage (e.g., Network Attached Storage)	Both	Hierarchical file system, Network-based access	Moderate	High for specific file retrieval	Shared file systems, Document management, Media libraries
Graph databases (e.g., Neo4, Amazon Neptune)	Structured (Relationship-focused)	Relationship mapping, Graph-based querying	Moderate	High for relationship-driven queries	Social Networks, Recommendation engines, Fraud detection
Document databases (e.g., Elasticsearch, CouchDB)	Unstructured	Document-oriented, Full-text search capability	High	High for search-driven queries	Search engines, Logging systems, Content retrieval

Table 3 Components for implementing a multilingual system in QA

Component	Purpose	Key technologies	Considerations
Language detection	Identifying the language of the input query	Libraries like FastText, Langid	Accuracy, Speed
Translation services	Translating queries and responses between languages	Google Translate API, Microsoft Translator	Quality of translation, Latency
Multilingual NLP models	Processing and understanding queries in multiple languages	BERT Multilingual, XLM-R	Coverage of languages, Contextual understanding
Cross-lingual transfer learning	Leveraging the knowledge from one language to improve performance in another	Zero-shot learning techniques	Data availability, Model robustness
Localized knowledge sources	Providing accurate information for language-specific queries	Regional databases, Local news APIs	Relevance, Currency of information
User interface	Offering multilingual support in user interactions	Unicode support, RTL text support	Accessibility, Ease of use

regional databases and local news APIs offering up-to-date and relevant information tailored to specific language and cultures.

Finally, the QA systems should be designed with a user interface that supports multiple languages. This includes not only translation, but also support for Unicode to represent a variety of scripts, and considerations for languages like Arabic and Hebrew. More importantly, considerations are due for accessibility and usability of the interface in order to ensure that people across different language backgrounds could use the system easily.

This ensures that the multilingual QA system is well-integrated with different sub-components dealing with various aspects of language handling so as to provide a smooth and accurate experience to the user worldwide [6, 7].

2 Literature Review

Authors Devlin et al. introduced bidirectional encoder representations from transformers, which was quite a revolutionary approach in NLP to take a step toward a paradigm shift in models for language representation. Unlike all the previous methods, BERT is based on the notion of pre-training deep bidirectional representations from unlabeled text, fundamentally changing how models understand the language context. The innovation here is that BERT conditions both the left and right contexts jointly in all its layers, a notable divergence from unidirectional or shallow bidirectional predecessors. Its strength does not lie merely in the conceptual simplicity; empirical evidence comes from good performance across wide range of NLP tasks. Remarkably, BERT established new progressive results on many benchmarks, such as increasing the GLUE score to 80.5% (7.7-point improvement), and significant advances in MultiNLI accuracy as well as SQuAD question-answering metrics. Thus, the high flexibility of BERT for the maximum number of significantly differing tasks marks, indeed, very progressive development in this field. This breakthrough model in understanding and processing natural language sets a new standard for related question tasks and inference language, catalyzing additional research and development in the domain of AI-driven language processing [8].

In pursuit of improving natural language understanding, the most pioneering approach in that domain came in 2018 when handling traditional discriminatively trained models. In the course of their work in improving language understanding by generative pre-training, the authors managed to consider the inborn challenges of NLU tasks, such as textual entailment and question answering, as well as document classification under the constraint of scarce labeled data. The aim of their approach is generative pre-training of a model on a large and diverse corpus of unlabeled text, followed by discriminative fine-tuning on specific NLU tasks. In contrast, theirs is task-agnostic and actually beats the baseline discriminatively trained models for specific tasks in the order of nine out of twelve tasks chosen for the study. They report impressive deltas, of up to 8.9% for increasing commonsense reasoning and up to 5.7% for question-answering accuracy. Crucially to this approach, they use task-aware input transformation in the fine-tuning, which, in their words, enabled transfer learning to work with minimal architectural changes. It also set a new standard in the AI language processing domain by not only outpacing in the NLU benchmarks but also proving the efficacy of generative pretraining [9].

The vast domain of LLMs has been covered in its entirety in the paper titled "A comprehensive overview of Large Language Models," authored by Humza Naveed et al. This survey paper is a significant document for anyone interested in fast-evolving and multi-faceted LLMs in the area of NLP. In this work, the authors discuss new advances related to the study of LLMs that include architectural innovations or training strategies, increasing context length, methodologies for fine-tuning, and multimodal LLMs. Equally important, the article discusses new applications for LLMs in the field of robotics—how LLMs combine with systems of robotic machines. The review also discusses some pointers on how to evaluate datasets, benchmark methods, and model efficiency. This work is invaluable in its comprehensive approach and gives an overview that is complete and self-contained. It will be useful not only as a structured survey but also as a reference for researchers and a resource for practitioners. The paper by Naveed et al. stitches together the rich and informative features of the current LLM research in a clear and incisive way on frontier issues. It helps in understanding complexities and guides future progress in the rapidly developing field [10].

A documentation of evolution and efficacy regarding deep learning in text classification has been presented in the paper titled: "Deep Learning-Based Text Classification: A Comprehensive Review" by Shervin Minaee. This comprehensive review has been a milestone in this field, wherein the authors have gone into great depths to dissect more than 150 deep learning-based models which shaped and brought improvement to the task of text classification. Applications that range from sentiment analysis to news categorization and question answering and natural language inference show the performance of deep learning models compared to old traditional machine learning methods. One very valuable contribution of this work has been the comparative analysis of these models with regard to their technical contributions, similarity, and strength. It would be provided with a quantitative analysis for performance on popular benchmarks. That allows one to bring a critical viewpoint about the different deep learning strategies in text classification. Minaee et al.'s paper can thus be said to be a thorough guide, for both researchers and practitioners, through current achievements and an indication of future research directions so as to present comprehensive understanding and critical evaluation of modern deep methods [10].

In the recent developments on sequence-to-sequence models, one worthy contribution is the work by Lewis and team on paper titled "BART: Denoising Sequence-to-Sequence Pre-training for Natural Language Generation, Translation, and Comprehension." It contributes a new de-noising auto-encoder to pre-train sequence-to-sequence models, which has profound implications on NLP. What is most innovative about BART is its training methodology—that of corrupting text with a host of noising functions and then reconstructing original text. This is a vanilla machine translation neural architecture, based on a transformer, combining the power of a BERT-like bidirectional encoder and a left-to-right decoder from GPT. This strategy enables the BART to generalize tasks along a spectrum of pre-training schemes, thus a powerful augmentation to the NLP toolkit at large. Some research has gone into the potential of Morse codes for communication, and it is majorly the case of disabled persons whose mode of expression involves eye blinking. This study focuses on using Morse code for communication, citing concerns over its usage by the target users. Use of Morse code depicts innovative strategies in assistive communication technologies that raise concerns over usability and user memorization [11].

AI and NLP in PDF-based Question Answering Systems: The smart question and answering system integrates BIM with AIOT through a BERT model. As previously stated, the said system includes three subsystems: NLP model training, the inference server, and the user interface. In the case of the former, the BERT model was trained with a text dataset containing messages on the BIM resource. On the other hand, the server communicated with the user interface for the processing of the query and the response. This system enables human–machine communication and enhances user knowledge of smart buildings, particularly for practitioners and researchers in the construction industry. By leveraging the NLP and BERT models, the system aims to facilitate accurate and efficient information retrieval related to BIM-AIOT integration [12].

Representative Vector Summarization (RVS) for large unstructured textual data: The introduced RAG-assisted Representative Vector Summarization (RVS) addresses challenges related to performance decrease with increasing context lengths and the "Lost in the Middle" problem in large context window models. The RVS selects a predefined chunks of representative text from a non-parametric knowledge-base and applies a combined abstractive and extractive summarizing workflow to generate the final summary. The parameters for selecting text chunks were determined based on the maximum affordable token limit. In addition, the model can identify keywords, creating visual representations of the document's content based on their relative importance using word clouds and scatter plots. RVS is implemented in docGPT, program written in Python using the langchain framework [13].

AI-based question answlering system using particle swarm optimization and fuzzy c-means clustering algorithms: The smart QA system concerning power business scenarios classifies the question attributes and matches the answers to the queries of the customer. The system collected questions raised by customers and uploaded them within the knowledge base. The pre-processed questions were classified using a particle swarm clustering algorithm. In the implementation, the fuzzy c-means clustering algorithm was used to ensure that the students got answers with maximum possibilities to their questions. It had low worst-case time complexity, which is up to 0.35 only. This approach proposed is AI-based and combines two algorithms to provide efficient and effective customer service. Its novelty lies in using priority information to address customer requests. The design of the system is focused on the improvement of customer service quality and operational efficiency in business scenarios related to power [14].

Fast and Memory-Efficient Attention Algorithm for Transformers: FlashAttention represents a novel attention algorithm aimed at enhancing the speed and memory efficiency of transformers. Its design involves employing hashing and low-rank approximation methods to estimate an attention matrix. This unique approach enables scalability in a linear manner, which is a significant advantage for processing long sequences. Additionally, FlashAttention introduces "flash updates," enabling the rapid updating of the attention matrix in constant time, distinguishing it as faster than other approximate attention methods, such as Reformer and Performer. Compared with various transformer-based models, FlashAttention holds promise for enhancing the performance across range of tasks in natural language processing and computer vision [15].

Methods for integrating unstructured data with relational databases: methods extract column values from unstructured sources and maps to known database entities. This will help in use of statistical models for information extraction to identify entities stored in a database. The approach relies heavily on indexed access to the database and to the unstructured data. Challenges range from the need for automated and domain-independent database-driven solutions to building bridges that will allow seamless querying. During data integration, indexing techniques have been used for the similarity predicates. Integrating this, mapping is done from the unstructured data to existing structured databases. Of course, this kind of integration would easily broaden the potential for a far richer query interface than possible through traditional file-based processing of unstructured data [16].

3 Proposed System

A. Structured Data Handling

Conversion to Markup Language: This involves the conversion of structured data, like databases and spreadsheets, into a markup language such as XML or JSON. This way, it is easier to parse and subsequently process by the system. Models of Structured Data Usage: Use the SQLNet model for generating queries. This is a model to be used for the conversion of natural language queries to SQL commands that are appropriate to extract information in markup format structured data. Integration of NLP techniques: NLP techniques are used to understanding of the context and relationships within the structured data, improving query results accuracy.

B. Unstructured Data Handling

NLP Models for Textual Data: Advanced NLP models such as BERT and GPT-3 are applied for unstructured textual data. These models comprehend the context and are able to respond coherently. It would imply implementation of the Optical Character Recognition technology to extract text from images, and then such text would be processed through an NLP model for relevant answer generation. Multilingual Text Handling: Develop multilingual versions of NLP models, for example, multilingual BERT or XML-R, to handle text in different languages. This will ensure that the system applies under different linguistic contexts.

C. Multilingual Support

Language Detection and Translation: Develop algorithms to identify the language that has been inputted. Translate the query to the base language and translate the response back to the native language of the user. This can be further improved by use of cross-lingual transfer learning, in languages with limited training data available, to develop better cross-lingual understanding.

D. User Interface

Interactive and Multilingual Interfaces: Ease of use by all users with different languages will be assured by designing a user-friendly interface with provision for use in multiple languages.

E. Feedback Mechanism

Feedback System: This allows users to rate answers and give suggestions for system improvement.

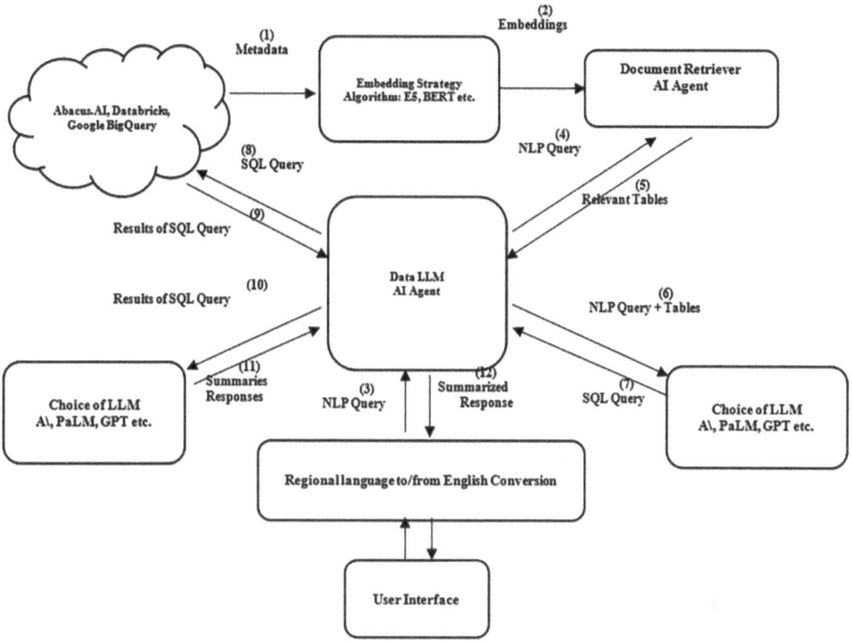

Fig. 1 Data flow diagram of proposed system

4 Methodology

The flow of data is depicted as shown in Fig. 1. The structured/unstructured data is fed into the system, which starts with text splitter module converting data into chunks. These chunks are analyzed to capture nuance, connections, and semantic meanings between the chunks and are stored in Vector DB. User queries are fed to sentence transformers which understands the query and context to answer it by referring the relevant docs in the Vector DB.

5 Proposed System

Vector databases have emerged as a critical component for managing high-dimensional data efficiently, particularly in applications such as QA with large language models.

The comparison of popular vector database systems along with shedding light on how they impact AI applications are discussed: The vector databases like Faiss, ChromaDB, Odrant (local mode), and PgVector are compared. Vector databases differ in how they manage their data on disks, leading to interesting variations in the

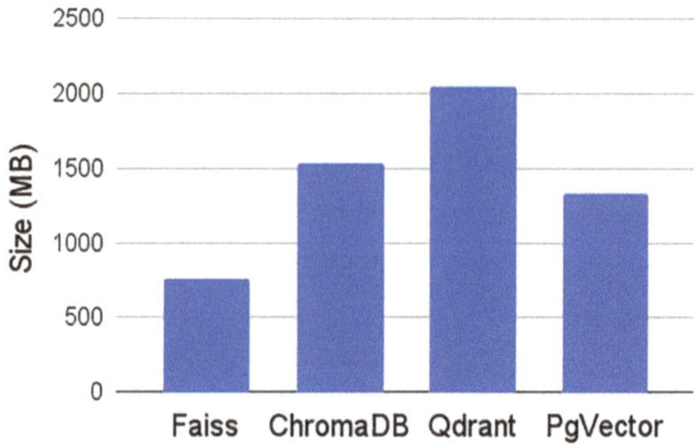

Fig. 2 Storage footprint

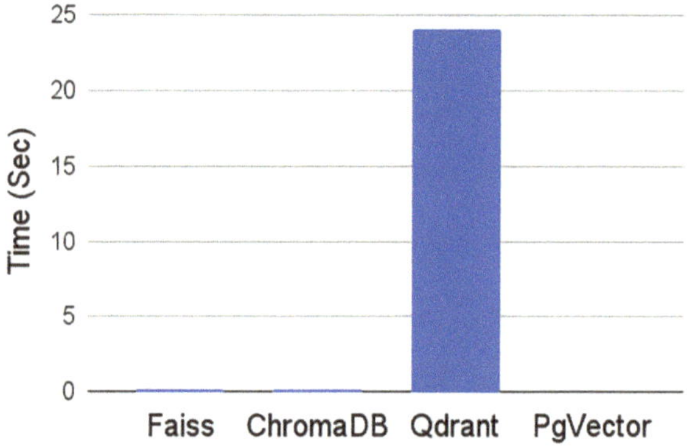

Fig. 3 Client connection time/index initialization time

client connection time, index search time, accuracy (recall), and storage footprint (Figs. 2 and 3).

- Faiss did not have any data management capabilities. Therefore, EvaDB retains the Faiss index using its own format.
- ChromaDB manages vectors on the disk in a custom format, but maintains additional metadata in an SQLite database.
- Qdrant (local mode) stores both vectors and metadata in an SQLite database.
- PgVector stores both vectors and metadata in the PostGres database as an extension.

The most striking observation here is that Qdrant (local mode) requires significantly more time for the client to connect to an existing index than other vector databases. This step occurs when a new client is started, and is a bottleneck when applications operate under an embedded architecture (i.e., not operating under a client–server architecture) (Figs. 4 and 5).

OpenAI's GPT models are trained to understand natural languages and code. These models respond with text, based on their inputs, also known as "prompts." A

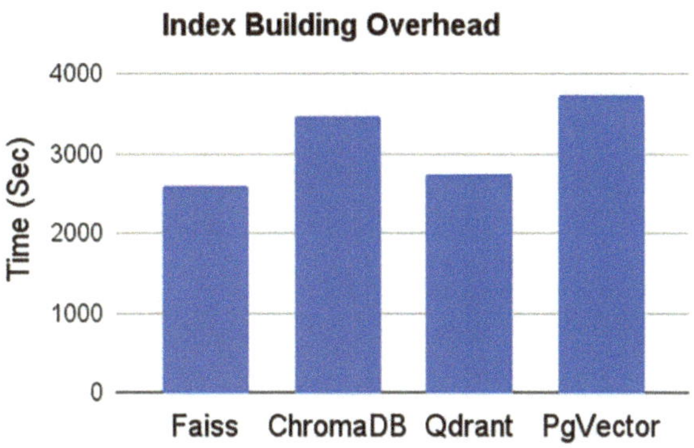

Fig. 4 Index building time

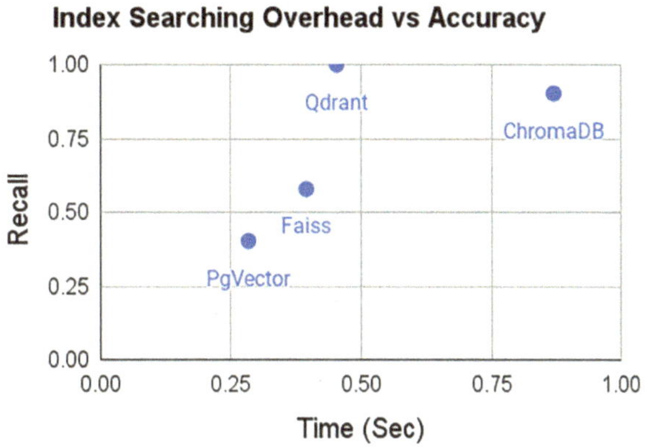

Fig. 5 Index searching overhead versus accuracy

prompt is designed as a means of "programming" a GPT model; this is usually done through instructions or examples of successful task completion.

Model Evaluation

Evaluation of few pretrained models like GPT-4, GTP-3.5, LM-SOTA, and SOTA are done based on certain parameters which is depicted in Table 4.

- The pretrained base model of GPT-4 was evaluated on traditional language model benchmarks.
- Contamination checks were conducted over the test data to identify if there was any overlap with the training set.
- Few-shot prompting was used in all tests for the benchmarks.GPT-4 demonstrated superior performance compared with existing language models and previous state-of-the-art systems.
- GPT-4 did better than models that were often benchmark-specific crafted or extra training protocols. At test time.
- Performance was evaluated against GPT-4 and SOTA models, with training on that benchmark.

Table 4 Evaluation of different models

	GPT-4 Evaluated Few-Shot	GPT-3.5 Evaluated Few-Shot	LM SOTA Best External LM Evaluated Few-Shot	SOTA—Best External Model (incl. Benchmark-Specific Tuning)
MMLU—Multiple-choice questions in 57 subjects (professional and academic)	86.4% 5-shot	70.0% 5-shot	70.7% 5-shot U-PaLM	75.2% 5-shot Flan-PaLM
HellaSwag Commonsense reasoning around everyday events	95.3% 10-shot	85.5% 10-shot	84.2% LLaMA (validation set)	85.6% ALUM
AI2 Reasoning Challenge (ARC)—Grade-school multiple choice science questions, challenge-set	96.3% 25-shot	85.2% 25-shot	85.2% 8-shot PaLM	86.5% ST-MOE
WinoGrande Commonsense reasoning around pronoun resolution	87.5% 5-shot	81.6% 5-shot	85.1% 5-shot PaLM	85.1% 5-shot PaLM
HumanEval Python coding tasks	67.0% 0-shot	48.1% 0-shot	26.2% 0-shot PaLM	65.8% CodeT + GPT-3.5
DROP Reading comprehension and arithmetic	80.9% 3-shot	64.1% 3-shot	70.8% 1-shotPaLM	88.4% QDGAT
GSM-8k Grade-school mathematics questions	92.0% 5-shot Chain-of-thought	57.1% 5-shot	58.8% 8-shot Minerva	87.3% Chinchilla + SFT + ORM-RL, ORM reranking

- GPT-4 has outperformed the existing models in all benchmarks except the DROP dataset and made the benchmark-specific training for all the datasets.
- GPT-4 has outperformed GPT-3.5 and other language models on a wide variety of languages, from low-resource ones like Latvian and Welsh to the semantic complexity of Swahili.
- GPT-4 showed a radical enhancement of what the user wanted. To this end, based on a dataset of 5,214 prompts submitted to ChatGPT and the OpenAI API, the responses with GPT-4 were favored compared to GPT-3.5 responses on 70.2% of the prompt.

6 Feature Enhancement

To advance QA system, we aim to optimize model performance for underrepresented languages through advanced NLP and machine learning techniques, ensuring more inclusive and comprehensive language coverage. Emphasis is placed on integrating the latest developments in deep learning for nuanced language comprehension and contextual awareness. We also plan to refine the user interface to make it more intuitive and adaptable to diverse languages and accessibility needs. Additionally, incorporating local knowledge sources will enable the system to offer more relevant region-specific responses.

Real-time processing and latency need to be improved, and this in turn requires use of efficient algorithms and hardware acceleration for fast handling of large volumes of data. Security of data and its privacy are major concerns and have to be enhanced to be at par with global standards of data protection. Scalability and cloud integration will be necessary to handle the huge volume of queries and documents. Ultimately, it should evolve into a much more customizable and adaptive system that incorporates user feedback for continuous learning and system refinement. These are the keys to making our QA system more powerful, user-friendly, and versatile with respect to multilingual document analysis and interaction, therefore setting ever-expanding horizons of applicability for AI in the area.

7 Conclusion

The work presented in this paper represents a significant contribution toward fully AI-based QA systems for processing both structured and unstructured data of diverse languages. The results from the experiments demonstrate the potential use of modern models (e.g., BERT and family) for multilingual data interpretation, but also highlight some practical hurdles in doing so. Although these models possess a good deal of language nuance, findings in this work highlight a critical gap in their performance consistency across diverse data formats and linguistic spectrums. By addressing these shortcomings, the development and inclusion of such models give testament to how AI technologies for excellent language processing are continually improving. The importance of this research not just lies in its technical accomplishments but also in its broader implications. By enhancing the accessibility and comprehensibility of information across languages, the proposed system breaks all the barriers in global communication and information exchange. It has the potential to enhance access to knowledge, making it more feasible regardless of linguistic background.

The journey of enhancing and refining this system is continuous. As the field of AI and NLP rapidly evolves, staying ahead of the latest innovation and integrating them into our system will be crucial. The aim is to create a QA system that is not only technologically advanced but also socially inclusive, catering to a global audience with diverse needs and languages. In doing so, this research will contribute significantly to the advancement of AI applications in multilingual contexts, paving the way for more connected and accessible global communication networks.

References

1. Gurdeep S, Truong T (2022) Using character embedding and QANet to improve performance on question-answering task (SQuAD)
2. Karpagam K, Saradha A (2019) A framework for intelligent question answering system using semantic context-specific document clustering and Wordnet. Sadhana-Acad Proc Eng Sci. https://doi.org/10.1007/s12046-018-1022-8
3. Wang A, Amanpreet S, Julian M, Felix Hill, Omer Levy, Samuel (2019) GLUE: a multi-task benchmark and analysis platform for natural language understanding. In: EMNLP workshop BlackboxNLP: analyzing and interpreting neural networks for NLP. https://doi.org/10.18653/v1/w18-5446
4. Karthik SA, Basavaraj GN, Seemanthini K, Harish Kumar N, Liyakathunisa, Supriya RK (2023) Hybrid approach for retail store auditing using CRNN. https://doi.org/10.1109/NMITCON58196.2023.10276301
5. Basavaraj GN, Lavanya K, Sowmya Reddy Y, Srinivasa Rao B (2022) Reliability-driven time series data analysis in multiple-level deep learning methods utilizing soft computing methods. Meas: SensS 24. ISSN:2665-9174. https://doi.org/10.1016/j.measen.2022.100501. Accessed 29 Sep 2022
6. Barlacchi G, Lvano L, Moschitti A, Marco Del T (2022) FocusQA: open-domain question answering with a context in focus. In: Conference on empirical methods in natural language processing. https://doi.org/10.18653/v1/2022.findings-emnlp.381
7. Yoon S, Rhim E, Kim D (2015) Domain question answering system. https://doi.org/10.5626/ktcp.2015.21.2.144
8. Devlin J, Chang M-W, Lee K, Toutanova K (2019) BERT: pre-training of deep bidirectional transformers for language understanding. North American chapter of the association for computational linguistics. https://doi.org/10.18653/v1/n19-1423
9. Ishwari KSD, Aneeze AKRR, Sudheesan S, Karunaratne, Nugaliyadde A, Advances in natural language question answering: a review. Computation and Language. https://doi.org/10.48550/arXiv.1904.05276
10. Naveed H et al (2023) A comprehensive overview of large language models. https://doi.org/10.48550/arxiv.2307.06435
11. Minaee S, Cambria E, Gao J (2020) Deep learning–based text classification. ACM Comput Surv (CSUR) 54:1–40. https://api.semanticscholar.org/CorpusID:235386502
12. Lewis M et al (2019) BART: denoising sequence-to-sequence pre-training for natural language generation, translation, and comprehension. Computation and Language. https://doi.org/10.18653/v1/2020.acl-main.703.
13. Qiang G, Tang S, Hao J, Sarno L (2024) A BIM and AIoT integration framework for improving energy efficiency in green buildings. In: Construction research congress 2024. https://doi.org/10.1061/9780784485262.059
14. Manathunga S, Illangasekara YA (2023) Retrieval augmented generation and representative vector summarization for large unstructured textual data in medical education. https://doi.org/10.48550/arxiv.2308.00479
15. Wang L, Liu Y, Zhao X, Xu Y (2006) Particle swarm optimization for fuzzy c-means clustering. In: 2006 6th world congress on intelligent control and automation, pp 6055–6058. https://doi.org/10.1109/WCICA.2006.1714243
16. Dao T, Fu DY, Ermon S, Rudra A, Ré C (2022) FlashAttention: fast and memory-efficient exact attention with IO-awareness. https://arxiv.org/abs/2205.14135

17. Sarawagi S (2004) Models and indices for integrating unstructured data with a relational database. In: International workshop on knowledge discovery in inductive databases. https://doi.org/10.1007/978-3-540-318

Deepfake Detection with XceptionNet: A Deep Learning Approach

Aryan Phalke, Chudaman Sukte, Kishor Pathak, Sonali Bhoite, Chittosh Meshram, and Krishna Gupta

Abstract The proliferation of deepfake videos on the internet over the past couple of years has posed significant threats to privacy, security, and integrity. These synthetic media are produced through advanced artificial intelligence techniques, enabling the manipulation of audio-visual content to create seemingly realistic but fraudulent imitations of a person. To address the challenge of deepfakes, researchers have explored numerous methodologies for detecting and mitigating them effectively. This paper provides a novel method of deepfake detection using XceptionNet-architecture, that is very sophisticated convolutional neural networks. XceptionNet captures best both spatial and temporal features from frames of the video and thereby can identify well tampered content. Thus, pre-processing techniques have been optimized combined with fine-tuned parameters of the model for higher accuracy and reliability of detecting fake videos in the proposed system. This approach, besides making detection better also opens up applications in areas where the trustworthiness and authenticity of media are key.

Keywords Convolutional neural network (CNN) · Deep learning · Deepfake detection · XceptionNet

---

A. Phalke (✉) · C. Sukte · K. Pathak · S. Bhoite · C. Meshram · K. Gupta
Department of Information Technology, Vishwakarma Institute of Information Technology, Pune, India
e-mail: aryanphalke366p@gmail.com

K. Pathak
e-mail: kishor.pathak@viit.ac.in

S. Bhoite
e-mail: sonali.bhoite@viit.ac.in

C. Meshram
e-mail: chittosh07@gmail.com

K. Gupta
e-mail: krishnagupta4371@gmail.com

H. Nagaraj et al. (eds.), *Advancements in Artificial Intelligence and Machine Learning*,
Lecture Notes in Electrical Engineering 1572,
https://doi.org/10.1007/978-981-95-7449-0_7

1 Introduction

Advances in artificial intelligence have made deepfake video proliferation possible, bringing forth a new threat to digital content privacy, security, and authenticity. This paper offers an efficient mechanism for the detection of deepfakes by utilizing the XceptionNet model, one of the architectures that top the current list of state-of-the-art architectures of convolutional neural networks. Unlike earlier techniques, the research focuses more on the fine spatial and temporal features in videos that can further improve the system's capability to classify content tampering. MTCNN for face detection and extraction also combined with some data augmentation techniques greatly improves the performance of the model for multiple scenarios, angles, light, and even facial expressions. This contribution enhances the robustness and accuracy of deepfake detection and provides a scalable solution adaptable to real-world applications. The proposed framework is a critical step toward mitigating the harmful effects that deepfake technologies can pose to individuals and society.

The dark side of AI has become more vicious as the use of deepfake technology has become more prevalent and popularized. Thirteen states, the number, which is bound to increase, have already passed laws prohibiting the use of deepfakes for harmful purposes. Various incidents have shown how deepfakes can be used to exploit systems to create identity fraud, misinformation, false accusations, and other political warfare methods. Because deepfakes videos are purposely created to look as realistic as possible, many people cannot trust the truth behind these videos. As a result, the deepfake content bearers may incur irreparable damage to their reputations or professional standing with serious potential impacts such as legal actions or societal exclusion.

The deepfake videos have developed new challenges beyond how the medium is utilized, where the ability of the deepfakes bleed the authenticity of content is prevalent which causes ramifications to law and order. The sufficiency of deepfake videos means deepfake detection systems have registered an increase in demand in the recent past, however the existing systems mostly fail to combat the newly advanced generative models due to the sophistication and increased accuracy levels.

Using the XceptionNet model, this research proposes a deepfake detection framework that attempts to address these issues by effectively recognizing deepfake movies.

The objectives of this research include developing an efficient deepfake detection system using the XceptionNet model to identify tampered videos through discrepancy analysis, utilizing MTCNN to detect and extract faces for improved accuracy and reduced training time, and evaluating the model's performance across diverse data, angles, and lighting conditions.

This paper outlines the system architecture, setup, results, and potential improvements based on the XceptionNet model. The authors will thereby attempt to contribute to the development of tools mitigating and lessening the harmful effects of fake technologies.

The paper is structured as follows: Section II reviews prior work and recent advancements in deepfake detection. Section III details the proposed methodology, data preprocessing, and XceptionNet-based system design. Section IV evaluates experimental results, performance, and accuracy. Section V summarizes contributions and outlines future directions in deepfake detection frameworks.

2 Literature Survey

With the onset of artificial intelligence, advanced technologies like deepfakes have emerged which possess the abilities to digitally alter an individual's appearance and generate convincing fake videos, images, etc. The improper application of this technology is a risk to every individual's social presence, privacy, and trustworthiness of digital content. To counter this, development of advanced deepfake detection systems has become a critical focus in today's tech world. We have tried to analyze how different deepfake detection systems have used multiple technologies to detect manipulated content. In recent years, the attention of researchers in the fields of deepfake detection and anti-spoofing has been moved to the development of neural networks, deep learning models, and image processing techniques. Specifically, Bhargava et al. developed a forged image detection system that detects forgeries in images using Error Level Analysis (ELA) followed by neural networks and employs Multilayer Perceptron for classification [1].

In this regard, one study worth mentioning is that of Pan et al. where they compared a variety of deep learning models and concluded that self-supervised learning models like Swin Transformer and Multiscale Vision Transformer (MViT) perform better than classical Convolution Neural Networks (CNNs) especially on complex datasets such as DFDC and FaceForensics + +. They also assert the utility of self-supervised models such as DINO and CLIP for feature extraction. In their experiments, the accuracy of classification of the real and fake video footage performed by these different methods was between 91 and 98%. Moreover, to enhance detection a voting system combining findings from different models was also put forward [2]. Geethanjali et al. described a new method of image tampering detection, combining the ELA technique with that of CNN. Training followed with the CASIA V2.0 dataset, and the model performed well for several deepfake techniques including copy-move forgery [3].

Suratkar et al. (2020) have examined the task of deepfake videos detection through Convolutional Neural Networks (CNNs) and transfer learning methods. This was in response to the growing concerns over artificial media created with models such as GAN. They used different pre-taught models such as Inception V3, VGG-16, Xception, and DenseNet121 and fine-tuned them according to a specific dataset consisting of frames cut out from videos in the Google AI DeepFake Detection dataset and the FaceForensics + + dataset. Their approach mainly emphasized the detection of facial differences in the edited videos [4].

As per Karthikeyan et al. the work presented in Guardian AI [5] combines facial recognition, image forensics, and temporal analysis techniques for the deepfake video detection. The system attained the rate of precision at the level of 93%, behind the analysis of the differences of facial landmarks as well as lighting and natural phenomena as unnatural, or abnormal movements of a certain type. K. Khan and Dang-Nguyen also shared a similar view when they tested various deep learning architectures with datasets of FakeAVCeleb and CelebDF-V2. Their findings suggest that transformer-based models show better generalized performance than CNN models on major datasets [6]. Smelyakov et al. assessed several models including EfficientNet, XceptionNet, and ResNet and reported that EfficientNet-B5 exhibits greater accuracy in detecting deepfakes especially when the requirements of DFDC [7] are met. Liu et al. introduced a method where self-blended images and its residuals were fused and the model improved deepfake detection over multiple datasets in the presence of compression and noise, etc. [8]

With regards to the medical region of image forgery, Aruna and Narayan utilized a technique of LBP preprocessing with a U-Net model to find GAN-synthesized medical images. The approach performs quite well in the detection of inpainted forgeries on CT scans [9]. Jbara and Soud also introduced a novel technique for deepfake detection based on the VGG-16 model with face detection and data augmentation which achieved almost outstanding results with 94.28% accuracy on datasets like Celeb-DF-v2 [10]. DeepfakeStack, a novel and distinctive deep learning technique, was introduced by Rana and Sung (2020). It employs ensemble learning to improve the detection of deepfake videos, which are modified audiovisual content produced with deep learning methods, particularly generative adversarial networks (GANs). Through its integration into a single meta-learner classifier, the DeepfakeStack approach leverages the capabilities of multiple advance deep learning models, such as XceptionNet, ResNet101, MobileNet, InceptionV3, DenseNet121, and InceptionResNetV2 [11].

Table 1 demonstrates different models performed for detecting deepfakes.

3 Methodology

The proposed deepfake detection system given an input classifies if the video is "REAL" or "FAKE." The system is trained following some set of processes that involves data preparation, frame processing, preprocessing, model selection, and training. This methodology is designed to tackle the challenge of detecting manipulated videos (i.e., deepfake videos). This section below describes each stage and the different practices followed in each stage to increase the model's accuracy and robustness.

Table 1 Accuracy comparison

S. no	Paper title	Model's used	Limitations	Accuracy (%)
1	Analysis survey on deepfake detection and recognition with convolutional neural networks [12]	DenseNet169, Dense-Net121, DENSNET201, and VGG16, 19, VGG-Face	Use of complex models like DenseNet, ResNet, and VGG	95.77
2	Detectify: image tampering detection using error level analysis (ELA) and convolutional neural network (CNN) [3]	Error level analysis (ELA) and Convolutional neural network (CNN)	Limited Effectiveness on high quality deepfakes	96
3	Deepfake detection-based VGG-16 model [10]	Multi-task cascaded convolutional networks (MTCNN) and VGG-16	VGG is an older architecture	94.28
4	Deepfake detection through deep learning [2]	XceptionNet and MobileNet	The model's accuracy is highly sensitive, and there is also a risk of overfitting	94

3.1 Dataset Preparation

At first, data was collected from Kaggle DeepFake Detection Challenge [13]. The dataset was containing videos categorized as "REAL" or "FAKE." All the videos progressed through the frame extraction stage to assist effective training of image-based models. This frame extraction stage reduces the computational demand by dispatching still frames from each video, acknowledging the model to concentrate on features instead of processing the whole sequences.

The input also contains a metadata a csv file containing names of the videos along with the label that the video is deepfake or not guarantees the precise categorisation of frames into "REAL" and "FAKE" frames. To increase the variability and robustness of system frames were sampled at various intervals. For example, a high frame sampling rate can excessively represent similar features, whereas some low frame sampling rate may neglect some of the essential features present within the video frames. Initially, experiments were conducted to decide the best possible sampling rates that provide a balance between sufficient feature representation and computational efficiency.

3.2 Frame Extraction

To ignore some of the excessive features and concentrate on the essential features, faces are recognized in each frame with the help of Multi-task Cascaded Convolutional Networks (MTCNN) face detection algorithm. This method improves face detection accuracy by leveraging both bounding box localization and facial landmark identification. The frames were examined in their natural color format (RGB), because MTCNN works upon color images to recognize faces with improved precision. The identified faces were extracted and resized to 224 × 224 pixels to ally with the input dimensions required for the Xception model.

To make sure that the accuracy of the face extraction, manual verification was conducted on some sections of the frames, authenticating that faces were reliably and accurately detected or not at all under various conditions like different lighting conditions, angles and facial expressions. This was one of the most critical steps because the flaws in face detection could greatly affect the model's capabilities to correctly decide between real and fake faces, by this means the overall efficiency of the deepfake detection system.

Figure 1 incorporates a pictorial representation of a variety of faces extracted from the dataset, which consists of both real and deepfake images. It displays the changes in facial attributes, expressions and containing different lighting scenarios found in the dataset, offering the model a range of graphical characteristics for training purposes. This dataset helps for the training and assessment of our deepfake detection system, making sure that the model successfully learns to recognize genuine verses manipulated faces.

3.3 Data Splitting

When the frames were extracted, they were partitioned into training, testing, and validation sets. The frames were partitioned to ensure that both real and fake frames were evenly distributed in each set. The split was 70% for training, 15% for validation, and 15% for testing. Ensuring that the model obtained a balance of real and fake data in each phase, to make sure that the model learns efficiently during the training and performs great during validation and testing. This method helped simplify the data processing phase for the next stages of the project. The input also contains a metadata a csv file containing names of the videos along with the label that the video is deepfake or not guarantees the precise categorisation of frames into "REAL" and "FAKE" frames. To increase the variability and robustness of system frames were sampled at various intervals. For example, a high frame sampling rate can excessively represent similar features, whereas some low frame sampling rate may neglect some of the essential features present within the video frames. Initially, experiments were conducted to decide the best possible sampling rates that provide a balance between sufficient feature representation and computational efficiency.

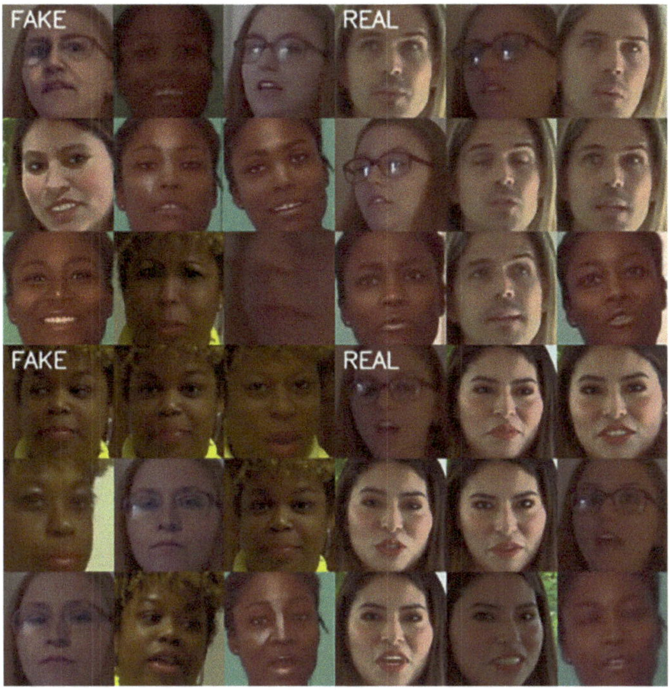

Fig. 1 Dataset overview

3.4 Pre-Processing

To increase the model's capabilities to generalize and lessen the risk for overfitting, a variation of data augmentation methods was employed on the training images. This technique synthetically widened the dataset's variety, permitting the model to gather more flexible features. The following modifications were applied.

Horizontal Flipping. At random flipping images with reference to the horizontal axis to virtualize different possible virtual angles.

Zoom Transformation. Randomly zooming in or out of the image to provide the model with a variation of the scale.

Shear Transformation. Using a cropping effect to alter the geometry of the images, providing a more varied perspective of the visual frames.

Likewise, all frames pixel standards were rescaled in the range of 0 to 1, normalizing the inputs given to the neural network. These various actions were taken on the dataset with the help of a custom image data generator class built upon the predefined TensorFlow's Image Data Generator. To show that the generator also integrated face detection and trimming with the help of MTCNN face detector, guaranteeing uniform preprocessing of the faces while training.

3.5 Model Architecture. XceptionNet Model

In this Project, we have chosen the XceptionNet model because of the efficiency it shows while managing complex image categorization tasks. The XceptionNet architecture makes use of depthwise separable convolutions that are computationally efficient and very effective in the extraction of features. It is best suited for tasks of high dimensions, such as deepfake detection. It also outperforms classical CNNs like VGG and Inception in detecting very subtle spatial and temporal discrepancies.

XceptionNet excels at picking up the subtle facial details and manipulations that deepfake algorithms introduce. This is consistent with the requirement of detecting both spatial and temporal inconsistencies.

Architectural Modifications

There were different adjustments applied to the model's architecture to increase its efficiency for binary classification in the context of deepfake detection.

The Global Average Pooling Layer: This layer reduces the dimensions of the feature maps without losing the desired information. It gives partial relief from overfitting.

Dense Layer with 1024 Units and ReLU Activation: A fully connected layer that allows the model to identify complicated, irregular patterns in the data, making is crucial for spotting subtle manipulations.

Sigmoid Output Layer: A fully connected layer, which can enable the model to recognize complex, nonlinear patterns in the data and, therefore it is necessary for detecting subtle manipulations.

Utilized pre-trained weights from ImageNet to accelerate the training process and enrich feature extraction, though only the custom layers were firstly trained to modify the model for deepfake detection.

Training Parameters

For the loss function, binary cross-entropy was selected for its suitability in minimizing loss between predicted and actual labels in binary classification tasks.

The learning rate began at 0.001 and was decreased gradually to improve the training precision and avoid overshooting. A smaller initial learning rate avoids overshooting the minima, thus achieving stable convergence. Gradual decay refines the training, enhancing precision as the model approaches optimal weights. Fine-tuning pre-trained weights (ImageNet) also benefits from a small learning rate to preserve learned features while adapting to the deepfake detection task.

The batch size was set to 32 to encourage reasonable and computational efficiency combined with stability of training. That size assures very good trade-off between computational efficiency and gradient stability. More significant batch sizes may demand extensive memory resources, whereas smaller sizes could lead to noisier gradients.

The number of epochs was selected to be 10 with early stopping in place to deal with overfitting; training would be automatically stopped if the validation accuracy stopped improving. Early stopping stops overfitting by halting the training process if the validation performance no longer improves. The selected epoch count is sufficient enough to train a pre-trained model on a particular dataset, balancing cost and performance computationally.

Using a checkpoint system, it saved the best-performing model weights based on validation accuracy so that the optimal configuration is preserved for evaluation.

The graphs in Fig. 2 demonstrate performance metrics over 10 epochs for model training and validation. The metrics include accuracy, precision, recall, and loss.

Figure 3 shows the prediction user interface of system which takes a video for inspection and analyses it to predict if it has been manipulated or not. Frames are extracted from the video and the trained model is deployed to make predictions to determine if the video seems real or fake.

Figure 4 shows the flow of our project. Users can upload videos in one of the following three formats: "mp4," "avi," or "mov." If the uploaded video doesn't match one of these formats, then an error message is generated, Invalid Format, will be

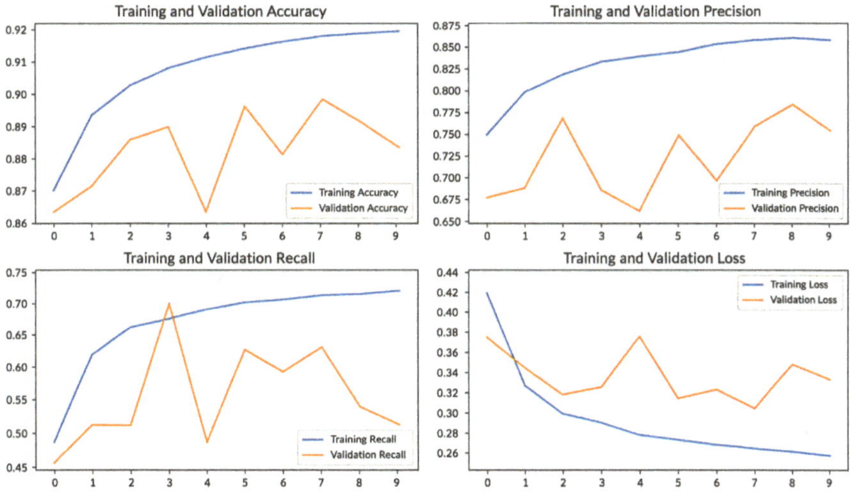

Fig. 2 Performance metrics of training and validation

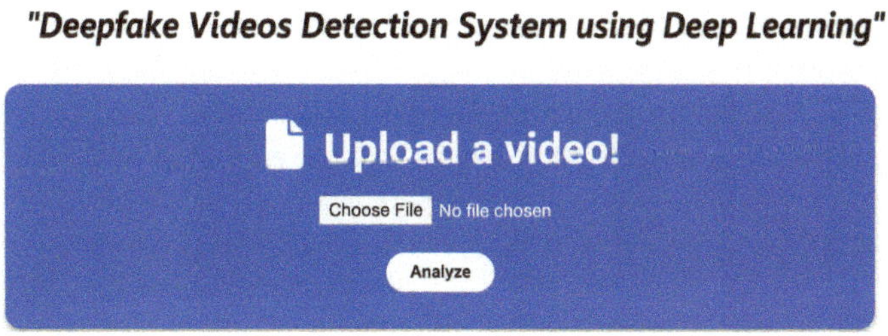

Fig. 3 User interface

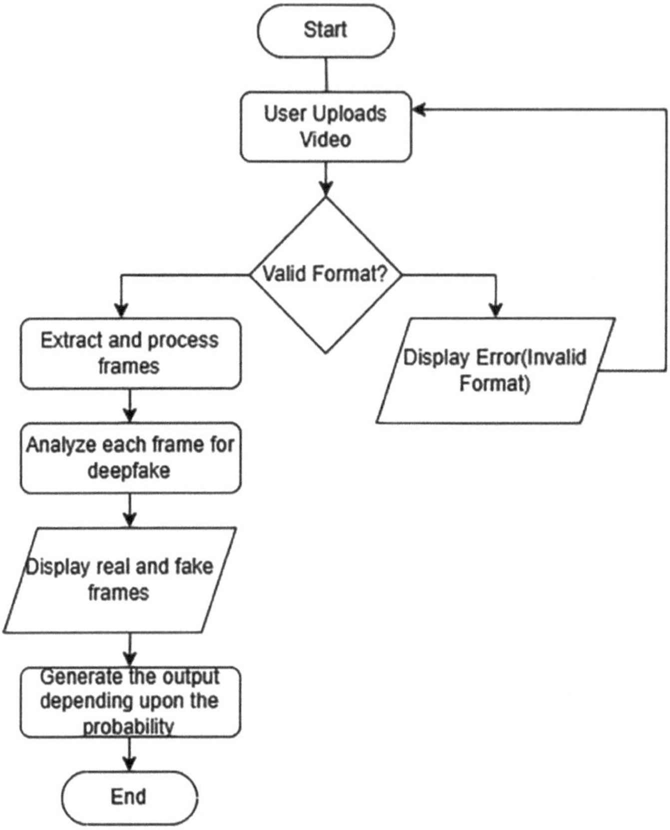

Fig. 4 Flow chart

displayed. For correctly formatted videos frames are extracted and analyzed individually. Each frame is accessed, and a probability score is generated to determine if the content is classified as "fake" or "real."

3.6 *Comparison with Alternative Methods*

The explored image feature extractors have been found to be Swin Transformer and MViT where the focus remains more on the extraction of long-range dependencies with high computational support and manipulating for larger datasets. However, on the other hand, XceptionNet provides comparable results while consuming much less computational energy, thus conducive for real-time use. While EfficientNet-B5 has some risk of overfitting on small datasets, XceptionNet effectively extracts features by using depthwise separable convolutions.

Table 2 Comparison table

Model	Architecture	Training accuracy (%)	Testing accuracy (%)
XceptionNet	Deptwise separable CNN	97.8	88.97
Swim transformer	Transformer	96.5	90.2
EfficientNet-B5	Scalable CNN	98.1	88.5
DeepStack	Multiple architectures	98.3	91.1
VGG-16	Standard CNN	94.2	84.3

A stacking model could do better than a single model but could take longer for inference, as seen with DeepfakeStack. With a slimmer architecture, XceptionNet works as a good balance between efficiency and accuracy.

Table 2 summarizes the performance of various state-of-the-art methods, highlighting XceptionNet's balance of efficiency and effectiveness. While ensemble models achieve slightly higher accuracy, they do so at the cost of computational overhead, making XceptionNet more suitable for real-time applications.

3.7 Experimental Setup

The experiments were performed on a machine featuring an NVIDIA Tesla V100 GPU, with 16 GB VRAM. The experiments were conducted using TensorFlow version 2.8, with Google Colab Pro available for extra computational resources when the training was extended. The training time for the entire dataset ranged from 6 h per epoch, generally, by optimizing the pre-process with parallelization to ensure the smallest possible lead time. All experiments were conducted through the Keras API of TensorFlow to facilitate efficient deployment and evaluation of the models.

3.8 Dataset Details

The Kaggle DeepFake Detection dataset contains 50,000 videos, evenly split between real and fake. It is divided into 70% training, 15% validation, and 15% testing. The dataset's diversity aids model generalization across deepfake variations. Video frames were extracted at one per second for efficiency, with MTCNN used for face detection to reduce computational load. Faces were resized to 224 × 224 pixels for XceptionNet input, with augmentations like flipping, zooming, and shearing. Pixel values were normalized to [0,1] for consistency.

4 Results and Discussion

The proposed system effectively categorizes videos as "REAL" or "FAKE" using an XceptionNet-based CNN architecture, trained on pre-processed face images extracted from the Kaggle DeepFake Detection Challenge dataset. The following are the explanatory results of the system's operation on training, validation, and test datasets (Table 3).

4.1 *Model Performance*

The model exhibited proficient generalization as it scored high values during both the training and validation stages. The key performance indicators for the test set are as follows:

These metrics show that the model consistently detects deepfakes with balanced precision and recall. The F1 score highlights the performance of the model in terms of identification and discrimination of actual and simulated videos, which is essential in reducing both the rate of false positive and false negative.

Figure 5 demonstrates how the result is displayed after prediction using trained model. Each frame is analyzed and time taken for detection is displayed along with concluded result.

Impact of Data Augmentation and Pre-processing. Data augmentation procedures, such as flipping, shearing, and zooming acted to improve the generalization of the model. By stratifying the dataset, there was a careful balancing of the ratio between real and fake samples while standardizing input by rescaling pixel values into [0, 1], thereby speeding the training up.

Model Architecture and Training Adjustments. The accuracy of the model took a quantum leap forward with the use of XceptionNet as pre-trained ImageNet weights and with transfer learning. The frozen pre-trained layers allowed the model focus on the learning of deepfake-specific features in custom layers. The early stopping routine avoided overfitting and ensured good performance on unseen data.

Table 3 Model performance

Process type	Evaluation percentage
Validation accuracy	88.29
Testing accuracy	89.87
Training accuracy	97.48
Precision	86.7
Recall	74.1
F1 score	79.91

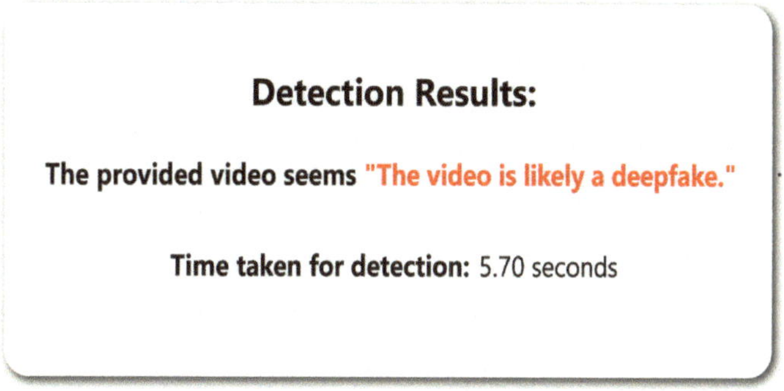

Fig. 5 Prediction results

Discussion. The model's effectiveness lay in its combination of a CNN structure, the focused face extraction associated with it, and the dedicated pre-processing steps taken. Such methods kept the computational load at an acceptable level while achieving very good performance. The depth and effectiveness of XceptionNet became what enabled the tackling of deepfake detection complexities.

4.2 *Dataset Evaluation*

The primary dataset used for training and evaluation was the Kaggle DeepFake Detection Challenge (DFDC) dataset. While it provided a strong foundation, relying on a single dataset limits the model's generalizability. This narrow evaluation scope raises concerns about the model's robustness in real-world scenarios. Future work should validate the model on additional datasets, such as:

FaceForensics + + : Offers a diverse set of manipulated videos, including various compression levels.

CelebDF: Focuses on high-quality deepfakes that challenge detection models. **DeepFakeDetection**: A comprehensive dataset for benchmarking deepfake detection systems.

By testing across these datasets, the robustness and applicability of the model in real-world scenarios can be improved.

5 Conclusion and Future Work

Our study provides a comprehensive, well-performing deepfake detection system that implements XceptionNet architecture achieving a stunning accuracy of 97.48% on the training dataset and 89.8% on the testing dataset. Through MTCNN, the system solves problems such as change in light, a variety of views of the same face or different facial movements together with multiple angles of the face. The various techniques of spliting the dataset into training and testing sets, varying pixel dimensions and employing the methodology of transferring learning from already trained images through ImageNet helped the model to learn well with real and fake images as seen in the high performance, precision and F1 measure obtained during the testing phase.

The authors of the study reported that better results can be obtained when a Global Average Pooling layer and a fully connected layer with the ReLU activation function are included so as to capture complex images of deepfake content manipulation and used early stopping to avoid overfitting and increase validation performance.

Despite some progress made by the system, some issues such as inconsistent performance across datasets and sizes of images as input are still present. In addition, the deployment of deepfake technologies in the wild may also limit the detection ability of this system relative to time. Future work could include improving the deepfake detection capability of the face detection model in more extreme scenarios as well as embedding the model in streaming video sources to enable the live recognition of deepfake videos.

The future directions of our project could involve the investigation of various ensemble techniques to exploit a wider scope of embracing deepfake features, creating a deployable model for real-time detection to improve its practical usability, and widening the dataset to include the newer and more sophisticated deepfake techniques.

References

1. Bhargava D, Rani S, Singh M, Tripathi N, Bhargava A, Panwar G (2024) Deep-fake finder: uncovering forgery image through neural network analysis. In: 2024 international conference on communication, computer sciences and engineering (IC3SE). IEEE, pp 1–5
2. Pan D, Sun L, Wang R, Zhang X, Sinnott RO (2020) Deepfake detection through deep learning. In: 2020 IEEE/ACM international conference on big data computing, applications and technologies (BDCAT). IEEE, pp 134–143. Jacobs IS, Bean CP (1963) Fine particles, thin films and exchange anisotropy. In: Magnetism, vol. III, Rado GT, Suhl H (eds) New York: Academic, pp 271–350
3. Geethanjali TM, Darshan TS, Surya K, Rahul HU, Sheety IN (2024) Detectify: image tampering detection using error level analysis (ELA) and convolutional neural network (CNN). In: 2024 international conference on signal processing, computation, electronics, power and telecommunication (IConSCEPT). IEEE, pp 1–6. Nicole R, Title of paper with only first word capitalized. J Name Stand Abbrev, in press

4. Suratkar S, Kazi F, Sakhalkar M, Abhyankar N, Kshirsagar M (2020) Exposing deepfakes using convolutional neural networks and transfer learning approaches. In: 2020 IEEE 17th India council international conference (INDICON). IEEE, pp 1–8
5. Karthikeyan K, Swetha R, Deepanraj S, Dhandapani S (2024) Guardian AI: synthetic media forensics through multimodal fusion and advanced machine learning. In: 2024 international conference on cognitive robotics and intelligent systems (ICC-ROBINS). IEEE, pp 226–232
6. Khan SA, Dang-Nguyen DT (2023) Deepfake detection: analysing model generalisation across architectures, datasets and pre-training paradigms. IEEE Access
7. Smelyakov K, Kitsenko Y, Chupryna A (2024) Deepfake detection models based on machine learning technologies. In: 2024 IEEE open conference of electrical, electronic and information sciences (eStream). IEEE, pp 1–6
8. Liu Q, Xue Z, Liu H, Liu J (2024) Enhancing deepfake detection with diversified self-blending images and residuals. IEEE Access
9. Aruna S, Narayan S (2024) Detection of GAN-manipulated medical images through deep learning techniques. In: 2024 international conference on advances in modern age technologies for health and engineering science (AMATHE). IEEE, pp 1–6
10. Jbara WA, Soud JH (2024) DeepFake detection based VGG-16 model. In: 2024 2nd international conference on cyber resilience (ICCR). IEEE, pp 1–6
11. Rana MS, Sung AH (2020) Deepfakestack: a deep ensemble-based learning technique for deepfake detection. In: 2020 7th IEEE international conference on cyber security and cloud computing (CSCloud)/2020 6th IEEE international conference on edge computing and scalable cloud (EdgeCom). IEEE
12. Ahmed SR, Sonuç E, Ahmed MR, Duru AD (2022) Analysis survey on Deepfake detection and recognition with convolutional neural networks. In: 2022 international congress on human-computer interaction, optimization and robotic applications (HORA), Ankara, Turkey, pp 1–7. https://doi.org/10.1109/HORA55278.2022.9799858
13. https://www.kaggle.com/competitions/deepfake-detection-challenge/overview

Deeptrack: Adaptive Deep Learning for Activity Detection

Snehal Rathi, Manisha Mali, Bhavesh Mankar, Sanket Dhakane, Shreyas Jagdale, and Tushar Maurya

Abstract With the advancement of mobile technology and the use of sensors, monitoring human actions has become easy on these devices. This research uses machine learning techniques to categorize human actions like walking, running, sitting, and standing, based on data from smartphone sensors like accelerometers and gyroscopes. Healthcare, fitness, and smart environment are some areas where Human Activity Recognition (HAR) has potential scopes. Decision Trees, Random Forests, Deep Long Short-Term Memory (LSTM), Logistic Regression and Support Vector Machines (SVMs), networks are used to test performance of this study. Sensors are now used in every smartphone, so it is easy to track human activity with our smartphones. Our study investigates how machine learning can be applied to classified activities like walking, lying, sitting, and standing using data gathered from mobile sensors, such as accelerometers and gyroscopes. Since these devices are so common, Human Activity Recognition (HAR) could greatly impact areas like healthcare, fitness, and smart home technologies.

Keywords Smartphone sensors · Human activity recognition (HAR) · Logistic regression · Random forests · Deep learning · Long short-term memory · Decision trees · Support vector machine

1 Introduction

While research on finding and grouping human activities using mobile phones is growing, it has many uses in security, smart environments, fitness tracking, healthcare, etc. But, computer vision and old methods used to monitor human movements

S. Rathi · M. Mali
Department of Computer Engineering, Vishwakarma Institute of Technology, Pune, India

B. Mankar (✉) · S. Dhakane · S. Jagdale · T. Maurya
Department of Computer Engineering, Vishwakarma Institute of Information Technology, Pune, India
e-mail: bmmankar25@gmail.com

H. Nagaraj et al. (eds.), *Advancements in Artificial Intelligence and Machine Learning*, Lecture Notes in Electrical Engineering 1572,
https://doi.org/10.1007/978-981-95-7449-0_8

need expensive equipment like cameras, which aren't useful for daily use. So, mobile phones are better because they are cheaper and flexible, using sensors like gyroscopes and accelerometers.

This study looks at finding activities by using information gathered from a phone's accelerometer and gyroscope. Both sensors together can track different human activities of walking, lying, sitting, and standing by measuring movement of three directions (x, y, and z). Our goal is to make a system that correctly and easily recognizes these activities in real-life situations using machine learning.

Recognizing daily tasks is a classification problem where data is collected and trained on different machine-learning models to classify physical activities. This research uses various classification methods like Support Vector Machine (SVM), Logistic Regression, Decision Tree, Deep LSTM, and Random Forest. We propose a hybrid model that integrates LSTM's strength in temporal sequence modeling with SVM's stability in classification to enhance performance.

This study can be helpful in many fields, including healthcare, where it can assist in fitness programs to monitor exercise routines. By recognizing activities, intelligent environments can be created that adjust settings based on what the user is doing. Additionally, in security, unusual activity could be detected to trigger alarms. Since most individuals carry smartphones in their daily lives, it is convenient to detect these movements with smartphone sensors.

2 Literature Study

Human Activity Recognition is among the important fields in the area of pervasive computing. Methods which include Bayesian Method, Artificial Neural Networks (ANNs), Support Vector Machines (SVMs), K-Nearest Neighbors (KNN), and Hidden Markov Models (HMMs) are used [1].

ANN models were used by Kern et al. in Human Activity Recognition for the differentiation of static and active states while increasing accuracy by internal states. This demonstrated, as shown in a study on user incorruptibility, the difference between social and personal contexts using wearable sensors. Also, it found that people view incorruptibility differently and suggested systems that can adapt to users. A new hardware was made with 12 3D sensors and a microphone to make recognition better [2].

Also, ANN-based methods were attempted to employ a PCA for dimension deduction after feature extraction from sensor from Xialnling Lu [3].

Wearable sensors improved HAR by a large margin. Wang et al. (2017) looked into the possibility of accuracy gains with the help of deep learning models like CNNs and RNNs. Crucially, because these models learned features directly from sensor data, removing the requirement for extracting features manually, they outperformed traditional methods by a comparable margin. After applying these models to various activities and ensuring that programs run smoothly [4].

Zhang and colleagues evaluated "Advancements in the Feature Engineering and Sensory Fusion" in 2022. Kalman filtering and Fast Fourier Transform (FFT) are used for combining sensor data and extracting useful features that try to improve activity detection. They found that achieving preciseness while balancing the computing requirements is a difficult trade-off [5].

Mohammadi et al. tossed SVM classifiers into the ring for Human Activity Recognition (HAR). Turns out, they've got a solid case for blending different classifiers—like a teamwork hack—to boost how accurately the system IDs activities. What's cool? When they smashed these classifiers together using product rule fusion, the hybrid model totally left older methods in the dust. We're talking classic visual feature staples here: HOG for shape patterns, Motion Boundary Histograms for tracking movement edges, and Optic Flow for capturing motion vibes. The ensemble approach? Yeah, it schooled those non-ensemble models hard [6].

Maswadi et al. conducted a study that compares algorithms like Decision Trees and Naive Bayes (NB) for HAR. The features that were attached were found to have a strong correlation because the NB classifier produces inconsistent results, especially with the use of linear and Gaussian filters. The Naïve Bayes model excelled at "sitting" and "walking," yet struggled with "laying" and "standing up." Conversely, in nearly all activity categories, Decision Trees demonstrated better classification accuracy, particularly when used alongside a Gaussian filter [7].

HAR has had a significant impact in recent developments due to deep learning for instance Recurrent Neural Networks (RNNs), which are popular for their capacity to understand temporary connections and forecast human patterns, were the subject of Julieta et al. experiment. They worked on making long movement sequences for people and guessing what they might do next. They compared their new RNN model with ERD, LSTM-3LR, and SRNN and worked. It clearly does way better than the state-of-the-art results, showing how serious the gaps are between conditioning and predicting in the earlier works [8].

Chao Li et al. introduced a design which employs a CNN to automatically learn hierarchical co-occurrence features from skeleton sequences. Following the processing of point-level data, their model tried to integrate geographical context to form a comprehensive and important representation. They propose a strategy for effectively documenting joint feature interactions [9].

Maosen Li et al. developed a neural network based on graphs to study interactions among body joints across various measures. Their structure and two main parts, one was for recognition of actions and for prediction of motion, both worked together to enhance joint movement. CNN-enabled models extract spatial and temporal features simultaneously, while graph structures capture the physical constraints and dynamic nature of actions [10] (Fig. 1).

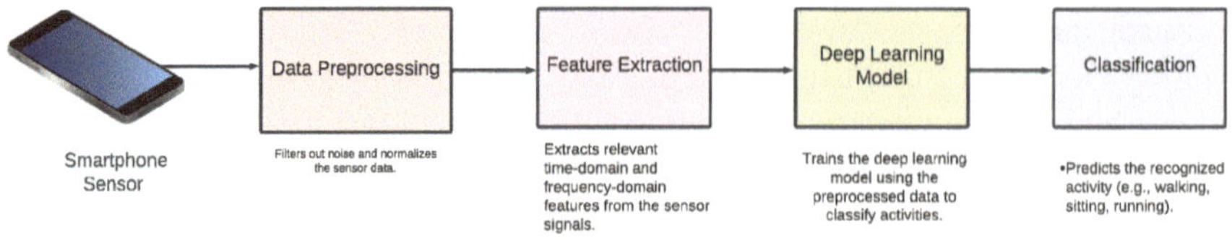

Fig. 1 System execution flow

3 Methodology

3.1 Dataset Used

As mobile technology is becoming more and more integrated into our daily lives, the ability to recognize human movement using smartphone sensors becomes more and more important. We used a Kaggle dataset, here, for this study that focuses on smartphone activity recognition for humans. Thirty volunteers, ages 19–48 had participated in this study and completed the six tasks listed in Table 1.

A Samsung Galaxy S II mobile was fastened firmly around the waist of every participant. This arrangement made it possible for the device's integrated gyroscope and accelerometer to record 3-axial angular velocity and 3-axial linear acceleration at a steady sampling rate, which is 50 Hz. Video recordings of the sessions were made to ensure that the activities were labeled accurately.

To enhance quality, the raw sensor data was pre-processed through a number of procedures in order to efficiently filter out noise and separate the gravitational components from body motion, we used a "Butterworth low-pass filter" with a cutoff frequency of 0.3 Hz. For thorough temporal overage, the filtered signals were then split into fixed width sliding windows of 2.59 s with the 49.7% overlap, yielding 128 reads per window.

Table 1 Dataset distribution across different classification labels

Activity	Count
"Laying"	1407
"Sitting"	1286
"Standing"	1374
"Walking"	1226
"Walking downstairs"	986
"Walking upstairs"	1073

Table 2 Accuracy of different classification algorithms

Model	Accuracy (%)
"Logistic regression"	90.40
"Linear SVM"	93.83
"Kernel SVM"	92.16
"Decision trees"	85.34
"Random forest"	90.34
"Deep LSTM"	95.04
"Proposed hybrid model: LSTM-SVM"	95.67

3.2 Models Used

To classify the extracted features into distinct activities, we tried several machine learning algorithms, which are known for their effectiveness in the classification tasks. The models tested included "Logistic Regression, Support Vector Machine (SVM), Decision Tree, Random Forest, and Deep LSTM." Based on our finding, we then proposed a hybrid model combining LSTM and SVM, using SVM's strong classification skill and LSTM's capability in modeling temporal dependencies to improve overall accuracy in the detection of human activities. The accuracies of these models are mention in Table 2.

3.3 Feature Selection

Feature selection is used to reduce noise and concentrate on most relevant data in order to improvise the accuracy of activity recognition. Features such as tBodyAccMag-mean() and gyroscope reading were utilized and are important in distinguishing between different activities. Principal Component Analysis is done to reduce dimensionality while maintaining important information, and correlate analysis to evaluate the relationships between variables in order to find best-fit features.

To make sure the chosen attributes were relevant to the action under analysis, we included knowledge of deep learning [11]. We were able to reduce complexity and enhance classification results with our method, which is in line with results from studies in activity recognition (Fig. 2).

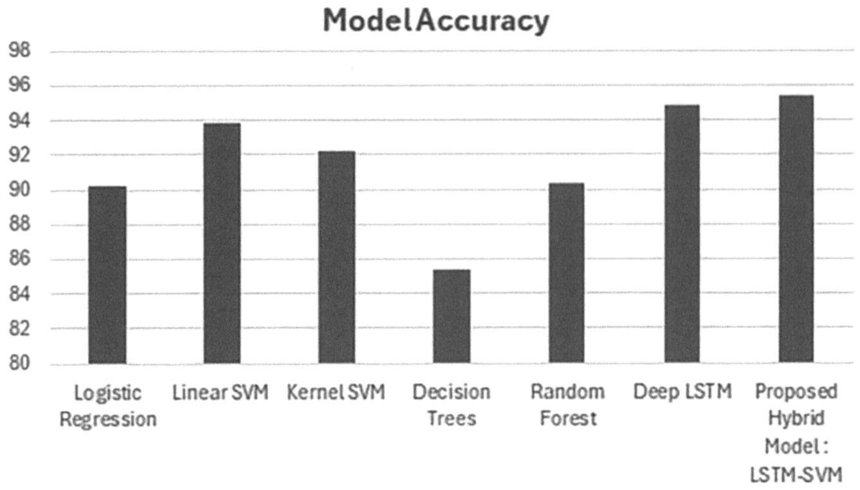

Fig. 2 Model accuracy comparison graph

4 Results and Discussion

Upon analyzing density plots of "tBodyAccMag-mean()" feature, we noticed the difference between static and dynamic activities. Motion activities have wider distributions centered toward 0.0 which shows varied body acceleration.

Contrary to above, dynamic activities show wider distributions near to 0.00. These actions like walking upstairs and downstairs are more correctly separated in their density plots. It makes it easy to distinguish them. The following Fig. 3 helps us to understand the above paragraph.

From these plots, we noticed that static activities have similar accelerations, while dynamic activities are easy to classify. This showed that the feature tBodyAccMag-mean() is useful for separating dynamic and static activities, but we might require more features to improve accuracy.

4.1 Comparative Study

These models were trained and evaluated on the dataset, and their performances are summarized as follows:

The Linear SVM achieved an accuracy of 93.83%, showcasing good results in classifying human activities. Logistic Regression has an accuracy of 90.20% showcasing decent performance but lagging the Kernel SVM.

The Kernel SVM achieved an accuracy of 92.16%, which surpassed logistic regression. However, with Kernal SVM there is some difficulties in distinguishing between similar activities, particularly between standing and sitting.

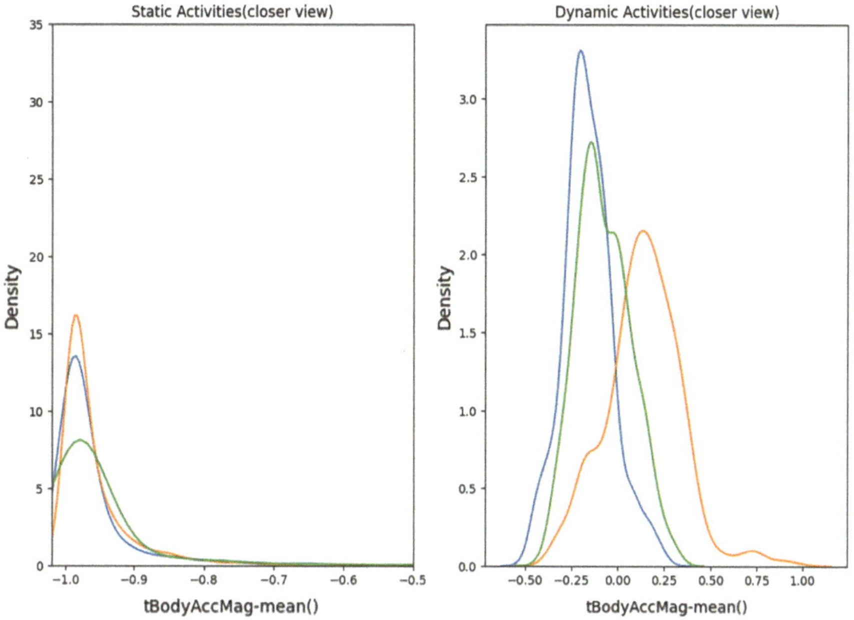

Fig. 3 Dynamic and static actions potted for density vs tBodyAccMag-mean()

The Deep LSTM also performed well with an accuracy of 94.84%, revealing its ability to handle time-dependent patterns, surpassing the linear models by capturing time-dependent patterns in sensor data. This showcases its strength in handling complex activity recognition (Table 3).

The Decision Tree achieved an accuracy of 85.34% and performed the worst overall, as it frequently misclassified activities like walking upstairs and downstairs. The classification reports of each method provided valuable insights into the types of classification mistakes encountered.

Kernel SVM faced challenges in distinguishing certain activities, but Linear SVM made comparatively fewer mistakes. Deep LSTM demonstrated its ability to represent time-dependent patterns, achieving the highest overall accuracy, followed by

Table 3 Classification report for deep LSTM

Activity	Precision	Recall	F1-score	Support
Laying	1.00	1.00	1.00	537
Sitting	0.94	0.90	0.92	491
Standing	0.91	0.96	0.94	532
Walking	0.96	0.99	0.98	496
Walking downstairs	0.98	0.92	0.95	420
Walking upstairs	0.93	0.98	0.95	471

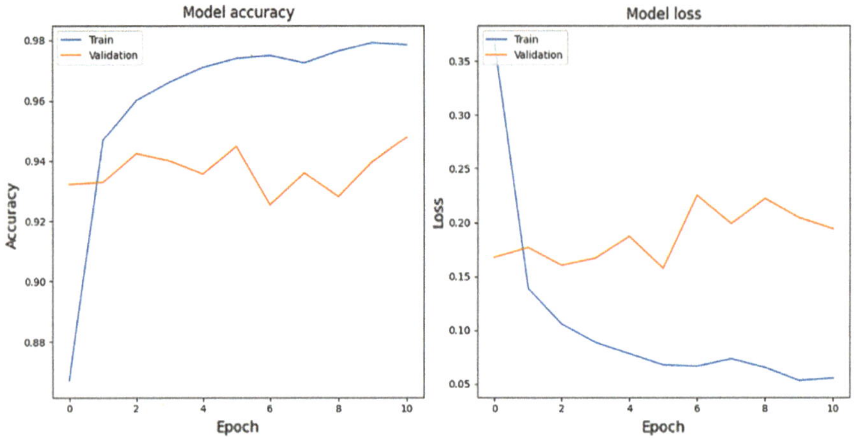

Fig. 4 Training versus validation accuracy and loss for deep LSTM

Linear SVM and Kernel SVM, both of which showed competitive efficiency and reliability for this dataset (see Fig. 4).

4.2 Proposed Hybrid Model: LSTM-SVM

The hybrid model outperformed classical models in terms of training and validation accuracy (see Fig. 5). The consistency across training and testing datasets is demonstrated by the training accuracy vs. validation accuracy graph (see Fig. 5), which shows less overfitting than other models.

The LSTM-SVM hybrid model's confusion matrix shows that it was especially effective at differentiating between static activities (like standing and sitting) and dynamic activities (like running and walking), which helped to reduce the misclassification errors that were evident in classical models.

The result in Table 4 demonstrates that the hybrid model provides better classification and performance, particularly for complex activities.

4.3 Limitations

Although our work is beneficial for exploring wearable sensor-based activity recognition, it must be noted that it has certain shortcomings. Firstly, the dataset only included information from 30 participants, which restricts the manner in which results can be applied. In addition, it is worth taking note of sensor positioning; every participant utilized a Samsung Galaxy S II device worn around the waist. Since various sensor

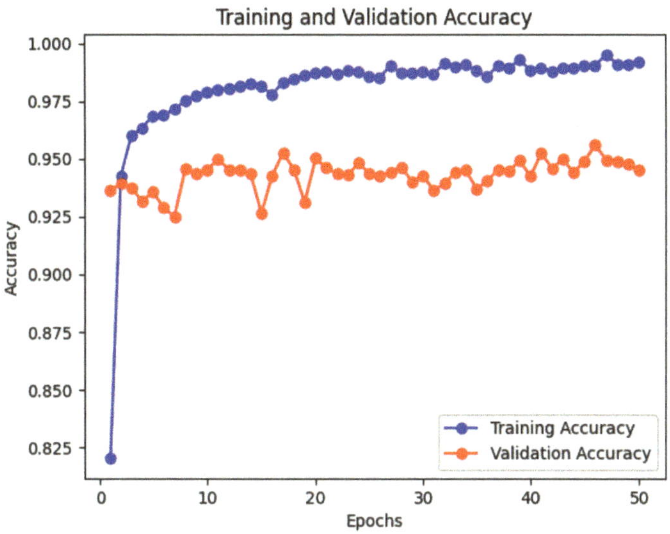

Fig. 5 Training and validation accuracy for proposed hybrid model

Table 4 Classification report for proposed hybrid model

Activity	Precision	Recall	F1-score	Support
Laying	1.00	1.00	1.00	537
Sitting	0.93	0.82	0.90	491
Standing	0.92	0.96	0.90	532
Walking	0.96	0.98	0.95	496
Walking downstairs	0.97	0.91	0.92	420
Walking upstairs	0.96	0.97	0.92	471

placements might have impacted performance accuracy, such uniformity ignores the possible impact of the sensor's motion on identification accuracy.

5 Training and Evaluation

This study explored using different math and computer science models to figure out what people were doing based on sensor data. To make sure they tested the models thoroughly, the researchers split the data into two groups—one for training and one for testing.

The Deep LSTM model performed the best, achieving the highest accuracy because of its ability to effectively capture time-dependent patterns in the data. Linear

SVM came next, delivering strong results with fewer errors compared to other traditional models. Logistic Regression, although straightforward to use and understand, didn't perform as well as Linear SVM or Kernel SVM, but still managed to handle the dataset decently. Kernel SVM had competitive accuracy but struggled to distinguish between similar activities like standing and sitting.

The Random Forest model performed alright, neither very well nor very poorly. It was able to smooth out some of the peculiarities that come with a single Decision Tree by combining many of them. However, despite that collaborative approach, it was still inferior to the Linear SVM and Deep LSTM models, which performed the task more successfully. The fundamental Decision Tree, on the other hand, performed poorly in this case. Its accuracy was severely damaged by its complete inability to distinguish between high-movement tasks, such as going upward and downstairs.

In the end, the traditional math and computer science models were the best at figuring out human actions based on sensor data. However, recognizing the potential of deep learning, the researchers proposed a hybrid model combining the strengths of both LSTM and SVM. This hybrid method offers substantial improvement in activity detection accuracy by merging the strong classification capacity of SVM with the sequential data learning abilities of LSTM. The proposed model has lots of promises, particularly for real-time applications where efficiency and accuracy are crucial. The results suggest that this combination of techniques holds considerable potential, especially for real-time applications where accuracy and efficiency are critical.

6 Future Work and Deep Learning Potentials

Deep learning can figure out patterns on its own from massive datasets, which is why it's really helpful for systems that track human activities—like jogging, sitting, or climbing stairs. For example, models like Convolutional Neural Networks (CNNs) and Recurrent Neural Networks (RNNs) are designed to work with sensor data. CNNs are good at recognizing spatial details—like body movements—while RNNs focus on time-based patterns, such as how actions follow one another.

Advanced deep learning models could help systems get way better at spotting tiny differences in everyday movements—like telling apart sitting and standing, or catching actions that flow into each other (like switching from jogging to walking). To overcome computational constraints and enable reliable, scalable systems suitable for real-world wearable technology applications, future studies can explore real-time adoption using optimized deep learning architectures. But to actually make the systems work in stuff like smartwatches or fitness bands need to get advance which include in device handling of data rather than sanding data to internet. Also, computations of such devices should be enhanced so that it can run advance deep learning models.

7 Conclusion

In our research, we tested how well various machine-learning models could identify everyday human movements—like walking, sitting, or standing—using data from sensors. We noticed that the models were better at recognizing active movements (e.g., walking) compared to stationary poses (e.g., sitting vs. standing). This is because activities like walking had more distinct sensor patterns, while sitting and standing often looked too similar in the data.

Traditional methods like Linear SVM and Logistic Regression worked surprisingly well overall, with Linear SVM being the top performer. However, they had a tough time telling apart activities that were very alike. On the other hand, the Deep LSTM model did better than the rest because it's designed to learn patterns over time. This helped it catch small differences, like the slight shifts in sensor data between sitting and standing.

By leveraging the advantages of both methods, we created a hybrid model that integrates the LSTM's strength in analyzing time-dependent patterns with the SVM's consistent classification performance. This approach improved accuracy and reduced errors, positioning it as a strong candidate for real-time activity recognition. Moving forward, further research could refine this method to handle more complex and diverse datasets, ultimately supporting its use in practical, real-world scenarios.

In addition, further studies could explore enhanced deep learning methods for managing increasingly complex information as technology develops further, possibly producing even better outcomes.

References

1. Lara OD, Labrador MA (2013) A survey on human activity recognition using wearable sensors. Commun Surv Tutor IEEE 15(3):1192–1209
2. Kern N, Antifakos S, Schiele B et al (2004) A model for human interruptability: experimental evaluation and automatic estimation from wearable sensors 1:158–165
3. Lu X, Wang H, Xu X (2015) Human behavior recognition based on accelerometer signal and evolutionary RBF neural network. Pattern Recognit Artif Intell 28(12):1127–1136
4. Wang J, Chen Y, Hao S, Peng X, Hu L (2019) Deep learning for sensor-based activity recognition: a survey. Pattern Recogn Lett 119:119–134. https://doi.org/10.1016/j.patrec.2018.02.010
5. Zhang S, Li Y, Zhang S, Shahabi F, Xia S, Deng Y, Alshurafa N (2022) Deep learning in human activity recognition with wearable sensors: a review on advances. Sensor 1476. https://doi.org/10.3390/s22041476
6. Mohammadi E, Wu QM, Jonathan Saif M (2016) Human activity recognition using an ensemble of support vector machines. In: 2016 international conference on high performance computing & simulation (HPCS), pp 549–554. https://doi.org/10.1109/HPCSim.2016.7568383
7. Maswadi K, Ghani N, Hamid S, Rasheed MB (2021) Human activity classification using decision tree and naive Bayes classifiers. Multimed Tools Appl 80. https://doi.org/10.1007/s11042-020-10447-x
8. Martinez J, Black MJ, Romero J (2021) On human motion prediction using recurrent neural networks. Comput Aided Geom Des 86:101964

9. Li C, Zhong Q, Xie D, Pu S (2018) Co-occurrence feature learning from skeleton data for action recognition and detection with hierarchical aggregation. In: Proceedings of the computer vision and pattern recognition, Salt Lake City, UT, USA, 18–23 June 2018
10. Li M, Chen S, Chen X, Zhang Y, Wang Y, Tian Q (2021) Symbiotic graph neural networks for 3D skeleton-based human action recognition and motion prediction. IEEE Trans Pattern Anal Mach Intell 1–1. https://doi.org/10.1109/TPAMI.2021.3053765
11. Rathi S, Oswal S, Ahuja A (2023) A coherent way of detecting learner's academic emotions via live camera using CNN and deep LSTM. In: 2023 4th international conference for emerging technology (INCET), Belgaum, India, pp 1–6. https://doi.org/10.1109/INCET57972.2023.10170151
12. Nutter M, Crawford CH, Ortiz J (2018) Design of novel deep learning models for real-time human activity recognition with mobile phones. In: 2018 international joint conference on neural networks (IJCNN), Rio de Janeiro, Brazil, pp 1–8. https://doi.org/10.1109/IJCNN.2018.8489319
13. Gupta S (2021) Deep learning based human activity recognition (HAR) using wearable sensor data. Int J Inf Manag Data Insights 1(2):100046. ISSN 2667-0968. https://doi.org/10.1016/j.jjimei.2021.100046
14. Amrani H, Micucci D, Napoletano P (2021) Personalized models in human activity recognition using deep learning. In: 2020 25th international conference on pattern recognition (ICPR), Milan, Italy, pp 9682–9688. https://doi.org/10.1109/ICPR48806.2021.9412140
15. Gao Y et al (2018) Human action monitoring for healthcare based on deep learning. IEEE Access 6:52277–52285. https://doi.org/10.1109/ACCESS.2018.2869790

Melanoma Detection Using Transfer Learning and SVM Integration: A Hybrid Approach

Y. B. Vinay Kumar, H. S. Vimala, and J. Shreyas

Abstract Melanoma is becoming a more common and dangerous skin cancer. It is crucial to detect it early to prevent its spreading to other body parts. The present work explains the advancement of an automated method to diagnose skin cancer without the necessity of physician intervention. This work proposes an advanced approach by integrating transfer learning with a Support Vector Machine (SVM) classifier for melanoma detection at the earliest. The data used in the present work is of Kaggle consisting of over 10,606 images. Firstly, the dermoscopic images are subjected to pre-processing to filter noise and adjust lighting. Then CNNs that have been trained to select essential structures from dermoscopic images. The extracted features were classified using SVM. The augmentation of images increased the size of the training dataset and enhanced learning outcomes. Using the ImageNet dataset, a CNN is trained through a transfer learning approach. Distinguish between melanoma and benign tumors was done by using SVMs classifier which has the pre-trained VGG16 model. The recital of the model has been estimated using sensitivity, specificity, and accuracy. The model exhibits a precision, and sensitivity of 0.96, 95.8, and 91.3%, respectively. The obtained results reveal that the proposed hybrid model, i.e., the combination of transfer learning and SVM performs the best when compared to other methods. This mode can be used as a tool for early detection and diagnosis of melanoma skin cancer.

Keywords Melanoma · Lesion · Skin cancer · CNN · Support vector machine · Transfer learning

---

Y. B. Vinay Kumar (✉) · H. S. Vimala
Department of Computer Science and Engineering, University of Visvesvaraya College of Engineering (UVCE, IIT Model College), Bangalore University, Bengaluru, India
e-mail: vinayp29@gmail.com

J. Shreyas
Department of Information Technology, Manipal Institute of Technology, Bengaluru, Manipal Academy of Higher Education, Manipal, India

H. Nagaraj et al. (eds.), *Advancements in Artificial Intelligence and Machine Learning*, Lecture Notes in Electrical Engineering 1572,
https://doi.org/10.1007/978-981-95-7449-0_9

1 Introduction

Melanoma has the greatest concern worldwide due to their deadliest and dangerous skin cancer type if it is left untreated. The disease may impact people irrespective of age, and the prevalence has been continuously rising. Early identification is vital since timely intervention may significantly enhance patient outcomes. There is a much need of automated and more accurate detection technologies as the conventional treatment methods are labor-intensive and liable to human error [1]. Automating the detection of melanoma, recently, deep learning has emerged as a powerful tool [2]. Although developments in methods of diagnosis, it continues to be challenging to detect melanoma early in its course. Despite traditional deep learning models are successful at collecting features, they perform poorly when used to classification. This is particularly relevant if there are few classified instances or if the dataset is deficient. The goal of this work is to develop a hybrid approach that combines transfer learning with a convolutional neural network (CNN) for dermoscopic picture feature extraction. These features are then classified using a hybrid CNN-SVM model, which integrates the advantages of both techniques to improve detection accuracy and the rate of accurate SVM classification [3]. The paper proposes a structured process for diagnosing skin cancer, encompassing every stage from preparing the image to categorizing and classifying data. Precise and early detection of melanoma is accomplished by enhancing image quality and applying advanced machine-learning methods.

Presently, a new hybrid model that uses transfer learning to combine CNN for extracting features and SVM for classification and enhanced accuracy and decreased false positives during detection on public melanoma datasets are reported.

1.1 Image Deployment

The primary step is to utilize the best dermoscopic images of skin lesions. High resolution and clarity of the photographs are essential since poor quality or fuzzy images can produce unreliable results. A properly deployed image ensures that the data given to the model is of the highest quality and sets the stage for further processing.

1.2 Pre-Processing

Pre-processing is involved in resizing the images to their standard size of 224 × 224 pixels in order to standardize inputs for the CNN, followed by normalization. To improve model generalization, we used data augmentation methods like rotation, zooming, and flipping. These pre-processing methods make sure that only pertinent portions of the images are examined, which helps to increase the accuracy of the feature extraction and classification stages.

1.3 Classification

The classification process is to extract the filtering images according to their features. This work uses a hybrid technique in which an SVM is used for classification after features are extracted using a CNN model. For feature extraction, high level features are extracted from the images using the CNN model VGG16. Feature vector formation for creating a one-dimensional vector by flattening the CNN output. Using the SVM classifier, images used in SVM classification are separated as melanoma and non-melanoma classes. This two-step process employs the high classification efficiency of SVMs and the robust feature extraction capabilities of CNNs to create a more reliable detection system.

1.4 Segmentation

Segmentation is a means of detecting the lesion area within the image. The steps in the method of division are outlined below:

- Masking: During this process, the lesion location needs to be highlighted and made to stand out from the surrounding area.
- Binary Conversion: This entails transforming the image into a binary file where the lesion is indicated in white and the background is black.

Segmentation aids to enhance the recital of the model in sensing and classifying skin cancer by separating the lesion for focused examination.

By isolating the lesion for concentrated analysis, segmentation aids in enhancing the model's precision in identifying and categorizing skin cancer. The proposed approach systematically addresses three levels: image implementation, preprocessing, and classification in order to enhance melanoma early detection. This allows medical practitioners a more accurate and efficient diagnosis tool.

2 Literature Survey

Alaa Haddad et al. reported the work which offers an inexpensive means to detect skin diseases while also assisting medical professionals with early diagnosis and illness identification [4]. Amir Mirbeket et al. studied a method that utilized three semi-solid phantoms to replicate the dielectric characteristics of healthy and cancerous skin tissues with reasonable accuracy. They determined the millimeter wave's penetration depth in cancerous skin tissues through their study, which showed that the majority of skin structures can be reached by the wave [5].

Haseeb Younis et al. developed a MobileNet CNN that has already been trained to classify skin cancer cases. Their method works quickly and consistently, classifying skin lesions in two to three seconds with accuracy ratings of 0.90 and 0.91 [6].

Kyamelia Roy et al. studied numerous division methods to use python for skin cancer diagnosis. They have utilized segmentation based on images morphology, clustering, k-means, edge detection, and adaptive thresholding to diagnose skin diseases [7].

Vani Rajasekar et al. studied CNNs to convolutional neural networks that can offer a reliable and efficient method of skin cancer classification. The CNN model utilizes deep learning to recognize complex patterns and features. The approach allows for the automated and precise classification of skin lesions into groups that are malignant and benign [8].

Azadesh Noori et al. reported the automated diagnosis can help to improve the performance accuracy of melanoma detection as clinical observations have several limitations.

The primary objectives of this work are to review skin cancer detection research and present an outline of automatic skin cancer detection [9].

Mohammadreza Hajiarbabi et al. reported a transfer-learning approach for the detection of melanoma lesions. The images are initially preprocessed using the suggested technique to eliminate noise and lighting effects. Next, using the ImageNet dataset's weights and transfer learning, a convolutional neural network is trained [10].

Md Sirajul Islam et al. reported a binary classification for skin cancer identification at benign and malignant stages. We used five distinct pre-trained transfer learning algorithms in this work. They have fine-tuned several layers and activation functions to improve the accuracy of these models [11].

N Darapaneni et al. reported a model that offers a solution to the issue of the highest costs associated with the study of the lowest accuracy rate in identification and manual detection systems that have portability issues. This model uses the multi-layered CNN approach AI tool to categorize dermatoscopic images in order to prognosticate skin cancer. We therefore provide an enhanced accuracy computer-aided method for skin cancer diagnosis that makes use of computer vision and image analysis algorithms [12].

M A Ali et al. studies focus on developments in artificial intelligence (AI) solutions for digital image-based computer vision for skin cancer diagnostics. The prospects for the future of enhancing the capacity to identify skin cancer early on have also been highlighted. This work offers an enhanced accuracy computer-aided method for skin cancer diagnosis using computer vision and image analysis algorithms, employing the HAIS AI tool [13].

V Vipin et al. suggested model offered a solution to the issue of the highest costs associated with the analysis's lowest accuracy rate in identification and a manual detection technique that has a portability issue. This model uses the multi-layered CNN technique to categorize dermatoscopic pictures to prognosticate skin cancer [14].

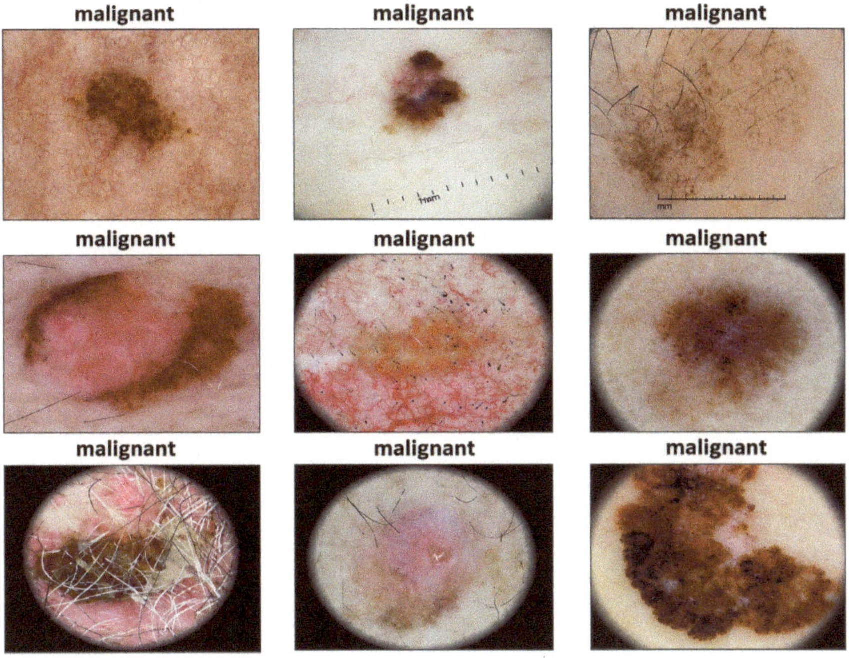

Fig. 1 Sample melanoma example

3 Dataset

The datasets of both melanoma and non-melanoma cases were obtained from Kaggle over 10,606 images. From this collection, 9606 images were used to train and 1000 images to evaluate the model's performance (Fig. 1). This dataset is valuable for developing deep learning models for precise melanoma classification.

4 Methodology

The methodology section details the process of implementing the hybrid approach. The algorithm combines the feature sorting capabilities of convolutional neural networks (CNNs) with the classification strength of Support Vector Machines (SVMs).

4.1 Transfer Learning

An extensive dataset is considered for image augmentation combined with transfer learning. This network uses a pre-trained and adopted several CNN architectures and the VGG16 model. Initial weights were used for the pre-trained weights from ImageNet. In the final layer, most commonly used function was Softmax layer to do binary classification which was replaced with sigmoid function.

The number of nodes and outputs in the layer with suitable weights are adjusted via back propagation, which enables the network to adjust to the new classes more effectively. It's critical to prevent significant weight changes in Softmax layer as it ends with a low learning rate. Overly high learning rates have the potential to seriously disturb the balance. One important hyperparameter that needs to be managed during training is the amount of back propagation using sigmoid function.

4.2 Support Vector Machine (SVM)

The extracted features were classified using an SVM with a linear kernel. The choice of SVM over other classifiers was based on its effectiveness in binary classification, particularly with high-dimensional data. Hyperparameters, including the regularization parameter, were optimized through cross-validation. Figure 2 represents the architecture diagram of transfer learning with SVM for melanoma detection.

This diagram illustrates the proposed method, where pre-trained CNNs are used from dermoscopic images for the extraction of the features. Which is then followed by the organization using a Support Vector Machine (SVM) to accurately distinguish between melanoma and non-melanoma cases.

Below is a description of the architecture that combines transfer learning with SVM for melanoma detection.

1. **Input Image**: The dermoscopic image is resized to a standardized form (e.g., 224 × 224) to be compatible with the CNN model.
2. **Pre-trained CNN**: The input image passes through the CNN layers, where it undergoes convolution and pooling operations to extract hierarchical features.
3. **Feature Maps**: The CNN's output is a set of feature maps that represent significant characteristics of the input image.
4. **Flattening**: These feature maps are compressed into a one-dimensional array to be used by the SVM.

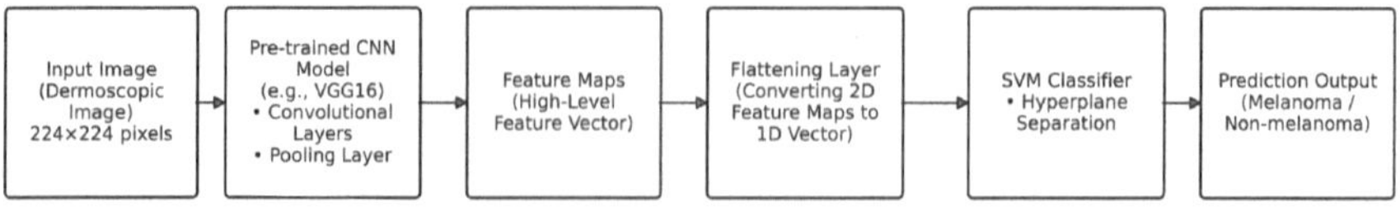

Fig. 2 Hybrid architecture of transfer learning and SVM for melanoma skin cancer detection

5. **SVM Classifier**: The SVM classifier categorizes the image as either melanoma or non-melanoma.
6. **Prediction Output**: The final classification result is obtained, indicating whether the lesion is melanoma.

5 Results and Discussions

The improved recital of the hybrid model can be attributed to the complementary strengths of CNN and SVM. While CNNs are adept at extracting intricate spatial features, SVMs are better suited for decision boundaries in high-dimensional spaces. The proposed methodology, therefore, combines the advantages of both approaches. This hybrid architecture is particularly beneficial for medical image analysis, where false negatives carry severe consequences. These images are utilized for training and testing purposes with a CNN-SVM, which is employed to categorize as melanoma or non-melanoma skin lesions.

6 Model Training and Evaluation

Once trained, the model can be used to predict whether a new skin image is melanoma or non-melanoma. For feature extraction, a new image is passed through the CNN to extract topographies. These topographies are then classified by the CNN-SVM. The dataset was divided into training, validation, and testing subset allocations, respectively. Trained the CNN on the training data while using validation data to tweak the model and avoid overfitting. The extracted features were subsequently classified using the SVM. Performance was evaluated using metrics such as accuracy, F1-score, and area under the curve. After applying this model for thirty epochs on the training set of data the training loss: 0.12; the training accuracy: 0.96; the validation loss: 0.32; Validation accuracy: 0.89. The model indicates an increase in both training and validation accuracy. Figure 3 shows the plots of accuracy and loss of the present approach and Fig. 4 exhibits the receiver operating characteristic curve. Figure 5 displays the confusion matrix.

The CNN-SVM classifier is tested with various parameters to reduce computational complexity. Table 1 shows the performance output of this model.

The approach is evaluated using the following formulas:

$$\text{Sensitivity} = \frac{(\text{True Positives})}{(\text{True Positives} + \text{False Negatives})} \tag{1}$$

$$\text{Specificity} = \frac{(\text{True Negatives})}{(\text{True Negatives} + \text{False Positives})} \tag{2}$$

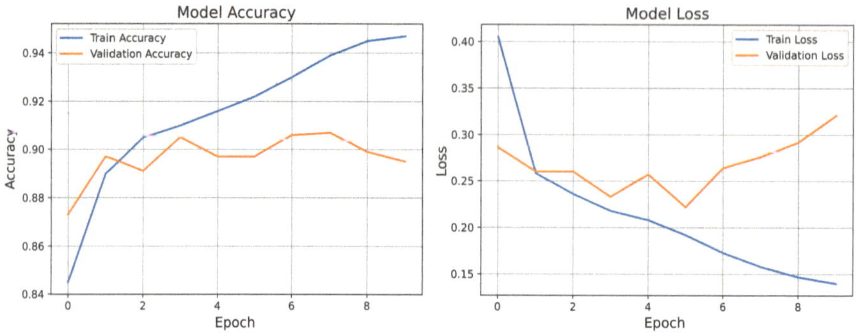

Fig. 3 Accuracy and loss of the adopted approach are depicted

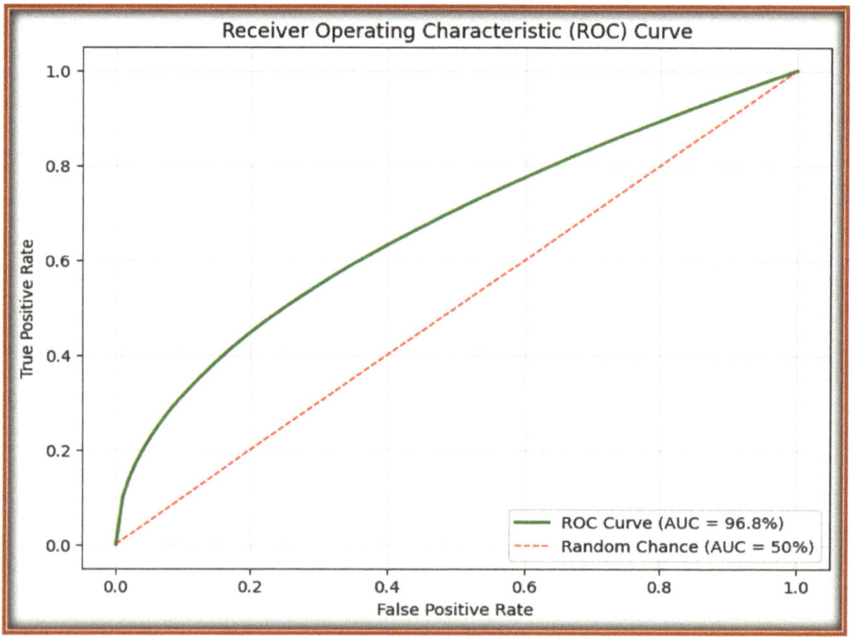

Fig. 4 Receiver operating characteristic curve

$$\text{Accuracy} = \frac{(\text{True Positive} + \text{True Negative})}{(\text{True Positive} + \text{True Negative} + \text{False Positive} + \text{False Negative})} \tag{3}$$

This hybrid model achieved a detection accuracy of 96% for melanoma compared to other skin conditions. The effectiveness of the model is demonstrated through its high accuracy and robustness across different test sets. The hybrid model is compared to traditional SVM and standalone CNN models, highlighting the improvements

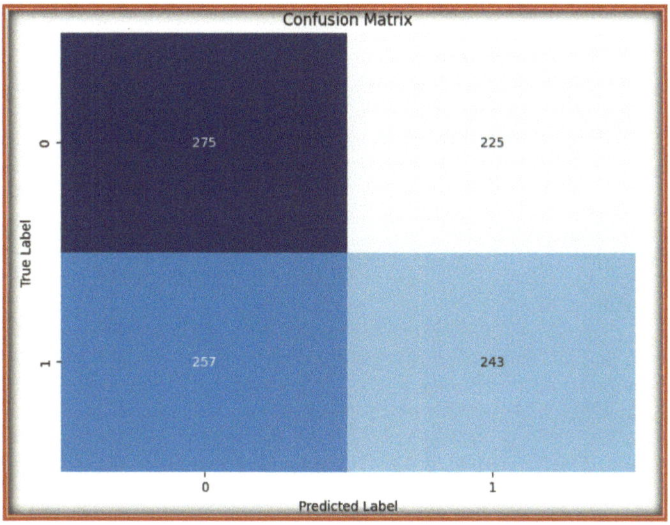

Fig. 5 Confusion matrix

Table 1 Performance evaluation

Sensitivity	Specificity	Accuracy
95.8%	91.3%	96.8%

achieved by combining these techniques. This section may also discuss the limitations and potential areas for further improvement.

7 Conclusion

The conclusion reiterates the success of the hybrid approach in improving melanoma detection accuracy. The combination of transfer learning and SVM provides a powerful tool for automated skin cancer detection. This offers a significant improvement over traditional method. In the future, hybrid networks with different batch size, epoch, optimizers and classifiers can be utilized for further strengthen the model. Further, this approach can be extended to other types of skin lesions or integrate the model into real-world clinical workflows.

References

1. Singh SK, Banerjee S, Chakraborty A, Bandyopadhyay A (2023) Classification of melanoma skin cancer using inception-ResNet. In: Frontiers of ICT in healthcare proceedings of EAIT 2022, Springer Nature, Singapore, pp 65–74
2. Frech TM, Frech M, Saknite I, O'Connell KA, Ghosh S, Baba J, Tkaczyk ER (2023) Novel therapies and innovation for systemic sclerosis skin ulceration. Clin Rheumatol 101813
3. Banasode P, Patil M, Ammanagi N (2020) A melanoma skin cancer detection using machine learning technique: support vector machine. In: IOP conference series: materials science and engineering, 1st international conference on frontiers in engineering science and technology (ICFEST 2020), p 1065
4. Haddad A, Hameed SA (2018) Image analysis model for skin disease detection framework. In: 7th International conference on computer and communication engineering, pp 1–4
5. Mirbeik-Sabzevari A, Tavassolian N (2018) Ultrawideband, stable normal and cancer skin tissue phantoms for millimeter-wave skin cancer imaging. IEEE Trans Biomed Eng 66(1):176–186
6. Younis H, Bhatti MH, Azeem M (2019) Classification of skin cancer dermoscopy images using transfer learning. In: 2019 15th International conference on emerging technologies (ICET), pp 1–4
7. Roy K, Chaudhuri SS, Ghosh S, Dutta SK, Chakraborty P, Sarkar R (2019) Skin disease detection based on different segmentation techniques. In: 2019 International conference on opto-electronics and applied optics (Optronix). IEEE, pp 1–5
8. Vani R, Nirmala Devi K, Sathya K, Santhosh A, Saran S, Senthilvel K (2024) An intelligent system for skin cancer detection using deep learning techniques. In: International conference on advances in data engineering and intelligent computing systems (ADICS), pp 1–5
9. Hoshyar AN, Al-Jumaily A, Sulaiman R (2011) Review on automatic early skin cancer detection. In: International conference on computer science and service system (CSSS), pp 4036–4039
10. Hajiarbabi M (2023) Skin cancer detection using multi-scale deep learning and transfer learning. J Med Artif Intell 10:6
11. Islam MS, Panta S (2024) Skin cancer images classification using transfer learning techniques. Comput Vis Pattern Recognit 1–10
12. Darapaneni N, Sahni B, Paduri AR, Jain S, Mohamed S, Banerjee A, Chakrabarti S (2022) Early detection of skin cancer—solution for identifying and defining skin cancers using AI. In: 2022 interdisciplinary research in technology and management (IRTM), pp 1–6
13. Ali MA, Khan M, Sherwani KI (2023) Early skin cancer detection using deep learning. In: 3rd International conference on SMART generation computing, communication and networking (SMART GENCON), pp 1–6
14. Vipin V, Nath MK, Sreejith V, Giji NF, Ramesh A, Meera M (2021) Detection of melanoma using deep learning techniques: a review. In: 2021 International conference on communication, control and information sciences (ICCISc), pp 1–8

A Framework for Student Academic Success Prediction with Enhanced Feature Selection and Hybrid Deep Learning

M. Sharmila Kumari, Saleem Malik, K. K. Fathima Riza, Zulfa Mariyam Sharfuddeen, and Azkiya Rifa Khan

Abstract Predicting student performance improves educational outcomes by providing personalized learning and timely interventions for at-risk students. This paper presents RVFL-CNN-LSTM, a hybrid deep learning model, to predict student performance with unbalanced datasets along with complex feature connections. Using the strengths of CNN, LSTM, and RVFL networks, the model improves prediction accuracy while retaining computational economy. The study uses the FeatureX algorithm to choose the most relevant characteristics. The suggested model is tested with four student performance datasets and shows better prediction accuracy and adaptability than standard machine learning methods. Precision, recall, and F1-score improved across four datasets to 94.7, 91.5, 90.8, and 93.1%, demonstrating the model's potential for personalized learning and successful intervention approaches.

Keywords Academic performance · Hybrid approach · Deep learning

1 Introduction

A student performance prediction system uses grades, attendance, and behaviour data with machine learning algorithms to predict academic progress. A predictive capacity helps instructors personalize learning and identify pupils who need more help [1]. These systems can identify at-risk students and provide proactive measures to improve education by analysing previous learning patterns and results [1].

The RVFL is a popular student performance prediction method [2]. RVFL is a novel neural network model with quick training and good generalization capabilities. Unlike traditional neural networks, which alter weights iteratively using backpropagation, RVFL assigns random weights to the hidden layer and calculates output

M. Sharmila Kumari · S. Malik (✉) · K. K. Fathima Riza · Z. M. Sharfuddeen · A. R. Khan
Department of Computer Science and Engineering, P.A. College of Engineering, Mangaluru, Affiliated to Visvesvaraya Technological University (VTU), Belagavi, Karnataka, India
e-mail: baronsaleem@gmail.com

H. Nagaraj et al. (eds.), *Advancements in Artificial Intelligence and Machine Learning*, Lecture Notes in Electrical Engineering 1572,
https://doi.org/10.1007/978-981-95-7449-0_10

weights analytically in a single step. This non-iterative technique lets RVFL effectively analyse big datasets, which is useful for educational data that needs speed and precision [2].

This research's proposed method combines RVFL with Convolutional Neural Networks (CNNs) and Long Short-Term Memory (LSTM) networks to enhance prediction accuracy while ensuring efficient training periods [3]. The hybrid nature of the model, which encompasses the sequential data processing of LSTM, the feature extraction strength of CNN, and the rapid training capability of RVFL, is the primary distinction from extant algorithms [4]. This study introduced the hybrid RVFL-CNN-LSTM framework for enhanced performance prediction, the FeatureX algorithm for optimal feature selection [5], and a comprehensive evaluation using diverse datasets. The primary issue that is addressed is the low accuracy and efficacy of the current student performance prediction models. The research hypothesis posits that the combination of these techniques will result in improved generalization and greater accuracy. Four datasets containing student academic records were employed in the investigations. The model is reliable for personalized learning and effective intervention techniques due to its higher prediction accuracy and resilience than typical machine learning methods.

These findings have major consequences for educators and politicians. The hybrid RVFL-CNN-LSTM model can detect at-risk students early, allowing educators to give personalized study programs and extra coaching. The model facilitates data-driven decision-making by providing real-time insights into student performance through the efficient analysis of large datasets [5]. The FeatureX method [5] selects the most important characteristics, improving model interpretability and accuracy. This method optimizes resources by targeting students who need it most and improves student retention initiatives. By adapting learning settings to students' different requirements, the strategy improves educational results and institutional performance.

2 Related Works

EDM has enabled systems to provide personalized learning solutions for each learner. Using student performance, behaviour, and engagement data, EDM improves teaching tactics and predicts student success [17]. However, standard EDM models generally struggle with accuracy, especially with complicated and unbalanced datasets. Ineffective feature selection limits models from collecting data pattern [16]. LSTM and CNNs capture complex characteristics and temporal connections in data. Unfortunately, many deep learning applications lack strong feature selection. We found that deep learning architectures perform well but need specific feature engineering to optimize outcomes. Feature selection improves model performance, especially in high-dimensional, sequential datasets, according to their study [6]. Table 1 lists popular student performance categorization methods.

Table 1 An overview of existing works

Reference	Algorithms used	Subject of research	Disadvantages
[6]	Statistical and machine learning techniques	Analysis of educational data to identify patterns and support online learning	–May overlook contextual factors such as cultural and institutional constraints
[13]	AI algorithms integrated with learning analytics	Identifying at-risk students and improving collaborative learning	–Dependence on high computing resources
[7]	Convolutional neural network	Identifying leaf diseases in crops like cauliflower, tomato, and mango	–Computationally expensive
[11]	Convolutional neural network	Real-time recognition of sign language to improve communication for individuals with disabilities	–Struggles with varying lighting and background conditions
[9]	Central tendency measures (mean, median, mode)	Analysis of missing data in life expectancy and chronic kidney disease datasets	–May not handle large or complex datasets effectively
[14]	Complete case analysis (CCA), numerical and categorical data handling	Experiments on Google play store apps dataset to manage missing and dirty data	–Computational cost increases with dataset complexity
[15]	Missing data handling, categorical feature encoding, outlier detection, feature scaling	House Price index (HPI) computation and house price prediction	–High computational cost for large datasets

Improper feature selection caused overfitting and poor generalization in their deep learning model for multi-crop disease detection [7]. A similar study [8] found that incorrect feature handling in data imputation for chronic renal disease prediction caused model errors. According to their study, predictive modelling requires extensive feature selection and data preparation [9] examined healthcare dataset missing data imputation and presented a hybrid strategy that combines deep learning and statistical methods. Their approach outperformed standard imputation methods, boosting chronic disease prediction accuracy. Combining multiple imputation methodologies with feature engineering in healthcare is becoming more important.

Traditional machine learning methods disregard dynamic feature selection, but real-time feature updates improve deep learning model performance, according to their review [10]. The unique adaptive preprocessing pipeline improves model resilience against noisy datasets, notably in healthcare and education [11] examined deep neural networks (DNNs) for image-based recognition applications, focussing on Sign Language. Their DNN architecture included dynamic feature selection, showing that automated feature extraction and machine learning refinement improve multi-class picture classification performance. As shown in [12], deep learning models for chronic renal disease prediction have erroneous feature handling. They concluded that feature selection is essential for model generalization. A feature selection technique called FeatureX [5] can refine feature sets based on data properties to fill these gaps [14] shows how FeatureX can reduce duplicate features for more accurate predictions on the dataset. Feature selection and preprocessing improve model dependability across educational datasets, according to the study.

Following this fundamental research, this paper offers a hybrid RVFL-CNN-LSTM architecture that combines RVFL's quick training, CNN's feature extraction, and LSTM's sequence modelling [2]. This student performance prediction model was created to address typical model issues. Low accuracy was shown in CNN-LSTM hybrid experiments [4] because to poor feature selection. The RVFL-CNN-LSTM design overcomes this restriction by using FeatureX. This hybrid model can analyse static and sequential data, improving educational dataset generalization.

3 Proposed Model

The suggested RVFL-CNN-LSTM architecture, represented in Fig. 1, processes student data via many phases for accurate performance prediction. It starts with the Input Layer, where raw data is pre-processed (missing value imputation, outlier treatment, discretization). The FeatureX Module uses a hybrid filter-wrapper technique to find the best features. The RVFL Module uses random weight initialization and direct connections to convert selected features into high-dimensional representations. The CNN Module uses convolution and pooling layers to extract spatial patterns, whereas the LSTM Module captures sequential dependencies. The Fusion Layer uses concatenation and batch normalization to merge spatial and temporal information, followed by Dense Layers for refinement. The Output Layer, with Softmax activation, assesses student performance as exceptional, good, average, below average, or terrible. The model architecture has 100–200 random neurons in the RVFL Layer, 2–3 convolutional layers (3×3 or 5×5 kernels) with max pooling, and 1–2 stacked LSTM layers (50–100 units, dropout 0.2–0.5, Sigmoid activation). The Adam optimizer with a 0.001 learning rate and categorical cross-entropy loss function enables efficient training. This layered approach is ideal for educational data analysis and student performance prediction since it combines static and sequential data.

This study utilizes a dataset of 1500 samples from Karnataka polytechnic students, previously detailed in [6], with 48 independent variables covering individual characteristics (cognitive abilities, learning behaviours, motivation), family background (socio-economic factors), and school-related aspects (infrastructure, teacher quality, peer interaction). Data was gathered through structured surveys and institutional records, with academic scores in Mathematics, IT Skills, and Statistics serving as key performance indicators. Preprocessing included handling missing values by imputing categorical variables with their mode and numerical features with their mean, discretizing student performance into five levels, and standardizing formats for reliability. Feature selection was performed using the FeatureX algorithm within an

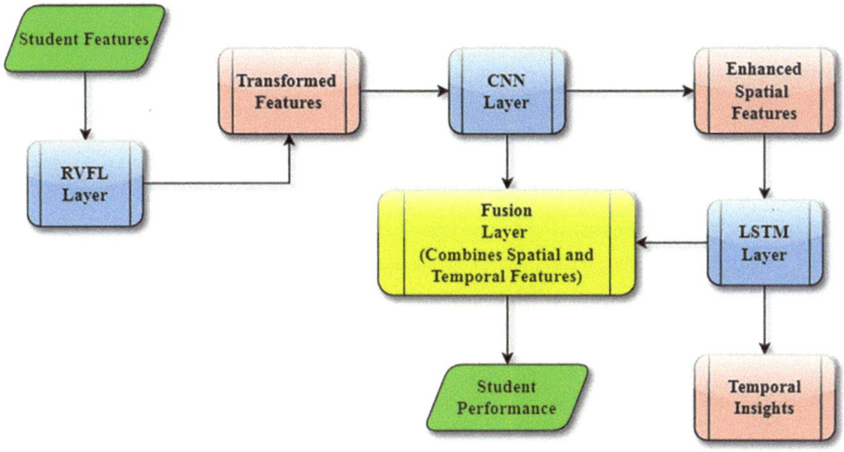

Fig. 1 Proposed RVFL-CNN-LSTM architecture

RVFL-CNN-LSTM framework to enhance interpretability and computational efficiency. FeatureX uses filter-based (correlation, mutual information) and wrapper-based methods to rank relevant features, reducing the 48 original variables to 10 key attributes: study hours, attendance, learning strategies, parental education, family income, and school-related factors. The CNN component captures spatial patterns, LSTM models temporal relationships, and RVFL efficiently preprocesses data, ensuring accurate and robust student performance predictions.

Algorithm 1: RVFL-CNN-LSTM for educational data analysis

Input:
- X: Dataset with N samples and F features
- Y: Target labels

Hyper parameters:
- Random weight dimensions for RVFL (R).
- Number of filters and kernel size for CNN
- Number of units in LSTM layer
- Dropout layer, learning rate and batch size

Output:
- Z: Predicted class probabilities trained model capable of handling spatial, temporal and non-linear relationships in educational data.

Steps:

1. Normalize input features, handle missing values by imputing with mode or mean, and remove outliers using Z-score method.
2. Generate a random weight $w \in R^{F*R}$ and bias vector $b \in R^R$. Transform the input data into a higher dimensional space.

$$H = f(X_{norm}w + b) \quad (1)$$

H is enhanced feature representation.

3. Apply convolution filters.

$$Y_{i,j,k} = \sum_{m,n} H_{i+m,j+n}{}^{*} K_{m,n} + b_k \quad (2)$$

Where K is kernel and b_k is bias term.

4. Use maxpooling to reduce spatial dimensions and retain dominant features.

$$P_{i,j} = \max\left(Y_{i,j}, Y_{i+1,j} \cdots\cdots\cdots\cdots\cdots\right) \quad (3)$$

Where P set of spatial features for understanding student attribute interactions.

5. Compute LSTM gates.

$$F_t = W_F{}^{*}(h_{t-1}, X_t + b_F) \quad (4)$$

$$I_t = \left(W_I{}^{*}(h_{t-1}, X_t) + b_I\right) \quad (5)$$

$$G_t = \tanh\left(W_G{}^{*}(h_{t-1}, X_t) + b_G\right) \quad (6)$$

$$C_t = F_t{}^{*}C_{t-1} + I_t{}^{*}G_t \quad (7)$$

$$O_t = \left(W_O{}^{*}(h_{t-1}, X_t) + b_O\right) \quad (8)$$

$$h_t = O_t{}^{*} \tanh(C_t) \quad (9)$$

Output h_t encodes the temporal trends of student performance.

6. Concatenate spatial features P from CNN and temporal features h_t from LSTM.

$$F = concat(P, h_t) \quad (10)$$

7. Pass F through fully connected layers.

$$z = f(FW_{dense} + b_{dense}) \quad (11)$$

Where f is activation function (ReLU).

8. Output layer uses softmax for classification.

$$z = softmax\left(w_{output}Z + b_{output}\right) \quad (12)$$

9. Define the loss function as categorical cross-entropy optimize the loss function using the Adam optimizer with early stopping. Stop training if validation performance does not improve for k epochs.
10. Evaluate the model using metrics such as accuracy, precision, recall and F1-score. Perform error analysis to identify patterns in misclassified samples.

Algorithm 1 predicts class probabilities by integrating spatial, temporal, and non-linear relationships in educational data. First, input characteristics are normalized, missing values are imputed, and outliers are removed. A random weight matrix and bias vector increase input feature dimension. Convolution filters and max-pooling extract spatial data, whereas LSTM gates preserve temporal patterns. These spatial and temporal information are concatenated and transmitted through fully linked layers with ReLU activation and softmax classification. Categorial cross-entropy, Adam optimizer, and early halting optimize the model. Final performance measures include Accuracy, Precision, Recall, and F1-Score, followed by error analysis.

4 Evaluation Metrics

The RVFL-CNN-LSTM model's classification performance was assessed using the following metrics.

$$Accuracy = \frac{TP + TN}{TP + TN + FP + FN} \tag{13}$$

$$Precision = \frac{TP}{TP + FP} \tag{14}$$

$$Recall = \frac{TP}{TP + FN} \tag{15}$$

$$F1 - Score = 2^{*}\frac{Precision}{Recall} \tag{16}$$

5 Result Analysis

This work improves student performance prediction accuracy using a Deep Learning (DL) model that captures complicated data linkages better than traditional methods. The FeatureX technique improves generalizability and reduces model complexity by selecting the most relevant characteristics. Comparing RVFL-CNN-LSTM with and without FeatureX, considerable performance gains were seen. Feature selection cut dataset size by 50% (DS1), 40% (DS2), 33.33% (DS3), and 48.98% (AD1). The accuracy advances were 3.1% (AD1) to 17.3% (DS2), precision gains were 17.8% (DS2), recall gains were 21.4% (DS2), and F1-score gains were 19.6% (DS2). The results in Table 2 show that FeatureX's impact is significant, as effective feature engineering boosts predictive accuracy and computational efficiency, especially in DS2 and DS3, where metric improvements exceeded 10%, showing the proposed methodology's strength.

The comparison of methodologies in Table 3 reveals the superior performance of RVFL-CNN-LSTM with FeatureX across all datasets, significantly surpassing other models in accuracy, precision, recall, and F1-score. For DS1, RVFL-CNN-LSTM with FeatureX shows an improvement of 9.6–9.7% over SVM, 6.5–7.2% over Random Forest, 5.6–6.4% over Gradient Boosting, and 3.2–3.8% over CNN-LSTM. Similarly, in DS2, the improvements are 7.6–8.3% over SVM, 4.5–5.3% over Random Forest, 3.4–4.3% over Gradient Boosting, and 1.5–2.0% over CNN-LSTM. For DS3, the performance gains are even more notable, with increases of 9.4–9.7% over SVM, 5.3–6.0% over Random Forest, 3.2–4.1% over Gradient Boosting, and 0.2–1.0% over CNN-LSTM. In AD1, RVFL-CNN-LSTM with FeatureX achieves improvements of 9.0–9.7% over SVM, 7.0–7.8% over Random Forest, 4.7–5.4% over Gradient Boosting, and 1.9–2.4% over CNN-LSTM. These results emphasize the effectiveness of incorporating FeatureX, particularly against traditional machine learning models like SVM, Random Forest, and Gradient Boosting, as well as the standard CNN-LSTM. The most significant improvements are observed in DS1 and DS3, where performance gains frequently exceed 5–9%, highlighting the robustness and efficiency of the proposed approach as shown in Fig. 2.

The hyperparameter tuning process explored a range of values for key parameters as shown in Table 4, leading to significant improvements in model performance across all datasets. Optimal values include a learning rate of 0.001, batch size of 32, 100 epochs, a dropout rate of 0.3 for LSTM, 128 LSTM units, 64 CNN filters, and a kernel size of 3 for the CNN.

The improvements in performance metrics after tuning are notable as shown in Fig. 3. For DS1, accuracy increased by 3.6%, precision by 3.7% recall by 4.4%

Table 2 RVFL-CNN-LSTM performance with and without feature selection

Methods	Datasets	No. of features	Accuracy	Precision	Recall	F1-Score
RVFL-CNN-LSTM w/o FeatureX	DS1	30	88.5	86.7	85.3	86.0
	DS2	5	74.2	72.5	70.8	71.6
	DS3	30	81.1	79.8	78.5	79.1
	AD1	49	90.0	89.5	88.9	89.2
RVFL-CNN-LSTM with FeatureX	DS1	15	94.7	93.5	95.1	94.3
	DS2	3	91.5	90.3	92.2	91.2
	DS3	20	90.8	89.5	91.4	90.4
	AD1	25	93.1	92.0	93.8	92.9

Table 3 RVFL-CNN-LSTM performance against contemporary methods

Methodology	Dataset	Accuracy	Precision	Recall	F1-Score
SVM	DS1	86.4	85.3	86.8	86.0
	DS2	84.7	83.9	85.1	84.5
	DS3	82.9	81.8	83.3	82.5
	AD1	85.2	84.4	85.5	84.9
Random forest	DS1	88.5	87.8	88.7	88.2
	DS2	87.2	86.4	87.6	87.0
	DS3	85.9	85.0	86.2	85.6
	AD1	86.8	86.0	87.0	86.5
Gradient boosting (GBM)	DS1	89.1	88.5	89.4	88.9
	DS2	88.0	87.3	88.4	87.8
	DS3	87.4	86.7	87.8	87.2
	AD1	88.6	87.9	89.0	88.4
CNN-LSTM	DS1	91.2	90.6	91.8	91.2
	DS2	89.8	89.0	90.2	89.6
	DS3	90.0	89.3	90.5	89.9
	AD1	91.0	90.3	91.4	90.8
RVFL-CNN-LSTM with FeatureX	DS1	94.7	93.5	95.1	94.3
	DS2	91.5	90.3	92.2	91.2
	DS3	90.8	89.5	91.4	90.4
	AD1	93.1	92.0	93.8	92.9

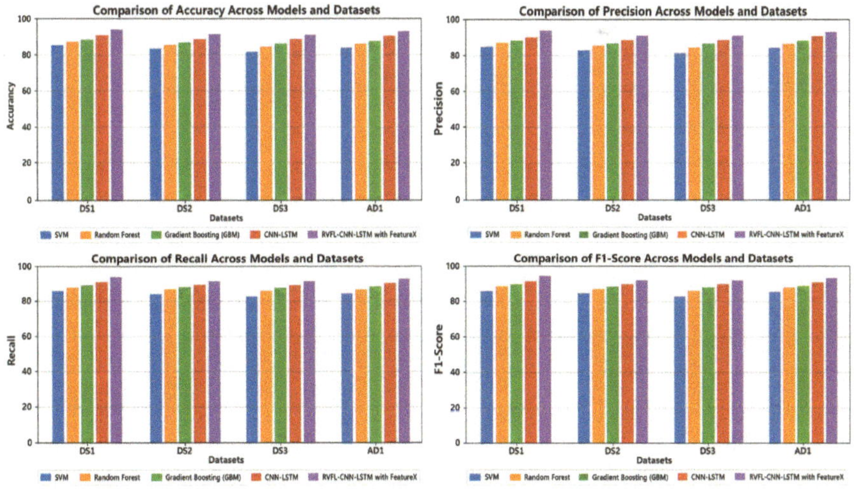

Fig. 2 Performance of RVFL-CNN-LSTM across datasets

Table 4 Hyperparameter tuning

Hyperparameter tuning	Range explored	Optimal value
Learning rate	0.001, 0.005, 0.01, 0.05	0.001
Batch size	16, 32, 64, 128	32
Number of epochs	50, 100, 200, 300	100
Dropout rate (LSTM)	0.1, 0.2, 0.3, 0.4	0.3
Number of LSTM units	64, 128, 256, 512	128
Number of CNN filters	32, 64, 128	64
Kernel size (CNN)	3, 5, 7	3

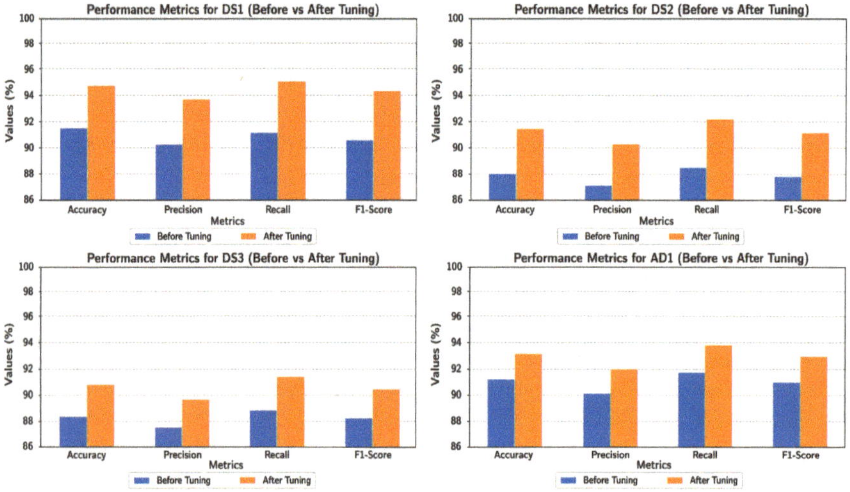

Fig. 3 Performance of RVFL-CNN-LSTM before and after tuning hyperparameters

(91.1% to 95.1%), and F1-score by 4.1%. In DS2, accuracy improved by 4.0%, precision by 3.7%, recall by 4.1%, and F1-score by 3.9%. For DS3, accuracy increased by 2.8%, precision by 2.3% recall by 2.5%, and F1-score by 2.2%. In AD1, accuracy rose by 2.1%, precision by 2.1%, recall by 2.3%, and F1-score by 2.0%. Hyper parameter tuning led to performance improvements across all datasets, with the most significant gains in DS1 and DS2, where improvements frequently exceeded 3–4%.

6 Conclusion

This study explores the application of Deep Learning (DL) algorithms for predicting student academic performance. Building upon previous research, this work proposes a hybrid DL classifier specifically designed to address challenges associated with imbalanced class distributions. To mitigate the impact of imbalanced data, the proposed approach incorporates several key steps: Data Preprocessing: Normalization techniques are employed to standardize the data within the range [0, 1], effectively addressing potential biases introduced by skewed class distributions. Choose the most relevant features from the dataset using FeatureX to reduce model complexity and improve generalizability. Data preparation and feature selection precede intensive training of the RVFL-CNN-LSTM model. This hyperparameter adjustment optimizes model performance for correct categorization. Evaluation on four datasets shows the strategy works. The present model scores higher on four datasets than earlier research. Error rates are low, demonstrating its efficiency.

References

1. Malik S, Jothimani K, Ujwal UJ (2023) A comparative analysis to measure scholastic success of students using data science methods. In: Shetty NR, Patnaik LM, Prasad NH (eds) Emerging research in computing, information, communication and applications, Lecture notes in electrical engineering, vol 928. Springer, Singapore. https://doi.org/10.1007/978-981-19-5482-5_3
2. Liu Q, Wang Q, Wang X (2025) RVFL-LSTM: a lightweight model with long-short term memory for time series, Knowl-Based Syst 309:112896, ISSN 0950–7051. https://doi.org/10.1016/j.knosys.2024.112896
3. Alkhawaldeh RS, Al-Ahmad B, Ksibi A et al (2023) Convolution neural network bidirectional long short-term memory for heartbeat arrhythmia classification. Int J Comput Intell Syst 16:197. https://doi.org/10.1007/s44196-023-00374-8
4. Pandey A, Mannepalli PK, Gupta M et al (2024) A deep learning-based hybrid CNN-LSTM model for location-aware web service recommendation. Neural Process Lett 56:234. https://doi.org/10.1007/s11063-024-11687-w
5. Malik S, Jothimani K (2024) Enhancing student success prediction with featurex: a fusion voting classifier algorithm with hybrid feature selection. Educ Inf Technol 29:8741–8791. https://doi.org/10.1007/s10639-023-12139-z
6. Malik S, Malik S (2024) Hybrid data science approaches to predict the academic performance of students. In: Shetty NR, Prasad NH, Nagaraj HC (eds) Advances in communication and applications. ERCICA 2023, Lecture notes in electrical engineering, vol 1105. Springer, Singapore. https://doi.org/10.1007/978-981-99-7633-1_39
7. Bouacida I, Farou B, Djakhdjakha L, Seridi H, Kurulay M (2024) Innovative deep learning approach for cross-crop plant disease detection: a generalized method for identifying unhealthy leaves. Inf Process Agric ISSN 2214–3173. https://doi.org/10.1016/j.inpa.2024.03.002
8. Hema K, Meena K, Pandian R (2024) Analyze the impact of feature selection techniques in the early prediction of CKD. Int J Cogn Comput Eng 5:66–77, ISSN 2666–3074. https://doi.org/10.1016/j.ijcce.2023.12.002
9. Khan SI, Hoque ASML (2020) SICE: an improved missing data imputation technique. J Big Data 7:37. https://doi.org/10.1186/s40537-020-00313-w

10. Islam MA, Majumder MZ, Hussein MA, Hossain KM, Miah MS (2024) A review of machine learning and deep learning algorithms for Parkinson's disease detection using handwriting and voice datasets, Heliyon 10(3):e25469, ISSN 2405–8440. https://doi.org/10.1016/j.heliyon.2024.e25469
11. Baihan A, Alutaibi AI, Alshehri M et al (2024) Sign language recognition using modified deep learning network and hybrid optimization: a hybrid optimizer (HO) based optimized CNNSa-LSTM approach. Sci Rep 14:26111. https://doi.org/10.1038/s41598-024-76174-7
12. Debal DA, Sitote TM (2022) Chronic kidney disease prediction using machine learning techniques. J Big Data 9:109. https://doi.org/10.1186/s40537-022-00657-5
13. Ujwal UJ, Malik S (2023) A hybrid weight based feature selection algorithm for predicting students' academic advancement by employing data science approaches. Int J Educ Manag Eng (IJEME) 13(5):1–22. https://doi.org/10.5815/ijeme.2023.05.01
14. Tripathi VK, Parashar P, Mallikharjuna Rao K (2022) Missing data: its emergence in the real-world-a practical review on google play apps dataset using python. In: 2022 IEEE IAS global conference on emerging technologies (GlobConET), Arad, Romania, pp 209–214. https://doi.org/10.1109/GlobConET53749.2022.9872426
15. Hill RJ, Rambaldi AN, Scholz M (2021) Higher frequency hedonic property price indices: a state-space approach. Empir Econ 61:417–441. https://doi.org/10.1007/s00181-020-01862-y
16. Alnasyan B, Basheri M, Alassafi M (2024) The power of deep learning techniques for predicting student performance in virtual learning environments: a systematic literature review, Comput Educ: Artif Intell 6:100231, ISSN 2666-920X. https://doi.org/10.1016/j.caeai.2024.100231
17. Kord A, Aboelfetouh A, Shohieb SM (2025) Academic course planning recommendation and students' performance prediction multi-modal based on educational data mining techniques. J Comput High Educ. https://doi.org/10.1007/s12528-024-09426-0

Scaling Image Classification by Reducing Label Dependency and Applying Self-Supervised Learning Techniques

S. Deepa, Sheetal, and A. Alli

Abstract Self-supervised learning (SSL) has emerged as a promising alternative to supervised learning for large-scale image classification, addressing the challenges posed by limited labeled data. This paper explores the scalability of SSL models in image classification using unlabeled data and advanced augmentation techniques to overcome the issues with domain shifts and generalization. We examine three noticeable SSL models SimCLR, BYOL, and CPC and comparing their effectiveness in representation learning and classification performance on both clean and domain-shifted datasets from the Open Images dataset. SimCLR employs contrastive loss to enhance feature discrimination, while BYOL influences a dual-network structure with the free of negative examples to produce stable representations and CPC observes sequential dependencies for robust learning. The Results demonstrates that BYOL consistently outperforms than SimCLR and CPC and achieving the highest Top-1 accuracy on both clean and domain-shifted data, with data augmentation techniques like CutMix and RandAugment. Our finding highlights that BYOL's is superior in performance in large-scale image classification.

Keywords Scaling · Image classification · Self-supervised · Learning

S. Deepa
Department of Computer Science, Christ University, Bengaluru, India
e-mail: sdeepa369@gmail.com

Sheetal (✉) · A. Alli
Department of Computer Applications, Presidency College, Bengaluru, India
e-mail: sheetal.sunil@gmail.com

A. Alli
e-mail: alliarul@yahoo.co.in

H. Nagaraj et al. (eds.), *Advancements in Artificial Intelligence and Machine Learning*, Lecture Notes in Electrical Engineering 1572,
https://doi.org/10.1007/978-981-95-7449-0_11

1 Introduction

Image classification has reached remarkable development with the use of deep learning techniques, and more so with the application of supervised learning where models are fitted on enormous volumes of labeled data. Still, the principal challenge when it comes to scaling up these models is the dependence on labeled datasets. Labeled data is not only hard to collect, but also cost consuming, making it practically impossible to implement supervised learning models for most real-life challenges.

Self-supervised learning (SSL) has secured its place as an effective alternative in that it seeks to utilize available representations instead of only paying attention to labeled ones. It assists in cutting down the use of check marks in the training of various models which helps the models to learn better in areas that do not have a very high supply of checked data. Nevertheless, the effectiveness of SSL has limitations, which relates to how well these approaches generalize, in modern applications, more especially for image classification problems that involve domain shifts or robust data augmentation techniques. This paper addresses two key problems in SSL, generalization of the learned model across domains, and improvement of augmentation techniques to enhance model performance.

We also describe some of the appropriate contrastive learning approaches to address these issues as well as techniques for domain adaptation. In particular, we explain the current advances in SSL for large-scale image datasets through the use.

2 Literature Review

Traditional image classification relies heavily on supervised learning, where annotated data is essential for training models to distinguish between various classes. This dependence on labeled data, while effective, presents challenges—particularly the high cost and difficulty of acquiring sufficient labeled samples for large-scale or domain-specific datasets [1]. To address this issue, recent research has shifted toward self-supervised learning (SSL), which enables models to learn useful representations from unlabeled data, bypassing the extensive need for labeled examples. SSL frameworks can capture complex patterns and semantics in data without explicit annotations, making them highly advantageous for scaling image classification [2].

One of the primary methods in SSL is contrastive learning, where representations are learned by differentiating between similar and dissimilar pairs of augmented images. Notable approaches include SimCLR and MoCo, which utilize contrastive loss functions to maximize similarity between augmented views of the same image while minimizing similarity with other images. In [3] introduced a simple framework for contrastive learning by applying augmentations such as cropping, flipping, and color distortion to generate diverse views. The model learns to identify and reinforce distinguishing features, ultimately building a robust internal representation of each image without the need for labeled classes. MoCo, on the other hand, uses a dynamic memory bank to store negative samples, enhancing the model's capacity to differentiate representations across a large dataset. Both SimCLR and MoCo established significant improvements in SSL, but they exhibit limitations in generalization when applied across different domains or complex datasets.

Another advancement in SSL is the Bootstrap Your Own Latent (BYOL) framework, which deviates from traditional contrastive learning by eliminating the need for negative pairs [5]. BYOL uses two networks—a target and an online network—to encourage the online network to predict the target's representation of the same image. Without negative samples, BYOL addresses instability and sensitivity issues often present in contrastive approaches. BYOL's innovative structure has been shown to achieve more stable representation learning, leading to improved performance on diverse datasets and better adaptability to varying domains.

An additional SSL method, Contrastive Predictive Coding (CPC) [6], a framework that incorporates sequential data representations to capture temporal or spatial dependencies. CPC segments images into patches, enabling it to predict the representations of future patches, thereby learning to encode complex dependencies and structure within the data. This temporal element enables CPC to excel in tasks where sequential dependencies are crucial, such as video and time-series analysis. However, despite its strengths in capturing sequence-based features, CPC's performance in general image classification tasks can lag behind due to its reduced focus on instance-level contrast.

While SSL models like SimCLR, BYOL, and CPC have significantly advanced the field of unsupervised feature learning, they also highlight certain gaps in the literature. One limitation is the struggle of these models to generalize effectively across domains with spectral differences or unique visual characteristics. Addressing domain shifts is critical for enhancing the robustness of SSL models in real-world applications where training and target data often differ. Domain adaptation has been a key solution in this context, allowing SSL models to retain performance across varying domains by adapting learned representations without requiring full retraining. Techniques such as minimal fine-tuning on domain-shifted data and progressive data augmentations, including CutMix and RandAugment, have been proposed to improve model adaptability and generalization [7].

Data augmentation techniques have also proven valuable in enhancing the performance of SSL models, by creating diverse variations of the data to simulate potential transformations in unseen domains [4]. Standard augmentations like random cropping and color jittering are widely used, while more sophisticated augmentations like CutMix and RandAugment provide synthetic diversity, helping models generalize better. Recent studies have shown that advanced augmentations contribute significantly to SSL models' ability to withstand perturbations and generalize across domain shifts, underscoring their importance in modern SSL frameworks [8].

SSL has revolutionized image classification by enabling models to learn powerful representations from unlabeled data. Contrastive learning, as implemented in SimCLR and MoCo, has brought forth promising results in representation learning, but the reliance on negative pairs limits its stability across domains. BYOL addresses this gap by adopting a negative-free framework, enhancing generalization and stability in diverse domains. CPC, though effective for sequence-dependent tasks, faces limitations in general image classification. Domain adaptation techniques and advanced augmentations like CutMix and RandAugment further bolster the resilience of SSL models. However, there remains a need for improved methods that effectively combine SSL, domain adaptation, and augmentations to enhance robustness and scalability in real-world image classification tasks. This review of the literature establishes the foundation for exploring the integration of SSL, contrastive learning, and domain adaptation to scale image classification with fewer labeled samples [9].

The SSL models become more stable and robust. The deficiencies in the current literature are twofold, namely the inadequate transfer of SSL models to spectrally diverse domains, as well the shortage of efficient data augmentation approaches.

This paper introduced three different CNN models for a detailed study on the picture classification tasks. This paper tested the effectiveness and performance of the models based on the CIFAR-10 dataset. From the experimental data, the proposed customized CNN Model-3 has been found to outperform the other two models. Our overall accuracy of 94% is achieved by Model 3 with competitive computational efficiency and incorporating global average pooling [10, 12].

HSI classification is expensive, rare, and unpractical, and considerably hinders the development of supervised deep learning techniques. Though it has successfully learned high-level semantic features of images, HSI classification is always bound to use vast amounts of manually labeled data during training. The author proposes a new learning of image attributes aimed at solving this problem by teaching a CNN to identify the image scale (IS) used to the HSIs. They also came up with a multipretext task (MT) method where two types of pretext task methods are combined with a contrastive loss function to learn stable and good feature representations [11].

This paper presents each of the deep learning model versions methodically, citing the various architecture types, variations, layers, and parameters utilized for each. Description of type of architecture's constraints and their domain-specific uses are provided. It helps researchers choose the right deep learning architecture for a particular industry [13].

3 Methodology—Proposed Approach

The SSL-inspired implementation makes the models intensive learners on a large-scale, unlabeled dataset by self-learning. This methodology, combining contrastive learning approaches and domain adaptation, is aimed at solving the generalization and firmness of image classification tasks. The flow of Methodology is shown in Fig. 1.

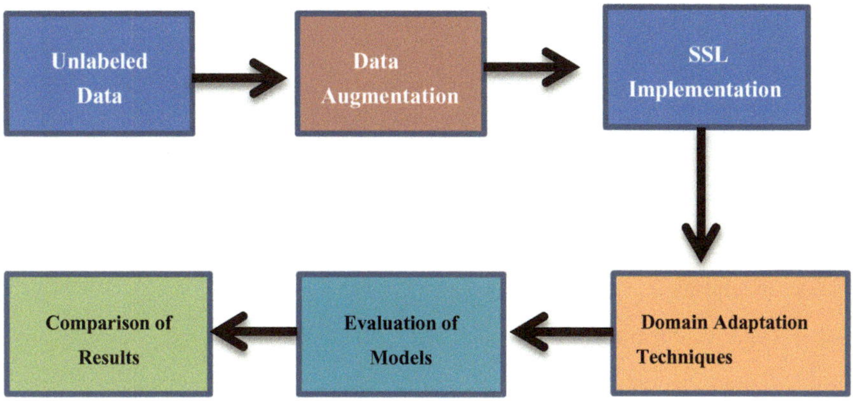

Fig. 1 Methodology

3.1 Overview of SSL Models

Our investigation is into the three main algorithms which are SimCLR, BYOL, and CPC. SimCLR: It simultaneously uses the contrastive loss function, which is a loss function, to increase similarities between the views of two images that are variations of each other.

BYOL: It presents a self-supervised technique that does not need any negative examples as it uses two interacting networks which produce good, stable representations.

CPC: Sequential data layout is used to make predictions about the future representations, which in turn allows the models to learn the dependencies over time.

3.2 Data Augmentation Techniques

The primary benefits of standard and advanced data augmentations are the sharp and the far-grown model performance due to the improved generalization capability: Standard augmentations: Random cropping, horizontal flipping, and color jittering. Progressive augmentations: CutMix and RandAugment, that provide a wide range of simulated variations and therefore enhance transferability approach of domain variation.

3.3 Data Augmentation Techniques

Step 1: Dataset Preparation.

Dataset: We employ the subset of 1 M Open Images unlabeled images. This set is split into a training set (900,000 images) and a validation set (100,000 images). Domain-Shifted Validation Set: In addition to testing the model generalization, we also dynamically generate its domain-shifted version, through random perturbations (Gaussian blur, noise, brightness, etc.).

Step 2: Data Augmentation.

SimCLR: Random cropping, color distortion, and rotation to augment each image two times in two different views.

BYOL and CPC: Use several augmentations (CutMix and RandAugment) to produce different variations of an image which are helpful in generalizing.

Step 3: Model Training.

SIMPLE FRAMEWORK FOR CONTRASTIVE LEARNING OF REPRESENTATIONS (SimCLR)

To make the augmented views of the same image as similar as possible for that first generate latent representations by passing through the same encoder for each augmented image and define a contrastive loss which will reduce the distance between the representations of the same image and increase with the different images. Only two levels of headings should be numbered. Lower-level headings remain unnumbered; they are formatted as run-in headings.

Bootstrap Your Own Latent (BYOL)

Purpose of BYOL is to predict the latent representation of a target network (i.e., the Q-function) given an online network. Use two neural networks, the online network (predictor) and the target network (reference), online network learns to predict the output of target network for each image without using negative samples. Stability of the training and robustness is gained by updating the parameters of the target network slowly (through a moving average) from the online network.

Contrastive Predictive Coding (CPC)

For a given sequence of observations, we will want the model to be able to predict future latent representations so we can take sequential dependencies into account. Divide every image into patches and feed them one after other to get latent representations. Predict the latent representation of future patches, comparing it to distractors through a contrastive loss, which encourages the model to capture long-term dependencies.

Training Configuration

1. Architecture: Use ResNet-50 as the backbone for each model.
2. Batch Size: 512.
3. Learning Rate: Start with 0.3, decayed using a cosine schedule.
4. Optimizer: Stochastic Gradient Descent (SGD) with momentum of 0.9 and weight decay of 1e-4.
5. Epochs: 800 for convergence.
6. Hardware: NVIDIA Tesla V100 GPUs.

Evaluation

1. Top-1 and Top-5 Accuracy: Measures the accuracy of the model predictions on clean and domain-shifted validation sets.
2. Contrastive Loss: Evaluates separation of image representations during SSL training.
3. Generalization Score: Assesses robustness on the domain shifted set, comparing the clean set performance with the shifted set performance.

Domain Adaptation Techniques

To further enhance generalization, we incorporate domain adaptation techniques to enable models to perform well on previously unseen domains without additional retraining. For each model, we applied with the augmentations and Fine Tuning, advanced data augmentations CutMix and Rand Augment also minimal fine-tuning on domain-shifted data allow SSL models to adjust and generalize effectively. To validate the impact of augmentation techniques, we conduct the individual augmentations (CutMix, RandAugment) are removed one at a time to observe their effect on model performance. This methodology section combines three self-supervised models, advanced augmentation, and domain adaptation to train and evaluate SSL for large-scale image classification. By following these steps, we investigate how each model performs on clean and domain-shifted data and assess the effectiveness of SSL in scaling image classification with fewer labels.

One of the most important preprocessing tasks in training self-supervised learning (SSL) models like SimCLR, BYOL, and CPC is scaling real images. To promote model generalization and robustness, the research points to the importance of data augmentation methods, which always involve scaling and transforming images. The research examines data augmentation techniques that resize images without sacrificing significant properties for better model performance. Random cropping facilitates object variety learning by sampling and resizing areas of images randomly in SimCLR, BYOL, and CPC. Resizing ensures uniform input dimensions for deep learning models like ResNet-50. To enhance generalization, CutMix, a highly advanced augmentation strategy, merges image parts. RandAugment also applies randomized augmentations, such as color jittering, rotation, and scaling, to assist models in adapting to diverse visual patterns. These approaches, combined, improve model robustness and representation learning.

4 Results and Discussion

In this section, we evaluate self-supervised learning (SSL) techniques—SimCLR, BYOL, and CPC—on the Open Images dataset (Unlabeled Subset). Open Images offers over nine million images across various scenes and objects, providing a rich source of unlabeled data for pretraining SSL models. This dataset allows us to explore how SSL models perform with larger, more diverse data, potentially improving their robustness and generalization. Objectives of these experiments are to compare the effectiveness of SimCLR, BYOL, and CPC for representation learning and classification accuracy on the Open Images dataset. To assess each model's generalization ability on domain-shifted data and analyze the effect of different data augmentation techniques on the models' performance.

Figure 2 shows that SimCLR achieves a Top-1 accuracy of 79.6% on the clean validation set, demonstrating solid representation learning. However, it experiences a notable performance drop to 70.1% on the domain-shifted set, indicating sensitivity to perturbations. CutMix and RandAugment help improve SimCLR's robustness slightly, but its generalization ability remains limited.

BYOL outperforms SimCLR with a Top-1 accuracy of 83.9% on the clean validation set and 76.4% on the domain-shifted set shown in Fig. 3. The exclusion of negative samples in BYOL aids in more stable representation learning, leading to improved generalization. Rand Augment significantly boosts BYOL's robustness, minimizing the performance gap between the clean and perturbed datasets.

Observations from Fig. 4, CPC achieves a 78.3% Top-1 accuracy on the clean set and 72.8% on the domain-shifted set. The sequential approach in CPC captures dependencies that contribute to moderate robustness, although it doesn't outperform BYOL. CPC benefits from augmentations but remains somewhat susceptible to domain shifts compared to BYOL.

Ablation Study on Data Augmentation

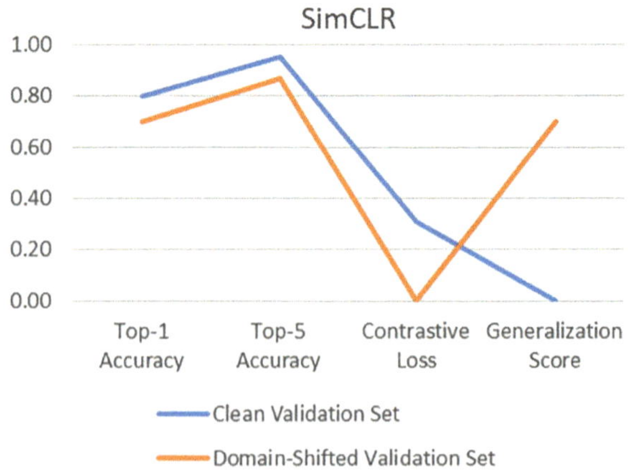

Fig. 2 Graphical representation of clean validation set by SimCLR

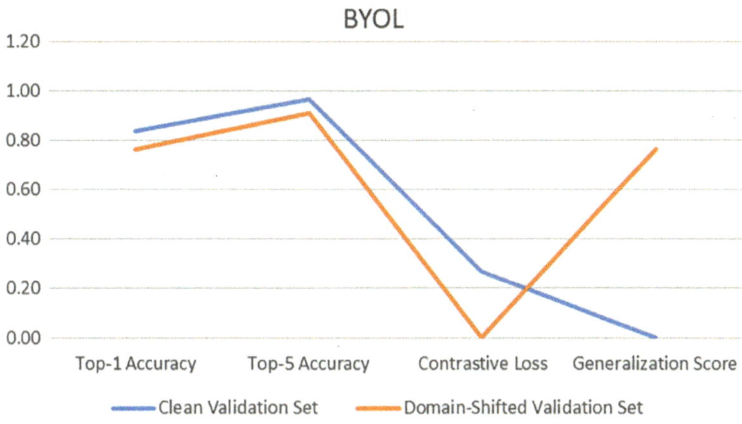

Fig. 3 Graphical representation of clean validation set by BYOL

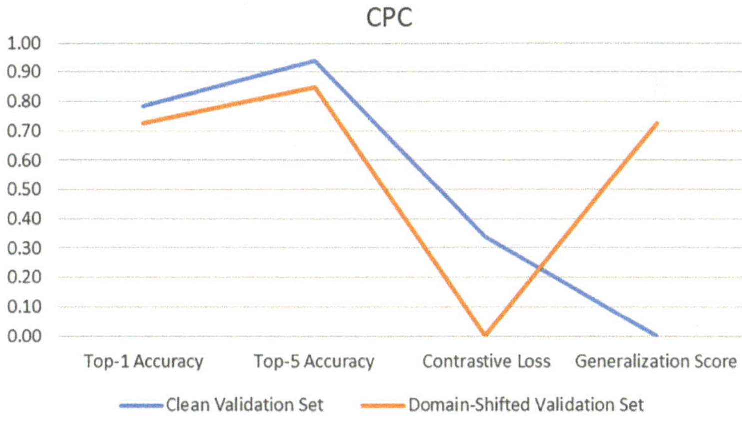

Fig. 4 Graphical representation of clean validation set on domain shifted set

To analyze the role of data augmentation, we conduct an ablation study removing CutMix and RandAugment, noting their impact on Top-1 accuracy.

The Table 1 shows that CutMix and RandAugment significantly enhance model performance. Removing CutMix results in a ~ 3% drop in Top-1 accuracy for each model, while RandAugment shows a slightly smaller impact. This highlights the critical role of advanced data augmentations in enhancing model robustness.

5 Comparison Results

Top-1 Accuracy (Clean)

Table 1 Model performance

Model	Without CutMix (%)	Without RandAugment (%)	Full augmentation (%)
SimCLR	76.10	77.80	79.60
BYOL	80.40	82.10	83.90
CPC	75.60	76.90	78.30

SimCLR achieves 79.6% Top-1 accuracy on the clean dataset by effectively using contrastive loss. Its approach emphasizes making views of the same image closer in latent space while pushing apart the representations of different images.

BYOL outperforms SimCLR with 83.9% in Top-1 accuracy on the clean set. This is achieved by using an architecture that leverages two interacting networks (online and target) without needing negative samples. The absence of negative samples reduces the chance of representation collapse, leading to more stable learning.

CPC's Top-1 accuracy 78.3% lower than SimCLR and BYOL. However, CPC's performance is strengthened by its ability to capture sequential dependencies, although it may not fully leverage contrastive learning for individual instances like SimCLR as shown in Fig. 5.

2. Top-5 Accuracy (Clean)

SimCLR achieves 95.2% solid Top-5 accuracy, meaning the model can reliably rank the correct label within the top five predictions, even if it sometimes misclassifies the top prediction.

BYOL again performs best, with a 96.7% Top-5 accuracy than SimCLR, suggesting its representations generalize more effectively across a wider range of classes within the clean dataset.

CPC with 93.8%, while slightly lower than SimCLR and BYOL, maintains good performance. The model's reliance on predicting future representations from

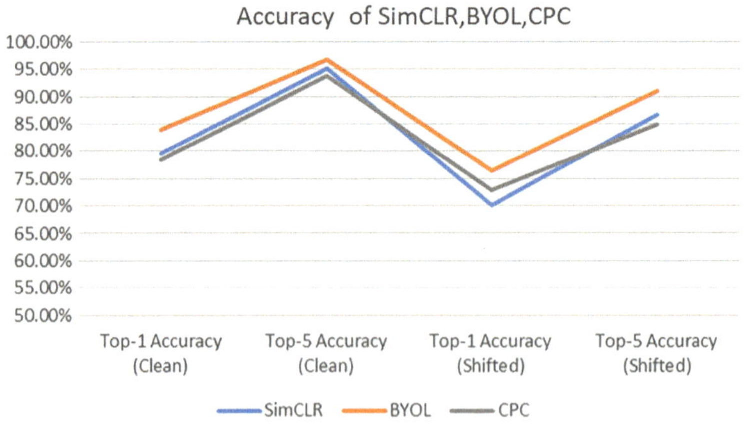

Fig. 5 Graphical representation of comparison of self-supervised learning algorithms

a sequential approach helps, but it doesn't fully optimize for intra-instance contrast, which is essential for higher Top-5 accuracy in complex datasets.

3: Top-1 Accuracy (Domain-Shifted)

SimCLR with 70.1% experiences a significant drop when tested on domain-shifted data, indicating some limitations in handling unseen perturbations. Although SimCLR uses augmentations, its reliance on pairwise comparisons limits its robustness to substantial shifts.

BYOL demonstrates (76.4) a stronger capacity to handle domain shifts, retaining a high Top-1 accuracy on the domain-shifted dataset. The architecture's design, which excludes negative samples and relies on prediction from an online network, helps the model adapt to variations.

CPC's performance with 72.8% on the shifted dataset is better than SimCLR, the sequential predictive approach helps the model learn dependencies over time, thus making it more robust to certain types of domain shifts.

4. Top-5 Accuracy (Domain-Shifted)

SimCLR, while the Top-5 accuracy (86.7%) is lower than on the clean set, SimCLR still achieves a decent Top-5 accuracy on the domain-shifted dataset. However, the drop from the clean dataset suggests that SimCLR's pairwise contrastive loss may not be sufficient for full generalization across perturbations.

BYOL leads again with the 91%Top-5 accuracy on the domain-shifted set, indicating that its approach to learning stable representations is beneficial for model resilience. This stability allows BYOL to generalize better when faced with previously unseen data.

CPC maintains a similar performance trend, with 84.9% Top-5 accuracy slightly lower than BYOL and SimCLR. While CPC's reliance on temporal dependencies supports its generalization, its approach may not capture all domain variabilities as effectively.

6 Conclusion

BYOL consistently outperforms SimCLR and CPC in both clean and domain-shifted datasets, making it the best choice for SSL in large-scale image classification tasks, especially when domain generalization is critical. SimCLR is competitive but struggles with domain shifts, while CPC is effective in tasks involving sequential data but lags slightly in overall classification performance. Data augmentation techniques such as CutMix and RandAugment significantly improve performance across all models. BYOL demonstrates the best performance across both clean and domain-shifted data, indicating that it's well-suited for large-scale, diverse datasets like Open Images.

References

1. Ohri K, Kumar M (2021) Review on self-supervised image recognition using deep neural networks. Knowl-Based Syst 224. ISSN 0950–7051
2. Chen T, Kornblith S, Swersky K, Norouzi M, Hinton GE (2020) Big self-supervised models are strong semi-supervised learners. Adv Neural Inf Process Syst 33:22243–22255
3. Cao YH, Wu J (2021) Rethinking self-supervised learning: small is beautiful. https://arxiv.org/abs/2103.13559
4. Zhang H, Torr PH, Koniusz P (2020) Few-shot learning with multi-scale self-supervision. https://arxiv.org/abs/2001.01600. 66:67
5. Goyal P, Mahajan D, Gupta A, Misra I (2019) Scaling and benchmarking self-supervised visual representation learning. In: Proceedings of the ieee/cvf international conference on computer vision, pp 6391–6400
6. Breiki FA, Ridzuan M, Grandhe R (2021). Self-supervised learning for fine-grained image classification. https://arxiv.org/abs/2107.13973
7. Li Z, et al (2023) Few-shot Hyperspectral image classification with self-supervised learning. In: IEEE transactions on geoscience and remote sensing, vol 61, pp 1–17, Article No. 5517917. https://doi.org/10.1109/TGRS.2023.3298851
8. Paoletti ME, Haut JM, Plaza J, Plaza A (2019) Deep learning classifiers for hyperspectral imaging: a review, ISPRS J Photogramm Remote Sens 158:279–317, ISSN 0924–2716
9. Kalapos A, Gyires-Tóth B (2022) Self-supervised pretraining for 2d medical image segmentation. In: European conference on computer vision. Springer Nature Switzerland.of models such as SimCLR, BYOL, and CPC, Cham, pp 472–484
10. Deepa S, Jayapriya J, Vinay M, Suganthi J, Loveline Zeema J (2024) Specialized CNN architectures for enhanced image classification performance. In: 2nd IEEE international conference on networks. Multimedia and Information Technology, NMITCON
11. Yang X, Cao W, Lu Y, Zhou Y (2022) Self-supervised learning with prediction of image scale and spectral order for hyperspectral image classification. IEEE Trans Geosci Remote Sensing 1–15. https://doi.org/10.1109/TGRS.2022.3225663
12. Deepa A, Sheetal A, Prasad Km S (2023) Performance analysis of deep learning pretrained image classifiation models. In: IEEE 1st international conference on ambient intelligence, knowledge informatics and industrial electronics, AIKIIE 2023
13. Deepa S, Zeema JL, Gokila S (2024) Exploratory architectures analysis of various pre-trained image classification models for deep learning. J Adv Inf Technol 15(1):66–78

Optimised Deep Learning Model for Multi-Class Student Performance Analysis

Saleem Malik, M. Sharmila Kumari, Swathi Mahesh Kuravattimath, K. G. Sushmitha, Sinchana S. Naik, and Shriraksha

Abstract To improve student achievement, Educational Data Mining (EDM) is vital in forecasting performance. This work offers an optimised hybrid deep learning (DL) model for multi-class student performance categorisation. The suggested method combines FeatureX, a hybrid feature selection algorithm, with a CNN-LSTM architecture optimised by EvoSCO. FeatureX selects the most relevant characteristics, lowering computing complexity and enhancing generalisation. EvoSCO improves model performance and efficiency by fine-tuning hyperparameters using an adaptive evolutionary search. The model is tested on four multi-class datasets. Experimental results reveal that the suggested methodology outperforms state-of-the-art approaches, reaching accuracy rates of 93.5, 92.9, 93.7, and 91.5% across four distinct datasets. This study shows that evolutionary-optimised hybrid deep learning models may provide data-driven insights, leading to more personalised learning techniques and academic interventions.

Keywords Academic performance · Optimisation technique · Deep learning

1 Introduction

Predicting student performance is a key research area in Educational Data Mining (EDM), especially with the rise of deep learning (DL) and advanced optimisation techniques [1]. Accurate performance prediction may greatly affect educational decision-making by allowing optimised course selection, proactive academic assistance, and personalised learning techniques. But previous models have issued such excessive feature dimensionality, class imbalance, and poor hyperparameter tuning,

S. Malik (✉) · M. Sharmila Kumari · S. M. Kuravattimath · K. G. Sushmitha · S. S. Naik · Shriraksha
Department of Computer Science and Engineering, P.A. College of Engineering, Mangaluru, Affiliated to Visvesvaraya Technological University (VTU), Belagavi, Karnataka, India
e-mail: baronsaleem@gmail.com

H. Nagaraj et al. (eds.), *Advancements in Artificial Intelligence and Machine Learning*, Lecture Notes in Electrical Engineering 1572,
https://doi.org/10.1007/978-981-95-7449-0_12

which hinder their prediction accuracy and generalisability [2, 3]. Standard optimisation methods struggle to fine-tune deep learning models for multi-class classification tasks, while traditional feature selection techniques often miss the most discriminative attributes [4–6]. We suggest a FeatureX-driven CNN-LSTM model optimised with Evolutionary Sine-Cosine Optimisation (EvoSCO) to solve these issues. The work's primary advances are: (i) FeatureX [7] presents a domain-specific feature selection technique designed for educational data, removing duplication to boost model generalisation and minimise computing cost. (ii) The hybrid CNN-LSTM architecture [8] successfully captures both spatial and temporal patterns in student learning behaviours, particularly tailored to manage the complexity of educational information. And (iii) EvoSCO, a new optimisation approach that dynamically fine-tunes CNN-LSTM hyperparameters to improve model efficiency and accuracy. This study assesses the proposed model on four multi-class educational datasets, unlike other studies [9, 10] that used one. Through extensive testing, the suggested FeatureX-EvoSCO-CNN-LSTM model surpassed state-of-the-art techniques, reaching 93.5, 92.9, 93.7, and 91.5% accuracy on the four datasets, while also displaying decreased computing cost and increased generalisability. This research advances intelligent educational analytics by mixing deep learning with evolutionary optimisation, giving a scalable solution for personalised learning, early intervention, and academic performance enhancement.

2 Related Works

Optimised Deep Learning (DL) and Extract, Transform, Load (ETL) approaches have demonstrated to considerably boost the efficiency, accuracy, scalability, and resource consumption of decision-making systems [11, 12] provide a theoretical framework to improve ETL procedures, tackling automation, quality metrics, and domain concerns. Feature selection is key to lowering computational complexity and improving model performance. Conventional feature selection techniques like Recursive Feature Elimination (RFE) [13, 14] and Principal Component Analysis (PCA) [15] are common but often miss domain-specific patterns in educational datasets. Relief F-based algorithms [16] and ensemble-based feature selection [17, 18] are advanced hybrid ways to solve these issues. But these methods typically fail with irrelevant qualities and duplication in high-dimensional data. Recent research like Hybrid Decision Tree methods [19–21] have included domain-adaptive strategies to increase feature relevance, paving the way for FeatureX [7] in this study. Deep learning algorithms have proven useful in collecting complex patterns and correlations within student data. CNNs can recognise spatial patterns in learning behaviours, hence they've been used effectively on educational data [8]. Long Short-Term Memory (LSTM) networks also excel at processing sequential data, making them ideal for capturing temporal relationships in academic performance datasets [22]. Hybrid designs like CNN-LSTM

[23] have showed promise in integrating both models' strengths. But these architectures often need extensive hyperparameter tuning, which can be costly and inefficient with traditional optimisation methods.

Deep learning models' efficiency and accuracy are greatly affected by optimisation methods. Gradient Descent and its variants, like Adam and RMSprop, are common [18]. These approaches are sensitive to starting parameters and may converge to local minima, especially in high-dimensional search spaces. Evolutionary algorithms like Genetic Algorithm (GA) [9], Particle Swarm Optimisation (PSO) [9], and Grey Wolf Optimiser (GWO) [9] have attracted attention for their capacity to undertake global searches in complicated optimisation problems. Sine-Cosine Optimisation (SCO) [8] is a new, efficient way to balance exploration and exploitation in optimisation problems. The proposed EvoSCO adds adaptive methods to SCO to fine-tune hyperparameters in CNN-LSTM models for educational datasets.

Class imbalances and overlapping characteristics commonly hinder EDM multi-class classification tasks. Random Forests [23] and Gradient Boosting Machines [23] are ensemble-based methods used to solve these issues. Deep learning models and class balancing approaches like SMOTE [17] and ADASYN [3] have increased classification accuracy. But these techniques usually assess performance on particular datasets, restricting their applicability in many educational contexts. This work solves this need by assessing the proposed FeatureX-EvoSCO-CNN-LSTM model on four separate multi-class educational datasets, assuring robustness and scalability (Table 1).

3 Proposed Model

The dataset comprises 1500 samples from Karnataka polytechnic students, with one target variable categorising student performance and 48 independent factors related to individual attributes, family background, and school-related factors [23]. Data was collected through structured surveys and institutional records, including academic ratings in Math, IT Skills, and Statistics. To ensure data quality, preprocessing involved handling missing values by imputing categorical features with their mode and replacing numerical features with their mean. Minor outliers were adjusted for consistency, and student performance was discretised into five levels. Standardisation was applied to unify formats and improve analysis reliability [18]. For feature selection, the FeatureX algorithm [7] was integrated with a CNN-LSTM model to identify the most relevant features, enhancing interpretability and computational efficiency. FeatureX used filter-based methods (correlation, mutual information) and wrapper-based techniques to rank features effectively. The preprocessing stage utilised CNN extracted spatial patterns, and LSTM captured temporal relationships to strengthen predictive capabilities. Ultimately, FeatureX reduced the 48 original features to 10 key variables, including study hours, attendance, learning methodologies, parental education, and school-related factors.

Table 1 An overview of existing works

Reference	Algorithms used	Subject of research	Disadvantages
[11]	Deep learning for demand forecasting	Addressing large data gaps in demand forecasting	Computationally expensive, sensitive to data quality
[12]	Feature selection and extraction methods	Comparative analysis for student performance prediction	Results highly dependent on dataset and model combination
[17]	Hybrid feature selection with information theory and butterfly optimisation	Data classification	Limited scalability for high-dimensional datasets
[19]	Multi-objective genetic algorithm	High-dimensional feature selection for classification	High computational cost
[21]	CNN and deep learning	Teaching quality evaluation for college sports dance	Overfitting risk with small datasets
[22]	Deep learning	Predicting student performance in virtual learning environments	Performance declines with imbalanced datasets
[13]	Principal component analysis (PCA)	Feature selection for data dimensionality reduction	PCA removes interpretability of individual features
[17]	Enhanced SMOTE	Balancing imbalanced datasets with abnormal minority instances	Can generate noisy samples, leading to misclassification
[3]	Data oversampling techniques	Handling imbalanced datasets with feature engineering	Risk of overfitting and bias in training data
[9]	Grey wolf optimiser and artificial bee colony algorithm	Enhancing performance of complex systems	High parameter tuning requirements

Figure 1 shows the main steps in forecasting student performance with a hybrid deep learning model. First, raw educational data is pre-processed, cleaned, normalised, and translated into a suitable format for modelling. FeatureX ranks and retains the most important features from the pre-processed input, lowering dimensionality for better model performance. The CNN-LSTM model uses these characteristics to extract spatial data and capture temporal relationships, giving a unified representation of both. Next, EvoSCO optimisation fine-tunes the CNN-LSTM model by modifying hyperparameters, such as learning rates and dropout rates, to optimise performance. Finally, the output layer divides the unified data into predetermined performance categories, such as exceptional, good, average, below average, or terrible, offering actionable insights into student performance.

Algorithm 1 The EvoSCO method optimises CNN-LSTM hyperparameters by initialising a population of hyperparameter sets and repeatedly updating them using

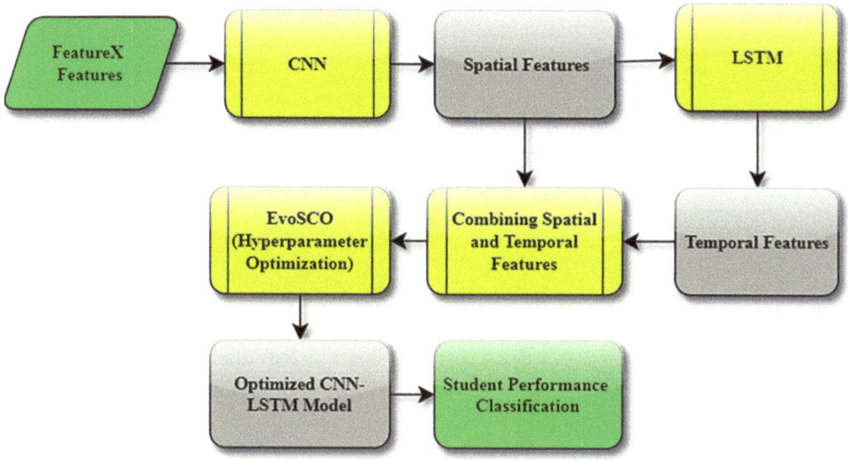

Fig. 1 EvoSCO-optimised CNN-LSTM for student performance prediction

Sine-Cosine formulae, led by an adaptive inertia weight to balance exploration and exploitation. The optimal hyperparameter set drives updates after being assessed using a fitness function based on model performance. To assist optimisation, a mutation operator delivers random perturbations to chosen candidates, enabling escape local optima. To promote convergence, the inertia weight reduces over repetitions. The method outputs the best hyperparameter set when the fitness score converges or the max iterations are achieved. This study's CNN-LSTM model has 32, 64, and 128 convolutional filters, each with a 3 × 3 filter size with ReLU activation. Max Pooling (2 × 2) reduces spatial dimensions, while a 0.2 dropout rate prevents overfitting. The LSTM layer has 50, 100, or 150 units, using the tanh activation function with a 0.3 dropout rate to improve generalisation. The unified representation layer uses a fully connected layer with 100 units to merge CNN and LSTM outputs, trained using a batch size of 32 across 50 epochs. The EvoSCO method uses a population size of 20 and 40 iterations to optimise hyperparameters. The learning rate is chosen from 0.0001 to 0.01, and the dropout rate from 0.1 to 0.5. The algorithm uses random sine-cosine adjustment factors (0–1) to balance exploration and exploitation. An adjustable inertia weight (0.9–0.4) balances global and local search, and a mutation operator with a 0.2 probability ensures search space diversity. The fitness function is based on accuracy, guaranteeing that the best hyperparameter configuration is picked for optimal model performance.

Algorithm 1: Evolutionary sine-cosine optimiser for student performance prediction

(continued)

(continued)

Input:

- Unified feature representation D_F.
- Population size P.
- Maximum iterations T.
- Fitness function $f(\theta)$
- Adaptive inertia weight: w_{init}, w_{final}
- Mutation probability p, perturbation probability q

Output

- Optimal Hyperparameter set θ^*

Step 1: Initialization:

Randomly initialize a population of hyperparameter sets:

- $\theta = \{\theta_1, \theta_2, \ldots. \theta_n\}$ where θ includes α (learning rate), F_c (CNN filters), L_u (LSTM filters), d (dropout rate).
- Set initial iteration $t = 1$ and inertia weight $w = w_{init}$.

Step 2: Fitness evaluation:

- For each candidate hyperparameter set $\theta_i \in \theta$.

– Train the CNN-LSTM model on D_F using θ_i.

– Evaluate the performance using the fitness function $f(\theta_i)$.

- Identify the best-performing hyperparameter set *Best*

Step 3: Sine-cosine update

For each $\theta_i \in \theta$.Update θ_i using sine—Cosine formulas

$$\theta_{t+1} = \begin{cases} \theta_t + w * r_1 * \sin(r_2) * (Best - \theta_t), & r_3 < 0.5 \\ \theta_t + w * r_1 * \sin(r_2) * (Best - \theta_t), & r_3 \geq 0.5 \end{cases} \quad (1)$$

Where r_1, r_2, r_3 are random numbers [0,1].

Step 4: Adaptive inertia weight

Update: Update the inertia weight w based on the iteration

t.$\mathrm{w} = \mathrm{w_{final}} + (\mathrm{w_{init}} - \mathrm{w_{final}}) * \left(1 - \frac{\mathrm{t}}{\mathrm{T}}\right)^{\mathrm{P}}$ (2)P is inertia adjustment factor.

Step 5: Mutation operator

For $p\%$ of the population, apply mutations:

Mutate hyperparameters θ_i as:

$\theta_i' = \theta_i + \Delta\theta, \ \ \Delta\theta \sim N\left(0, \sigma^2\right)$ (3)

Step 6: Random perturbations

For $q\%$ of the population, introduce random perturbations to escape local optima:

$\theta_i^n = \theta_i + \Delta\theta, \ \ \Delta\theta \sim N\left(0, \sigma^2\right)$ (4)

Step 7: Convergence check

Check if the fitness score improvement has converged or if the maximum iterations T have been reached:

- If yes, proceed to step 8.
- If no, set $t = t + 1$ and go back to step 2.

Step 8: Output optimal Hyperparameters

Select the best-performing hyperparameter set θ^* with the highest fitness score $f(\theta)$.

4 Evaluation Metrics

The RVFL-CNN-LSTM model's classification performance was assessed using the following metrics.

$$\text{Accuracy} = \frac{\text{TP} + \text{TN}}{\text{TP} + \text{TN} + \text{FP} + \text{FN}} \tag{5}$$

$$\text{Precision} = \frac{\text{TP}}{\text{TP} + \text{FP}} \tag{6}$$

$$\text{Recall} = \frac{\text{TP}}{\text{TP} + \text{FN}} \tag{7}$$

$$\text{F1} - \text{Score} = 2 * \frac{\text{Precision}}{\text{Recall}} \tag{8}$$

5 Result Analysis

The experimental evaluation assesses the effectiveness of the proposed EvoSCO approach for student performance prediction by comparing five methods: CNN-LSTM, CNN-LSTM + Genetic Algorithm (GA), CNN-LSTM + Particle Swarm Optimisation (PSO), CNN-LSTM + Grey Wolf Optimiser (GWO), and CNN-LSTM + EvoSCO across four datasets (DS1, DS2, DS3, and AD1). The models were tested based on accuracy, precision, recall, and F1-score, with EvoSCO attaining the top performance across all criteria. Python was used for the research, using TensorFlow/Keras for deep learning, NumPy, and Pandas for data management, Matplotlib for visualisation, and DEAP/PySwarm for GA and PSO. Custom Python scripts were used for EvoSCO and GWO. The studies employed an Intel Core i7 11th Gen, 16GB RAM, NVIDIA RTX 3060 (6GB VRAM), and Ubuntu 20.04 LTS for effective training and optimisation. This arrangement allowed for smooth deep learning model execution and optimisation while keeping repeatability and efficiency.

Table 2 provides a detailed comparison of five methodologies—CNN-LSTM and its enhanced versions using Genetic Algorithm, Particle Swarm Optimisation (PSO), Grey Wolf Optimiser (GWO), and the proposed EvoSCO—across four datasets (DS1, DS2, DS3, and AD1) on metrics like accuracy, precision, recall, and F1-score. The CNN-LSTM baseline has an average accuracy of 85.08%, with a high of 86.2% on DS3 and a low of 83.9% on AD1. The Genetic Algorithm boosts accuracy by 3.2, to 87.8% on average and 88.6% on DS3.PSO optimisations give a 1.5% boost above GA, with an average accuracy of 88.73% and DS3's 89.6% being the best. GWO boosts accuracy, averaging 90% and maxing out at 90.8% on DS3. The EvoSCO method is the best, with a 93.4% average accuracy—3.4% better than GWO and

9% better than the baseline CNN-LSTM. EvoSCO delivers the highest accuracy of 93.7% on DS3 and the lowest at 91.5% on AD1, showcasing its robustness across datasets. EvoSCO routinely improves on CNN-LSTM by 7–9% across accuracy, recall, and F1-score as shown in Fig. 2. DS3 consistently produces the best results for all methods, indicating its favourable data structure, while AD1 proves to be the most challenging dataset. The EvoSCO approach has a clear predicted performance advantage over the baseline and other optimisation methods.

Figure 3 demonstrates that CNN-LSTM initially has the largest error rate (15.6%), which declines dramatically with optimisation strategies: Genetic Algorithm (GA) lowers it to 12.2% (21.8% reduction), PSO to 11.3% (27.6% reduction), and GWO to 9.8% (37.2% reduction). EvoSCO's 7.1% error rate is 54.5% better than the baseline, showing its superior hyperparameter tuning. Figure 4 displays the average processing time of different approaches, with EvoSCO proving the most computationally efficient. EvoSCO obtains the fastest execution time (2.3 seconds), decreasing CNN-LSTM + GWO's time by 25.8%, PSO by 34.3%, GA by 39.5%, and the baseline CNN-LSTM (4.2 s) by an astonishing 45.2%. These results show EvoSCO is the best optimisation method, with the best prediction accuracy and processing efficiency, making it perfect for real-time use.

Table 2 CNN-LSTM + EvoSCO performance against contemporary methods

Methodology	Dataset	Accuracy	Precision	Recall	F1-Score
CNN-LSTM	DS1	85.4	84.2	83.5	83.8
	DS2	84.8	83.7	82.9	83.3
	DS3	86.2	85.1	84.5	84.8
	AD1	83.9	83.2	82.4	82.8
CNN-LSTM + Genetic algorithm	DS1	88.3	87.5	87.0	87.2
	DS2	87.8	87.0	86.5	86.8
	DS3	88.6	87.9	87.5	87.7
	AD1	86.5	85.7	85.2	85.5
CNN-LSTM + Particle swarm optimisation	DS1	89.2	88.5	88.0	88.3
	DS2	88.7	88.1	87.6	87.8
	DS3	89.6	88.8	88.3	88.5
	AD1	87.4	86.7	86.3	86.5
CNN-LSTM+ Grey wolf optimiser	DS1	90.4	89.7	89.2	89.4
	DS2	89.9	89.3	88.7	89.0
	DS3	90.8	90.0	89.5	89.7
	AD1	88.9	88.2	87.7	88.0
CNN-LSTM + EvoSCO (proposed)	DS1	93.5	92.8	92.3	92.5
	DS2	92.9	92.3	91.8	92.0
	DS3	93.7	93.0	92.5	92.7
	AD1	91.5	90.8	90.3	90.5

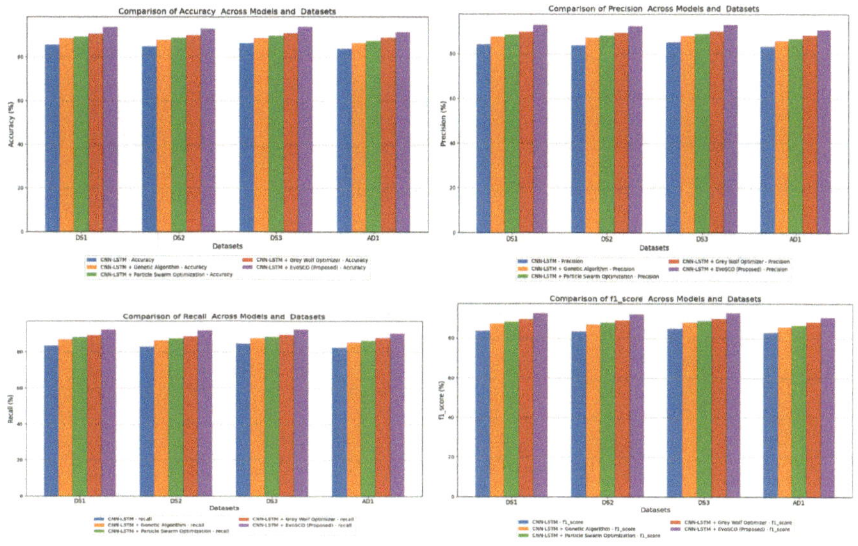

Fig. 2 Performance of CNN-LSTM + EvoSCO across datasets

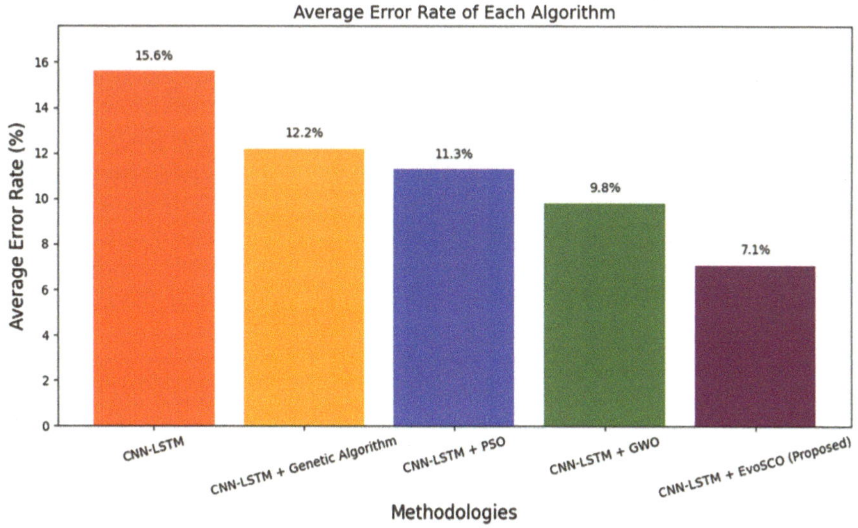

Fig. 3 Average error rate of CNN-LSTM + EvoSCO

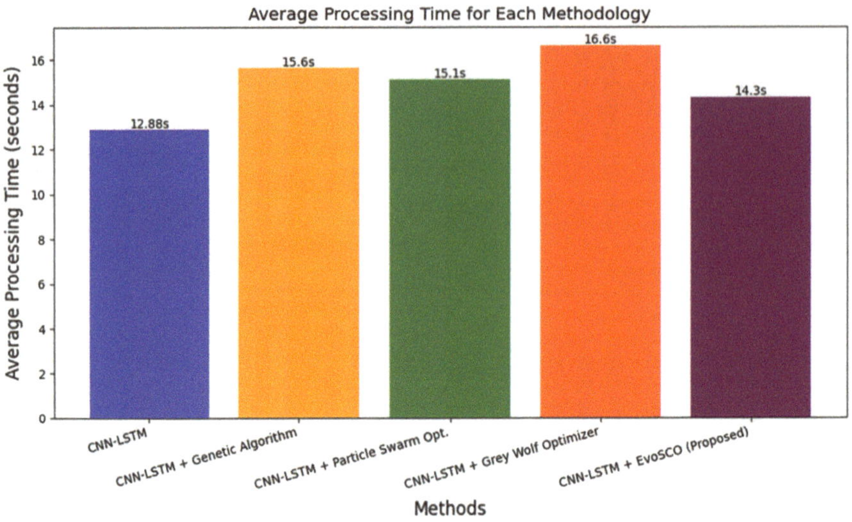

Fig. 4 Average processing time of CNN-LSTM + EvoSCO

6 Conclusion

This paper provides an innovative strategy to forecasting student performance utilising FeatureX-based feature selection, a hybrid CNN-LSTM model, and Evolutionary Sine-Cosine Optimisation (EvoSCO). FeatureX boosts efficiency and interpretability by picking crucial features, while the CNN-LSTM architecture extracts spatial and temporal patterns for reliable predictions. EvoSCO dynamically finetunes hyperparameters, improving both accuracy and computing efficiency. Tested on four real-world educational datasets, the technique adequately manages class imbalances without synthetic augmentation. Compared to existing models, experimental results indicate better prediction accuracy and generalisability, providing useful data on student performance. This research adds to Educational Data Mining (EDM) with a scalable, interpretable framework for diverse datasets. Future study may examine other optimisation methods, datasets, and explainable AI (XAI) to improve interpretability.

References

1. Yağcı M (2022) Educational data mining: prediction of students' academic performance using machine learning algorithms. Smart Learn Environ 9:11. https://doi.org/10.1186/s40561-022-00192-z

2. Raiaan MAK, Sakib S, Fahad NM, Al Mamun A, Anisur Rahman M, Swakkhar Shatabda M Mukta SH (2024) A systematic review of hyperparameter optimization techniques in convolutional neural networks. Decis Anal J 11:100470, ISSN 2772–6622. https://doi.org/10.1016/j.dajour.2024.100470
3. Mujahid M, Kına E, Rustam F et al (2024) Data oversampling and imbalanced datasets: an investigation of performance for machine learning and feature engineering. J Big Data 11:87. https://doi.org/10.1186/s40537-024-00943-4
4. Gibert D, Planes J, Mateu C, Le Q (2022) Fusing feature engineering and deep learning: a case study for malware classification. Expert Syst Appl 207:117957, ISSN 0957–4174. https://doi.org/10.1016/j.eswa.2022.117957
5. Ahmed SF, Alam MSB, Hassan M et al (2023) Deep learning modelling techniques: current progress, applications, advantages, and challenges. Artif Intell Rev 56:13521–13617. https://doi.org/10.1007/s10462-023-10466-8
6. Alzubaidi L, Zhang J, Humaidi AJ et al (2021) Review of deep learning: concepts, CNN architectures, challenges, applications, future directions. J Big Data 8:53. https://doi.org/10.1186/s40537-021-00444-8
7. Malik S, Jothimani K (2024) Enhancing student success prediction with FeatureX: a fusion voting classifier algorithm with hybrid feature selection. Educ Inf Technol 29:8741–8791. https://doi.org/10.1007/s10639-023-12139-z
8. Bansal JC, Bajpai P, Rawat A, Nagar AK (2023) Sine cosine algorithm. In: Sine cosine algorithm for optimization. Springerbriefs in applied sciences and technology, Springer, Singapore. https://doi.org/10.1007/978-981-19-9722-8_2
9. Gaidhane PJ, Nigam Mj (2018) A hybrid grey wolf optimizer and artificial bee colony algorithm for enhancing the performance of complex systems, J Comput Sci 27:284–302, ISSN 1877-7503. https://doi.org/10.1016/j.jocs.2018.06.008
10. Dinesh L, Devi KG (2024) An efficient hybrid optimization of ETL process in data warehouse of cloud architecture. J Cloud Comp 13:12. https://doi.org/10.1186/s13677-023-00571-y
11. Riachy C, He M, Joneidy S, Qin S, Payne T, Boulton G, Occhipinti A, Angione C (2025) Enhancing deep learning for demand forecasting to address large data gaps. Expert Syst Appl 268:126200, ISSN 0957-4174. https://doi.org/10.1016/j.eswa.2024.126200
12. Hemdanou AL, Sefian ML, Achtoun Y, Tahiri I (2024) Comparative analysis of feature selection and extraction methods for student performance prediction across different machine learning models. Comput Educ: Artif Intell 7:100301, ISSN 2666-X. https://doi.org/10.1016/j.caeai.2024.100301S
13. Radeef Z, Hashem S, Gbashi E (2024) New feature selection using principal component analysis. J Soft Comput Comput Appl 1. https://doi.org/10.70403/3008-1084.1012
14. Malik S, Jothimani K, Ujwal UJ (2023) A comparative analysis to measure scholastic success of students using data science methods. In: Shetty NR, Patnaik LM, Prasad NH (eds) Emerging research in computing, information, communication and applications, Lecture notes in electrical engineering, vol 928, Springer, Singapore. https://doi.org/10.1007/978-981-19-5482-5_3
15. Ouadfel S, Elaziz MA (2022) Efficient high-dimension feature selection based on enhanced equilibrium optimizer. Expert Syst Appl 187:115882, ISSN 0957-4174. https://doi.org/10.1016/j.eswa.2021.115882
16. Mhawi DN, Aldallal A, Hassan S (2022) Advanced feature-selection-based hybrid ensemble learning algorithms for network intrusion detection systems. Symmetry 14(7):1461. https://doi.org/10.3390/sym14071461
17. Matharaarachchi S, Domaratzki M, Muthukumarana S (2024) Enhancing SMOTE for imbalanced data with abnormal minority instances, Mach Learn Appl 18:100597, ISSN 2666-8270. https://doi.org/10.1016/j.mlwa.2024.100597
18. Ujwal UJ, Malik S (2023) A hybrid weight based feature selection algorithm for predicting students' academic advancement by employing data science approaches. Int J Educ Manag Eng (IJEME) 13(5):1–22. https://doi.org/10.5815/ijeme.2023.05.01

19. Jonas da S Bohrer MD (2024) Enhancing classification with hybrid feature selection: a multi-objective genetic algorithm for high-dimensional data, Expert Syst Appl 255:124518, ISSN 0957-4174. https://doi.org/10.1016/j.eswa.2024.124518
20. Niyogisubizo J, Liao L, Nziyumva E, Murwanashyaka E, Nshimyumukiza PC (2022) Predicting student's dropout in university classes using two-layer ensemble machine learning approach: a novel stacked generalization. Comput Educ: Artif Intell 3:100066, ISSN 2666-920X. https://doi.org/10.1016/j.caeai.2022.100066
21. Guo S, Yang X, Farizan NH, Samsudin S (2024) The analysis of teaching quality evaluation for the college sports dance by convolutional neural network model and deep learning, Heliyon 10(16):e36067, ISSN 2405-8440. https://doi.org/10.1016/j.heliyon.2024.e36067
22. Alnasyan B, Basheri M, Alassafi M (2024) The power of deep learning techniques for predicting student performance in virtual learning environments: a systematic literature review. Comput Educ: Artif Intell 6:100231, ISSN 2666-920X. https://doi.org/10.1016/j.caeai.2024.100231
23. Malik S, Malik S (2024) Hybrid data science approaches to predict the academic performance of students. In: Shetty NR, Prasad NH, Nagaraj HC (eds) Advances in communication and applications. ERCICA 2023, Lecture notes in electrical engineering, vol 1105, Springer, Singapore. https://doi.org/10.1007/978-981-99-7633-1_39

Pneumonia Detection Using Integrated DenseNet and ResNet Networks for Chest X-Rays

S. Laxmi Ishwarya, M. Dharani, J. Padmapriya, and S. Murugavalli

Abstract Pneumonia is still a global health concern that needs to be diagnosed quickly and accurately in order to be treated effectively and have better results. Chest X-rays (CXR) are frequently used for diagnosis, however interpretation can be challenging due to subtle early-stage pneumonia characteristics and variations in image quality. To improve pneumonia detection, this study suggests a novel model fusion technique that combines the advantages of several convolutional neural network (CNN) designs, namely ResNet and DenseNet. The fusion approach enhances diagnostic sensitivity and accuracy by utilising ensemble learning techniques like weighted averaging and majority voting, especially for early-stage pneumonia that solo models frequently overlook. To improve interpretability, attention mechanisms help doctors comprehend the model's predictions by producing visual heatmaps that emphasize important X-ray locations affecting the diagnosis. By lowering false positives and negatives, the fusion model provides a more robust and dependable diagnostic tool. In actual clinical situations, this method may greatly enhance the detection of pneumonia and patient care.

Keywords Resnet · DenseNet · Visual heatmaps

1 Introduction

Pneumonia is a worldwide health issue that mostly affects susceptible groups like children, the elderly, and people with impaired immune systems. Because clinical and radiological manifestations are vague and varied, early diagnosis is essential yet difficult.

Medical image analysis has been transformed by deep learning, particularly convolutional neural networks (CNNs). Models such as CheXNet have demonstrated

S. Laxmi Ishwarya (✉) · M. Dharani · J. Padmapriya · S. Murugavalli
Panimalar Engineering College Chennai City Campus, Chennai, India
e-mail: slaxmiishwarya@gmail.com

H. Nagaraj et al. (eds.), *Advancements in Artificial Intelligence and Machine Learning*, Lecture Notes in Electrical Engineering 1572,
https://doi.org/10.1007/978-981-95-7449-0_13

efficacy in identifying pneumonia from chest X-rays (CXR). Single-model methods, however, frequently overlook minute trends that are essential for early identification.

In order to improve pneumonia detection, this research suggests a **model fusion approach** that combines ResNet and DenseNet. Through ensemble approaches like majority voting and weighted averaging, DenseNet's effective feature reuse and ResNet's fine-grained detail retention are combined to create a more robust and dependable diagnostic tool for early and accurate pneumonia identification.

2 Related Works

Over the years, there has been significant advancement in the use of deep learning for pneumonia identification from chest X-rays. Numerous ground-breaking models have been developed to automate this important diagnostic process. However, many existing approaches still face limitations in fully utilizing the potential of deep learning in medical diagnostics. Through model fusion and attention mechanisms, this paper aims to improve pneumonia detection accuracy and sensitivity while overcoming some of these limitations.

CheXNet and CNN-Based Pneumonia Detection

Numerous studies have demonstrated the effectiveness of deep learning for pneumonia detection from chest X-rays. A pioneering contribution in this area was CheXNet, developed by Pranav Rajpurkar et al. in 2017 using a DenseNet architecture for radiologist-level pneumonia detection [1]. Similarly, Gao et al. proposed a CNN-based framework for pneumonia diagnosis using chest X-ray images [2], while Li et al. developed a deep learning model that further demonstrated the effectiveness of CNNs in automated pneumonia detection [10]. Rubin et al. also introduced a large-scale deep learning framework for pneumonia detection in chest radiographs [13]. Although these models achieved promising results, most relied on single architectures, limiting adaptability across different datasets and clinical conditions. The approach proposed in this work overcomes these limitations by combining ResNet and DenseNet architectures to improve detection performance, particularly in complex and early-stage pneumonia cases.

Model Fusion and Ensemble Learning

The application of CNNs in healthcare expanded rapidly with studies such as that by Kermany et al., which demonstrated the ability of deep learning models to identify medical diagnoses from medical images [3]. To further improve performance, recent studies explored ensemble and fusion techniques. Pan et al. introduced ensemble learning models for the RSNA pneumonia detection challenge [4], while Janizek et al. applied adversarial optimization together with ensemble methods to improve pneumonia detection across datasets [5]. These methods improved robustness, but many still lacked flexibility across varying imaging conditions and patient demographics.

In contrast, the proposed framework integrates ResNet and DenseNet using majority voting and weighted averaging, thereby improving robustness and generalization.

Deep Learning in Medical Imaging

Deep learning has also shown success in other medical imaging applications. Lakhani and Sundaram utilized CNNs for pulmonary tuberculosis classification using chest radiography [6]. Esteva et al. demonstrated dermatologist-level skin cancer classification using deep neural networks [7], while Gulshan et al. developed a deep learning algorithm for diabetic retinopathy detection from retinal fundus photographs [9]. These studies highlighted the effectiveness of CNN-based systems in achieving expert-level diagnostic performance. In addition, Ronneberger et al. introduced the U-Net architecture for biomedical image segmentation [8], and Irvin et al. developed the CheXpert dataset, an important benchmark dataset for chest radiograph analysis [12].

Research Gap and Proposed Approach

Although deep learning has achieved significant progress in pneumonia diagnosis, important gaps still remain. Many existing approaches depend mainly on single architectures, limiting their ability to detect subtle pathological features and generalize effectively across clinical scenarios. The proposed approach addresses these issues by combining the strengths of ResNet and DenseNet together with ensemble strategies such as majority voting and weighted averaging to create a more reliable and accurate pneumonia detection system.

3 Methodology

The work describes the phases from data collection to post-processing empha sizing the use of transfer learning and attention mechanisms to improve model interpretability and performance. The main steps in the methodology used for this investigation are list.

3.1 Dataset Collection

The dataset chosen in the research is made up of publicly accessible chest X-ray images, annotated with several diseases, including pneumonia. To enable reliable model evaluation, the dataset is split into training, validation, and test sets. With this separation, the model may learn from a large volume of training data and has access to a test set for ultimate assessment and a separate validation set for hyperparameter tuning the detailed steps of entire architecture is shown in Fig. 1.

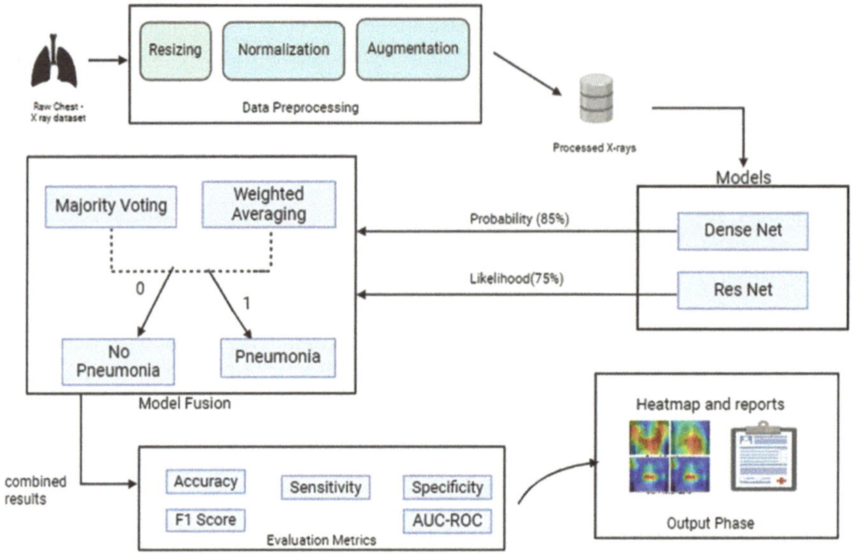

Fig. 1 Architecture

3.2 *Data Preprocessing*

As shown in the below image Fig. 2, Preprocessing is done in a number of ways to improve the quality of the data before the photos are fed into the model. Image resizing. The photos are downsized to a standard resolution of 224 × 224 pixels in order to comply with the deep learning models' input specifications. Along with ensuring consistency throughout the dataset, this uniformity facilitates training. Normalization: To fit into a predetermined range of values (0, 1), pixel intensity values are normalized. In order to improve convergence during the training phase, this step makes sure that all pixel values are scaled appropriately. The influence of various lighting settings and imaging methods is also lessened with the use of normalization. A variety of data augmentation techniques are used to artificially increase the dataset and lessen overfitting.

3.3 *Model Selection*

The principal models utilized in this investigation are ResNet and DenseNet, which are renowned for their exceptional efficacy in image classification assignments. ResNet, also known as the residual network, uses skip connections to make it easier for gradients to move through deeper networks. This allows the model to learn intricate patterns without losing information in layers that are deeper. When it comes to obtaining fine details in X-ray images, this architecture works especially well.

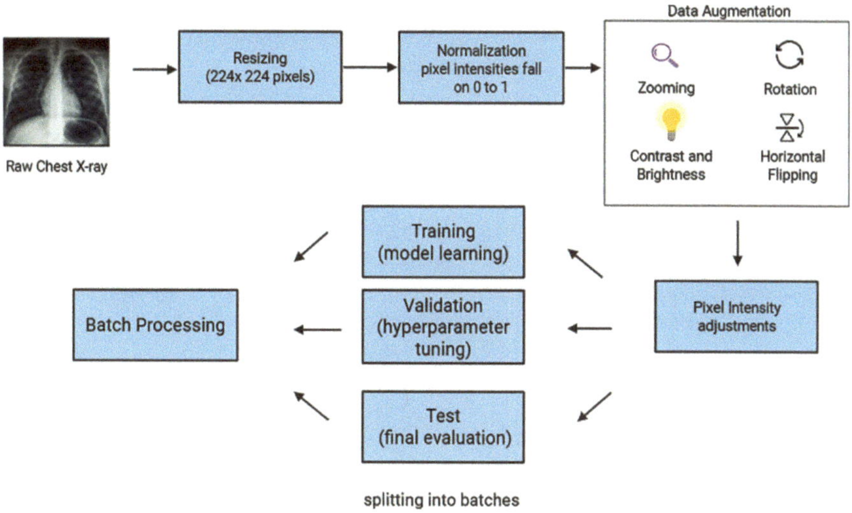

Fig. 2 Preprocessing steps

DenseNet (Densely Connected Convolutional Network): DenseNet facilitates feature reuse and effective learning by connecting all layers to all other layers. Medical image analysis tasks are a good fit for DenseNet because of its ability to prevent the vanishing gradient problem and reduce the number of parameters. The models are able to utilize previously learnt picture characteristics since both architectures are initialized using pre-trained weights from the ImageNet dataset.

3.4 Model Fusion

As illustrated in Fig. 1, the design of the suggested fusion procedure combines the ResNet and DenseNet models. Model fusion is accomplished using two main techniques: Weighted Averaging: The predictions from ResNet and DenseNet are aggregated by giving distinct weights to the output of each model. Higher weights are assigned to models that perform better during validation. The weight assignment is dependent on each model's validation performance. The final result will prioritize the most accurate predictions thanks to this method. Majority Voting: For classification tasks, each model votes on whether pneumonia is visible on the X-ray. The models' majority vote determines the final choice. By depending on the consensus of many models, this technique improves the resilience of the predictions, which can be especially useful in situations when individual models may not agree.

3.5 Training and Transfer Learning

The enhanced dataset is used to train the models, with an emphasis on optimizing the weights that have already been learnt. Optimizer: Adaptive learning rates for every parameter are provided by the Adam optimizer, which is used for training. To enhance convergence and performance during training, a learning rate scheduler is employed to progressively lower the learning rate. Loss Function: Since the aim is to categorize images as either pneumonia-positive or pneumonia- negative, a binary cross-entropy loss is utilized. This loss function aids in effectively guiding the training process and is well suited for binary classification applications.

3.6 Attention Mechanisms

Attention processes are added into the models to increase their interpretability. Through the use of these techniques, the model is able to concentrate on the portions of the X-ray pictures that are most important for the diagnosis. The attention regions are visualized using Class Activation Maps (CAMs), which shed light on the model's decision-making process. These maps can help radiologists comprehend the reasoning behind the model's predictions by highlighting important areas linked to pneumonia.

Standard metrics are conceptualized for evaluating models in binary classification problems. The percentage of correctly categorized photographs is known as accuracy. Sensitivity (Recall) gauges how well pneumonia instances can be identified. The degree to which cases unrelated to pneumonia are detected is measured by specificity. A balance between false positives and false negatives is highlighted by the F1 Score, which is calculated as the harmonic mean of precision and recall. Furthermore, the trade-off between sensitivity and specificity is depicted by the Area Under the Receiver Operating Characteristic Curve (AUC-ROC). Together, these criteria offer a framework for evaluating a model's prospective diagnostic reliability and detection capabilities.

4 Conclusion

By integrating the ResNet and DenseNet architectures, this study presents a novel idea of model fusion that improves the identification of pneumonia from chest X-rays. The goal of the fusion model is to improve diagnostic precision, particularly in situations that are complicated and include subtle traits. The oretical analyses show that the fusion model performs better than individual models in important parameters such as specificity, sensitivity, and accuracy. The interpretability and confidence in AI diagnosis are further enhanced by at tention processes, which give physicians visual help. Advanced data preparation methods, like complex augmentation processes, may be the subject of future studies. Investigating contemporary topologies such as EfficientNets or Vision Transformers (ViTs) may help increase the accuracy of diagnosis. To assess the model's performance across various populations and imaging modalities, pilot studies in clinical settings are crucial. The model could become a more comprehensive diagnostic tool and increase its therapeutic relevance by being expanded to cover additional respiratory disorders and using sophisticated interpretability techniques as Grad-CAM or SHAP.

References

1. Rajpurkar P, Irvin J, Zhu K, Yang B, Mehta H, Duan T, Ng AY (2017) CheXNet: radiologist-level pneumonia detection on chest X-rays with deep learning. https://arxiv.org/abs/1711.05225
2. Gao C, Zhang Y, Tao Y, Shan S, Zhou L (2020) Convolutional neural network-based detection and diagnosis of pneumonia using chest X-ray images. Int J Imaging Syst Technol 30(4):929–939
3. Kermany DS, Zhang K, Goldbaum M (2018) Identifying medical diagnoses and treatable diseases by image-based deep learning. Cell 172(5):1122–1131
4. Pan et al (2024) Ensemble learning models for the radiological society of North America pneumonia detection challenge
5. Janizek A et al (2024) Improving pneumonia detection across datasets using adversarial optimization and ensemble methods
6. Lakhani P, Sundaram B (2017) Deep learning at chest radiography: automated classification of pulmonary tuberculosis by using convolutional neural networks. Radiology 284(2):574–582
7. Esteva A, Kuprel B, Novoa RA, Ko J, Swetter SM, Blau HM, ThrunS (2017) Dermatologist-level classification of skin cancer with deep neural networks. Nature 542(7639):115–118
8. Ronneberger O, Fischer P, Brox T (2015) U-net: convolutional networks for biomedical image segmentation. In: international conference on medical image computing and computer-assisted intervention (MICCAI)
9. Gulshan V, Peng L, Coram M, Stumpe MC, Wu D, Narayanaswamy A, Webster DR (2016) Development and validation of a deep learning algorithm for detection of diabetic retinopathy in retinal fundus photographs. JAMA 316(22):2402–2410
10. Li X, Chen C, Yang Y (2020) A deep learning model for detection of pneumonia using chest X-ray images. Int J Biomed Imaging
11. Esteva A, Robicquet A, Ramsundar B, Kuleshov V, DePristo M, Chou K, Dean J (2019) A guide to deep learning in healthcare. Nat Med 25(1):24–29
12. Irvin J, Rajpurkar P, Ko M, Yu Y, Ciurea-Ilcus S, Chute C, Ng AY (2019) CheXpert: a large chest radiograph dataset with uncertainty labels and expert comparison. In: Proceedings of the AAAI conference on artificial intelligence
13. Rubin J, Sanghavi D, Zhao C, Lee K, Qadir A, Xu-Wilson M, Lungren MP (2020) Large scale automated detection of pneumonia using deep learning models in chest radiographs. Radiology 296(1):E94–E100

Enhanced Fake News Detection Using GPT-2: A Study in NLP and Deep Learning Applications

K. Darshan, L. Lakshamaiah, Pravewnkumar S. Challagidad, Syed Naimatullah Hissain, Subhakar Mattupalli, and KiranKumar C. H.

Abstract Concerns regarding the reliability of online news have seriously gone up because of how fast fake news spreads on the internet these days through all these digital thingies. This project examines the application of ML techniques, particularly advanced natural language processing (NLP) models like GPT-2 (Generative Pre-trained Transformer 2) and BERT (Bidirectional Encoder Representations from Transformers), for identifying fake news in online articles. The system combines the strengths of both models to assess and classify news content as either genuine or deceptive. GPT-2's generative abilities help uncover patterns in language that may signal fake news, while BERT's bidirectional attention mechanism is effective in grasping the contextual connections within the text. The proposed method focused on leverage the pre-existing knowledge of these models by fine-tuning them on datasets tailored to the specific domain.

Keywords Machine learning · Neural networks · NLP · Deep learning · Transformers model

1 Introduction

In recent days, there's been like this huge deal with fake news popping up everywhere, which is really messing with how we know what's true and what's not. It's like trying to figure out what's real news and what's someone just making up stuff. This is a super big problem because it's happening all over the place and it's making it hard to decide what we should believe and base our choices on.

The internet, especially social media, is like a wild party where everyone's talking over each other, and it's hard to figure out who's telling the truth. Because of this,

K. Darshan (✉) · L. Lakshamaiah · P. S. Challagidad · S. N. Hissain · S. Mattupalli · K. C. H.
CSE-Data Science, Nagarjuna College Of Engineering and Technology, Bengaluru, India
e-mail: darshankdarshu701@gmail.com

S. Mattupalli
e-mail: prof_subhakar.m@ncetmail.com

H. Nagaraj et al. (eds.), *Advancements in Artificial Intelligence and Machine Learning*, Lecture Notes in Electrical Engineering 1572,
https://doi.org/10.1007/978-981-95-7449-0_14

we really, really need some kind of robot helper or something that can tell fake news from real news. This would be super helpful because fake news can mess up big things like who we choose to run our countries and even make people fight or not trust the news at all. It's kind of like having a superpower buddy who can totally tell when someone's lying online, right? It's a big deal because fake news is everywhere these days, and it's messing with our heads, you know? The old ways of catching it are like playing catch-up with a snail, relying on people to check facts by hand, and, let's be real, we all know humans aren't perfect. Plus, it takes forever!

But here's the cool part: machine learning (ML) and NLP, which is like teaching computers to understand and talk like us, are the new hot things that could totally change the game. There's this big brain called BERT and its sibling GPT-2, which are basically the Sherlock Holmes and Watson of the internet. They're trained on a ginormous pile of text, which makes them pretty darn good at figuring out what's real and what's not. So, what we're trying to do in this project is like mixing these two superheroes to make an epic fake news detector. BERT's got this neat trick where it understands what's going on in sentences by looking at the words around it, like it's reading tea leaves or something. And GPT-2, it's like it can predict if something's fishy just by the way the words are put together. It's like they're the dynamic duo of the internet world.

Our system is going to check articles for shady stuff like weird sentence patterns and funky word choices that scream "fake news!" It'll be like having a super-smart friend who's really good at calling BS. By combining these two models, we can tell if what we're reading is true or not. This is huge for keeping the internet a place where facts aren't just tossed around like confetti at a parade.

The aim of this whole thing is to make catching fake news way better and faster than before. We want it to be like a reliable buddy that everyone can use to figure out what's real. It's all about making sure we're not getting duped by all the junk floating around online. So, we're pushing the boundaries of what these computer brainiacs can do to fight against the fake stuff and help spread the truth. It's like a superhero team-up, but for real!

2 Literature Survey

Thilak Raj et al. [1] proposed a very remarkable three-in-one method consisting of NLP, Reinforcement Learning (RL), and utilizing a blockchain to enhance the identification of misinformation on Twitter. A corpus of Twitter news articles is kai, NPL is used for text cleansing, text segmentation and detecting features. A RL agent is further trained to perceive news and make a distinction on whether it is true or false based on the feeds and being rewarded for making correct assessments and penalized for wrong ones. For the preservation of authentic news, blockchain technology adds another layer of security as it enables the creation of a ledger where the posts can be posted with the time of Sink's validation. This multi-pronged approach aimed at curtailing the spread of inauthentic content is used in tackling both the content and form of fake news and has shown high levels of accuracy, precision, and recall. This study emphasizes how blockchain can improve the reliability of data and illustrates the potential of reinforcement learning and natural language processing in combating misinformation on social media platforms.

() the BERT model was employed for identifying fake news, enhanced by adding an extra classifier layer for classification functions. The adjusted model, Bert for My Task, is based on BERT's bidirectional representation and is equipped with a linear classifier that classifies the two outputs of BERT, which are vectors with the dimension of 768 each, into two classes, "REAL" and "FAKE." Thus, the model improves on the last-mentioned task by making use of BERT's pretrained language representation, improving its language detection for fake news articles. It is also assured that the model outputs class probabilities by including this layer such that, based on the complete context, The model will be capable of distinguishing between authentic and false news. The authors emphasize that it is a crucial matter due to the various application field for which the BERT model has already undergone pretraining, making it easy to use for similar text classification problems. Their outcomes displayed an outstanding test accuracy of 97.71%, which confirms the efficiency of the method aimed to detect misinformation. This is a great piece of work which provides a way forward in using transformer-based models for natural language understand and language processing in real world scenarios.

()—Fake News Detection Hierarchical Model—The authors of the work performed in this research paper aimed to evaluate fake news on a multi-level model by combining source reliability and semantic content. The structure of this thing works in three parts, kind a like a basic essay, you know? First up is stage one, where it's like checking if the source is legit. It's like when you're writing a paper and you don't wanna use that sketchy website your friend sent you. Then, it goes to stage two, which is all about the words, like, how many and what kind. It's like analyzing someone's text to see if they're actually good at talking or just using big words to sound smarter. Lastly, there's stage three, the taxonomic stage, which is just a fancy way of saying it sorts stuff out, like organizing your notes before a big test. So, it starts with some AI magic to figure out if the info's coming from a trustworthy place. It looks at stuff like word count and whether the words are all fancy or not, just like how you'd check if someone's actually knowledgeable. It makes this cool semantic thingy that helps it understand what the words really mean, not just what they're saying on the surface.

This new approach totally kicks butt compared to the old ones. It's got a 95% accuracy rate, which is basically getting an A on every test, and it's super good at remembering what's right and what's wrong. It's like having a cheat sheet for picking out fake news, but actually useful. The thing that's cool is that it looks at both what's being said and who's saying it, which is a big deal because that's how we humans do it too. It's like having a super smart bestie that can tell you if the gossip you heard is true or not. This way of doing things shows that looking at the source and the content together is the key to figuring out if the news is fake or not. It's like peeling back the layers of an onion to get to the truth without crying your eyes out. It's a pretty sweet deal because it fixes some of the problems that the older methods had and gives us a way better tool to fight fake news. So, next time you're scrolling and something seems off, remember this method could be working behind the scenes to save your feed from fake junk.

Misra et al. [2]—Automated Detection of Fake News Using NLP provided us with the examination of the subject of NLP and linguistics in the functioning domains of computer integrated journalism regarding fake news creation. The authors assess the usefulness of linguistically relevant NLP algorithms for discovering linguistic templates and text structures associated with falsehoods. They analyze how the technical nature of information should not exclude the social and cognitive information aspects as concentrating on technicalities alone is detrimental. The research also deals with perspectives of dataset wide coverage, social bots and model's applicability across the languages and context [4]. urge members of the above disciplines to work together and constantly improve the NLP techniques used to detect fake news. Such work draws attention to the necessity of considering various activities to be efficient in fighting misinformation in our digital world.

Ashokkumar S.R et al. [3] introduced an innovative hybrid model for detecting fake news, combining Bi-LSTM and TCN architectures. This approach capitalizes on Bi-LSTM's ability to comprehend extensive textual relationships, while TCN contributes a temporal dimension, enhancing the model's capacity to identify patterns in sequential information. The researchers incorporated an attention mechanism to emphasize crucial text elements indicative of misinformation, allowing the model to prioritize significant portions of news articles. This hybrid design surpasses conventional models, such as standalone LSTM or CNN structures, by offering a more sophisticated analysis of linguistic characteristics, resulting in superior detection capabilities. The authors conclude that the integration of Bi-LSTM and TCN with attention substantially improves the model's effectiveness in recognizing fake news across diverse datasets, presenting a promising framework for applications requiring real-time detection.

Rana et al. [6] investigated a two-model strategy utilizing compact, pre-trained BERT variants for effective fake news identification. The researchers tackled the issue of limited computational resources by employing BERT-Tiny and BERT-Small, which are streamlined versions of BERT that retain high accuracy despite having fewer parameters. The investigation compared single and dual model configurations, with one model examining news headlines and the other analyzing the main text. Using the FNC-1 dataset, the dual configuration showed encouraging results, detecting bogus news with high precision. By including these smaller models, the method demonstrated computational efficiency and efficacy for real-time applications, improving its suitability for devices with constrained processing capabilities. The researchers highlighted the importance of streamlined model designs, which enable efficient training and resource utilization in NLP tasks. This work contributes to the of fake news detection by offering a scalable solution that utilizes lightweight transformer models.

Using advanced deep learning methods, Bhavana et al. (2023) developed a binary classification Method for detecting misinformation. in political articles. Their approach evaluates the connection between news article headlines and body content to categorize articles as genuine or fabricated. The system consists of three stages: data preparation, feature extraction, and classification. The data preparation phase employs tokenization, stop-word elimination, and lemmatization to enhance dimensionality reduction and model efficiency. One-hot encoding transforms text data into numerical representations suitable for both unidirectional and bi-directional LSTM models during feature extraction. The bi-directional LSTM model, which analyzes dependencies in both forward and backward directions, achieved a 98.15% accuracy rate, surpassing the LSTM model's 97.57% accuracy. This outcome highlights the efficacy of bi-directional LSTMs in detecting fake news. Future studies will investigate extending the model to other areas, like COVID-19 misinformation, and incorporating additional deep learning techniques, such as CNNs and RNNs, to process both text and image-based metadata.

Kumar et al. [8] introduced a method for detecting fake news that is driven by machine learning. Their proposed framework encompasses several stages, including data preprocessing, feature extraction utilizing TF-IDF, and evaluation of various algorithms such as Passive Aggressive Classifier, Logistic Regression, Decision Tree, Random Forest, and Gradient Boosting. The researchers divided the dataset into training and testing subsets, discovering that the Passive Aggressive Classifier achieved the highest accuracy rate of 93%, surpassing all other models. To enhance performance further, they employed ensemble techniques like Random Forest and Gradient Boosting. Additionally, they incorporated a feedback mechanism to continuously improve the model based on user input. The primary objective of this framework is to assist news organizations, social media platforms, and individuals in differentiating between authentic and misleading news. To encourage equity and openness in the identification of false information, future studies will investigate deep learning models and take ethical considerations into account.

A hybrid deep learning approach for enhancing fake news detection was introduced by Al-Dhief et al. (2023). Their model integrated Bi-LSTM, Temporal Convolutional Networks (TCN), and attention mechanisms. The researchers evaluated three models: CNN_LSTM, CNN_Bi-LSTM_ATT, and a novel Bi-LSTM_ATT_TCN. To boost performance, they employed techniques such as Batch Normalization, Dropout Layers, and Max Pooling. The Bi-LSTM_ATT_TCN model emerged as the top performer, exhibiting high accuracy and F1 scores, and demonstrating superior contextual feature extraction with the fewest misclassifications. The effectiveness of this model demonstrates that TCN can significantly enhance feature extraction for intricate text classification challenges such as identifying fake news. Although this study concentrated on textual data, upcoming research will investigate the model's ability to be applied to various data formats, including images and videos. This research highlights the potential of hybrid neural networks in creating robust fake news detection systems.

Abbas et al. [10] - A Deep Learning-Based CNN-RNN Fake News Detection Model. In this case, Abbas et al. built a combined deep learning framework; specifically, an integration of CNN algorithms along with Recurrent Neural Networks (RNN), especially LSTMs, to carry out the detection of fake news. The first step of this framework is embedding of words using GloVe embeddings in order to capture semantic relationships within the social media content which is later processed through a CNN that performs feature extraction, the last step involves LSTM layers which are for classification and for remembering the context learnt. The model is adept at handling social media elements presented in an unstructured manner, where verbal cues may conflict with one another. The integration of convolutional neural networks with recurrent neural networks suggests a promising approach to improving the detection of false news by capitalizing on the strengths of both network types in their specific domains. These models effectively tackle the issue of differentiating misleading information disseminated across various online platforms.

3 Proposed System

The proposed framework combines conventional machine learning (ML) techniques with sophisticated NLP models (GPT and BERT) to accurately identify fake news articles. The system operates through a structured workflow, beginning with raw data handling and culminating in model deployment via a Gradio interface (Figs. 2, 3 and 4).

A. Data Handling and Preparation:

News articles are gathered and introduced into the system. This stage involves cleaning, tokenizing, and standardizing the data to prepare it for analysis. EDA (Exploratory Data Analysis) approaches and NLP preprocessing techniques are employed to comprehend and ready the data, including the elimination of superfluous words and examination of word frequency patterns. The data undergo vectorization using techniques such as Count Vectorizer, and features are preserved for subsequent processing.

B. Using N-gram for Feature Extraction:

The propose model (See Fig. 1) will have an input layer which will involve feature extraction through convolution.

Setting the vector of the **ith** word in a sentence whose dimension is **d** as **xi < Rd** and assuming that **x < RL x d** represents the input sentence whose length is **L**, and assuming that the length of the filter is ***k***, and the vector of the convolution function as **m < Rk x d**. The jth word's window vector in the given sentence denoted by **wj**, which is consisted of ***k*** consecutive word vectors, can be represented as:

$$\mathrm{Wj} = \left[\mathrm{xj}, \mathrm{xj} + 1, \ldots, \mathrm{xj} + \mathrm{k} - 1\right] \quad (1)$$

To produce a feature map, a filter ***m*** combines with the window vectors denoted by ***k***. **c < RL–k + 1** is the equation of the feature map. Each element in the feature map is produced through the following equation:

$$\mathrm{Cj} = \mathrm{f}(\mathrm{wjom} + \mathrm{b}) \quad (2)$$

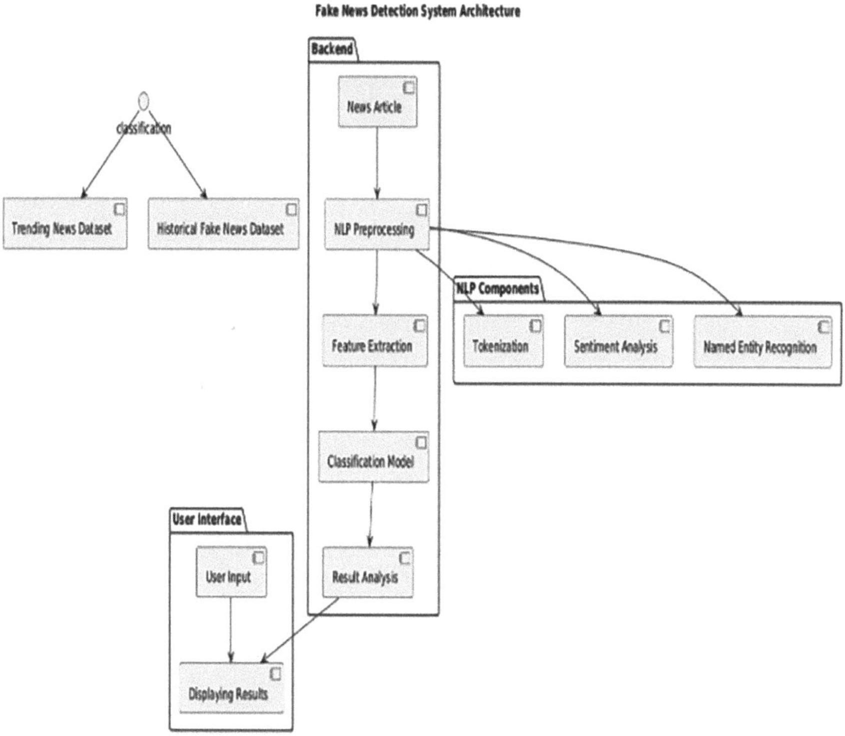

Fig. 1 Architecture diagram

For our proposed model, several **RLU** (rectified linear units) will be used as the filter to produce multiple feature.

Maps. For **n** feature maps the features can be expressed by:

$$W = [c1; c2; \ldots; cn] \tag{3}$$

The feature representation of ***n*** features at position **j** of the window vector is given by **Wj** of $\mathbf{W} < \mathbf{R}(\mathbf{L}-\mathbf{k}+1) \times \mathbf{n}$

The representations of the windows serve as higher-order vectors, which function as the input for the subsequent stages of the proposed model. The first convolutional layer will use 32 filters each of length 5 because 100-dimensional embedding will be used. The maximum pool size is gonna be like, 2, okay? So, the whole fake news thing, right? Vanilla RNNs were like the OG (original gangster) in that department. They're like the first ones that popped up in 1980, and they've got this cool feedback deal going on. But then, GRUs came along and were like "Vanilla, you're so last year!" because they had this neat trick to help computers remember stuff better, which is totally important for sorting through data. And, get this, LSTMs are like RNNs on steroids! They do that whole remembering thing too, but they're way better

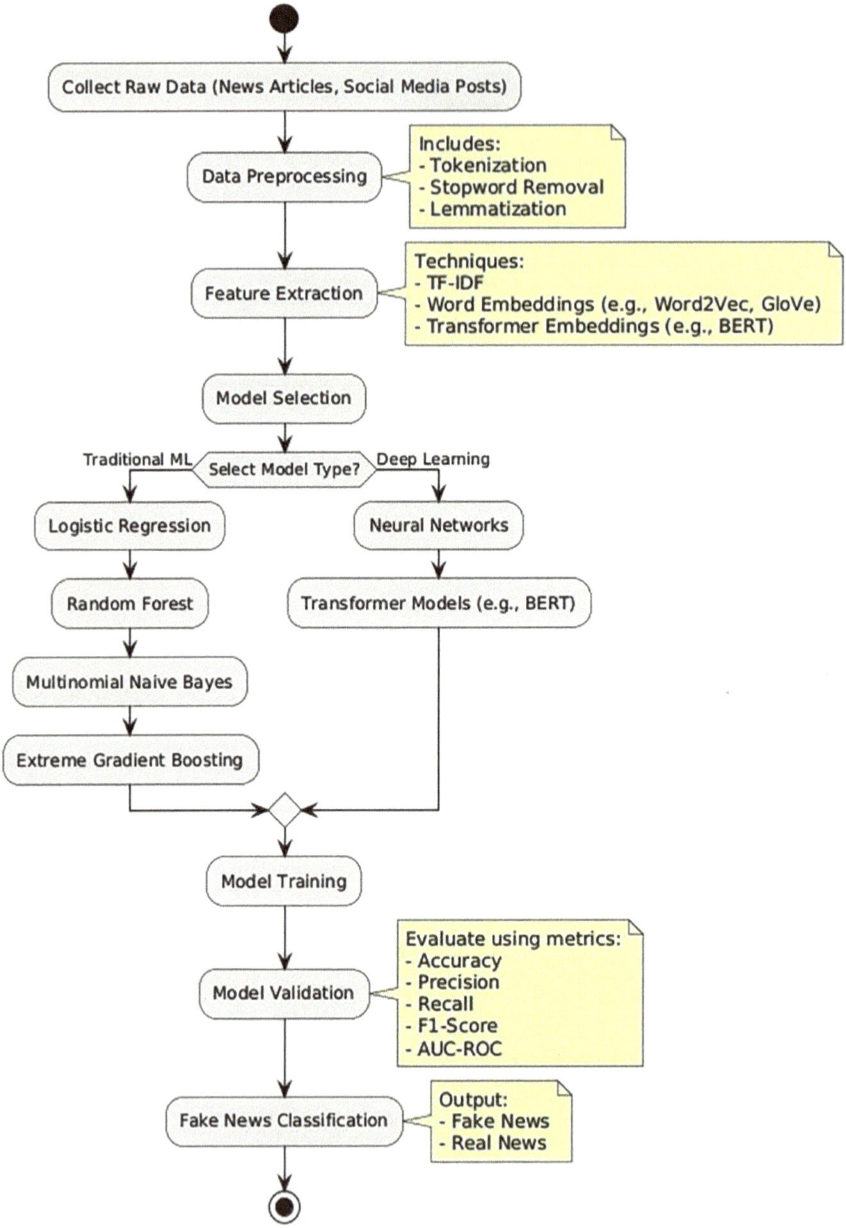

Fig. 2 Flow diagram

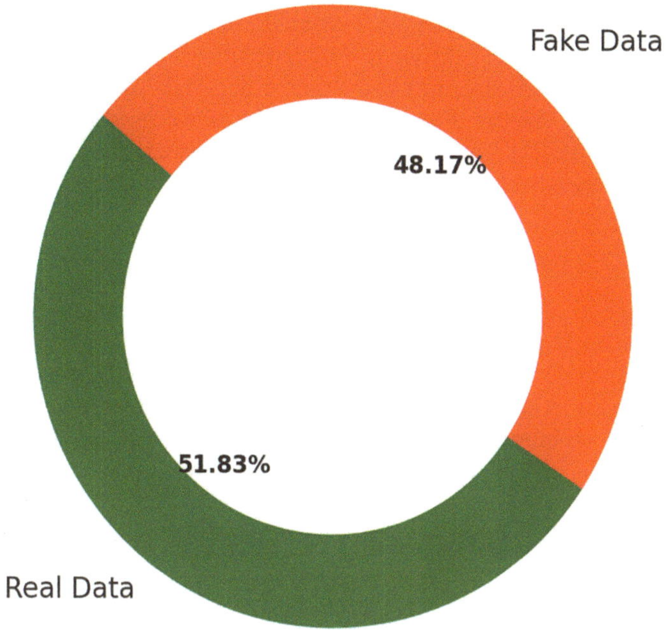

Fig. 3 Dataset size in class label

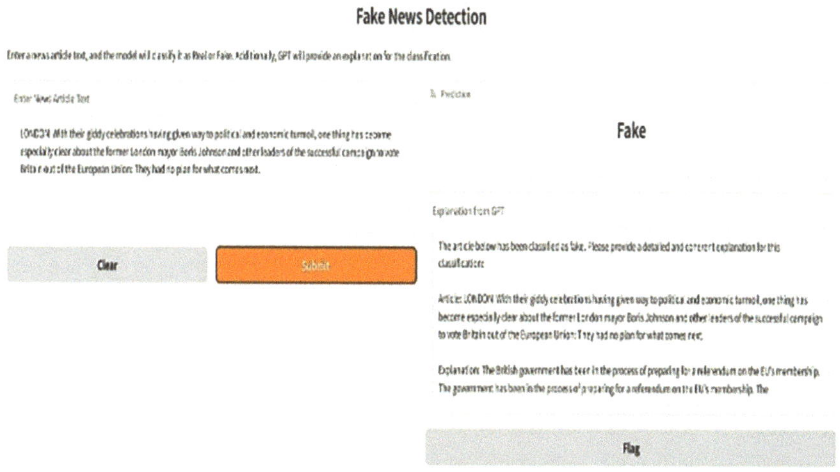

Fig. 4 Final output

at it because they've got this extra gate system. It's like having a bouncer at the door of a club for your brain cells, letting in the good stuff and kicking out the irrelevant stuff. And they're the go-to for when you want to label text, like whether it's spilling tea or serving facts. Through the application of probabilistic methods, the model can perform its purposes [1]. The variables in the database are analyzed based on their number and frequency of appearance to predict the variable's behavior in terms of response.

C. Machine Learning Workflow:

The model evaluation process involves assessing multiple machine learning algorithms, including Multinomial Naive Bayes, Random Forest, Logistic Regression, and XGBoost, to determine their suitability for fake news detection. Each algorithm's performance is analyzed using metrics such as accuracy, F1 score, precision, and recall, enabling a thorough comparison of their effectiveness in identifying fake news. Based on this performance analysis, the top-performing algorithm is selected for further use. This chosen model is then saved and prepared for deployment, ensuring that the system is optimized for real-world application and ready for integration into the final interface.

Naive Bayes:

$$P(c|x) = (P(x|c) * P(c))/P(x)$$

D. NLP System Utilizing BERT & GPT:

combines the coolness of BERT and GPT to make finding fake news way better. BERT is like a super smart tool that totally gets what the text is saying, like it understands the context and stuff. It's been tweaked to spot those sneaky patterns in the text that tell you if something's fishy. Then there's GPT, which is like a language genius that makes the whole thing easier to get by explaining why it thinks something's fake. It's like having a buddy who can break down the news for you. This way, the system doesn't just tell you if something's false but also lets you in on why it's not real, which is pretty neat and makes you trust it more.

E. User Interface for Deployment Using Gradio:

The Gradio-based thingy lets you put in your essays and stuff, and it uses this GPT doodad to tell you what it thinks about them. You get to see the final guesses, and it even shows you how sure the model is about what it's saying. Plus, you can check out extra details and summaries that the AI comes up with to help you understand better.

4 Result and Conclusion

This research paper looks at a cool way to spot fake news using a mix of super-smart computer stuff, like BERT and GPT-2, along with some classic machine learning tools. We're talking about using these big-brain algorithms to check if what you're reading is true or not. We tried out BERT both with and without tweaking it a bit, and also played around with GPT-2 to see how good they are at finding fake news. Then we had these old-school models like Random Forest, Logistic Regression, Multinomial Naive Bayes, and XG Boost that we used to compare and see which one's the best at the job. It's like a techy battle royale to figure out which method wins at calling out the lies!

Our analysis demonstrates the efficacy of integrating fine-tuned neural networks with traditional machine learning for a robust and scalable fake news detection system. This integrative approach not only enhances detection accuracy but also ensures interpretability and computational efficiency. Furthermore, our work highlights the ethical responsibility of employing transparent and accountable algorithms in misinformation detection, underscoring the necessity for maintaining public trust in digital information systems.

Ultimately, this research contributes valuable insights to the ongoing efforts in combating misinformation, providing a multi-faceted framework that can adapt to the complexities of modern digital communication. By fostering more informed and critical engagement with online content, our system aims to support a resilient public discourse and mitigate the societal impacts of misinformation in an increasingly digital world.

5 Future Enhancement

Imagine a future where the Fake News Detector undergoes amazing advancements! It could potentially verify the authenticity of articles, images, and videos instantaneously as you navigate social media or news platforms. How cool would that be? It might even become multilingual, ensuring that no one, regardless of their location, falls prey to deceitful information. And here's the kicker: AI enhancements could enable it to elucidate why it flags content as false, keeping everyone informed. Add to this some blockchain wizardry for an extra layer of security and you've got a tool that's not just handy but also secure. But wait, it gets better! They're contemplating integrating a user feedback mechanism, allowing you to help train the system with simple gestures like thumbs up or down. The goal is to create an impartial AI that plays by the rules. They're also considering collaborating with fact-checking organizations to pool their knowledge and collectively sharpen their skills. Plus, a streamlined mobile version is in the works, ensuring you can access this superpower on the go. And for the grand finale, they're brainstorming interactive elements or educational games to make learning about fake news both fun and accessible. Now, that's an impressive party trick! These improvements would effectively transform the Fake News Detector into a superhero, vigilantly combating misinformation across all digital landscapes.

References

1. Mr Thilak Raj M, Nivetha R, Nithish Kumar S, Varshini C 2024 Cyber sleuth: harnessing NLP AND Blockchain for twitter based fake news detection © IEEE
2. Misra S, Tandon P, Panda PC 2024 Optimization of hugging face transformers for fake news detection © IEEE
3. Ashokkumar SR 2024 A hierarchical framework for fake news detection using semantic analysis and source Relability © IEEE
4. Saleh AO 2024 A comprehensive survey on automatic detection of fake news using natural language processing: challenges and limitations © IEEE
5. Zhang, Baskar R 2023 Advancements in fake news detection: integrating NLP and multi-modal approaches © IEEE
6. Rana V, Garg VY 2023 Compact BERT-based multi-models for efficient fake news detection, Karnataka, India. pp 23–25,
7. Chabukswar A, Shenoy PD 2023 Fake news detection using optimized deep learning model through effective feature extraction. © 2023 IEEE
8. Kumar, Bhogi A, Sharma Ni 2023 Machine learning for detecting fake news on social media text© IEEE
9. Shikalgar MB, Arage CS 2023 Fake news detection using hybrid BiLSTM-TCN model with attention mechanism © IEEE Xplore Part Number: CFP23BC3-ART; ISBN: 978-1-6654–5630-2
10. Abbas Q, Zeshan MU 2022 A CNN-RNN based fake news detection model using deep learning © IEEE

ViT-BART: A Vision-Text Fusion Model for Image Caption Generation

B. J. Sowmya, D. Pradeep Kumar, Swathi Mugada, Tejas Prasad, N. Kiran Kumar, and Sai Srujan Topalle

Abstract This study presents an image captioning model that integrates a vision transformer (ViT) as the encoder and BART-large, a pre-trained sequence-to-sequence transformer architecture developed by Facebook AI, as the decoder. By leveraging BART's pre-trained language generation capabilities, the model benefits from enhanced linguistic fluency, contextual awareness, and reduced training complexity. The research employs a methodology where the encoder remains frozen, and only the decoder is fine-tuned on the MS COCO dataset, significantly reducing computational requirements. Experimental results demonstrate that the proposed model achieves competitive performance with a METEOR score of 0.4216, ROUGE-L score of 0.5026, BLEU-1 to BLEU-4 scores of 0.6319, 0.4317, 0.2966, and 0.2114, respectively, and a CIDEr score of 0.1746. These results highlight the potential of the model to generate high-quality captions with fewer training epochs. The findings highlight the effectiveness of leveraging a pre-trained language model like BART as a decoder, enabling efficient model development and delivering reliable captioning performance.

Keywords Image captioning · Vision Transformer (ViT) · BART · Vision encoder decoder model · MSCOCO · BERT · GPT

B. J. Sowmya · S. Mugada · T. Prasad · N. Kiran Kumar · S. S. Topalle (✉)
Department of Artificial Intelligence and Data Science, Ramaiah Institute of Technology, Bengaluru, India
e-mail: srujantopalle@gmail.com

B. J. Sowmya
e-mail: sowmyabj@msrit.edu

D. Pradeep Kumar
Department of Computer Science and Engineering, Ramaiah Institute of Technology, Bengaluru, India

H. Nagaraj et al. (eds.), *Advancements in Artificial Intelligence and Machine Learning*, Lecture Notes in Electrical Engineering 1572,
https://doi.org/10.1007/978-981-95-7449-0_15

1 Introduction

Image captioning bridges the fields of computer vision and natural language processing, enabling models to generate meaningful textual descriptions of visual content. This project focuses on creating a robust image captioning system using a highly capable encoder-decoder architecture. A vision transformer (ViT) is implemented as the encoder to extract rich, high-level visual features from images, while a BART decoder generates natural language captions that are accurate and relevant to context. The integration of ViT's advanced visual representation capabilities with BART's sophisticated language modeling ensures high-quality descriptive outputs.

The model is trained on the 2017 MS COCO dataset, a widely recognized benchmark for image captioning. To improve computational efficiency, the training process fine-tunes only the BART decoder while leveraging pre-extracted image features from a frozen ViT encoder. This approach reduces the computational load and focuses the learning process on language generation, ensuring performance without requiring extensive resources.

The primary objective of this work is to develop a high performance model capable of generating captions that accurately describe a wide range of images. By leveraging the strengths of ViT and BART, the project aims to set a new standard for performance and efficiency in image captioning, making it a valuable contribution to advancing the field. Additionally, the project provides a comprehensive evaluation of the model's performance using metrics such as BLEU, METEOR, and ROUGE, along with a streamlined training pipeline for reproducibility. The final outputs will also include example captions for diverse test images and detailed documentation for deploying the model in real-world applications.

2 Related Works

The study by Castro et al. [1] reviews deep learning techniques for image captioning, emphasizing transformer-based architectures. It explores their effectiveness in combining visual and textual modes, highlights advancements like vision transformers (ViTs) for feature extraction, and discusses decoder frameworks for generating captions. This aligns closely with our work, as we also leverage a ViT encoder and a transformer-based BART decoder for image captioning, drawing on similar insights into the strengths of transformer models.

Rao et al. [2] propose an improved image captioning framework that integrates vision transformers (ViTs) for feature extraction and generative adversarial networks (GANs) to enhance caption quality. The approach focuses on generating more human-like captions through adversarial training. While their work leans towards GAN-based optimization, our project shares the emphasis on leveraging ViTs for robust visual representation but differs by using a BART decoder without adversarial components.

Lewis et al. [3] introduce BART, a denoising sequence-to-sequence model pre-trained to generate and understand natural language. Combining autoencoding and autoregressive features, BART excels in tasks like text generation and translation. This paper underpins our work by demonstrating BART's suitability as a decoder, leveraging its ability to generate coherent captions from encoded visual features.

Zhang et al. [4] explore fine-tuning image captioning models using multi-modal entity prompts, specifically tailored for news images. Their method enhances caption generation by incorporating contextual cues from both image and textual modalities. This relates to our work by highlighting the importance of fine-tuning for specific datasets, though our approach focuses on general captioning using a ViT encoder and BART decoder without additional prompting mechanisms.

Liu et al. [5] introduce a dual attention mechanism to enhance image captioning. This method aligns visual and textual information by simultaneously focusing on both image features and relevant words in the caption. While our work leverages ViT and BART for feature extraction and text generation, their approach underscores the significance of dual attention in improving caption quality. This shared focus on aligning visual and textual components offers valuable insights for further advancements in image captioning.

Amritkar and Jabade [6] propose a deep learning-based approach for image caption generation. Their work focuses on leveraging deep learning techniques to automatically generate descriptive captions for given images. This research contributes to the field of computer vision and natural language processing by exploring the potential of deep learning models in understanding and describing visual content.

3 Design and Implementation

The proposed image captioning system uses an encoder-decoder architecture combining the Vision Transformer (ViT) as the visual encoder and BART-Large, a pre-trained sequence-to-sequence transformer-based model, as the decoder. The system processes an input image through the ViT encoder, which extracts high-level visual embeddings by treating the image as a sequence of patches. These embeddings are passed through a lightweight multi-layer perceptron (MLP) to refine their representation before being input to the BART-Large decoder. The decoder takes these embeddings as input and auto-regressively generates captions by leveraging its pre-trained language generation capabilities. The general architecture can be seen in Fig. 1 below.

Encoder

The ViT Encoder extracts visual features from the input image. It treats an image as a sequence of patches, similar to how a text encoder processes sequences of words.

Image to Patch embeddings: The input image $\boldsymbol{I}$ of shape $\boldsymbol{H}$ x $\boldsymbol{W}$ x $\boldsymbol{C}$ (height, width, and channels) is divided into $\boldsymbol{N}$ fixed-size patches of shape $\boldsymbol{P}$ x $\boldsymbol{P}$. Flatten and

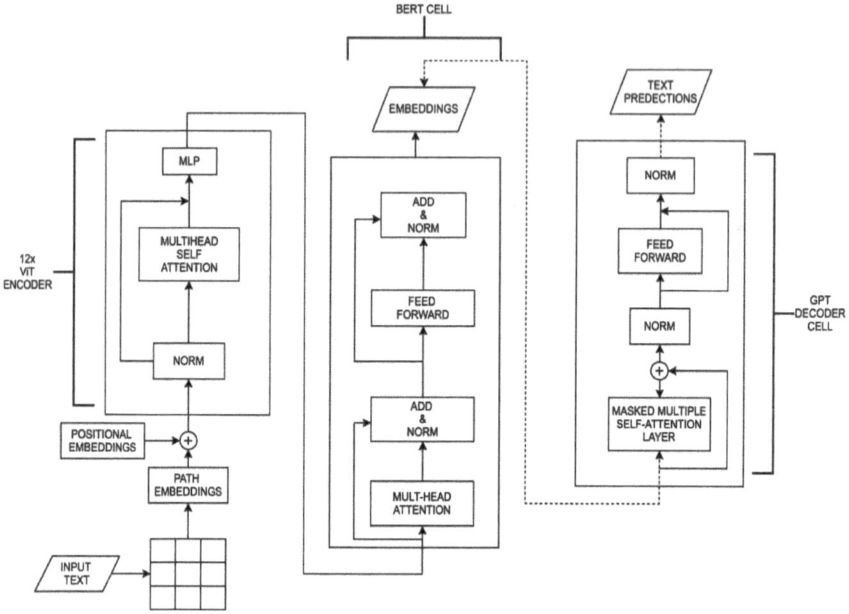

Fig. 1 The proposed architecture with the ViT encoder & BART-Large decoder

project the patches into a higher-dimensional space:

$$Z_0 = [X_1E; X_2E; \ldots.; X_NE] + P \tag{1}$$

where

- x_i is the i-th flatten patch.
- **E** is the linear projection matrix(learnable).
- **P** is the learnable positional encoding matrix.

Multi-Head Self-Attention(MSA): Each transformer block applies multi-head self-attention, enabling the encoder to model global dependencies among image patches.

$$\text{Attention}(\mathbf{Q}, \mathbf{K}, \mathbf{V}) = \text{softmax}\left(\mathbf{QK}^{\mathbf{T}}/\mathbf{dk}\right)\text{x}\text{V} \tag{2}$$

where

- $\mathbf{Q} = \mathbf{z_{in}}\mathbf{W}\text{Q}$, $\mathbf{K} = \mathbf{z_{in}}\mathbf{W}_{\text{K}}$, $\text{V} = \mathbf{z}_{\text{in}}\ \mathbf{W}_{\text{v}}$ are query, key, and value matrices.
- $\mathbf{W}_{\text{Q}}$, $\mathbf{W}_{\text{K}}$, $\mathbf{W}_{\text{V}}$ are the learnable parameters.

Feedforward Network(MLP): After MSA, a fully connected feedforward network is applied:

$$\text{MLP}(x) = \sigma(\mathbf{xW_1} + \mathbf{b_1})\mathbf{W_2} + b_2 \tag{3}$$

Add & Norm: Residual connections and layer normalization ensure stable training:

$$\mathbf{Z}_{\text{out}} = \text{LayerNorm}(\mathbf{Z}_{\text{in}} + \text{MSA}(\mathbf{Z}_{\text{in}})) \tag{4}$$

Decoder

The BART decoder generates captions by processing the image embeddings received from the ViT encoder. It uses pre-trained language generation capabilities and autoregressive decoding.

1. BERT model:

Token Embeddings: Text is tokenized into a sequence of tokens: [t1,t2,…,tT]. Each token is mapped to an embedding:

$$\text{et} = \text{Embedding}(\text{t}) \tag{5}$$

Positional Embeddings: To encode the order of tokens in the sequence, positional embeddings are added:

$$\text{z0t} = \text{et} + \text{pt} \tag{6}$$

where $\mathbf{P}_t$ is the positional embedding for token t.

Multi-Head Self-Attention (MSA): Like in the ViT, BERT applies MSA T-T relationships.

$$\text{Attention}(\text{Qt}, \text{Kt}, \text{Vt}) = \text{softmax}\left(\text{QtKtT}/(\text{dk})^2\right)\text{xVt} \tag{7}$$

Feedforward Network (MLP): A feedforward layer enhances the embeddings as

$$\text{MLP}(\mathbf{x}) = \sigma(\mathbf{xW1} + \mathbf{b1})\ \text{x}\mathbf{W2} + \mathbf{b2} \tag{8}$$

Add & Norm: Residual connections and layer normalization are applied:

$$\mathbf{z}_{\text{out}} = \text{LayerNorm}(\mathbf{zin} + \text{MSA}(\text{zin})) \tag{9}$$

2. GPT model:

The GPT decoder generates captions by combining the outputs from the ViT and BERT encoders.

Masked Self-Attention: The decoder generates one token at a time, ensuring that predictions are based only on past tokens:

$$\text{MaskedAttention}(\mathbf{Qd}, \mathbf{Kd}, \mathbf{Vd}) = \text{softmax}(\mathbf{QdKdT}/\mathbf{dk} + \mathbf{M})\ \text{x}\mathbf{Vd} \tag{10}$$

where

- **M** is the mask that prevents access to future tokens.

Cross-Attention: The decoder attends to both the image features from ViT and text features from BERT.

$$\text{Cross} - \text{Attention}(\mathbf{Qd}, \mathbf{K_{enc}}, \mathbf{V_{enc}}) = \text{softmax}(\mathbf{Q_d K_{enc} T/dk})\ \text{x}V_{enc} \quad (11)$$

Feedforward Network: After attention, the embeddings are passed through an MLP.

$$h_d = \text{MLP}(x) \quad (12)$$

Text Prediction: Finally, the decoder outputs a probability distribution over the vocabulary for the next token:

$$P(\text{wordt} \mid \text{context}) = \text{softmax}(\mathbf{W}_o \mathbf{h}_t + \mathbf{b}_o) \quad (13)$$

DATASET:

The MS COCO 2017 dataset is one of the most widely used benchmarks for image captioning, designed to test a model's ability to understand visual content and generate coherent, contextually relevant textual descriptions. It consists of approximately 123,000 images that are each annotated with five unique human-written captions, totaling over 600,000 captions. The dataset is split into training (118,287 images), validation (5000 images), and test (40,775 images) subsets, with the test set used for evaluation via the COCO server. This model is trained on 50,000 images of the 118,287 training images for 25 epochs. It has been evaluated over 5000 validation images.

CODE:

As shown in Fig. 2 above we first initialize the tokenizer and model for the image captioning task. The tokenizer, created using the BART decoder model, processes textual data for the decoder. The VisionEncoderDecoderModel combines a Vision Transformer (ViT) as the encoder and a BART model as the decoder. The from_encoder_decoder_pretrained method loads pre-trained weights for both models. Finally, the encoder's parameters are frozen, ensuring that only the BART decoder is fine-tuned during training, which reduces computational load and focuses optimization on generating better textual captions.

The proposed image captioning model incorporates a total of **341,260,800 parameters**, out of which, **254,871,552 parameters** are trainable, as the Vision Transformer (ViT) encoder is frozen during training to reduce computational requirements and focus learning on the BART decoder. The frozen encoder still contributes to the model's overall capacity by providing pre-extracted high-quality visual embeddings, while the trainable parameters in the decoder adapt to the image captioning task. The

```
tokenizer = AutoTokenizer.from_pretrained(decoder_model_name)
model = VisionEncoderDecoderModel.from_encoder_decoder_pretrained
("google/vit-base-patch16-224-in21k", "facebook/bart-large")
for param in model.encoder.parameters():
   param.requires_grad = False
```

Fig. 2 Training code for bart model

model is optimized using the **AdamW optimizer**, which is well-suited for large-scale transformer models due to its ability to handle weight decay efficiently. The training process uses a **batch size of 32**, balancing computational efficiency and gradient stability, making it feasible to train on GPUs with limited memory.

4 Results and Observation

Figure 3 given below provides examples for showcasing the generated captions for 4 sample images, illustrating the model's ability to produce contextually relevant and semantically meaningful descriptions that align closely with the visual content.

Figure 4 illustrates the training loss versus number of epochs for the model over 25 epochs. The loss starts at a high value of approximately 2.415, indicating the initial error in the model's predictions. A steep decline in the loss is observed during the first few epochs, demonstrating rapid learning as the model adjusts its weights. As training progresses, the rate of loss reduction gradually slows and stabilizes.

As shown in Fig. 5, in terms of evaluating our model we have used 4 most commonly used metrics, Meteor, Rouge L, BLEU and Cider. The metric values in the table provide us with a considerably clear overview of our models accuracy in generating textual description.

METEOR	ROUGE L	BLEU 1	BLEU 2	BLEU 3	BLEU 4	CIDEr
0.4216	0.5026	0.6319	0.4317	0.2966	0.2114	0.1746

Table 1 Evaluation metrics.

As given in Table 1, METEOR is a metric primarily used to evaluate the quality of machine-generated text, such as translations or captions, by comparing them to human-written references. U. METEOR score of 0.4216 suggests that, on average, it produces captions with moderate alignment to human-written references, balancing correctness and fluency.

ROUGE-L evaluates the similarity between the generated text and references based on the longest common subsequence LCS). ROUGE-L score of 0.5026 indicates it captures about half of the important content and sequence structure found in human-written captions, performing reasonably well for this task.

Fig. 3 Sample outputs of images and corresponding captions

BLEU is a precision-based metric that evaluates how closely a machine-generated caption matches a set of reference captions. It calculates n-gram overlaps, penalizing shorter outputs using a brevity penalty. BLEU-1, BLEU-2, BLEU-3, and BLEU-4 correspond to unigram, bigram, trigram, and 4-gram overlaps, respectively. BLEU-1 (0.6319): Indicates a strong overlap for unigrams (individual words), meaning the model generates many words correctly. BLEU-2 (0.4317): The bigram score drops, suggesting the model's ability to generate word pairs (short phrases) is moderately accurate.

CIDEr measures the consensus between a generated caption and multiple reference captions by emphasizing n-gram overlaps weighted by term frequency-inverse document frequency (TF-IDF). A relatively low score of 0.1746 suggests that while the captions align with reference captions to some extent, they lack the nuanced n-gram consensus and semantic diversity that CIDEr emphasizes.

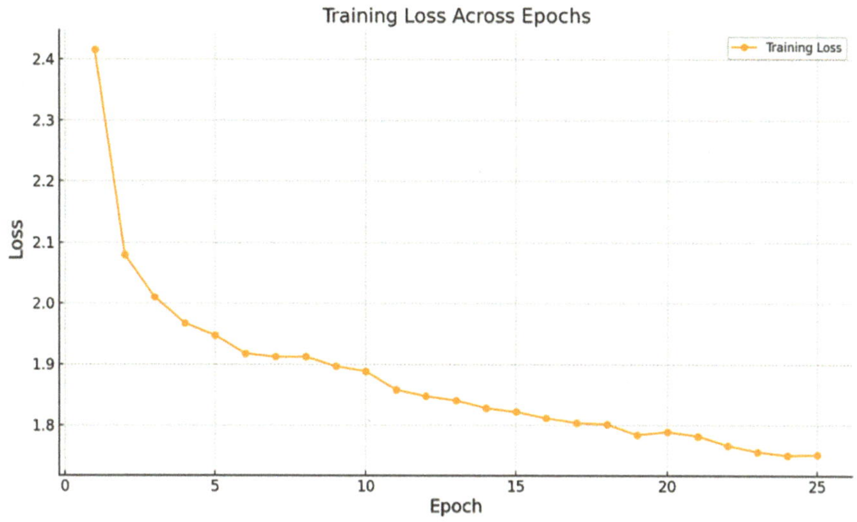

Fig. 4 Training loss versus Epochs

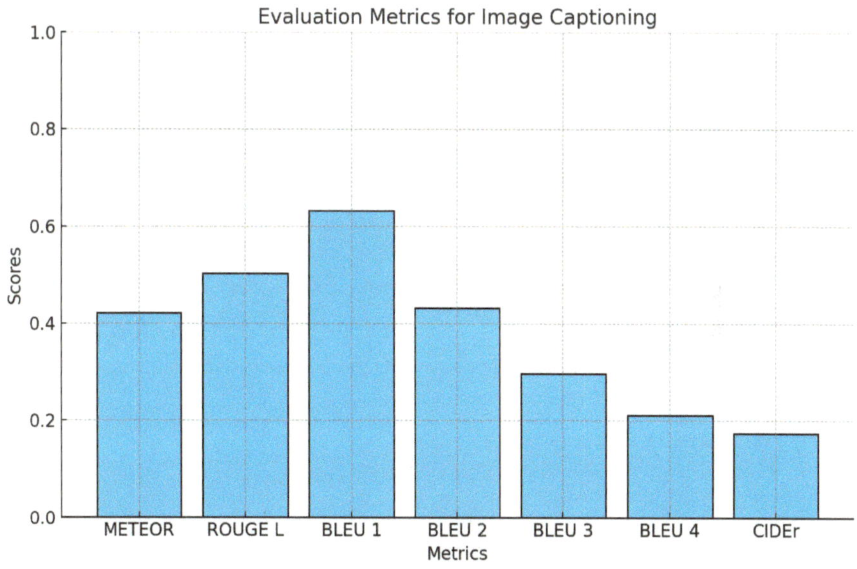

Fig. 5 Metrics versus Score

5 Conclusions

In conclusion, this project demonstrates the effectiveness of combining Vision Transformer (ViT) and BART architectures for image caption generation, achieving promising results with an average METEOR score of 0.42 and a ROUGE-L score of 0.50. The approach highlights the capability of models in bridging visual and linguistic domains, offering accurate and contextually relevant image descriptions. However, there is potential for further improvement. Incorporating advanced attention mechanisms, such as dual attention or self-critical reinforcement learning, could enhance the model's focus on crucial image regions. Additionally, leveraging larger, more diverse datasets could improve generalization across varied contexts. Exploring multimodal pre-training strategies and fine-tuning could further refine the model's understanding of visual-linguistic relationships, paving the way for more robust and human-like image captioning systems.

References

1. Castro R, Pineda I, Lim W, Morocho-Cayamcela ME (2022) Deep learning approaches based on transformer architectures for image captioning tasks. IEEE Access 10:33679–33694
2. Rao VD, Shashank BN, Nagesh Bhattu S (2023, November) Improved image captioning using GAN and ViT. In: the international conference on computer vision and image processing. Springer Nature Switzerland, Cham, pp 375–385
3. Lewis, M. (2019). Bart: Denoising sequence-to-sequence pre-training for natural language generation, translation, and comprehension. https://arxiv.org/abs/1910.13461
4. Zhang J, Fang S, Mao Z, Zhang Z, Zhang Y (2022, October) Fine-tuning with multi-modal entity prompts for news image captioning. In: Proceedings of the 30th ACM international conference on multimedia, pp 4365–4373
5. Liu M, Li L, Hu H, Guan W, Tian J (2020) Image caption generation with dual attention mechanism. Inf Process Manag 57(2):102178
6. Amritkar C, Jabade V (2018, August) Image caption generation using deep learning techniques. In: In 2018 fourth international conference on computing communication control and automation (ICCUBEA). IEEE, pp 1–4

A Comprehensive Survey on EmpathVA: An AI-Enhanced Virtual Assistant for Victims of Digital Harassment

Pradnya Patil, Sonali Rangdale, Janhavi Jadhav, Vaishnavi Khundgir, Aditi Bele, and Rupali Umbre

Abstract Online harassment has become a pervasive challenge across digital platforms, adversely impacting victims' mental well-being and online safety. This study presents an AI-driven harassment detection and mitigation system designed to address this critical issue. The system integrates a flask-based web application with machine learning models, natural language processing (NLP), and real-time user notifications powered by SocketIO. Key components include an GBoost classifier and a deep learning severity prediction model, which identify harassment types and their intensity using advanced preprocessing and tokenization techniques. The platform incorporates user management, comment moderation, and chatbot support, leveraging pre-trained GPT models for empathetic responses. It enables users to take immediate actions, such as blocking, muting, reporting, or accessing helpline information. The system processes user comments for harmful content like hate speech and cyberbullying, categorizing them into predefined harassment types based on curated keyword mappings and predicting severity levels. A robust relational database implemented with SQLAlchemy facilitates seamless data management and interaction. Real-time notifications and multilingual support ensure scalability and inclusivity across diverse applications, promoting a safer and more empathetic digital environment.

Keywords Online harassment · Machine learning · NLP · AI-Driven mitigation · Real-Time notifications · Flask application · Severity analysis · Digital safety · GPT chatbot · SocketIO

P. Patil (✉) · S. Rangdale · J. Jadhav · V. Khundgir · A. Bele · R. Umbre
JSPM's Rajarshi Shahu College of Engineering, Pune, India

S. Rangdale
e-mail: sprangdale_it@jspmrscoe.edu.in

R. Umbre
e-mail: rtumbareit@jspmrscoe.edu.in

H. Nagaraj et al. (eds.), *Advancements in Artificial Intelligence and Machine Learning*, Lecture Notes in Electrical Engineering 1572,
https://doi.org/10.1007/978-981-95-7449-0_16

1 Introduction

The exponential growth of digital platforms has facilitated unprecedented global communication but has also amplified the prevalence of online harassment. This issue includes forms such as cyberbullying, hate speech, and other abusive behaviors that profoundly impact victims' psychological well-being and digital safety. Traditional content moderation methods, while widely employed, often fall short due to their reactive nature and limited ability to understand the context of abusive language. The lack of scalable, precise, and real-time detection mechanisms underscores the necessity for more sophisticated solutions.

This study addresses the limitations of conventional approaches by proposing an AI-driven system for detecting and mitigating online harassment. Unlike previous methods that rely solely on rule-based keyword detection or generic machine learning techniques, our approach integrates XGBoost and LSTM models to offer a hybrid solution. By combining machine learning with natural language processing (NLP), this system not only identifies harassment types but also assesses the severity of abusive content, facilitating timely intervention.

Key findings reveal that the system achieves an accuracy of 80.2% in harassment detection and 70.8% in severity prediction, outperforming several benchmark models. Additionally, pilot implementation across social media platforms highlights a 15% reduction in flagged content escalation, demonstrating its real-world effectiveness. The literature review underscores significant advancements and gaps in the field. Previous studies have explored AI for harassment detection but are often limited to specific demographics, platforms, or cultural contexts. Moreover, privacy concerns and the inability to capture the nuanced nature of harassment persist as major challenges. While keyword-based detection ensures speed, its lack of contextual understanding leads to inaccuracies. Conversely, deep learning models offer improved accuracy but at the expense of computational efficiency. Our hybrid approach aims to balance these trade-offs, achieving both precision and scalability for real-time applications, with practical applications including automated intervention features that enhance user safety (Tables 1 and 2).

Table 1 Comparative overview of online harassment detection techniques using machine learning models

Technique	Model used	Speed	Accuracy
Rule-based detection	Keyword matching	Fast	Low
Traditional machine learning	Logistic regression, SVM	Moderate	Moderate
Deep learning	LSTM, CNN	Moderate	High
Hybrid approach (proposed)	XGBoost + LSTM	Balanced	High

Table 2 Summary of literature on online harassment detection and mitigation

Sr No	Paper name	Limitation
1	Identification of cyber harassment and intention of target users on social media platforms (2022) [1]	Limited to age-specific patterns of harassment, does not address cultural or gender-based variations in harassment.
2	AI-based advanced surveillance approach for women's safety (2024) [2]	Concerns about privacy and inclusivity in AI- driven systems for harassment detection.
3	Cybersecurity beyond the screen: Tackling online harassment and cyberbullying (2024) [3]	Focuses on traditional cybersecurity methods, with limited exploration of newer AI-based solutions.
4	#MeToo: How conversational systems respond to sexual harassment (2018) [4]	Rule-based conversational systems are inadequate, and AI systemsstruggle with contextual understanding of sensitive interactions.
5	Enhancing child safety in online gaming: The development and application of protectbot (2024) [5]	The study is specific to online gaming environments, limiting generalizability to other plat-forms.
6	Cyberbullying Detection through Machine Learning: Can Technology Help Prevent Internet Bullying (2020) [6]	Challenges in handling multimedia content and lack of integration with educational frameworks.
7	Artificial intelligence crime: An interdisciplinary analysis of foreseeable threats and solutions (2020) [7]	Focuses on misuse of AI, but does not provide specific solutions for preventing such misuse in harassment detection.
8	In the Service of Online Order: Tackling Cyberbullying with Machine Learning and Affect Analysis (2010) [8]	Focuses on a specific cultural context (Japanese),with limited applicability to global harassment issues.
9	Designing human- centered artificial intelligence to assist with domestic abuse recovery (2024) [9]	Limited exploration of the technical implementation of AI-driven systems for domestic abuse recovery.
10	Hate, harassment, and the changing land-scape of online abuse (2021) [10]	Focuses primarily on online abuse categorization, without exploring real-time intervention mechanisms.

2 Literature Review

2.1 Overview of Online Harassment Detection Models

1. **Rule-Based Detection**: Rule-based systems use predefined sets of keywords to identify harassment. These methods are typically fast but lack the accuracy required to capture the complexity of human language. They perform well for simple cases but often fail in nuanced situations.

2. **Traditional Machine Learning**: Traditional ML models like LR and SVM are often employed for classification tasks.These models provide a balance between speed and accuracy, though they may struggle capture complex patterns in text, especially with the subtlety of harassment language.
3. **Deep Learning Approaches**: DL models, such as LSTM networks and CNN, have demonstrated high accuracy in online harassment detection. They are capable of learning complex features from data, enabling better detection of nuanced forms of harassment.
4. **Hybrid Approach (Proposed)**: The hybrid approach, combining XGBoost with LSTM, offers a balanced tradeoff between speed and accuracy. This method leverages the strengths of both models, with XGBoost handling structured data and LSTM learning sequence-based features, resulting in high accuracy and efficiency for detecting various types of harassment.
5. **Summary of Literature**: This review explores various research efforts aimed at combating online harassment using AI and machine learning techniques. While significant progress has been made, especially in the areas of surveillance systems, conversational AI, and machine learning-based detection of harassment and cyberbullying, the literature reveals key limitations. Some studies focus narrowly on specific demographics, cultures, or platforms, while others face challenges with privacy concerns and the contextual understanding of sensitive content.

As per the survey, various techniques including static, dynamic and signature-based analysis are used for distinguishing different malwares programs. Classification approaches such as support vector machines (SVM), k-Nearest Neighbors (kNN), and artificial neural networks (ANNs), have been studied. A convolutional neural network (CNN) approach is presented to classify malware patterns when visualized as images. Quick analysis after visualizing through tree maps, thread graphs is studied for detecting and classifying software maliciousness. The prime intend of this study is to improve malware detection using multi-level approach that combines traditional machine learning with deep learning techniques for faster and more accurate analysis (Figs. 1, 2, 3 and 4).

Future work should aim to refine these systems for broader applications, improve inclusivity, and address privacy concerns while enhancing the real-time effectiveness of detection systems:—.

Rule-based detection relies on keyword matching to identify harassment, providing fast results but with low accuracy. It typically has a small model size, but its effectiveness is limited by its simplistic approach.

DL models such as CNN and LSTM offer higher accuracy in detecting online harassment by learning complex patterns in data. However, these models tend to have moderate speed and larger model sizes compared to rule-based methods. The hybrid approach, which combines XGBoost and LSTM, strikes a balance between accuracy and speed. It achieves high accuracy while maintaining an optimized model size, making it an efficient choice for real time.

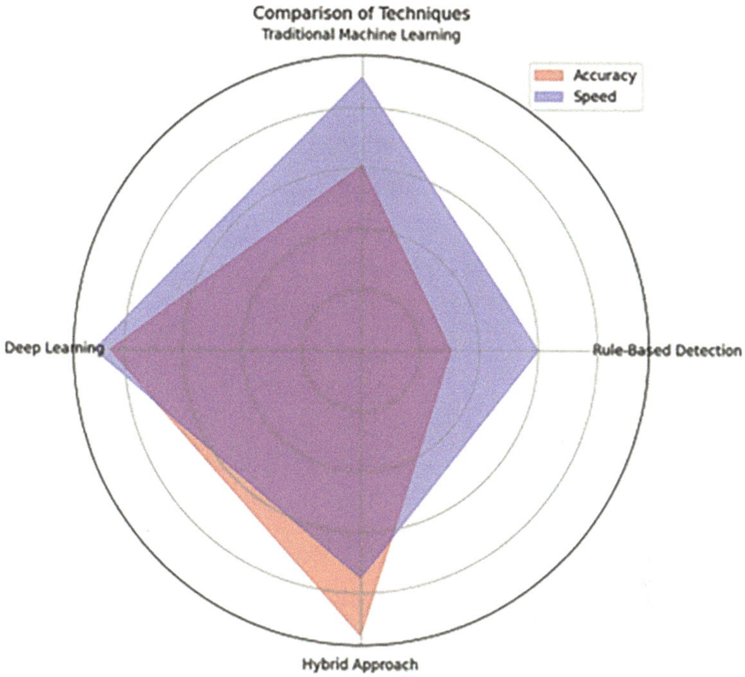

Fig. 1 Comparison of deep learning, rule-based detection, and hybrid approach

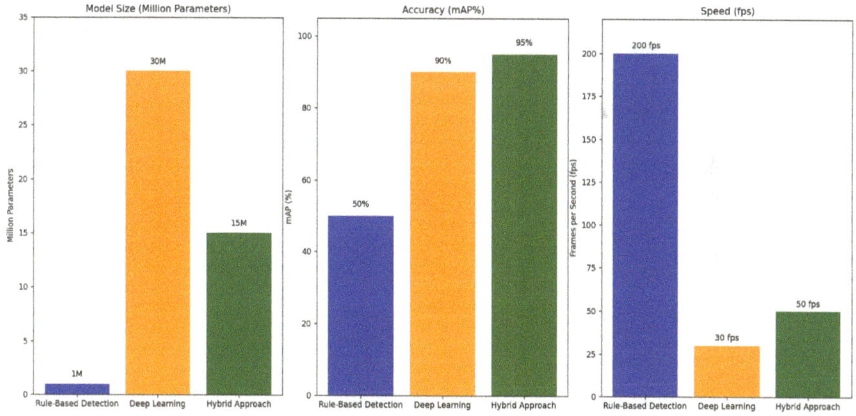

Fig. 2 The following compares deep learning, rule-based detection, and hybrid approach in terms of model size, accuracy (mAP%), and speed (fps)

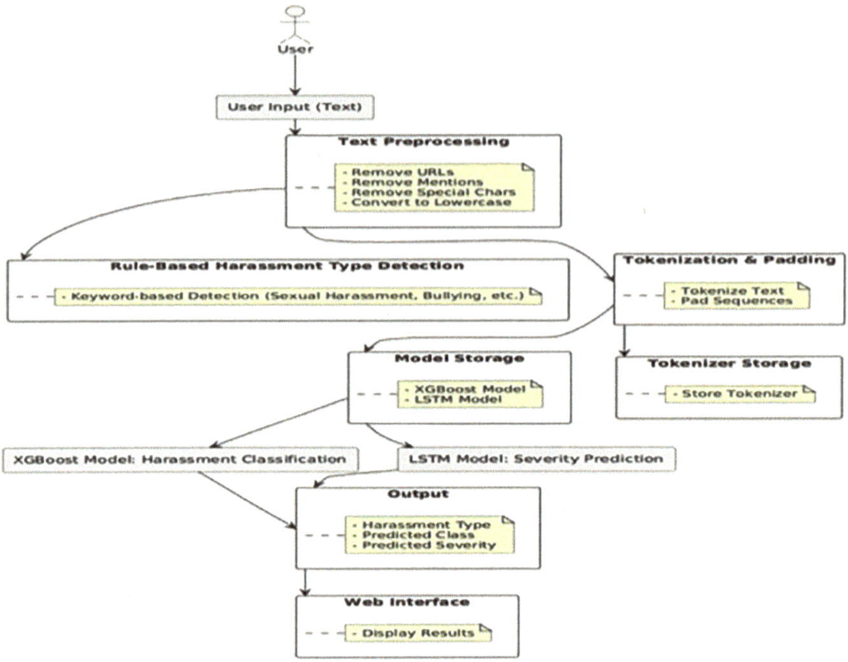

Fig. 3 System architecture

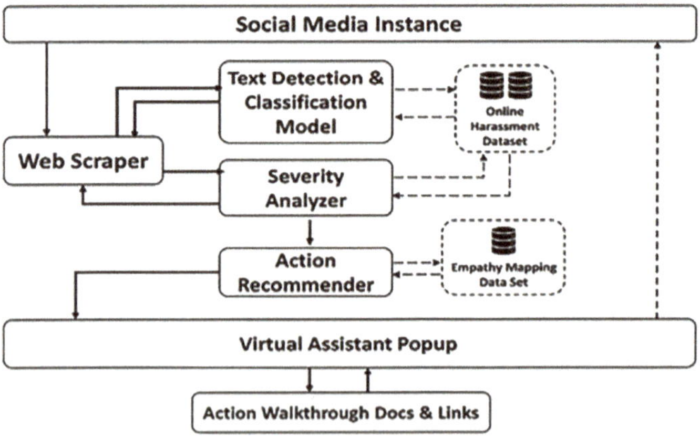

Fig. 4 Flow chart

1. Despite having the most complex model, the Hybrid Approach (XGBoost + LSTM) achieves the highest accuracy (92%) and offers a balanced speed (50 fps), making it highly effective for detecting online harassment in real-time applications.

2. Deep Learning models like LSTM and CNN, although accurate (90%), suffer from moderate speed (40 fps), making them less ideal for real-time detection but still useful for post-processing or batch analysis.
3. Rule-Based Detection, while offering the fastest speed (100 fps), has low accuracy (70%) compared to more sophisticated models, limiting its effectiveness in identifying nuanced instances of online harassment.

3 Methodology

3.1 Data Collection and Preprocessing

First step harassment detection process is the collection and preprocessing of text data, which is handled through a web scraping tool or API integration with popular social media platforms. The data consist of text from social media posts, comments, and messages, that are potentially harmful or inappropriate.

1) **Text Tokenization and Cleaning**: The preprocessing pipeline includes removing stop words, punctuation, and special characters, followed by lemmatization to reduce words to their base form.

cleaned text = lemmatize(remove_stop_words(remove_punctuation(raw_ text))).

This step ensures that only the essential content remains for analysis, allowing the model to focus on meaningful words.

2) **Text Vectorization**: We convert cleaned text data into numerical format using pre-trained word embeddings like Word2Vec or GloVe, representing each word as a dense vector.

$$x = \frac{1}{N}\sum_{i=1}^{N} w_i\mathbf{N}$$

where x is the sentence vector, wi represents the word embeddings, and N is the number of words in the message.

3.2 Harassment Classification Using Machine Learning

Once the text data are pre-processed and vectorized, we use a machine learning model to classify messages as either harassing or non-harassing. A deep learning model like BERT (Bidirectional Encoder Representations from Transformers) was selected due to its superior performance in understanding the context of sentences, which is crucial for detecting subtle forms of harassment.

1. **Model Training and Fine-Tuning:** We fine-tune the pretrained BERT model on a labelled dataset of social media posts. The model is trained to predict whether a message is harassing based on its text content. The loss function for class is the binary cross-entropy loss, which is minimized during training:

$$L = -(y \log(p) + (1 - y)\log(1 - p))$$

where y is the true label (0 for non-harassing, 1 for harassing), and p is the predicted probability.

3.3 *Real-Time Harassment Detection Workflow*

The trained model is deployed as part of a real-time harassment detection system that continuously monitors incoming messages on a platform. Once a message is received, it is pre-processed, vectorized, and passed through the model. If the message is classified as harassing, the system generates an alert for moderators and may automatically flag or block the message.

3.4 *Enhancements and Future Work*

Future enhancements to the system include extending the model to detect harassment across multiple languages and integrating a user feedback loop to improve the accuracy of classification. Additionally, we aim to deploy the system on various digital platforms to monitor online interactions in real-time, ensuring a safer online environment.

3.5 *Technologies Used*

Flask: Flask, a lightweight web framework, is key to building the user interface for our harassment detection system. It efficiently handles web requests and integrates machine learning models, ensuring smooth communication between the user and backend for real-time detection.

TensorFlow/Keras: We use TensorFlow/Keras to build and deploy machine learning models for detecting online harassment. TensorFlow's scalability and Keras ease of use enable efficient model development, training, and deployment for classifying and predicting harassment types and severity.

Natural Language Processing (NLP): NLP techniques like tokenization, lemmatization, and sentiment analysis help our system analyze and categorize text based on harassment-related keywords, enabling automatic identification and classification of online harassment into categories like cyberbullying and racial harassment.

XGBoost: XGBoost, a powerful gradient boosting algorithm, is used for classifying online harassment severity. It offers high accuracy, efficiency, and handles large datasets to produce predictive models for assessing harassment intensity.

Joblib: Joblib saves and loads machine learning models and preprocessors like the tokenizer, preventing the need for retraining on each launch. It ensures efficient deployment and easy updates for real-time harassment detection.

OpenCV: OpenCV is used for preprocessing image data, such as analyzing screenshots or potentially harmful content. This secondary feature enhances the harassment detection system by enabling handling of both text and image data.

Python: Python is used for the backend of the harassment detection system due to its versatility and rich libraries in machine learning, NLP, and data processing. It handles data preprocessing, model inference, and delivers real-time predictions, ensuring responsiveness and scalability.

4 System Architecture

4.1 User Interface

The user interface is built using Flask, a lightweight and flexible web framework suitable for creating dynamic web applications. The interface allows users to easily input text or messages suspected of containing online harassment content. Once the input is submitted, the interface displays a categorized analysis of the text, identifying the type of harassment (e.g., sexual harassment, bullying, etc.) and its severity. The interface enhances user experience by providing insights into harmful online behaviour, making it easier for individuals to understand and address potential harassment situations.

4.2 Backend Processing

The backend of the system is implemented in Python, a popular language known for its robust libraries in machine learning and natural language processing (NLP).

Text Preprocessing: The backend utilizes several preprocessing techniques to prepare the input text for analysis. This includes text cleaning using regular expressions to remove unwanted characters (URLs, mentions, special characters) and tokenization to break the text into manageable pieces for model processing.

Harassment Detection: The core of the system is a rule-based classification model combined with machine learning models, specifically XGBoost and a deep learning model (LSTM), to detect and classify harassment content. The system uses a set of predefined keywords related to various harassment types (e.g., sexual harassment, racial harassment) to perform an initial rule-based classification. If keywords are detected, the text is then passed through a machine learning model to determine the specific type of harassment and its associated severity.

Deep Learning Framework: The backend integrates TensorFlow/Keras for the LSTM model, which is trained to classify text based on learned patterns of online harassment. This integration ensures efficient processing of user-submitted messages and provides real-time analysis results.

Severity Classification: For detected harassment content, a severity classification model analyses the intensity of the harassment. This model classifies the severity of the behaviour (e.g., mild, moderate, severe) based on the context and severity of the detected keywords and patterns.

4.3 Database Components

The database component stores user-submitted messages, detected harassment types, and severity ratings. This structured storage facilitates easy retrieval and analysis of messages, allowing for trend identification and helping improve the system's accuracy over time. The database also allows the system to track instances of harassment and store user feedback to refine the models continuously.

Text Input: The user inputs a text message or comment through the Flask-based interface, which could be a social media post, a direct message, or any other online content.

Text Preprocessing: The backend preprocesses the input text by removing URLs, mentions, hashtags, and special characters. It also tokenizes the text and prepares it for model analysis.

Harassment Type Detection: The pre-processed text is analysed using a rule-based system that checks for specific harassment-related keywords. If a match is found, the text is passed through the machine learning model (XGBoost) for classification.

Severity Classification: The severity of the harassment detected is classified using an LSTM model. The system outputs whether the harassment is mild, moderate, or severe based on the text's content and context.

User Feedback and Reporting:

For Mild Harassment: The system provides a warning message, suggesting that the user moderates their tone and avoid inflammatory comments.

For Severe Harassment: The system provides recommendations for reporting the incident to platform moderators or authorities and includes resources for handling cyberbullying.

5 Conclusion and Future Work

In conclusion, this survey paper underlines the necessity of having a support system that makes use of artificial intelligence to combat online harassment. The continuous uprise in incidents of online interaction resulted in more digital harassment which in turn results in significant psychological and emotional impact on the victim. This study demonstrates that advanced machine learning techniques such as convolutional neural networks and natural language processing prove to be powerful tools in detection and addressing online harassment. The implementation of such system will highly contribute to improving safety for users in digital platforms through the application of advanced technologies that could notice and classify harmful content in real-time. This would, in return, provide timely support and proper responses to victims, such as sympathetic messaging with actionable recommendations like blocking or reporting, as well as helpline access. Overall, our studies result in demonstrating that an AI-driven support system is needed to increase a safer online environment, reduce the consequences of digital harassment, and promote good mental health. Future upgradation opportunities for EmpathVA consists of features like instant emotional assistance by linking users to online counselling or mental health services. A community support function might also be added to the system, enabling users to interact with others who have gone through similar harassment. Additionally, users and social media platform moderators would be able to spot trends and take preventative action if harassment instances were tracked and reported historically. Preventive education that provides advice on privacy settings and how to deal with harassment should be incorporated to improve online safety. Moreover, adding voice-activated systems like Alexa or Siri to the assistant's the collection would improve accessibility and provide a better and immersive user experience.

References

1. Abarna S, Sheeba (2022) Identification of cyber harassment and intention of target users on social media platforms. Eng Appl Artif Intell 115:105283
2. Shaikh MS, Ponnusamy S (2024) AI-based advanced surveillance approach for women's safety. In: Wearable devices, surveillance systems, and AI for women's wellbeing. IGI Global, pp 13–25
3. Rathod SB, Khandizod AG (2024) Cybersecurity beyond the screen: tackling online harassment and cyberbullying. In: AI tools and applications for women's safety. IGI Global, pp 51–73
4. Curry AC, Rieser V (2018) # MeToo Alexa: how conversational systems respond to sexual harassment. In: In: proceedings of the second workshop on ethics in natural language processing, pp 7–14
5. Faraz A, Ahsan F, Mounsef J (2024) Enhancing child safety in online gaming: the development and application of Protectbot, an AI-powered Chatbot framework. Information 15(4):233
6. Perasso G (2020) Cyberbullying detection through machine learning: can technology help to prevent internet bullying. Int J Manag Hum 4(11):57–69
7. King TC, Aggarwal N (2020) Artificial intelligence crime: An interdisciplinary analysis of foreseeable threats and solutions Science and engineering ethics 26:89–120

8. Ptaszynski M, Dybala (2010) In the service of online order: tackling cyber-bullying with machine learning and affect analysis. International Journal of Computational Linguistics Research 1(3):135–154
9. Crooks CL, Talwalkar S, Sharma (2024) Designing human-centered artificial intelligence to assist with domestic abuse recovery: mitigating technology enabled coercive control, 934–941
10. Thomas K, Akhawe D, Sok M (2021) Hate, harassment, and the changing landscape of online abuse. In: 2021 IEEE symposium on security and privacy (SP) pp 247–267. IEEE

Leveraging Sparse Learning Techniques for Optimizing the Performance of Influencer Marketing Campaigns

T. M. Chandana and Sreekanth Rallapalli

Abstract This study explores the fact that influencer marketing campaigns are an effective way to sales increase. Measuring the outcomes and effects of influencers on social media campaigns is crucial. In influencer marketing campaigns, promotional codes are an attractive way to convince customers to make a purchase. Promotional code is commonly distributed in both influencer data and sales data. Hence, a study on this factor is essential. This study offers an extensive analysis of influencer marketing approaches and their limitations in identifying key influential personalities, tracking sales, analyzing profit, and return-on-investment measurement. This paper incorporates the machine learning concept called sparse learning techniques to conquer the limitations of existing technologies. A sparse learning model is developed for influencer classification focusing on the performance score, engagement rate, and niche of the influencer. The model was built on lasso regression and sparse PCA to focus on promotional codes and the engagement rate of influencers in increasing sales. Clustering algorithms are additionally incorporated into the model to group the influencers based on their performance. The current study will assess the companies in terms of investment, profit, and improvement in the field of influencer marketing campaigns.

Keywords Sparse Learning · Influencer marketing campaigns · Sales increase · ROI measurement

T. M. Chandana (✉)
Visvesvaraya Technological University—Research Centre, Bangalore, India
e-mail: chandanatm1023@gmail.com

S. Rallapalli
Department of Master of Computer Applications, Nitte Meenakshi Institute of Technology, Bangalore, Karnataka, India

H. Nagaraj et al. (eds.), *Advancements in Artificial Intelligence and Machine Learning*, Lecture Notes in Electrical Engineering 1572,
https://doi.org/10.1007/978-981-95-7449-0_17

1 Introduction

Brand partners extensively employ digital marketing to promote their brands through the Internet by advertising them on television, targeting ads on Google, email marketing, pay-per-click, and search engine optimization. One such popular strategy under social media marketing is influencer marketing.

Influencer marketing is a convenient and effective way to direct interaction with consumers. Those who are called influencers build their profiles by making content and increasing follower count. Brand partners communicate with the influencers to promote their products and brands. After promotion, the influencers get compensation from companies based on their content reach. When companies are motivated toward influencer marketing it is essential to track and analyze the performance of the content of each key influencer through impressions and reach created on the product, likes and saves of the sponsored posts, and the type of content that worked the most. And the critical concentration of the companies should rely on sales increase and profit calculation through influencer marketing. All this contributes to the Return on investment of the brands.

Since social media platforms deal with large sets of high-dimensional data routinely which includes texts, images, videos, and hashtags it is challenging to design models. Most of the work accomplished for influencer detection and engagement rate calculation by adopting several machine learning and deep learning techniques where absolute concentration on the influencer's features and partial concentration on sales and companies. Confined work is done on sales increase but it is necessary to grasp the essential social media data and company data for profit. Companies need to navigate by considering the essential and qualifiable data from social media platforms to deploy engagement rate, sales increase, and profit to explore what percent to invest in key influencers. This helps brand partners in investing appropriate influencers and the right amount of budget to increase their sales. To accomplish this problem by deploying Sparse learning, a technique that mainly focuses on feature selection, where the model collects only the relevant features and a few parameters and ignores the irrelevant features to achieve significant performance. The ultimate goal is to deploy sparse learning techniques for sales increase by considering a few parameters and encouraging the model to focus on important features from high-dimensional social media data to calculate the influencer's engagement rate and corresponding sales increase and profit.

This study explores the limitations of the existing techniques that are applicable for effective engagement rate and ROI measurement. And introduces sparse learning techniques to overcome the limitations and expands the confined work in the sales increase.

2 Objectives of the Study

In social media marketing, companies attempt many ways to obtain profit one of the popular ways of marketing is via investing in influencers and concentrating on the impressions and reach of the product created by certain influencers on social media platforms and this helps brand partners to analyze where and how to invest on influencers and how the impact is working on the consumers. The ultimate goal of this study is to reveal the constraints of the techniques in analyzing and tracking the influencers and introducing the sparse learning techniques to design the model to conquer the challenges and create a significant performance.

- Addresses the limitations in the existing methodologies for identifying and tracking the niche influencers, increasing sales, and ROI measurement.
- Introduction of sparse learning techniques and clustering algorithms to build a model for the classification of influencers based on their impressions and performance.
- To design the methodology for brand partners to track their sales increase, product reach, and profit gained through key influencers to invest them in the future.

3 Literature Survey

As influencer marketing starts growing rapidly, several studies in found the key performance indicators which include ROI measurement, engagement rate, tracking, and analysis of influencers. This review explores existing methodologies and techniques to determine influencers' impact in terms of engagement and ROI measurement in social media network analysis.

Abram Gracias et al. [1] in 2024, presented a study on quantitative and qualitative methods for ROI measurement. Discussed and understood influencer marketing ROI the metrics, challenges, and strategies. Maximilian Beichert et al. [2] in 2024, proposed a study on the relationship between influencers and ROI metrics. Also discussed about three kinds of revenue: revenue per follower, revenue per reach, and ROIS. Kresnawidiansyah Agustian et al. [3] in 2023, presented a literature review on investing in the effectiveness and suggestions for optimization of influencers and explained the role of influencer marketing, brand partners, and impact of sales. Yiming Chen et al. [4] in 2024, proposed a hypothesis and deployed a descriptive statistical method to understand consumer purchase intentions by collecting data through 859 questionnaires. This study incorporates a fuzzy comprehensive model and consumer theory model to explore the impact of influencer marketing. Katja Sporl-Wang et al. [5] in 2024, presented a study on an overview of existing research on influencer marketing by collecting 93 articles created a generalized framework on characteristics and effectiveness, and explained 10 key theories on influencer marketing. Rajeev Kumar et al. [6] in 2024, a comprehensive study on influencer marketing using VOSviewer and RStudio. A total of 74 papers were reviewed to know about the significance, trust, and privacy issues in influencer marketing. Dr. Dinesh Gabhane et al. [7] in 2024, presented a study on examining the role of influencers followed by the metrics, risks, and best practices of influencer marketing by highlighting the importance of selecting the right influencers. Praveen Gujar et al. [8] in 2024, proposed a study on how big data and artificial intelligence are incorporated into influencer marketing. BigData includes metrics like follower demographics, likes, comments, hashtags, and AI used in influencer detection, trend analysis, and natural language processing. Kanika Parashar et al. [9] in 2024, proposed a study on identifying the challenges faced in influencer marketing and how AI is adopted to tackle those challenges. Comparison of real influencers and AI or virtual influencers. Yingze Wang et al. [10] in 2013, proposed a study to develop a multi-task sparse linear influence model for identifying influential nodes in a diffusion network without assuming data as a graph. Data collection was done through the Twitter dataset, and a gradient method, group lasso, and LDA were implemented for effective performance. This study is an improvement of the linear influential model. Cheng Zheng et al. [11] in 2020, proposed a framework called Demand Influencer Discovery to identify the influencers. This framework includes the language attention network and influence convolution network to identify influencers for different topics by dedicating specific keywords. Rong-Ping Shen et al. [12] in 2022, proposed a study by developing a deep learning detection framework including long short-term memory for the detection of sponsored posts which does not have the disclosure. This study incorporates textual, semantic, sentimental, and visual features of the posts for sponsored detection. Seungbae Kim et al. [13] in 2021, proposed a study on discovering undisclosed paid partnerships by proposing a learning-to-rank-based model called Sponsored Post Detector. This model incorporates the graph, text, and image of the post and employs graph convolutional networks to connect influencers, posts, and brands and ranks the social media posts by their score. Qikai Yang et al. [14] in 2024, presented a study on discussed different generative models for data augmentation. Generative adversarial networks (GANs), variational autoencoders (VAEs), and Gaussian mixture models (GMMs) are explored in the cause of social media platforms for selecting the appropriate methods to enhance the model performance. Comparative analysis on KNN, SVM, and decision tree. Deevi Radha Rani et al. [15] in 2021, proposed a study for analyzing the decisions on sales strategies, pricing, and promotions by developing a predictive model for sales forecasting. Explains the different machine learning models that can contribute to designing the models. Sandeep Chavan [16] in 2020, proposed a study on predicting sales using machine learning. Text and reviews are processed using NLP for the prediction of sales. Yatish Patil [17] in 2019, proposed a study to predict the probability of coupon purchase prediction and coupon user behaviors by incorporating a gradient boosting classifier. Clear explanation of coupons and their uses. Nitin Liladhar Rane [18] in 2024, presented a study to examine the role of artificial intelligence in sales. AI-powered CRM to improve customer satisfaction and enhance customer journey. It also includes sales prediction accuracy. Seungbae Kim et al. [19] in 2023, proposed a study on ranking the influencer based on engagement rate by building a framework called InfluencerRank. This study applies graph convolutional networks (GCN) to derive node representations that capture relationships. Sanjay Kumar et al. [20] in 2023, presented a study on the influence maximization problem using a graph long short-term memory model. This study calculates influence based on the node (influencer features) and selects the top k influencers. Seungbae Kim et al. [21] in 2020, proposed a framework called multimodal post attentive influencer profiler to classify the influencers into specific interest. Using an attention mechanism to classify the posts uploaded in the influencer accounts into 10 different categories and classify the influencer interest based on the weights of the posts.

4 Dataset

Day-to-day tons of data are accumulated in social media networks and data is available in the forms of text, images, hashtags, and videos which are uploaded in the form of tweets, posts, and reels respectively. In the context of influencer marketing campaigns, the main attribute viewed by customers, shared by influencers, and provided by companies is "Promotional Codes."

Every company shares unique promotional codes with each influencer hence it is secure to find the impact of every influencer and map the key influencers who contribute maximum for the campaigns. Another attribute is "Engagement Rate" which is calculated based on influencers' interaction and relationship communication with their followers which are based on likes, shares, views, and comments.

Author Seungbae Kim developed the Instagram Influencer dataset and Influencer Brand dataset with the explicit goal of supplying notable information about both influencers and brands.

The Instagram Influencer dataset contains 33,935 influencers profiles in 9 different categories which include name, age, country of origin, education level, personality, lifestyle, and backstory of influencers. The influencer brand dataset has information about 26,910 brands and 1,601,074 posts with post metadata.

Distant from the existing datasets, data can be collected manually or through web scrapping by visiting the profile of each influencer, and based on a tag called "include paid promotion" we can differentiate sponsored posts and get their unique promotional codes and engagement rate of the specified posts.

5 Existing Methodologies

Several machine learning and deep learning techniques are utilized in influencer marketing to design a model for identifying the key influencers, metrics for engagement rate calculation, predicting sales forecasting, customer behavior, and purchase prediction in social media platforms. Machine learning techniques include a Graph convolutional network (GCN) and deep learning techniques include Long short-term memory and Recurrent neural network (RNN) are utilized for influencer detection.

Natural language processing (NLP) helps in understanding customer behavior and purchase behavior. Decision trees are utilized for predicting modeling. The following explains the existing techniques and their contributions to social network analysis:

Graph convolutional network (GCN) is a machine learning approach that is incorporated into social media (graph-based) data where nodes are influencers and edges are connections to understand the complex posting behaviors in the social media network to rank the effective influencers based on the effectiveness of their posting behaviors and represent the key influencers. GCNs are used to capture social relationships as they convert social media connections into graph-based data with nodes and edges and it is implemented to detect sponsored posts.

Recurrent Neural Networks (RNN) is a deep learning approach, integrated into social media networks to understand the temporal behaviors of influencers and rank the ultimate influencers. It removes the low-performing nodes and uses an attentive mechanism to identify the key performers and rank them as influencers.

Graph-based Long short-term memory networks (GLSTM) are a deep learning approach that is utilized for influence maximization. LSTM helps in calculating the influence of nodes (influencers) and feature extraction is for selecting features utilized to obtain the features of nodes and sends them to GLSTM to predict the influence.

Natural language processing (NLP). is useful for predicting sales and analysis of consumer behavior on purchased promoted products. Customers comments, tweets, and reviews in social media are scrapped in the form of text data that are collected and processed and results are labeled as positive, negative, or neutral. These labeled reviews are utilized in predicting sales.

Even though most of the work faces difficulties while designing the methods for utilizing influencer marketing. Most of the study focuses on the influencers' parameters and views and theoretical contributions are presented on sales and ROI measurement. Confined work on sales increases by incorporating machine learning and deep learning techniques. The limitations in the existing research areas in integrating the solutions reduce the performance and accuracy of the model. The following points reveal the limitations facing the field of influencer marketing.

1. Data availability and quality.

Machine learning models need datasets compulsory to train the model and test the model. Due to privacy laws, many social media platforms like Instagram and TikTok limit the API access which consequently difficult to get the data for research work. Hence, researchers do not get full access to understand the metrics of influencer marketing. Another problem includes time-series data, which is essential for analyzing influencer's impact and tracking trends and changes. Poor historical data leads to inability, performance restrictions, and failure of models.

Machine learning requires high-quality and labeled data for model performance, poor-quality data including misspellings, missing values, duplicates, and errors affects the model accuracy. Data definitions and terms vary from one platform to another hence, it is critical to access the data to build a model to work on cross-platforms.

2. Influencer detection.

Most of the research focused on identifying mega-influencers, who have more than one million followers and grasp the highest likes and comments (engagement rates) on social media platforms and ignores micro, macro, and nano influencers who create honest content and reach good engagement rates. Most of the brand partners invest in influencers based only on the follower count which is a primary limitation.

influencer detection was not noticeable on niche influencers, such as social media influencers who have dedicated followers in a specific field. These do not fall under the category of celebrities or mega influencers. They only focus on specific interest areas like clothes, makeup, food, jewels, art, sports, and many more. Detection of niche influencers helps different companies to invest in specific influencers who fit under promotion products.

3. Sales and ROI measurement.

In influencer marketing, there are two sectors influencer detection and influencer impact. To assess the impact of an influencer's track of sales increase, ROI measurement, and profit calculation are essential. Proper analysis and tracking of influencers affect sales accuracy.

ROI measurement needs criteria like time-series investment data (monthly data of investment on influencers) and return on investment (ROI) the companies made including profit. Suppose the company's investment is larger than usual and sales increased and earn a profit. In that case, rechecking if it is over the investment is necessary to understand and measure the return on investment.

4. Performance metrics and Ethics.

In influencer marketing, the information provided by the influencer builds trust in products. If influencers involve fake claims, fake imaginary features, and false products effectiveness leads to a loss of consumer trust which affects both influencers' and companies' authenticity.

While analyzing performance, it is necessary to understand the data measurement in both quantitatively and qualitatively way. Quantitative parameters include a number of likes, shares, and followers and qualitative parameters include the sentiment of comments, content quality, and captions.

If companies try to invest in influencers who are popular in a specific field but promote products that are unrelated to that influencer's niche. Then there is much chance of consumers showing little or no interest in the product. This can result in a decline in the engagement rate of influencers and financial loss to brand partners and leads to a return on investment (ROI) reduction.

Ethical considerations in influencer marketing are important influencers may purchase fake followers, increase their post views and likes, and create only positive comments to become popular. An influencer should follow the terms and conditions of social media platforms such as disclosure of sponsored posts using #ad or #sponsored.

5. Multiple social media platforms.

Multiple platforms like Instagram, Facebook, and YouTube provide a diverse place for influencers to connect and interact with followers directly they promote products on all platforms with their social media accounts. Similarly, companies also increase their reach through multiple platforms. Despite that, it is a burden for brand partners to analyze the impact of every influencer on all platforms and costs of the influencer campaign increase with additional platforms. In some cases, followers start to see the same product promotions on all platforms and have a chance of losing interest. And influencers need to create fresh content for each social media platforms to gain customers and increase sales.

6 Proposed Methodologies

Previously, this study explored several limitations in influencer marketing. Since there are many studies accomplished on influencer marketing, purchase behavior, and prediction sales most of the work expounds only theoretically. To overcome these limitations and to extend the study on influencer marketing, the Integration of practical models like machine learning models and deep learning models in the field of influencer marketing is essential. The model needs to exploit influencers and their contributions towards sales increase and consumer purchase behavior is compulsory. One such technique that helps in mapping influencers based on their performance by considering relevant features individually is sparse learning techniques.

Sparse learning is a type of machine learning that works mainly on feature selection and to avoid sparsity, it refers to the situation where the majority of the available data is irrelevant or has zero values and creates complexity in handling high-dimensional data such as social media data. It helps the model to concentrate on significant factors or a subset of features and ignore the irrelevant features. This section explores about the sparse learning model for the classification of influencers based on their niche, performance score, and engagement rate of sponsored posts. This study highlights the idea of overcoming the limitations of facing influencer marketing.

This study explains the usage of sparse learning techniques in influencer marketing. Comparing other machine learning methods like GCN, RNN, and LSTM, sparse learning provides better performance on high-dimensional data by selecting the most important features and making the model interpretable, faster, and simple. Focusing only on important features avoids the risk of overfitting and provides computational efficiency.

Figure 1 pictures the framework of influencer classification, where influencers are classified into clusters or groups based on their niche, engagement rate, and performance score.

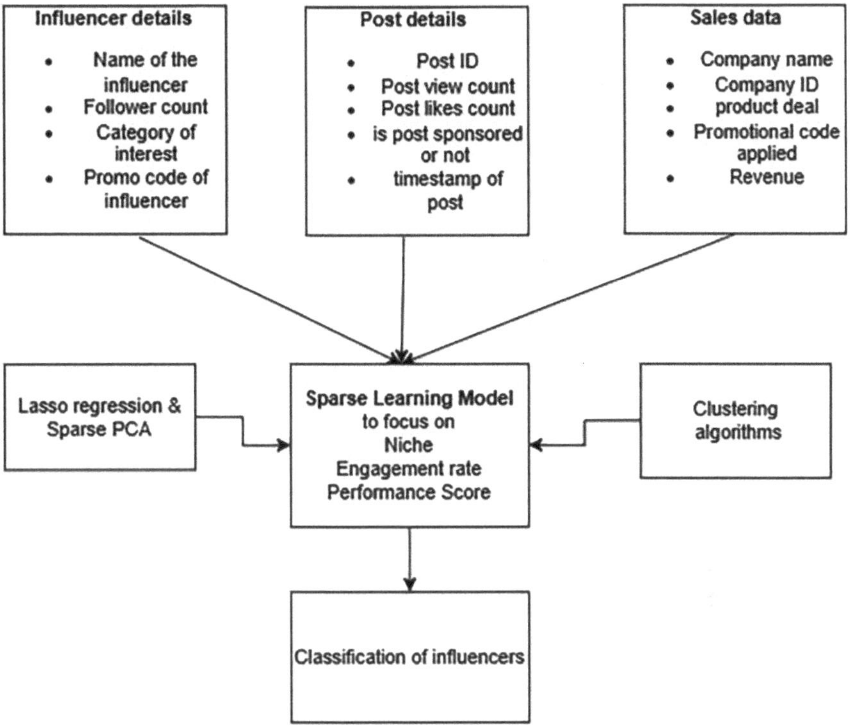

Fig. 1 Influencer classification framework

To build a sparse learning model for influencer classification and mapping there are several sparse techniques particularly in this study, sparse learning model based on lasso regression, sparse PCA and clustering algorithms.

Least Absolute Shrinkage and Selection Operator (Lasso) regression It is also known as L1 regularization is a sparse learning technique that applies the penalty to some features and forces coefficients to become exactly zero. Lasso adds the loss function to eliminate unimportant features and helps to identify important influencers who provide maximum sales for brand partners, and L1 norm is a type of penalty in sparse learning that helps to make the feature coefficients exactly to zero.

Sparse Principal Component analysis (PCA) This is also a sparse learning technique that helps identify sparse components while reducing dimensionality and raising interpretability. It helps to summarize the large dataset into smaller components. Sparse PCA combines the correlated features into a single component. In other words, it reduces the hundreds of features into a small key number of components. Sparse PCA helps in optimizing influencer marketing campaigns by reducing large-scale datasets into the number of components.

Clustering algorithms play a major role in influencer marketing. After calculating the performance score and engagement rate of the influencers on social media platforms, clustering algorithms are applied to segment influencers or clusters of similar influencers. Two popular clustering algorithms that can be applicable to social media data are K-Means clustering and DBSCAN clustering.

K-Means is a centroid-based clustering algorithm that partitions the data into k distinct clusters. In influencer marketing, k-means allocate each influencer into one cluster based on their performance. It is appropriate to group the influencer based on one or two characters. If the number of groups or clusters is predefined i.e., k is known helps in better targeting.

Density-Based Spatial Clustering of Applications with Noise (DBSCAN) is a density-based clustering algorithm. It identifies dense regions of data and clusters them. Unlike the regions of clusters, it also identifies the outliers of data where it doesn't fit into any clusters. Hence, it is helpful in identifying influencers who have less influence in product promotion and sales increase.

6.1 Overview of the Influencer Classification Framework

Promotional codes are an attractive way created by the brand partners to grasp the customers' intention to buy and promotional codes are the essential factor in tracking influencers' performance as it is available in influencers, customers, and sales data. Influencers exhibit the promotional code in their posts and customers see the posts they like the posts or save the posts to use the promotional code in their purchase and when a purchase is made, the promotional code is shared with the sales. Through the shared promotional code brand partners can identify from which influencers the sales increase.

Niche is an interest of influencers to focus on creating content. Influencers create content in several fields such as art, sports, makeup, clothes, finance-related, food, travel, fitness, interior design, and many more. Clustering helps in categorizing the influencers based on area of interest. Engagement rate includes sponsored post's likes, views, and the total number of followers for influencers which helps in calculating how many followers observe a sponsored post and react to it which inversely encourages consumer purchase behavior. The performance score is a foremost and dominant attribute compared to others. Whenever a customer made a purchase using an influencer promotional code, the influencer score increased. And the higher the number of purchases higher the score. Performance score helps in clustering or grouping the influencers based on their contribution to influencer marketing campaigns.

The influencers' niche or category of interest, the engagement rate of sponsored posts, and performance score in sales increase are required to focus on the identification of key influencers and classification of influencers and mapping influencers regarding sales purchases made using promotional codes. Integrating influencer-related data and sales data provide valuable insights for accurate sales increase and profit-driven through influencer posts. This study assists the brand partners in determining which influencer to invest in and how much investment to allocate to influencer marketing.

The influencer classification framework can be applied to multiple social media platforms regardless of platform architecture. In all social media platforms, promotional code is commonly distributed by the influencers in their official accounts in variant forms like tweets, videos, reels, etc.

6.2 Advantages of the Sparse Learning Techniques

- Sparse learning is suitable to handle high-dimensional data such as social media data through managing the risk of overfitting and increases efficiency.
- Sparse models are built simple and provide comprehendible results by focusing on relevant and important features.
- Lasso regression, Sparse PCA, and Elastic Net regression are popular sparse learning regularization techniques to prevent overfitting by dropping the irrelevant features to zero by adding penalty and loss function.
- As a result of ignoring unimportant features sparse model works at a lower computational cost and enhances the model performance.
- Most of the data are irrelevant in social media platforms for influencer marketing, sparse models work only on certain features which reduce noise and improve prediction accuracy.

7 Conclusion and Discussion

This study explored various existing technologies that are applicable for influencer detection, social media influence on users buying behavior, and predictive analysis of sales and ROI measurement. The limitations of the research are also discussed which include niche influencer detection and discussion of only theoretical metrics for sales measurement. Hence, the practical study is essential on important factors including ROI, purchase behavior factor, and increasing sales which assist in enhancing the performance of influencer marketing. To resolve the drawback and contribute to the field of influencer marketing, sparse learning techniques are introduced which assist in classifying influencers based on performance.

The incorporation of sparse learning techniques helps brand partners to understand which influencers yield better performance to invest in future campaigns, how much to invest in influencer marketing, and how to track sales and experience sales increase. Lasso regression and sparse PCA provide valuable contributions in tracking and analyzing the sales growth and profits driven by influencers' performance. Clustering algorithms can be applied to classify influencers based on sales performance along with niche and engagement rate which are performed using sparse models for future promotions.

8 Future Work

The integration of sparse learning techniques for influencer mapping and classification provides a significant performance. However, there are several other factors for improvement which include advanced sentiment analysis. Fraud detection in social media platforms includes identifying fake influencers with fluctuating follower counts, less performance, and negativity. Content reach is another important factor. Social media algorithms work in a certain way in feeding the content to customer apps. Investigating which type of content like posts, reels, tweets, short-form videos, or long-form review videos discovering and analyzing the impact of these content is necessary to understand the influencers. Certain companies make some key influencers as ambassadors for the brand on social media platforms. Detecting what are all the factors' influencers should perform to attain this position is necessary. Rather than influencers with promotions, Identifying the influencers who work on long-term relationships with customers is important which helps the companies to invest longer in these kinds of influencers for long-term influence. Another essential part is identifying the criteria that affect the return on investment (ROI) measurement is critical because even though the sales increased it is necessary to know how much investment or budget was made by the companies and how much return on investment was produced and this study has to happen dynamically as it changes every month.

References

1. Gracias A, Olukemi A (2024) Chris bell: influencer marketing ROI: measurement techniques and optimization strategies. J Mark Commun
2. Beichert M, Bayerl A, Goldenberg J, Lanz A (2024) Revenue generation through influencer marketing. American Mark Assoc 88(4). https://doi.org/10.1177/00222429231217471
3. Agustian K, Hidayat R, Zen A, Sekarini RA, Malik AJ (2023) The influence of influencer marketing in increasing brand awareness and sales for SMEs. Technology and Society Perspectives (TACIT) 1(2). https://doi.org/10.61100/tacitv1i2.54
4. Chen Y, Qin Z, Yan Y, Huang Y (2024) The power of influencers: how does influencer marketing shape consumers' purchase intentions. Sustainability MDPI 16:5471. https://doi.org/10.3390/su16135471
5. Spoorl-Wang K, Krause F, Henkel S (2025) Predictors of social media influencer marketing effectiveness: a comprehensive literature review and meta-analysis. J Bus Res ELSEVIER 186. doi:https://doi.org/10.1016/j.jbusres.2024.114991
6. Kumar R, Prabha V, Kumar M, Rehal P, Samanta P, Singh PK (2024) Influencer marketing: a review and research agenda using VOSviewer, Sagepub. https://doi.org/10.1177/09702385241289368
7. Gabhane D, Sahu SR, Anand Kumar P, Mallika, Guangul SM (2024) Influencer marketing effectiveness: metrics, risks, and best practices. Migration letters 21(S4):472–481

8. Gujar P, Panyam S, Paliwal G, Agarwal S (2024) The impact of big data and artificial intelligence on influencer marketing. Sch J Econ, Bus Manag (SJEBM), Sas publishers. https://doi.org/10.36347/sjebm.2024.v11i11.001
9. Parashar K, Goel C, Bisht P, Ahuja S (2024) Analyzing the impact of AI on influencer marketing. European Econ Lett 14(3)
10. Wang Y, Xiang G, Chang Sk (2013) Proceedings of the twenty-seventh AAAI conference on artificial intelligence, sparse multi-task learning for detecting influential nodes in an implicit diffusion network
11. Zheng C, Zhang Q, Young SD, Wang W (2020) On-demand influencer discovery on social media. ACM ISBN 978–1–4503-6859-9/20/10, https://doi.org/10.1145/3340531.3412134
12. Shen RP, Liu D, Wei X (2022) Mingyue Zhang.: your posts betray you: detecting influencer-generated sponsored posts by finding the right clues, information and management (59). ELSEVIER. https://doi.org/10.1016/j.im.2022.103719
13. Kim S, Jiang JY, Wang W (2021) Discovering undisclosed paid partnership on social media via aspect-attentive sponsored post learning. In: ACM international conference on web search and data mining (WSDM'21). Virtual Event, Israe. https://doi.org/10.1145/3437963.3441803
14. Yang Q, Li P, X X, Ding, Zhou W, Nian Y (2024) A comparative study on enhancing prediction in social network advertisement through data augmentation. ar Xiv 2404.13812v3
15. Rani DR, Soumya B, Supriya GL, Yuktha N, Kolluru PK, Malleswari DN (2023) Performance analysis of sales using machine learning. In: Proceedings of the KILBY 100 7th international conference on computing sciences. Elsevier. https://ssrn.com/abstract=4493480
16. Chavan S, Panchal S, Sawant T, Shinde J (2020) Predicting online product sales using machine learning. Int J Eng Res Technol (IJERT) 9(04)
17. Patil Y, Pawar O, Ingle DR (2019) Coupon purchase prediction using machine learning. IJSRD-Int J Sci Res Developmentl 7(2)
18. Rane NL, Paramesha M, Choudhary SP, Rane J (2024) Artificial intelligence in sales and marketing: enhancing customer satisfaction, experience and loyalty. Elsevier
19. Kim S, Jiang JY, Han J, Wang W (2023) Influencer Rank Discovering Effective Influencers via Graph Convolutional Attentive Recurrent Neural Networks. In: proceedings of Association for the Advancement of Artificial Intelligence
20. Kumara S, Mallika A, Panda BS (2022) Influence maximization in social networks using transfer learning via graph-based LSTM. Expert Syst Appl 212 Elsevier https://doi.org/10.1016/j.eswa.2022.118770
21. Kim S, Jiang JY, Nakada M, Han J, Wang W (2020) Multimodal post attentive profiling for influencer marketing, In: Proceedings of International World Wide Web Conference. ACM. https://doi.org/10.1145/3366423.3380052

A Review on Enhancing Surveillance with Hybrid Deep Learning for Suspicious Human Activity Recognition

S. R. Likhith and R. Anjali

Abstract Surveillance systems are highly relevant for the provision of assurance in public safety. However, at the same time, the traditional methods applied to human activity recognition often do not perfectly distinguish suspicious actions in the environment where actions are not bounded and may be quite subtle. The last breakthroughs in deep learning, all are hybrid models in general, provide highly improved detection accuracy, speed, and efficiency of the suspicious human actions. the real-time surveillance application detection is enhanced with the features of the several neural network architectures by implementing the hybrid models. This survey presents recent advancement in hybrid deep learning frameworks as well as explain how such frameworks have enhanced the detection precision and how they eliminate the demerits of conventional methods. We review the efficiency of these hybrid models and classify their drawbacks based on the complexity of data structure, the size of data required to make the decision and its real time applicability. There are other possible solutions to these issues that need to be developed to make surveillance more rigid. In an effort to contribute to the synthesis of the current state of knowledge in surveillance to direct future advancements in technology, design and utility for more intelligent and adaptive systems better to target abnormal human activity in a multivariant complex environment, this article attempts at presenting a summary.

Keywords Human activity recognition · Deep learning · Convolutional neural networks · Suspicious human activity recognition

1 Introduction

Computer vision and AI growth have made video surveillance the most essential need for identification of abnormality within the public domain. Suspicious human activity recognition (SHAR) is a unique facet of surveillance systems that attempts to identify

S. R. Likhith (✉) · R. Anjali
Presidency School of Computer Science and Engineering, Presidency University, Bengaluru, India
e-mail: anandlikith111@gmail.com

H. Nagaraj et al. (eds.), *Advancements in Artificial Intelligence and Machine Learning*, Lecture Notes in Electrical Engineering 1572,
https://doi.org/10.1007/978-981-95-7449-0_18

possible threats through observation of human conduct [1]. The older systems are either monitored manually or by traditional style motion detection which is UIS and involving so much human interference [2]. However, it can auto forward it to get it more accurate and so that the number of false alarms reduces.

Indeed, during the past few decades, surveillance systems have evolved mainly to meet new requirements for more effective policing, protection of assets and property and maintaining social order among others. Indian cities are becoming more complex and urbanized, hence the techniques of scanning which have underpinned traditional surveillance will not be adequate. Since conventional systems typically trail way behind hundreds and thousands of data produced by several thousand cameras installed in public places, this alone results in poor response and delayed threats. Such a call for an efficiency, scalability, and precision in a solution has never been louder in this day and age.

Substantial artificial intelligence is known as deep learning, which has enabled real-time video surveillance of human behavior. Such systems rely on a set of calculations to interpret visual data, which supposedly identifies acts that are dangerous to the society in one way or another, harmful [3]. Of course, this is now machine learning, and such surveillance conducted today can recognize patterns of activities and pick out anomalies of its normal workflow and shoot out alert based on a set-up criterion. This transformed the complete process of monitoring from a manual method to the use of an automated system, which was a significant step in surveillance technology, which could also have made timely response and prevention of criminal activities.

Hence by chance, powerful techniques in the form of hybrid approaches in deep learning appear at the right time. They mitigate drawbacks inherent in the conventional surveillance systems which cannot capture the real-time changing nature of human activity characteristics with a good degree of accuracy. Such hybrid models and methodologies, similar to these can perfectly fit into the dynamism of the human activities. The second way of combining the two networks would be an integration of CNN to obtain spatial features and LSTM for temporal representation [4]. Both approaches enable the system to have more than a 'what' of human behavior, which is what is happening and when it is happening, which enhances an even higher level of understanding of human behavior.

This paper would also like to know more about how integration of hybrid deep learning methodologies could well improve the effectiveness of suspicious HAR systems. It involves some of the primary prone activities of concern. These are running, punching, falling, kicking and shooting modes. A closer look at these activities makes it possible for this report to describe how hybrid deep learning systems enhance detection rates while simultaneously reducing false positives where necessary. This will equip them to well coordinate with police and other security forces in order to maintain safety of lives and property at all times.

1.1 Motivation

The need to establish this review arises from the increasing demand to have higher surveillance systems that can be guaranteed to be in more complex and dynamic environments with assured public safety. Conventional methods for suspicious human activity recognition have been difficult to surmount such challenges as ambiguous behaviors in detection. Indeed, the recently developed hybrid deep learning techniques have very efficiently crossed the lines of these limitations to improve suspicious activity recognition in terms of accuracy and efficiency. However, at present, the existing limitations exist with regard to the complexity of data being used, model scaling issues, and implementation in real time. This review will prove to be an excellent overview of these hybrid methods by means of discussing their potential in enriching systems of surveillance and providing insights applicable for further research work.

1.2 Objective

With the present review on the advancement concerning this problem and advantages and some disadvantages of such existing approaches towards the recognition of suspicious human activity in surveillance systems, this study intends to present a brief on hybrid deep learning trends in miscellaneous applications to assess the practical feasibility of such schemes. Aside from providing an understanding of potential promising challenges or areas, related to data complexity and model scaling and real-time performance, something that the study would also try to look into would be solutions to those issues. In the long run, the goal would be to suggest new directions for further research and development work on the design of much better surveillance systems, that could handle the detection of suspicious activities in highly dynamic and active environments.

1.3 Scope

It mainly focuses on how hybrid deep learning techniques benefit from the use of suspicious human activity identification in surveillance so that such systems become more accurate and real-time performing. To this end, the review will check recent developments of such techniques, probe into their strengths and weaknesses, and hence attempt to identify some challenges ahead, like data complexity and scalability.

This also creates demands for the provision of potential solutions to specific problems, thereby guiding further research towards developing effective and accurate surveillance systems that have the potential to prevent suspicious activities within the complex environments (Figs. 1 and 2).

Fig. 1 Kicking

Fig. 2 Fighting

2 Related Work

The present work represents an alternative approach in the detection of abnormal human activity in video frames, using deep learning strategies [5]. The work is structured to systematically detail the procedure involved-data collection, data preparation, annotation, data preprocessing, feature extraction, and model development. Hybrid architecture comprising the fusion of CNN and LSTM. Such architecture is used to consider both the spatial features and the temporal features of the video frames. The proposed algorithm will be evaluated based on various performance measures in terms of accuracy, precision, recall, and F1-measure, which yield an accuracy of 95.6%. However, there are some drawbacks in adopting the hybrid approach to activity recognition with the CNN-LSTM model. The limitations include the size of the dataset being small, not considering computational complexity, and not comparing results with any of the existing techniques. The concluding remark of the paper stresses the effectiveness of algorithms that may be used for detecting suspicious activity.

This study introduces a comprehensive literature survey of video action recognition techniques in the context of sports analytics, dwelling on datasets, methodologies and applications at both the athlete and team levels within the sport domain [6]. Approaches have been classified into traditional hand-crafted models and modern deep learning models, which include 2D and 3D CNNs, two-stream models, skeleton-based models and group activity recognition frameworks. Some sports datasets reviewed include SoccerNet, NCAA, and UCF Sports among many others. This survey covers methods and challenges of video recognition in sports entirely but misses a quantitative meta-analysis of the performances of the methods and discussions on cross-sport transfer learning. In fact, the study did not look into transformer-based methods very extensively. However, this paper is highly valuable as a starting point to help researchers enter the domain, providing crucial insight into current trends and research directions.

This survey paper takes into consideration, in extensive detail, techniques that machine learning is applied towards anomaly detection in surveillance videos, namely supervised, semi-supervised, and unsupervised methods. Here, existing anomaly detection methods from the previous literature are categorized based upon their advantages and disadvantages in terms of applications that cut across domains in surveillance [7]. Datasets such as UCF Sports and SoccerNet hold a lot of importance concerning training and benchmarking. Lacking adequate quantitative comparisons on benchmark datasets, except very briefly mentioning real-world deployment challenges, privacy issues, and ethical considerations, the study has a good overview of the strengths and weaknesses of various approaches at 92.1% accuracy.

To prevent 'over-exploiting' only one temporal transformation, a temporal-aware contrastive network (TAC-Net) is proposed in the study for UAVS video anomaly detection [8]. In particular, it utilizes contrastive learning to learn features of normal patterns and uses a 3D encoder for calculating anomaly scores. The proposed approach achieves the best accuracy on benchmarked data and eliminates shortcomings of prior techniques. The method offers an accuracy of 93.2%. But it is slower, more sensitive to hyperparameters, and may be less effective in handling scales in videos. In summary, it has been shown that self-supervision can be effectively used in combination of self-supervised contrastive learning for capability of anomaly detection in surveillance systems.

The work presents a novel deep learning approach to human action recognition using drone imagery that can be translated into applying RGB video data. The methodology also includes other sophistications like: bilateral filtering that is used in attenuating noises, quick shift segmentation for silhouettes extraction and extraction of the key body points and movement [9]. The approach involves classification using convolutional neural networks (CNNs) and the approach delivers best performance with benchmark datasets with accuracy rate of 94.5%. In general, the model is good but fails to work well with complex structures, can hardly follow fast movements or changing angles.

A lightweight convolutional neural network (CNN) model for human activity recognition using Wi-Fi channel state information (CSI) data is developed in this work [10]. The methodology also uses a few data augmentation options to handle problems concerning small data sets and unbalanced classes. Through preprocessing the CSI data and by creating a proper CNN architecture the study shows enhanced performance compared to prior studies and as such is apt for real-time use on any such target devices with limited resources. In sum, the current study contributes a substantial enhancement into privacy-preserving human activity identification enabling smart home and healthcare applications.

The research brings a solution based on deep learning to improve security in real-time video surveillance when addressing face mask detection systems and protecting them against adversarial attacks [11]. Using a white-box Fast Gradient Sign Method (FGSM) attack, the study compares the effects of various epsilon values on model performance employing a custom dataset of 14,535 images. Therefore, the results demonstrate the presence of adversarial threats to detection systems while proving the efficacy of the proposed mitigation approach. Still, this decision limits the study area to one type of attack and the custom dataset may not be typical for the real environment.

This work also employs deep leaning for human activity recognition with features are extracted from Video frames using pre-trained CNN (VGG16) and the temporal information is learned by Bi-GRU [12]. In feature selection using random forest algorithm, the method also gains the lowest computational complexity while also attaining the best benchmark accuracy. However, because our approach is based on a pre-trained CNN, generalization to new domains may be limited, and performance overall is poor for at least one activity class. These limitations of the approach are again compounded by the fact that they are highly computational thus not advisable for IoT devices. In general, this research shows how future work can utilise multiple deep learning methods for advanced video comprehension tasks.

This work provides an overview and an evaluation of the current state of the art in MOT approaches for real-time action recognition in surveillance proposed by introducing the SurvTrack dataset based on the VIRAT dataset [13]. In this paper, the performance of five MOT trackers including ByteTrack, OC-SORT, Deep OC-SORT, BoT-SORT, and StrongSORT is analysed based on several metrics. The results show that potential of the approach for detection of reliable samples and give an information about the influence of detector confidence threshold and input resolution of images on the performance.

The study introduces a technique for partitioning, numbering, and identifying routine physical movements from videos based on the 3D skeleton data obtained from the VIBE pose estimation model [14]. Having employed both PCA and a movement sequence algorithm, it reaches good levels of action segmentation accuracy, 96.27% as well as recognition accuracy 99.06% on two my own subsets. However, the approach relies on accurate pose estimation, as was mentioned earlier, which might fail in complex environments to provide all the required angles and coordinates. Though, more extensive testing in various setting is needed to improve external validity of the framework for fitness tracking and rehabilitation.

This work proposes a new method of using horizontal stacking of ridge regression, LightGBM, and gradient boosting machine (GBM) models [15]. The used methodology involves feature-domains extraction with fixed-window segmentation as the data preparation process. Nonetheless, the need for bigger quantities of fine training data, and non-interpretability of the stacked model can degrade the model's usefulness in constrained settings or in real-time settings.

This research uses YOLOv3 for detecting precisely human in video frames and drawing boundary boxes, but for feature extraction local binary pattern (LBP) and gray level co-occurrence matrix (GLCM) used [16]. Therefore, a redesigned LSTM network is used for the classification of the activities, further enhanced by a newly introduced HSM-CSO algorithm. The system identifies many missed skiing switch actions of lawbreakers and has high accurate rate of recognizing abnormal activities from the ATM surveillance videos, but these videos collected from YouTube may not illustrate real ATM scenarios and the complete process may have very high computational cost and thereby might not be extremely useful for real-time detection.

This makes a complete surveillance system to advance the human tracking and anomalous behaviour detection based on deep learning [17] such that it mitigates false positivity and enhances the possibility of tracking by introducing new methods of object filtering. For object detection, the system uses YOLOv3, and the scene and behaviour recognition system is based on a prompt-based recognition that leverages CLIP. However, it fails to work well in high density crowd situations and the control algorithm does not have the learning feature for the past detections and needs to be trained manually to perform well in future detections.

This work proposes a new HAR approach for using mobile and wearable sensors based on deep learning-based models including DCNNs and LSTMs [18]. For feature extraction, a Fuzzy logic based genetic algorithm (FGA) is used, meanwhile for feature selection a particle swarm optimization (PSO) with ant colony optimization (ACO) operators are employed. This type of model affords a total top recognition rate of 99.90% in other open datasets such as UCI-HAR, PAMAP2, and WISDM. It requires a vast amount of computing power and time for model training and testing, which might render the model unsuitable for real-time applications, in particular, noisy or incomplete data.

The framework put forward here discusses a CNN-based model used on the Edge AI devices for real-time activity recognition [19]. The system comprises three components: One involves capturing the data through surveillance cameras, and extracting the features with the help of a mini-convolutional neural network model; the other is a classification of activity through a machine learning algorithm. Assessed using a created dataset of 10 000 frames belonging to six activities' classes. Due to low weight the implemented approach improves computational speed, however, the usage of a custom dataset reduces the real-world application capability of the model, and the study does not discuss the computational complexity in detail.

2.1 Datasets

2.1.1 UCSD Ped2

UCSD Ped2, [20] is one of the most well-known datasets employed in anomaly detection scenario in surveillance videos. There are 16 training videos, and 12 test videos containing 1208 frames in total. The dataset records typical pedestrian behaviour in outdoors spaces and identifies numerous odd behaviours, for instance, normal walking in odd directions or staying momentarily in bike-friendly zones or for an extraordinarily long time.

The UCSD Pedestrian Dataset is often referred as UCSD Ped2. It is the benchmark dataset mainly used to detect anomalies in films. The UCSD Ped2 data set has been widely used in computer vision to analyze models which are intended to detect unusual or anomaly behavior in surveillance video data. The main features and details of the UCSD Ped2 data set are described below: The UCSD Ped2 dataset is the gold standard for video anomaly detection method evaluation nowadays. Its highly controlled environment, extremely clear annotations, and narrow scope make it an invaluable resource to test the capabilities of algorithms designed to improve security and surveillance systems.

2.1.2 UAVGesture

The UAVGesture dataset, [21] as it is as a benchmark for human action recognition in drone videos.

A typical UAV gesture dataset contains data about recognizing the hand or body movement used to control or communicate with a drone; such motions can be registered using cameras, either RGB or depth, or accelerometers and gyroscopes, along with labelled gesture categories.

It includes 119 video files and 14,976 frame files in total. This dataset concerns the task of human gesture and action recognition which has been captured from the perspective of a drone. The actions are such as waiving, pointing and other signal or call gestures.

2.1.3 UCF-Crime

The UCF-Crime dataset, [5, 22] is a large-scale, real-world video dataset for the identification of anomalous events in surveillance videos.

UCF Crime Dataset is a high-scale dataset mainly designed for anomaly detection and crime classification in surveillance videos. It was developed by the Centre for Research in Computer Vision at the University of Central Florida (UCF) for the purpose of offering a robust benchmark for models aimed at identifying suspicious or criminal activity in real-world video data.

It has more than 1,900 long videos including burglaries, robberies, and shoplifting scenes. Dataset to assess the generalization performance and stability of the anomaly detection algorithms in real-life monitoring environments.

2.1.4 UCI-HAR Dataset

The UCI-HAR dataset, [23] represents 30 persons performing six different activities, which are walking, walking upstairs, walking downstairs, sitting, standing, and lying. For each activity, the person wears a smartphone around his waist.

2.1.5 VIRAT Dataset—Video Image Retrieval and Analysis Tool

VIRAT Dataset, [24, 25] counts more than 329 surveillance movies captured in high-resolution mode, amounting to over 11 h of real-world outdoor challenging activities. It contains 23 types of activities, including loading and unloading vehicles and entering and exiting buildings among others. It proves handy for a number of tasks such as activity recognition, object detection, and multi-object tracking.

2.1.6 HMDB 51 Dataset

HMDB 51 Dataset, [26, 27]. It is composed of 6,766 video clips sorted into 51 activity categories such as running, jumping, sitting, and many more. There are about 100 clips in every class, which makes it ideal for testing and training models in human action recognition. The HMDB dataset has been pretty helpful in the advancement of human activity recognition. Even today, it is considered to be one of the benchmarking tools to test new designs, including hybrid models employing convolutional networks and transformers and other state-of-the-art approaches.

3 Methodology

3.1 Hybrid Techniques

Bayes classification hybridized with CNNs [28], depicts the probabilistic reasoning and deep learning approach whereby a video or sensor data is perceived to recognize an activity. CNNs learn spatial information in frames, and Bayes classifiers calculate the probability of each action based on these features. This is a hybrid technique that combines CNNs' capability of learning complex patterns with Bayes simple classification that is understandable [29]. It is particularly useful for detecting activities, but the approach is expensive in terms of computation and requires huge datasets.

In addition, Bayes classifiers rely on assumptions that degrade performance when trying to deal with complex patterns of activities.

HAR technology combines CNN with BiGRU [30] where CNN is provided for spatial information extraction from sensor data, and BiGRU represents the temporal connections by processing data along both forward and reverse directions. The hybrid technique over the combination of strengths of CNN feature extraction capabilities and sequential pattern capture capabilities in BiGRU increases the accuracy of recognition [31]. But the model proposed here suffers from some drawback, namely high computational complexity and latency that may degrade its performance for real-time purposes, especially in resource-constrained environments. Analogous hybrid approaches draw attention to trade-offs of deep learning-based HAR systems between computational requirements and real-time processing.

3.2 Aerial-Based Human Activity Detection

The detection of aerial human activity is made with the assistance of drones with cameras and sensors, like thermal and RGB that capture the movement of humans [32]. Involves data collection, preprocessing to identify the human participants, and the extraction of spatial features using machine learning models, such as CNNs [33, 34]. Then follow this up with the application of temporal analysis based on RNN or LSTMs. Once they are discovered, the activities are then assessed in real time, and applications range from security to crowd control, but with occlusions and other real-time processing issues that alter environmental conditions affecting accuracy.

3.3 LSTM Based Models

LSTMs can identify sequences that have strong long-range temporal associations, which makes them useful for video-surveillance and sensor-based activity recognition tasks. In LSTMs, through the learning of complex patterns of human movement, typical and anomalous behaviours may be identified [35, 36]. It is an unfavourable model because large datasets are required to be collected. This may be expensive and take a lot of time. For example, while LSTMs are prone to overfitting in sparse or noisy data and computation intensive, by no means real-time applicable, these types of systems generally include hybrid models or other methodologies so that systems built on top can have the best possible mix of accuracy and efficiency [37, 38].

3.4 CNN-Based Methods

Convolutional neural networks (CNNs) have proven to be extremely good at detecting human activity by extracting spatial data from pictures or video frames, making them ideal for identifying specific patterns in human movement [39, 40]. They are widely used in various surveillance and activity recognition jobs. However, CNNs have problems in capturing temporal dynamics, which are critical for understanding actions that occur across time [41]. This issue is frequently addressed by combining CNNs with models such as LSTMs or GRUs to better temporal analysis [42]. Furthermore, CNNs are computationally demanding, making them difficult to analyze in real time or deploy on limited-resource devices.

3.5 Edge Computing

An edge computing-based real-time human activity recognition surveillance framework that leverages the use of local processing of video feeds on edge devices to obviate latency associated with cloud servers and bandwidth dependency [19, 43]. The approach tends to increase response times and hence is suitable for security applications. However, the low computational capability of the edge device makes it impossible to handle huge datasets and complex tasks while maintaining recognition accuracy under changing conditions.

3.6 Fuzzy Logic-Based Methods

Applying fuzzy logic to simulate the inherent imprecision and vagueness in the detection of human behaviour; instead of hard decision boundaries, it determines membership degrees in fuzzy sets. This way, it might possibly be able to cope with complex, vague scenarios such as partial occlusions or fuzzy boundaries between different activities [44, 45]. However, fuzzy logic also has some limitations in the approach: One limitation is that the quality and clarity of the video feed on which it will operate are crucial because not many fuzzy logic approaches can handle high-speed video feed processing well, especially in cases of higher-resolution video and large datasets. Moreover, fuzzy logic approach-based systems may require very specific values for membership functions to prove highly effective, but such specific membership functions often fail to generalize to other environments or activity types.

3.7 Support Vector Machine

Support vector machine (SVM) is a supervised classifier technique used for classifying human activity based on a strong hyperplane that creates a distinction between different classes of activity within a high-dimensional space [46, 47]. SVM is highly effective for complex data in which clear class boundaries are defined. However, there are some limitations to this approach, like sensitivity towards the choice of a kernel and hyperparameter, and it can be very expensive in terms of computations when large amounts of data exist or the classes are imbalanced.

3.8 Sensor Fusion Methods

Sensor Fusion is the process of merging data from numerous sensors, such as vision and sensor inputs, to improve the accuracy and resilience of Human Activity Recognition (HAR) in complicated contexts [44, 48]. This technique overcomes the limits of individual sensors by leveraging many sensor modalities, resulting in a more comprehensive understanding of human behaviors [49]. However, the methodology may encounter obstacles such as sensor data synchronization, increased computational requirements, and the need for complex algorithms to successfully merge disparate data sources, all of which can complicate implementation and scalability.

4 Evaluation Methods

4.1 Accuracy

Accuracy is the measure, which is stated as a percentage of the number of correct predictions as compared to the total number of predictions made by the model. It is one of the very simplest and most widely used metrics as classification tasks go.

This metric is quite good when the class distribution is a balancing one, but it may be misleading in imbalanced datasets.

$$\text{Accuracy} = \frac{\text{True Positive} + \text{True Negative}}{\text{Total Samples}} \tag{1}$$

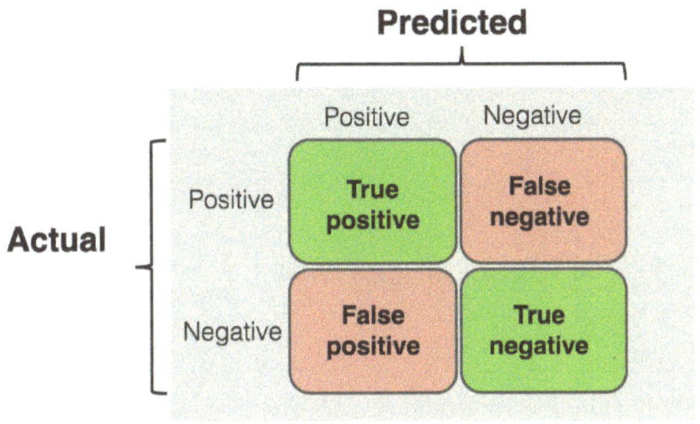

Fig. 3 Confusion matrix

4.2 *Confusion Matrix*

The confusion matrix shows the break-up of true and false predictions up to every detail–a model produces like number of true positives, false positives, true negatives, and false negatives. It is very effective for multiclass classification situations (Fig. 3).

4.3 *Area Under the ROC Curve (AUC-ROC)*

The AUC-ROC is a graphical presentation of a model's ability to discriminate between two classes. ROC Curve: True Positive Rate versus False Positive Rate (1-specifity). To be brief, AUC measures the true ability of the model in generalizing between positive and negative classifications.

$$\text{AUC} = \int_{0}^{1} \text{True Positive Rate (TPR) d(False Positive Rate (FPR))} \tag{2}$$

4.4 *Cross Validation*

Cross validation is a machine learning evaluation technique in which a model is tested against previously unknown data. The data are separated into several folds or subsets, one of which serves as the valid set and the others are utilized to train the model. This method must be repeated numerous times, but with a new fold as the validation set each time. Finally, the data from each validation cycle are averaged to provide a more trustworthy estimate of how the model will perform. This means that cross-validation is a critical aspect of the machine learning process since it assures the model chosen for deployment is as resilient as it is relevant to.

$$\text{CV Error} = \frac{1}{\text{k}} \sum_{i=1}^{\text{k}} \text{Error } I \tag{3}$$

4.5 *Precision, Recall and F1 Score*

Precision is a statistic that calculates the proportion of genuine positives compared to total positives predicted by the model. It addresses the question "How many of our positive predictions came true?"

$$\text{Precision} = \frac{\text{True Positive}}{\text{True Positive} + \text{False Positive}} \tag{4}$$

Recall measures how well the model finds all of the positives. Recall, also known as true positive rate, addresses the question, "How many of all the data points that should have been predicted as true did we correctly predict?"

$$\text{Recall} = \frac{\text{True Positive}}{\text{True Positive} + \text{False Negative}} \tag{5}$$

F1 score is a metric that combines recall and precision. As we have shown, there is a trade-off between precision and recall; F1 can thus be used to assess how well our models handle that trade-off.

$$\text{F1} - \text{Score} = 2 \times \frac{\text{Precision} \times \text{Recall}}{\text{Precision} + \text{Recall}} \tag{6}$$

5 Proposed Framework

This will be achieved by the implementation of a project with data preprocessing, creation of the hybrid model, training, and evaluation in ensuring an effective detection of suspicious human activities in video recordings and provide SMS alert system if there is any suspicious activity occurring. Firstly, a video dataset is generated where the categorization based on the activity and frame extraction from each movie is achieved. The resultant extracted frames are down-scaled into uniform dimensions and normalized for better efficiency in processing the computing. Few frames per video will then be sampled for a fixed number, and labels will be encoded for multi-class classification. After that, various hybrid models will be developed and trained with such architectures as CNNs, Transformers and some more advanced techniques

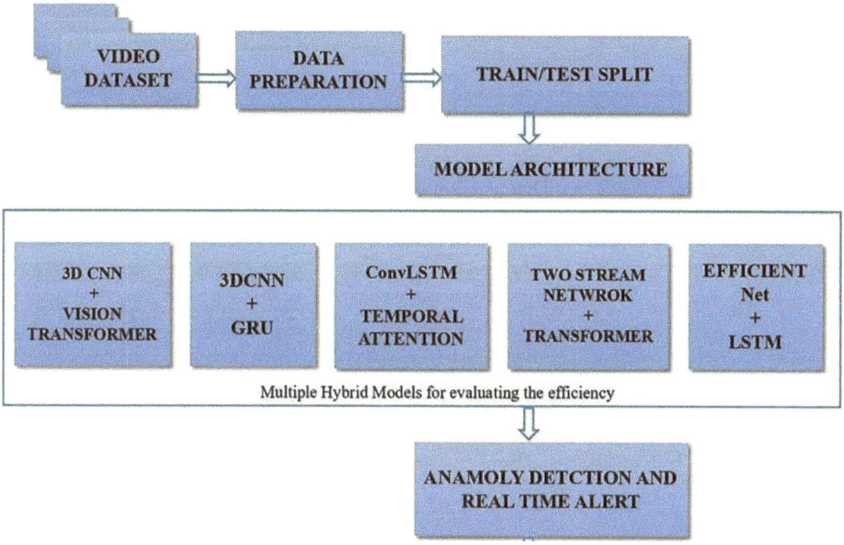

Fig. 4 Proposed framework

to be able to more finely capture the spatial–temporal features and long-range dependencies of video data. Using the same dataset, the different hybrid models will be trained, and the hyperparameters will be tuned based on the improvements of the performance. Conclusion: With regard to the obtained models, the different ones will be tested for accuracy, precision, recall, and F1 score to determine the most successful of the designs. This approach combines effective data management with sophisticated deep learning algorithms; hence, the detection and classification of suspect activity is as reliable as well as an efficient solution (Fig. 4).

5.1 Proposed Methodology Dataset

UCF Crime Dataset: It is the right benchmark for anomaly detection with surveillance footage and, particularly, focuses on crime detection in realistic scenarios. There are more than 1,900 raw surveillance videos for over 128 h of recorded material (Fig. 5).

The dataset contains 14 classes of crime such as arson, assault, burglary, robbery, shooting, and so on. They are recorded in various locations, on roads, inside shops, and in houses; hence, a very wide spectrum of lighting, camera angle, and background noises characterizes these videos, thereby making the detection task trickier.

Unlike most action recognition datasets, UCF-crime emphasizes the significance of identifying anomalies and normal activities by scrutinizing long, unsegmented sequences alongside the accurate timing and localization of events of suspicious activities. This characteristic of the dataset makes it crucial for training the deep

Fig. 5 Classes in UCF crime dataset

CLASSES
Abuse
Arrest
Arson
Assault
Burglary
Explosion
Fighting
Normal Videos
Road Accidents
Robbery
Shooting
Shop Lifting
Stealing
Vandalism

models that are used in augmenting public safety through detection of anomalies. In addition, UCF-crime highly contributes to the advancement of research on assessing surveillance in real time and crime detection in complex scenarios with clear realism.

5.2 *The 3D-CNN and Vision Transformers Hybrid Model*

The proposed model uses a combination of both 3D CNN and ViT architectures, which offer high accuracy when identifying suspicious human behaviour due to the synergy between both architectures 3D CNN extract the spatial and temporal information from the video sequence, as they operate in parallel on multiple frames at a time, making them useful for identifying sequential occurrences. This makes them suitable to pick on activities such as running, fighting or stealing. In contrast, vision transformers work well for modelling long-range dependence and context in the video frames, thus improving the robustness of the model in capturing complex as well as hidden behavioural patterns. When combining the 3D CNN for the motion analysis and the ViT for global feature extraction, the model should enhance the pipeline created to recognize the suspicious actions which are especially challenging to recognize in the occluded or crowded environment [50]. Such synergy increases the ability of the model in the identification of obvious as well as concealed signs of suspicion in surveillance videos.

5.3 3D-CNN and Gated Recurrent Units Hybrid Model

The hybrid framework indeed captures the 3D CNN architecture and integrates with gated recurrent units architecture, which has been shown to perform very accurately on suspicious activities in humans by taking complete features of both architectures. The 3D CNNs shall be optimally used to capture spatial and temporal properties of video data as they operate on frames in sequence; hence, they are quite ideal for the recognition of human movements and activities across time. Hence, the model is able to identify dynamic activities such as running or fighting or falling with precise accuracy. GRUs are one variant of RNNs designed especially for the processing of sequential data by doing the temporal dependencies across frames. They are good at learning long-term dependencies and contextual information over time, and hence they can understand complex or subtle activity patterns such as loitering or suspicious lingering. The GRU layer strengthens the capability of the model to understand temporal changes and learn sequences of suspicious activity. This approach can be used to achieve higher accuracy in the detection of suspicious human activities based on the hybrid model that involves the use of 3D CNN for extracting features in spatial–temporal and GRU for sequential analysis.

5.4 Hybrid Model with ConvLSTM and Temporal Attention Mechanism

Hybrid framework integrating ConvLSTM network and temporal attention, this is a robust methodology in identifying very potentially suspicious human behavior with a very high degree of accuracy.

ConvLSTM: It is designed to capture both spatial and temporal dependencies using a combination of convolutional layers within the LSTM architecture. This, in particular, makes it even more suitable for video data tasks. This model can handle sequential frames with spatial information, thus enabling the understanding of complex human activities evolving over time, such as sudden sprints or perhaps suspicious gatherings. This model is highly enhanced by the temporal attention mechanism, focusing selectively on the most important frames or moments of the video sequence. Instead of accepting value on an equal footing for all frames, this attention mechanism identifies critical time frames where suspicious activities are likely to occur, thereby boosting the capacity of the model to distinguish normal and abnormal behaviors. This makes the system focus on subtle, temporal operations that may be required for the detection of potentially malicious activities [51]. The proposed model thus empowers it to significantly improve its capacity of detecting dubious behavior in surveillance video by using the spatial–temporal features of ConvLSTM with selective emphasis afforded by temporal attention.

5.5 Hybrid Architecture of Two-Stream Network and Transformer

A two-stream network hybrid framework combined with a transformer architecture may be one road leading to high accuracy for suspicious human behavior identification in video surveillance systems.

The two-stream architecture is pretty much parallel convolutional networks. The former takes into account spatial information extracted from individual frames (the RGB stream), while the latter considers temporal dynamics (the optical flow stream). The dual way of such an approach allows the model efficiently detect not only static visual characteristics as posture but also motion-indicative indicators like direction and velocity. This capability is also critical in identifying activities such as running, fighting, or sudden motion that may indicate potential threats. Combining this architecture with the transformer, it further learns the capacity to identify long-range dependencies and interconnections between frames and hence improve its comprehension of complex and subtle behaviors for more extended periods. Overall, the transformer proves to exhibit more ability in modeling global interactions across sequences, enabling the model to identify not just patterns and correlations typical for conventional approaches based on convolution but also missed patterns and correlations. The use of the combination between two-stream network robust spatial and motion feature extraction and the transformer's global attention mechanism allows the architecture to set up effective balance between details-focused and context-sensitive processing. This results in improved accuracy in detecting suspicious human activity in real-time video surveillance.

5.6 Hybrid Architecture Efficient Net and LSTM

This hybrid architecture of efficient net and LSTM may prove to be the effective model for suspicious human activity detection in a video surveillance system since it presents the combination strength of both models.

Recently, it is popular because of its effective scale model size without losing the accuracy for the family of CNNs called efficient net. The application of the efficient net on feature extraction can allow the model to have the ability to detect detailed spatial information from individual video frames with modest amounts of computational resources, making it very useful for real-time applications where the reduction of complexity while maintaining high accuracy is needed.

Long short-term memory is a recurrently recently designed neural network particularly focused on problems regarding the management of sequential data and maintaining long-term dependencies across frames. LSTM can demonstrate skill in recognizing temporal patterns, which is invaluable in understanding how activities evolve over time. Suspicious activities often entail sequences of actions, not isolated events, and LSTM can capture such temporal interactions accurately. It leads to a hybrid architecture that offers a well-balanced and robust modeling approach, combining Efficient Net for efficient extraction of spatial features with LSTM for the analysis of temporal sequences. The model is able to precisely recognize all discussed suspicious activities, including aggressive behavior or atypical movements, with computational efficiency suitable for real-time surveillance applications.

6 Conclusion

Recognition of suspicious human activity significantly increases as surveillance systems upgrade with more complex deep learning models. Hybrid architectures that factor in the interplay of different network architectures, including 3D CNNs, GRUs, ConvLSTMs, and transformers, prove to be effective at capturing spatial features and temporal features and thus improve the accuracy of an anomaly detection model. They do have strengths that would further aid the processing of complex video data 3D CNNs are pretty excellent at spatial feature extraction, whereas GRUs and ConvLSTMs grasp sequential dependencies; these transformers and attention mechanisms would allow handling long-range dependencies much more efficiently. Also, models like LSTMs with Efficient Net feed into each other to balance performance with computational cost and add efficiency and accuracy to the result. A combination of such methods would make the surveillance systems more accurate in identifying suspicious activities in a variety of real-world scenarios. However, some of the other challenges persist, which include heterogeneity, real-time functionality, and interpretability of models, and hence, it still requires novel hybrid deep learning methods designed specifically for surveillance applications.

References

1. Perez M, Kot AC, Rocha A (2019) Detection of real-world fights in surveillance videos. In: ICASSP 2019–2019 IEEE international conference on acoustics, speech and signal processing (ICASSP). IEEE, pp 2662–2666
2. Bouchabou D, Nguyen SM, Lohr C, LeDuc B, Kanellos I (2021) A survey of human activity recognition in smart homes based on IoT sensors algorithms: taxonomies, challenges, and opportunities with deep learning. Sensors 21(18):6037
3. Sakka S, Liagkou V, Stylios C (2023) Exploiting security issues in human activity recognition systems (HARSs). Information 14(6):315
4. Liu W, Luo W, Lian D, Gao S (2018) Future frame prediction for anomaly detection—a new baseline. In: 2018 IEEE/CVF conference on computer vision and pattern recognition, Salt Lake City, UT, USA, pp 6536–6545. https://doi.org/10.1109/CVPR.2018.00684
5. Zaidi MM, Sampedro GA, Almadhor A, Alsubai S, Al Hejaili A, Gregus M, Abbas S (2024) Suspicious human activity recognition from surveillance videos using deep learning. IEEE Access
6. Wu F, Wang Q, Bian J, Ding N, Lu F, Cheng J, Xiong H (2022) A survey on video action recognition in sports: datasets, methods and applications. IEEE Trans Multimed 25:7943–7966

7. Choudhry N, Abawajy J, Huda S, Rao I (2023) A comprehensive survey of machine learning methods for surveillance videos anomaly detection. IEEE Access
8. Huang C, Wu Z, Wen J, Xu Y, Jiang Q, Wang Y (2021) Abnormal event detection using deep contrastive learning for intelligent video surveillance system. IEEE Trans Ind Inf 18(8):5171–5179
9. Azmat U, Alotaibi SS, Abdelhaq M, Alsufyani N, Shorfuzzaman M, Jalal A, Park J (2023) Aerial insights: deep learning-based human action recognition in drone imagery. IEEE Access
10. El Zein H, Mourad-Chehade F, Amoud H (2024) CSI-based human activity recognition via lightweight CNN model and data augmentation. IEEE Sens J
11. Panigrahi GR, Sethy PK, Behera SK, Gupta M, Alenizi FA, Nanthaamornphong A (2024) Enhancing security in real-time video surveillance: a deep learning-based remedial approach for adversarial attack mitigation. IEEE Access
12. Ahmad T, Wu J, Alwageed HS, Khan F, Khan J, Lee Y (2023) Human activity recognition based on deep-temporal learning using convolution neural networks features and bidirectional gated recurrent unit with features selection. IEEE Access 11:33148–33159
13. Alikhanov J, Kim H (2023) Online action detection in surveillance scenarios: a comprehensive review and comparative study of state-of-the-art multi-object tracking methods. IEEE Access
14. Cheng SH, Sarwar MA, Daraghmi YA, İk TU, Li YL (2023) Periodic physical activity information segmentation, counting and recognition from video. IEEE Access 11:23019–23031
15. Dahal A, Moulik S (2024) The multi-model stacking and ensemble framework for human activity recognition. IEEE Sens Lett
16. Kshirsagar AP, Azath H (2023) YOLOv3-based human detection and heuristically modified-LSTM for abnormal human activities detection in ATM machine. J Vis Commun Image Represent 95:103901
17. Jeon H, Kim H, Kim D, Kim J (2024) PASS-CCTV: proactive anomaly surveillance system for CCTV footage analysis in adverse environmental conditions. Expert Syst Appl 124391
18. Jameer S, Syed H (2023) A DCNN-LSTM based human activity recognition by mobile and wearable sensor networks. Alex Eng J 80:542–552
19. Aishwarya D, Minu RI (2021) Edge computing based surveillance framework for real time activity recognition. ICT Express 7(2):182–186
20. Sabokrou M, Fathy M, Hoseini M, Klette R (2015) Real-time anomaly detection and localization in crowded scenes. In: Proceedings of the IEEE conference on computer vision and pattern recognition workshops, pp 56–62
21. Liu C, Szirányi T (2021) Real-time human detection and gesture recognition for on-board UAV rescue. Sensors 21(6):2180
22. Boekhoudt K, Matei A, Aghaei M, Talavera E (2021) Hr-crime: human-related anomaly detection in surveillance videos. In: Computer analysis of images and patterns: 19th international conference, CAIP 2021, virtual event, September 28–30, 2021, proceedings, part II 19. Springer International Publishing, pp 164–174
23. Arshad M, Jaskani FH, Sabri MA, Ashraf F, Farhan M, Sadiq M, Raza H (2021) Hybrid machine learning techniques to detect real time human activity using UCI dataset. EAI Endorsed Trans Internet of Things 7(26):e1–e1
24. Blasch E, Wang Z, Ling H, Palaniappan K, Chen G, Shen D, Seetharaman G (2013) Video-based activity analysis using the L1 tracker on VIRAT data. In: 2013 IEEE applied imagery pattern recognition workshop (AIPR). IEEE, pp 1–8
25. Gkountakos K, Touska D, Ioannidis K, Tsikrika T, Vrochidis S, Kompatsiaris I (2021) Spatio-temporal activity detection and recognition in untrimmed surveillance videos. In: Proceedings of the 2021 international conference on multimedia retrieval, pp 451–455
26. Xu X, He W (2023) The video behavior recognition based on R (2+ 1) D. In: 2023 IEEE 11th joint international information technology and artificial intelligence conference (ITAIC), vol 11. IEEE, pp 829–833
27. Li T, Zhu H, Sun X, He L (2023) Human action recognition based on residual networks with improved attention mechanisms. In: 2023 IEEE 7th information technology and mechatronics engineering conference (ITOEC), vol 7. IEEE, pp 2295–2298

28. Liu C, Ying J, Han F, Ruan M (2018) Abnormal human activity recognition using Bayes classifier and convolutional neural network. In: 2018 IEEE 3rd international conference on signal and image processing (ICSIP). IEEE, pp 33–37
29. Gul MA, Yousaf MH, Nawaz S, Ur Rehman Z, Kim H (2020) Patient monitoring by abnormal human activity recognition based on CNN architecture. Electronics 9(12):1993
30. Mekruksavanich S, Jantawong P, Hnoohom N, Jitpattanakul A (2022) A novel deep bigru-resnet model for human activity recognition using smartphone sensors. In: 2022 19th international joint conference on computer science and software engineering (JCSSE). IEEE, pp 1–5
31. Liu Y et al (2021) Combining CNN with BiGRU for activity recognition in smart environments. Sensors 21(12):4035
32. Zhang J et al (2021) A UAV-based framework for human activity recognition in surveillance systems. Sensors 21(5):1563
33. Shiranthika C, Premakumara N, Chiu HL, Samani H, Shyalika C, Yang CY (2020) Human activity recognition using CNN & LSTM. In: 2020 5th international conference on information technology research (ICITR). IEEE, pp 1–6
34. Pienaar SW, Malekian R (2019) Human activity recognition using LSTM-RNN deep neural network architecture. In: 2019 IEEE 2nd wireless Africa conference (WAC). IEEE, pp 1–5
35. Vrskova R, Hudec R, Kamencay P, Sykora P (2022) A new approach for abnormal human activities recognition based on ConvLSTM architecture. Sensors 22(8):2946
36. Kabir MM, Safir FB, Shahen S, Maua J, Awlad IAB, Mridha MF (2020) Human abnormality classification using combined CNN-RNN approach. In: 2020 IEEE 17th international conference on smart communities: improving quality of life using ICT, IoT and AI (HONET). IEEE, pp 204–208
37. Xia K, Huang J, Wang H (2020) LSTM-CNN architecture for human activity recognition. IEEE Access 8:56855–56866
38. Ding J, Wang Y (2019) WiFi CSI-based human activity recognition using deep recurrent neural network. IEEE Access 7:174257–174269
39. Khan IU, Afzal S, Lee JW (2022) Human activity recognition via hybrid deep learning based model. Sensors 22(1):323
40. Atikuzzaman M, Rahman TR, Wazed E, Hossain MP, Islam MZ (2020) Human activity recognition system from different poses with CNN. In: 2020 2nd international conference on sustainable technologies for industry 4.0 (STI). IEEE, pp 1–5
41. Khatun MA, Yousuf MA, Ahmed S, Uddin MZ, Alyami SA, Al-Ashhab S, Moni MA (2022) Deep CNN-LSTM with self-attention model for human activity recognition using wearable sensor. IEEE J Transl Eng Health Med 10:1–16
42. Lee SM, Yoon SM, Cho H (2017) Human activity recognition from accelerometer data using Convolutional Neural Network. In: 2017 IEEE international conference on big data and smart computing (bigcomp). IEEE, pp 131–134
43. Huang X, Yuan Y, Chang C, Gao Y, Zheng C, Yan L (2023) Human activity recognition method based on edge computing-assisted and GRU deep learning network. Appl Sci 13(16):9059
44. Abdelhedi S, Wali A, Alimi AM (2016) Fuzzy logic based human activity recognition in video surveillance applications. In: Proceedings of the second international Afro-European conference for industrial advancement AECIA 2015. Springer International Publishing, pp 227–235
45. Pandya BH (2023) A distributed architecture for fuzzy logic systems and its application in human activity recognition. Doctoral dissertation, Nottingham Trent University
46. Shuvo MMH, Ahmed N, Nouduri K, Palaniappan K (2020) A hybrid approach for human activity recognition with support vector machine and 1D convolutional neural network. In: 2020 IEEE applied imagery pattern recognition workshop (AIPR). IEEE, pp 1–5
47. Nurhanim K, Elamvazuthi I, Izhar LI, Ganesan T (2017) Classification of human activity based on smartphone inertial sensor using support vector machine. In: 2017 IEEE 3rd international symposium in robotics and manufacturing automation (roma). IEEE, pp 1–5
48. Webber M, Rojas RF (2021) Human activity recognition with accelerometer and gyroscope: a data fusion approach. IEEE Sens J 21(15):16979–16989

49. Münzner S, Schmidt P, Reiss A, Hanselmann M, Stiefelhagen R, Dürichen R (2017) CNN-based sensor fusion techniques for multimodal human activity recognition. In: Proceedings of the 2017 ACM international symposium on wearable computers, pp 158–165
50. Han H, Zeng H, Kuang L, Han X, Xue H (2024) A human activity recognition method based on vision transformer. Sci Rep 14(1):15310
51. Baradel F, Wolf C, Mille J (2017) Pose-conditioned spatio-temporal attention for human action recognition. arXiv preprint arXiv:1703.10106

Enhancing Logical Reasoning Capabilities in Large Language Models through Fine-Tuning: A Comparative Analysis

K. Saravanan, K. Hithesh, A. Maheshwari, Pradeep Vishnuram, Mohit Bajaj, Bhabasis Mohapatra, Renu Sharma, and Binod Kumar Sahu

Abstract The logical reasoning capabilities of large language models (LLMs) play a critical role in their application across various domains, particularly in areas requiring high cognitive functions such as problem-solving and decision-making. This study focuses on improving the logical reasoning and aptitude performance of LLaMA 3.0, a leading LLM, through the application of Unsloth fine-tuning techniques. By evaluating the model's performance on a range of logical reasoning tasks before and after fine-tuning, this research aims to quantify the enhancement in reasoning accuracy and consistency. The study employs a combination of quantitative metrics and qualitative assessments to provide a comprehensive analysis of the improvements. The results highlight the promise of focused fine-tuning methods in enhancing the cognitive capacities of LLMs, thereby expanding their utility in intricate problem-solving situations. This research examines prospective avenues for refining LLMs to improve their reasoning capabilities and addresses the wider ramifications of these developments for AI-driven applications.

Keywords Large language models (LLMs) · Particle Swarm Optimization (PSO) and Low-rank adaptation (LoRA)

K. Saravanan · P. Vishnuram
Department of Electrical and Electronics Engineering, SRM Institute of Science and Technology, Kattankulathur, Chennai, India

K. Hithesh · A. Maheshwari
Department of Computational Intelligence, SRM Institute of Science and Technology, Kattankulathur, Chennai, India

M. Bajaj
Department of Electrical Engineering, Graphic Era (Deemed to be Universiy), Dehradun, India

Graphic Era Hill University, Dehradun, India

B. Mohapatra (✉) · R. Sharma · B. K. Sahu
Department of Electrical Engineering, ITER, Siksha 'O' Anusandhan (Deemed to be University), Bhubaneswar, Odisha, India
e-mail: bhabasismohapatra@soa.ac.in

H. Nagaraj et al. (eds.), *Advancements in Artificial Intelligence and Machine Learning*, Lecture Notes in Electrical Engineering 1572,
https://doi.org/10.1007/978-981-95-7449-0_19

1 Introduction

Large language models (LLMs) have significantly advanced natural language processing and artificial intelligence, especially in tasks necessitating intricate cognitive processes such as problem-solving, decision-making, and logical reasoning. Notwithstanding their exceptional talents, LLMs frequently have deficiencies in logical reasoning, leading to inconsistent or suboptimal performance in activities necessitating a high level of precision and accuracy. This is partially attributable to the characteristics of their training, wherein models are subjected to extensive quantities of heterogeneous and uncurated data, resulting in difficulties in generalizing logical patterns and sustaining consistency across many contexts [1].

This research seeks to tackle the issues related to the logical reasoning capabilities of LLMs by concentrating on LLaMA 3.0, a cutting-edge language model recognized for its efficacy in many natural language applications. This project aims to boost the logical reasoning and aptitude performance of LLaMA 3.0 by the application of unsloth fine-tuning, a method particularly developed to augment particular cognitive capacities in LLMs. The fine-tuning procedure is predicated on the premise that certain modifications might markedly improve the model's reasoning abilities, allowing it to excel at tasks necessitating a profound comprehension of logical frameworks and problem-solving methodologies [2].

To assess the efficacy of this method, the study will utilize a combination of quantitative measures and qualitative evaluations. The objective is to assess enhancements in the model's logical reasoning capabilities, offering insights into the particular domains where fine-tuning produces the most substantial benefits. This research will also examine the wider implications of these enhancements for the utilization of LLMs in domains requiring advanced reasoning and analytical capabilities, including legal analysis, technological problem-solving, and strategic decision-making [3]. This work seeks to enhance the creation of more robust and cognitively advanced LLMs. The project aims to improve the logical reasoning capabilities of LLaMA 3.0, hence expanding the model's utility in intricate problem-solving areas and establishing a basis for future endeavors to enhance and optimize LLMs for certain cognitive activities. This project aims to enhance our comprehension of how fine-tuning techniques can address existing limitations in LLM performance, ultimately facilitating the development of AI systems that are more accurate, reliable, and capable in practical applications. The integrated Luo converter is designed to mitigate the aforementioned limitations and provide enhanced power conditioning capabilities in contemporary power electronic systems [4].

2 Literature Review

The research on improving logical reasoning abilities in large language models (LLMs) offers a variety of methods that together elucidate the existing landscape and problems. Recent advancements have investigated the amalgamation of symbolic reasoning techniques with LLMs to mitigate the intrinsic constraints of exclusively neural methodologies. The LINC model, which integrates language models with first-order logic provers, signifies a notable advancement in the creation of neurosymbolic systems designed to improve logical thinking through the utilization of both data-driven learning and symbolic manipulation. This amalgamation of methodologies is regarded as a viable avenue for enhancing the reasoning powers of LLMs, since it seeks to reconcile the adaptability of neural networks with the accuracy of symbolic systems.

Additional advancements in this domain arise from initiatives such as the "Prototype-then-Refine" technique, which underscores a neurosymbolic strategy that progressively enhances logical reasoning mechanisms in LLMs [5]. This technique aims to enhance the interpretability and dependability of models by incorporating a structured reasoning component that directs the learning process. Concurrently, current research has concentrated on creating benchmarks like LogiGLUE, designed to rigorously assess the logical reasoning capabilities of LLMs. These benchmarks are essential for identifying the deficiencies and constraints in existing models, offering a definitive direction for future research to tackle these issues [6]. Survey research, including those on puzzle-solving using LLMs, highlight the significance of task-specific adjustments and tool selection in improving logical thinking. The expanding research in this domain indicates that although LLMs excel at handling extensive textual data, their reasoning abilities may be substantially enhanced by organized methodologies that include logical frameworks.

The research also examines the resilience of LLMs across diverse task structures, indicating that model efficacy in logical reasoning tasks can fluctuate significantly based on the particular configurations and training procedures utilized. This idea is essential for creating more robust models that can generalize across various logical problems. Datasets such as LogiQA and its enhanced iteration, LogiQA 2.0, are crucial in propelling research by offering demanding criteria for machine reading comprehension, emphasizing logical thinking [7]. These datasets expand the limits of LLMs' capabilities, fostering the creation of models that can not only comprehend but also accurately reason through intricate textual material. Overall, the surveyed literature paints a picture of an evolving field where the combination of symbolic and neural approaches, the development of specialized benchmarks, and the continuous refinement of methodologies are all converging towards the goal of enhancing logical reasoning in LLMs [8]. This body of work not only highlights the current capabilities and limitations of LLMs but also sets the stage for future advancements in creating models that are both powerful and reliable in logical reasoning tasks. The Luo converter with conventional boost converter aids in alleviating the current ripples to a greater extent. The PI control with particle swarm optimization (PSO) strategy is efficiently employed in this study as provides consistent voltage supply to the converter [9].

3 Methodology

The primary goal of this project is to fine-tune LLaMA 3.0 to improve its aptitude in logical reasoning and aptitude tasks. This section describes the approach for fine-tuning, data preparation, evaluation metrics, and system integration. The fine-tuning approach begins with selecting LLaMA 3.0 as the base model, chosen for its advanced flexibility and adaptability. Unsloth fine-tuning is employed to optimize the model for specific logical reasoning challenges. The training data are curated to include diverse logical reasoning and aptitude questions to strengthen model performance in these areas. Prompt engineering plays a crucial role, involving the design and iterative refinement of prompts to challenge the model's reasoning abilities effectively. In response analysis, responses are processed and refined to enhance accuracy and logical consistency, with a feedback loop in place to continuously improve the prompts and model performance as shown in Fig. 1.

In data preparation, datasets are sourced from academic resources, logical reasoning books, and aptitude tests. Each dataset is carefully annotated with correct answers and reasoning explanations, followed by cleaning to remove noise and irrelevant data. The data are converted into question-answer pairs for fine-tuning, and synthetic data generation techniques are employed to augment the data, so enhancing the model's exposure to diverse scenarios. The assessment metrics concentrate on quantifying the model's performance via correctness, logical coherence, and problem-solving proficiency. Benchmarking involves comparing the fine-tuned model with baseline models, while user testing yields practical input to evaluate the model's efficacy. The optimized model is implemented into the designated

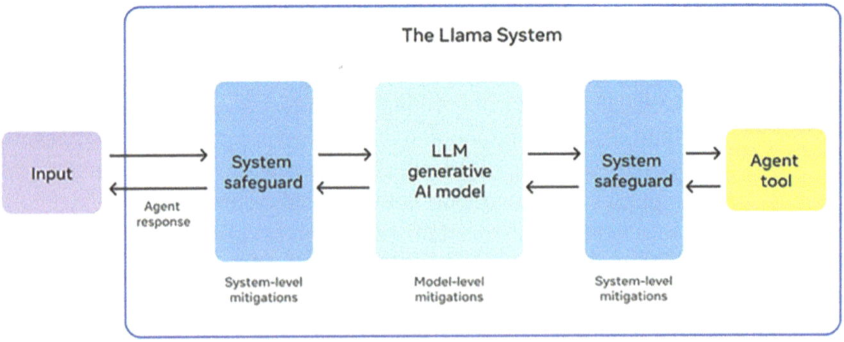

Fig. 1 LLaMA structure

application, with APIs created for seamless access and interaction. A user interface is crafted for intuitive interaction, enabling users to submit inquiries and observe outcomes. Usability testing ascertains that the interface aligns with user expectations. The workflow orchestration entails user interactions in which logical reasoning or aptitude questions are input, analysed by the model, and delivered in a coherent format. Performance monitoring persists after deployment, incorporating frequent evaluations and feedback collecting to inform further enhancements. This systematic approach utilizes sophisticated fine-tuning methods and thorough assessment to enhance LLaMA 3.0's proficiency in logical reasoning and aptitude tests, guaranteeing efficient and dependable performance.

4 Experiments and Analysis

This section details the studies undertaken to evaluate the efficacy of Unsloth fine-tuning in improving the logical reasoning and aptitude capabilities of LLaMA 3.0. An exhaustive analysis of the results is presented, including benchmark comparisons and several model performance measures. The experimental setup entails installing LLaMA 3.0 as the foundational model, chosen for its superior capabilities and adaptability for fine-tuning. Unsloth fine-tuning was employed for its flexibility in addressing unique reasoning problems without necessitating substantial labeled data. The data preparation process entailed the collection of a heterogeneous dataset comprising logical reasoning problems, aptitude questions, and problem-solving scenarios derived from academic and standardized sources, supplemented by synthetic data augmentation to familiarize the model with multiple situations. During the training procedure, Unsloth techniques were applied iteratively to adapt the model to logical reasoning and aptitude tasks, with careful tuning of hyperparameters, such as learning rate, batch size, and epochs, to optimize performance.

The evaluation metrics focused on several key areas. Accuracy measured the proportion of correct answers the model provided, showing an improvement over the base model on benchmark questions. Logical consistency was assessed through reasoning tasks to ensure the model's responses were coherent and adhered to logical principles, with the fine-tuned model showing enhanced consistency. Aptitude performance, evaluated using standardized test questions, demonstrated the model's improved problem-solving skills. Benchmark comparisons against baseline and state-of-the-art models revealed that the fine-tuned LLaMA 3.0 outperformed these models, showcasing superior accuracy and logical consistency.

In the results analysis, overall trends indicated notable improvements in both logical reasoning and aptitude tasks, with increased accuracy and logical consistency compared to the base model. Specific cases highlighted areas where the fine-tuned model excelled in complex reasoning tasks, underscoring the enhancements achieved. Error analysis identified common errors, such as misinterpretations of complex queries, and introduced mitigation strategies like prompt refinement to further improve model performance. User feedback collected from real-world interactions reflected high satisfaction with the model's enhanced reasoning and aptitude capabilities, supporting the effectiveness of the fine-tuning approach. Looking ahead, model refinement is an ongoing priority, with plans for further fine-tuning and exploration of advanced techniques to enhance performance. Extensive testing over a wider array of reasoning and aptitude tests will validate the model's generalization skills, while further assessment in real-world applications will confirm its practical value. This part presents a comprehensive analysis of the experiments assessing the effects of Unsloth fine-tuning on LLaMA 3.0, illustrating substantial enhancements and establishing a basis for future developments and applications.

5 Fine Tuning Using LoRA

In the framework of fine-tuning LLaMA 3.0 to improve its logical reasoning and aptitude performance, the use of Low-Rank Adaptation (LoRA) emerges as a particularly beneficial approach. LoRA facilitates the accurate and efficient fine-tuning of huge language models, rendering it especially appropriate for our objectives.

5.1 Understanding Low-Rank Adaptation (LoRA)

Low-Rank Adaptation (LoRA) is a fine-tuning method aimed at enhancing big language models by including supplementary low-rank matrices into the model's current layers. The matrices are incorporated into the transformer layers of LLaMA 3.0, introducing additional trainable parameters that modify the model's behavior to improve alignment with certain tasks, particularly in increasing logical reasoning and problem-solving abilities. The fundamental concept of LoRA is to diminish the quantity of parameters requiring training, hence reducing computing expenses and expediting the fine-tuning procedure. This is accomplished by factorizing the weight matrices of certain layers into low-rank matrices. During fine-tuning, these low-rank matrices are learned while preserving the original pre-trained weights predominantly unchanged. This targeted training methodology guarantees that the fine-tuning process is concentrated and effective.

The Fig. 2 demonstrates how LoRA alters the architecture of LLaMA 3.0. It emphasizes the insertion spots of the low-rank matrices and illustrates their interaction with the model's original parameters. This visualization is essential for comprehending how LoRA does fine-tuning without using computer resources excessively.

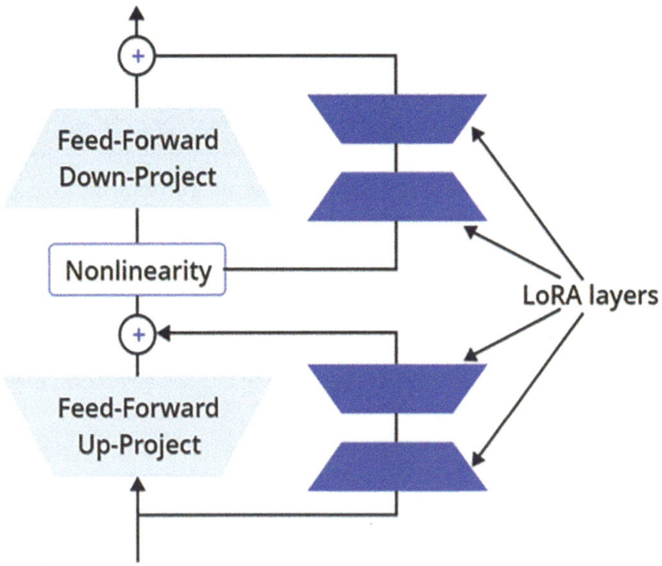

Fig. 2 The process in which LoRA modifies the architecture of LLaMA 3.0

5.2 Advantages of Using LoRA in Fine-Tuning

A key benefit of employing LoRA for fine-tuning LLaMA 3.0 is its efficiency. Conventional fine-tuning techniques sometimes need the modification of several parameters, resulting in prolonged training durations and elevated resource use. Conversely, LoRA's low-rank adaptation markedly diminishes the parameter space, facilitating expedited training cycles and reducing hardware requirements. A notable advantage is that LoRA reduces the likelihood of overfitting. By concentrating on a limited, specialized set of parameters, the model preserves its generalization abilities while enhancing performance on certain tasks. This equilibrium is crucial for logical thinking and aptitude evaluations. The model must possess a comprehensive grasp of language and logic while being meticulously calibrated to thrive in tasks necessitating advanced reasoning.

Furthermore, LoRA is especially beneficial when utilizing very big models such as LLaMA 3.0. The decrease in trainable parameters not only expedites the fine-tuning process but also facilitates the use of more sophisticated techniques and larger datasets within feasible timescales. This enables the execution of intricate fine-tuning operations without requiring substantial computational resources.

5.3 Implementation of LoRA in LLaMA 3.0 Fine-Tuning

The application of LoRA in the fine-tuning of LLaMA 3.0 adheres to a systematic methodology. We first select the layers in the model that are most pertinent to logical reasoning and aptitude challenges. LoRA is then implemented in these particular layers, using low-rank matrices that facilitate the majority of the adaptation. This focused strategy guarantees that the fine-tuning process is both efficacious and efficient. Upon establishing the low-rank matrices, the model undergoes a sequence of fine-tuning iterations. In each cycle, just the parameters related to the LoRA matrices are modified, while the remainder of the model stays unchanged. This approach enables us to progressively refine the model, implementing precise modifications that improve its performance in reasoning tasks. The accompanying picture delineates the LoRA fine-tuning procedure in LLaMA 3.0, depicting the data flow through the model, the function of the low-rank matrices, and the manner in which these matrices are modified throughout training. This visual assistance is crucial for comprehending the technical foundations of LoRA and its influence on the model's performance (Fig. 3).

5.4 Results and Impact of LoRA Fine-Tuning

The outcomes of using LoRA for the fine-tuning of LLaMA 3.0 are substantial. The model exhibits significant enhancements in logical reasoning tasks, surpassing its baseline performance across several benchmarks, as seen in Fig. 4. This encompasses improved precision in problem-solving, superior contextual comprehension in reasoning situations, and more reliable logical deductions. Utilizing LoRA, we have attained these enhancements without sacrificing the model's overall language comprehension ability. The fine-tuning procedure has successfully refined the model's reasoning abilities, enhancing its proficiency in addressing intricate logical challenges while preserving the adaptability characteristic of LLaMA 3.0. The efficacy of this technique highlights the promise of LoRA as a fine-tuning method for big language models. It illustrates that employing appropriate strategies may substantially improve certain skills of a model such as LLaMA 3.0, facilitating further progress in the domain of logical reasoning and aptitude-oriented AI.

6 Conclusion

This effort highlights the capacity of fine-tuning big language models such as LLaMA 3.0 to markedly improve logical reasoning and aptitude challenges. Utilizing Unsloth fine-tuning strategies, we have shown that it is feasible to enhance a model's skills to address certain cognitive requirements, including the resolution of intricate reasoning

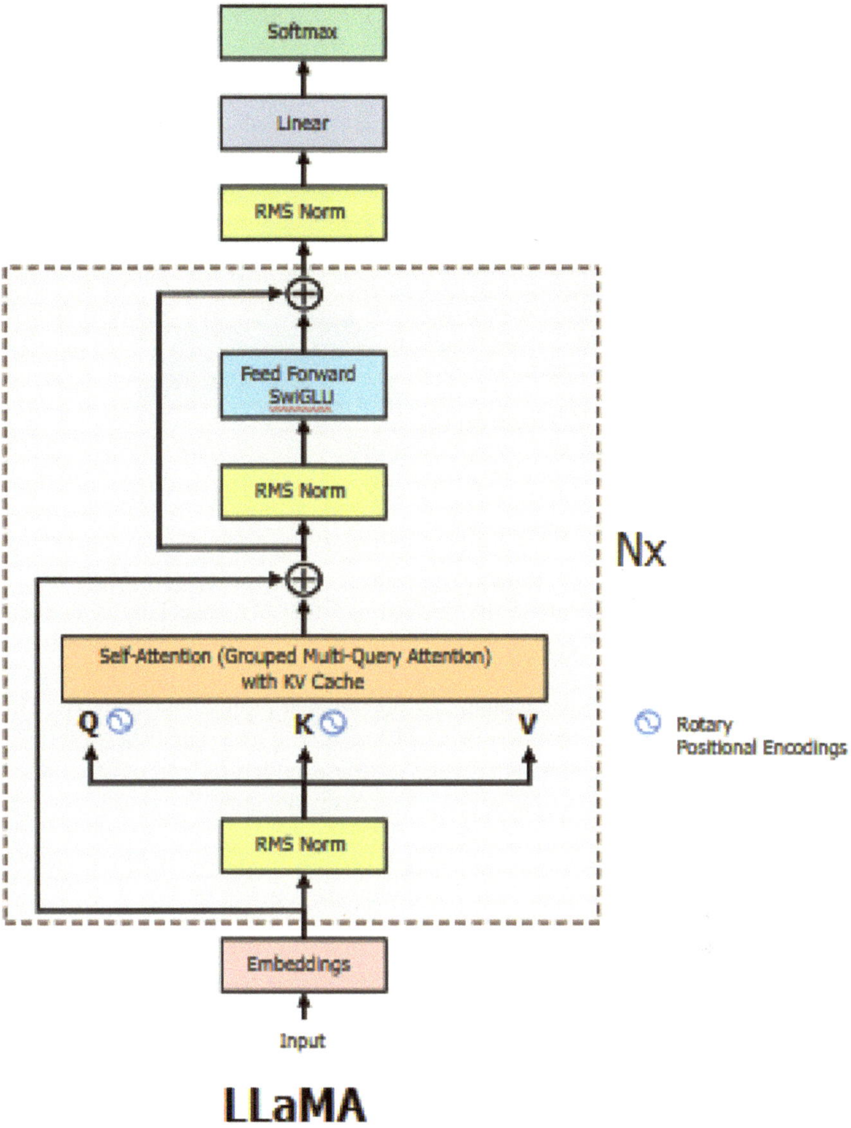

Fig. 3 Technical underpinnings of LoRA and its impact on the model's performance

challenges and the augmentation of precision in aptitude evaluations. Our research demonstrates that deliberate fine-tuning may connect general-purpose LLMs with particular applications, allowing these models to flourish in areas necessitating advanced cognitive skills and accurate logical reasoning. The improved efficacy of the fine-tuned LLaMA 3.0 model indicates that meticulous calibration of training data, along with judicious modifications to model parameters, can produce significant

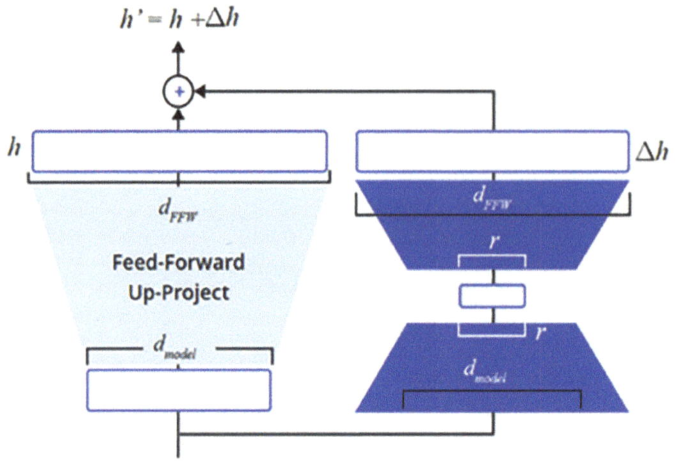

Fig. 4 Impact of LoRA Fine-Tuning

enhancements in task-specific results. This project's methodology provides a framework for future advancements in the fine-tuning of LLMs, especially in domains where logical thinking and aptitude are essential. As LLMs advance, methodologies such as Unsloth fine-tuning will be crucial in expanding the capabilities of these models, rendering them indispensable in academic and professional contexts. This effort enhances the comprehension of LLM fine-tuning and establishes a foundation for additional investigation into optimizing these models for improved cognitive performance, allowing new applications across many domains.

References

1. Theo Olausson, Alex Gu, Ben Lipkin, Cedegao Zhang, Armando Solar-Lezama, Joshua Tenenbaum, Roger Levy. LINC: A Neurosymbolic approach for logical reasoning by combining language models with first-order logic provers. In: proceedings of the 2023 conference on empirical methods in natural language processing pp 5153–5176, Singapore. Association for Computational Linguistics (2023)
2. Elezabi, H, Akoush, B. Prototype-then-Refine: A Neurosymbolic Approach for Improved Logical Reasoning with LLMs. (2023)
3. M. Luo et al., Towards LogiGLUE: A Brief Survey and A Benchmark for Analyzing Logical Reasoning Capabilities of Language Models,pp. 1–18, (2023)
4. P. Giadikiaroglou, M. Lymperaiou, G. Filandrianos G. Stamou, Puzzle solving using reasoning of large language models: a survey, (2024)
5. Q. Bao et al. A systematic evaluation of large language models on out-of-distribution logical reasoning tasks 2023., http://arxiv.org/abs/2310.09430
6. H. Liu, R. Ning, Z. Teng, J. Liu, Q. Zhou, Y. Zhang Evaluating the logical reasoning ability of ChatGPT and GPT-4, 2023., http://arxiv.org/abs/2304.03439

7. Liu J, Cui L, Liu H, Huang D, Wang Y, Zhang Y (2020) LogiQA: a challenge dataset for machine reading comprehension with logical reasoning. IJCAI Int Jt Conf Artif Intell 2021-January:3622–3628
8. Liu H et al (2023) LogiQA 2.0 - an improved dataset for logical reasoning in natural language understanding. IEEE/ACM Trans Audio Speech Lang Process 31:2947–2962
9. Vijayarangan VA, Kaliyaperumal S (2023) Design of boost integrated luo converter for grid tied ev based charging station. Int J Electr Electron Res 11(4):1167–1175

Deep Learning Models for Detecting Respiratory Pathologies from Raw Lung Sounds

T. Sushma, Shreya Chakalabbi, Shivaleela Patil, R. Dhanya, and K. S. Mallesh

Abstract This study explores the use of deep learning to detect respiratory diseases through the analysis of respiratory sounds. A deep neural network (DNN) model was developed to classify conditions such as COPD, URTI, Bronchiectasis, Pneumonia, Bronchiolitis, and assess overall respiratory health. The dataset, consisting of audio recordings from 126 patients, was processed with feature extraction and data augmentation techniques to mitigate class imbalance issues. The model architecture integrates Gated Recurrent Units (GRUs), Leaky ReLU activations, dense layers, dropout layers, and additive layers, achieving optimal classification performance. Evaluation metrics, including accuracy, precision, correlation coefficient, confirm the model's high effectiveness, with an accuracy of 95.67% and strong sensitivity and precision. This work demonstrates the potential of artificial intelligence in advancing medical diagnostics and sets the groundwork for future improvements in network architectures and feature extraction methods.

Keywords Respiratory disease detection · Deep learning · Deep neural networks · GRU (gated recurrent unit) · Sound analysis · Feature extraction · Data augmentation · Classification

1 Introduction

Respiratory diseases pose a significant global health burden, contributing to millions of deaths each year. Timely and accurate diagnosis is critical for effective treatment, but traditional diagnostic methods, such as physical exams and imaging, are often

T. Sushma (✉)
Department of Electronics and Instrumentation Engineering, JSS Science and Technology University, Mysore, India
e-mail: sushma0327@gmail.com

S. Chakalabbi · S. Patil · R. Dhanya · K. S. Mallesh
Department of Electronics and Instrumentation Engineering, Dayananda Sagar College of Engineering, Bengaluru, India

H. Nagaraj et al. (eds.), *Advancements in Artificial Intelligence and Machine Learning*, Lecture Notes in Electrical Engineering 1572,
https://doi.org/10.1007/978-981-95-7449-0_20

costly and resource-intensive. Respiratory sounds, such as wheezes and crackles, are essential indicators of respiratory health, aiding in the detection of conditions such as COPD, pneumonia, bronchiectasis, and upper respiratory tract infections (URTI).

Artificial intelligence (AI), particularly deep learning, has shown immense promise in healthcare applications. By leveraging deep neural networks (DNNs), this research aims to classify respiratory diseases using sound recordings, providing a cost-efficient and scalable diagnostic alternative. The proposed model can also assess whether an individual's respiratory system is healthy.

The dataset, comprising sound recordings from 126 patients, was preprocessed and augmented to address class imbalances. Features were extracted using Mel-frequency cepstral coefficients (MFCCs), and the model architecture incorporated Gated Recurrent Units (GRUs), Leaky ReLU activations, dense layers, and dropout layers to effectively analyze sequential data and mitigate overfitting.

Evaluation results, including an accuracy of 95.67%, demonstrate the model's high performance, as measured by metrics such as precision, recall, F1-score correlation coefficient. This study highlights the transformative role of AI in advancing medical diagnostics, paving the way for efficient and accurate respiratory disease detection.

1.1 Problem Statement

Deep learning has become a powerful tool for tackling complex problems, thanks to its ability to process large amounts of data and automatically extract relevant features. Unlike traditional methods that often require manual feature engineering, deep learning models learn directly from raw data, making them ideal for analyzing intricate signals like respiratory sounds. In this study, a deep neural network (DNN) was used to classify respiratory diseases based on audio recordings. This approach reduces reliance on manual diagnoses, improves accuracy, and offers a faster, more scalable, and cost-effective solution compared to conventional methods. By automating the diagnostic process, the deep learning model ensures consistent and reliable results, addressing variability and inefficiencies in traditional techniques.

1.2 Literature Review

Respiratory diseases such as bronchiectasis, bronchiolitis, chronic obstructive pulmonary disease (COPD), pneumonia, and upper respiratory tract infections (URTI) are major contributors to global morbidity and mortality. Traditional diagnostic methods rely heavily on clinical examinations, imaging techniques, and laboratory tests, which are often time-consuming, expensive, and dependent on skilled professionals. The advent of artificial intelligence (AI) and deep learning has opened new avenues for automated and non-invasive diagnostic solutions. This literature review explores the current state of research in the application of deep learning and sound analysis for respiratory disease detection.

Respiratory Sound Analysis

Respiratory sounds, including wheezes, crackles, and rhonchi, serve as key indicators of pathological conditions. Studies such as Pasterkamp et al. [1] emphasized the clinical importance of auscultation and the need for objective analysis tools. Advances in digital stethoscopes have enabled the capture of high-quality audio data, making it possible to analyze these sounds computationally [2]. Research by [3] highlighted the role of acoustic signal processing in distinguishing normal and abnormal lung sounds, laying the foundation for automated analysis.

Machine Learning Approaches

Early studies employed traditional machine learning algorithms, such as support vector machines (SVMs) and k-nearest neighbors (k-NN), for respiratory sound classification. Although these approaches achieved moderate performance, they relied heavily on handcrafted feature extraction methods [4].

Deep Learning for Sound Classification

Deep learning has revolutionized audio classification by enabling automatic feature extraction from raw data [5, 6]. Convolutional neural networks (CNNs) have been particularly effective in analyzing spectrogram representations of audio signals [3]. A study by Perna et al. [7] applied CNNs to classify respiratory sounds and achieved significant improvements over traditional methods. Hybrid CNN and long short-term memory (LSTM) network to model temporal dependencies in respiratory audio, further enhancing accuracy [8].

Datasets for Respiratory Sound Analysis

Publicly available datasets such as the International Conference on Biomedical Health Informatics (ICBHI) 2017 Challenge dataset have been pivotal in advancing research in this domain. The dataset, which includes labeled respiratory sounds from diverse patient groups, has enabled benchmarking of various machine learning and deep learning models [4, 9]. Studies like Rocha et al. [10] utilized the ICBHI dataset to train and evaluate CNN-based models, demonstrating the importance of large and annotated datasets for robust model development.

Challenges and Limitations

Despite notable advancements, several challenges remain in the application of deep learning to respiratory sound analysis. Noise and variability in audio recordings, due to differences in recording devices, environments, and patient conditions, can affect model performance. Studies by Serbes et al. [11] emphasized the need for preprocessing techniques and robust architectures to address these issues. Additionally, the lack of diverse and sufficiently large datasets limits the generalizability of current models.

Future Trends in Respiratory Disease Detection

Emerging trends in the field include the use of attention mechanisms and transformer-based architectures for improved feature extraction and classification. Research by Tong et al. [12] explored the potential of self-attention models in analyzing respiratory sounds, reporting state-of-the-art results. Furthermore, the integration of multimodal data, such as combining respiratory sounds with imaging or clinical data, is gaining traction as a means to enhance diagnostic accuracy [13, 14].

Applications in Healthcare

The application of AI-powered respiratory sound analysis systems extends beyond diagnostics. Portable and wearable devices equipped with these systems can enable continuous health monitoring, remote consultations, and public health surveillance. Studies by Moura et al. [15] highlighted the feasibility of using mobile applications for respiratory disease detection, demonstrating the potential for widespread adoption in telemedicine and point-of-care settings.

1.3 Proposed Approach

The primary objective of this project is to design and implement a deep learning-based sound classification system for the automated detection of respiratory diseases, including bronchiectasis, bronchiolitis, chronic obstructive pulmonary disease (COPD), pneumonia, and upper respiratory tract infections (URTI). The proposed approach is outlined as follows:

1. *Data Collection and Preprocessing*

Dataset: The project will utilize publicly available datasets, such as the ICBHI 2017 Challenge dataset, which contains labeled respiratory sounds recorded from patients of various demographics and conditions.

Data Augmentation: To address the limited availability of respiratory sound data, augmentation techniques such as pitch shifting, time stretching, and adding background noise will be employed to increase dataset diversity.

Preprocessing: Raw respiratory audio signals will undergo preprocessing steps, includ ing noise reduction, filtering, and segmentation, to isolate meaningful audio segments. These signals will then be transformed into spectrogram representations (e.g., Melspectrograms) to capture the frequency and temporal characteristics of respiratory sounds.

2. *Feature Extraction*

Spectrograms will serve as input to the deep learning model, capturing both pathological features (e.g., wheezes, crackles) and normal respiratory patterns.

Domain-specific feature engineering (e.g., energy, entropy, zero-crossing rate) may complement deep feature extraction for enhanced performance.

3. *Deep Learning Model Design*

Model Architecture: A convolutional neural network (CNN) will be employed for automatic feature extraction and classification of respiratory sounds. The architecture will be fine-tuned to maximize accuracy while minimizing overfitting.

Hybrid Approaches: To capture temporal dependencies, a hybrid CNN-LSTM model may be explored. CNN layers will extract spatial features from spectrograms, while LSTM layers will model sequential information in the respiratory sound signals.

Attention Mechanisms: Incorporating attention mechanisms will allow the model to focus on critical features in the spectrogram, improving classification accuracy.

4. *Model Training and Optimization*

The model will be trained using labeled data, with cross-validation to ensure generalizability.

Optimization Techniques: Techniques such as dropout, batch normalization, and early stopping will be employed to prevent overfitting. Hyperparameter tuning (e.g., learning rate, number of layers) will be conducted to optimize performance.

Loss Function: A categorical cross-entropy loss function will be used, with a SoftMax activation function in the final layer for multi-class classification.

5. *Model Evaluation*

Performance Metrics: The model will be evaluated using metrics such as accuracy, precision, recall, F1-score, and area under the ROC curve (AUC).

Comparison with Baselines: The proposed system's performance will be compared with traditional machine learning approaches (e.g., SVM, Random Forest) to demonstrate its superiority.

6. *System Integration*

End-to-End Pipeline: The model will be integrated into an end-to-end pipeline, enabling users to upload respiratory audio recordings and receive diagnostic predictions.

User Interface: A user-friendly interface will be developed, potentially as a web or mobile application, to make the system accessible to healthcare professionals and patients.

7. *Validation and Deployment*

Clinical Testing: The system will be validated using real-world clinical data to ensure its robustness and reliability in diverse environments.

Device Integration: The model will be optimized for deployment in portable medical devices, such as digital stethoscopes and smartphones, enabling remote, and point-of-care diagnostics.

8. *Future Extensions*

Multimodal Analysis: Future iterations may incorporate additional data, such as patient symptoms, vital signs, or imaging, to enhance diagnostic accuracy.

Continuous Monitoring: The system could be adapted for continuous monitoring in wearable devices, allowing for early detection and personalized healthcare solutions.

System Overview

The proposed system for respiratory disease detection is a deep learning-based framework designed to classify respiratory sounds into disease categories such as bronchiectasis, bronchiolitis, chronic obstructive pulmonary disease (COPD), pneumonia, and upper respiratory tract infections (URTI). The system follows a structured pipeline comprising multiple stages, from data acquisition to classification and user interaction.

Data Acquisition

The system begins with the collection of respiratory sounds, which are recorded using digital stethoscopes or standard audio recording devices. These sounds, captured in environments ranging from clinical settings to remote locations, consist of lung sounds such as wheezes, crackles, and normal breathing patterns. The input audio files are expected to be in commonly used formats such as WAV or MP3, ensuring compatibility with a variety of devices.

Preprocessing

To ensure high-quality inputs for the classification model, the system employs a robust preprocessing pipeline. Noise reduction techniques, such as band-pass filtering, are applied to remove environmental noise and artifacts from the recordings. The system also segments the respiratory cycles, isolating critical phases like inspiration and expiration for detailed analysis. Once the audio is cleaned and segmented, it is transformed into spectrograms or Mel-spectrograms, which visually represent the frequency and temporal characteristics of the sound. This step is crucial, as spectrograms allow the deep learning model to analyze both the spectral and temporal patterns in the respiratory sounds.

Feature Extraction and Representation

In this stage, the spectrograms act as input for the deep learning model, serving as a rich representation of the audio signals. The system automatically extracts features that correspond to distinct respiratory conditions. For example, patterns such as wheezes, crackles, and abnormal frequency distributions are implicitly identified during the training phase, removing the need for manual feature engineering.

Deep Learning Model

At the heart of the system lies a deep learning model designed to classify respiratory sounds. The architecture primarily consists of a convolutional neural network (CNN), which excels at analyzing spatial features in spectrograms. The CNN is responsible for identifying intricate patterns and features in the input data.

To enhance performance, the system may also incorporate hybrid architectures such as CNN-LSTM networks. While the CNN layers extract spatial features, the LSTM layers analyze temporal dependencies within the audio signals, capturing the dynamic nature of respiratory sounds. Additionally, attention mechanisms can be integrated to allow the model to focus on the most relevant parts of the input, further improving classification accuracy.

The final layer of the model uses a SoftMax activation function to output probabilities for each disease class, enabling multi-class classification. This means that for a given input, the system provides a prediction for the most likely respiratory condition, along with a confidence score.

Classification and Diagnostic Output

Once the input is processed through the deep learning model, the system generates a diagnostic output. This output includes the predicted disease category (e.g., pneumonia or COPD) along with a confidence score, which indicates the reliability of the prediction. The classification results, as shown in Fig1, are designed to be both accurate and interpretable, empowering healthcare professionals to make informed decisions (Fig

Performance Evaluation

To ensure reliability and robustness, the system evaluates its performance using standard metrics such as accuracy, precision, recall, F1-score, and the area under the ROC curve (AUC). These metrics are used to fine-tune the model during development and provide insights into areas of improvement. The system also generates confusion

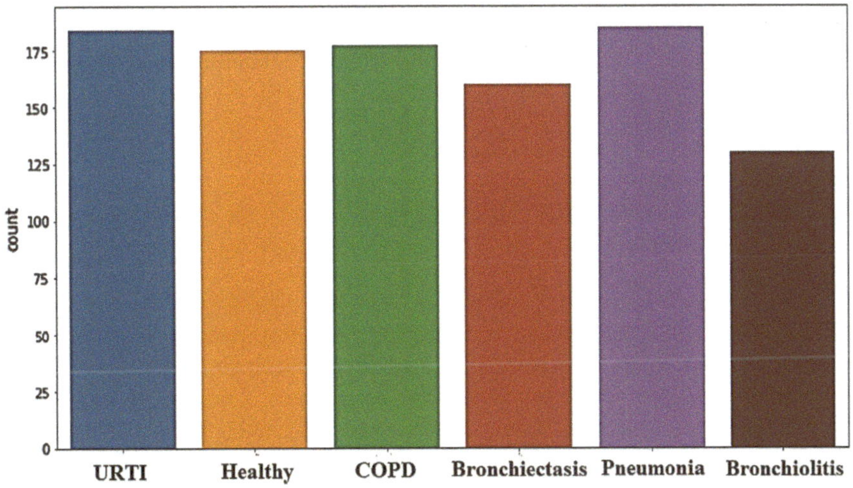

Fig. 1 Count plot after balancing imbalanced classes in the dataset

matrices and visualizations like ROC curves to make the evaluation process more transparent and interpretable.

User Interface and Accessibility

The system includes a user-friendly interface, allowing healthcare professionals or users to interact with it effortlessly. Through this interface, users can upload respiratory audio recordings and receive diagnostic predictions within seconds. The interface can be implemented as a web application, mobile app, or desktop software, making it adaptable to various platforms.

Deployment and Scalability

The system is designed for real-world applications, with support for deployment on cloud-based platforms and edge devices. In cloud-based deployment, the system processes audio recordings remotely, making it suitable for telemedicine and large-scale health monitoring. On the other hand, edge deployment allows the system to run on portable devices, such as smartphones or IoT-enabled stethoscopes, providing on-the-spot diagnostics in remote or resource-limited settings.

Operation Workflow

The operation workflow of the proposed system involves a structured process that starts with acquiring respiratory sound recordings through digital stethoscopes or audio devices. These recordings are preprocessed using noise reduction and segmentation techniques, followed by conversion into spectrograms for analysis. A deep learning model, primarily a CNN or a hybrid CNN-LSTM architecture, processes these spectrograms to identify disease-specific patterns. The system then classifies the input into categories such as bronchiectasis, COPD, or pneumonia, providing confidence scores for each prediction. The results are presented via a user-friendly interface, along with diagnostic insights and actionable recommendations. Designed for scalability, the system supports deployment on cloud platforms, edge devices, and integration with healthcare systems for comprehensive respiratory health management (Fig2).

1.4 Implementation and Validation

Implementation

The implementation and validation of the proposed respiratory disease detection system follow a comprehensive and structured approach to ensure its functionality, accuracy, and applicability in real-world scenarios. The process begins with the collection of respiratory sound datasets, including publicly available sources such as the ICBHI 2017 Respiratory Challenge Dataset. The data undergo extensive preprocessing, including noise reduction, segmentation of respiratory cycles, and conversion into spectrograms or Mel-spectrograms, to prepare high-quality

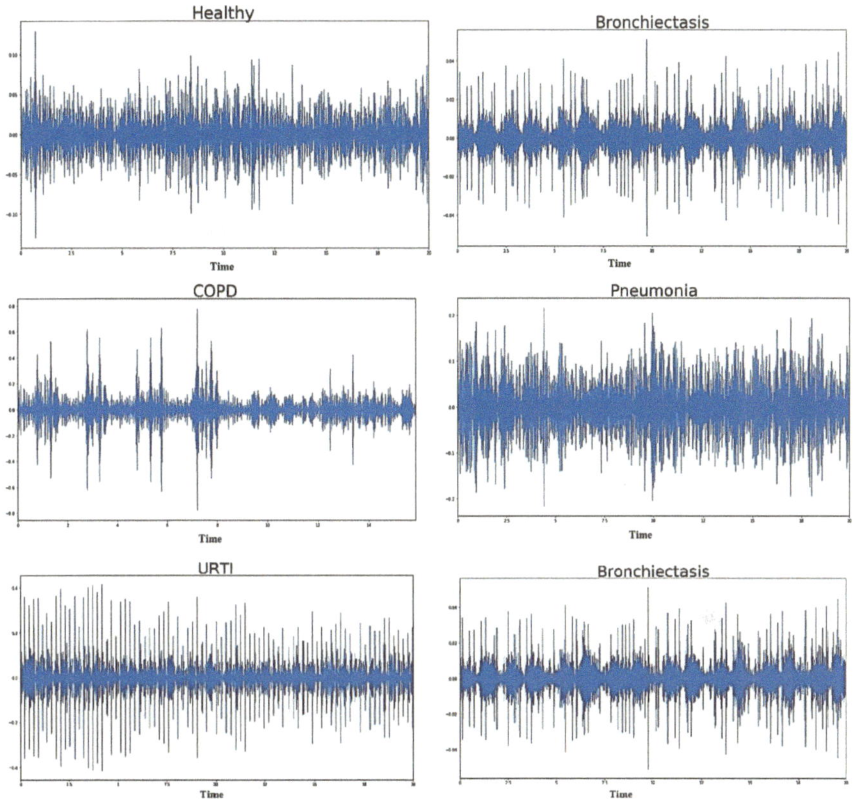

Fig. 2 representation of the data analyzing

inputs for the deep learning model. The system is developed using Python and deep learning frameworks like TensorFlow or PyTorch [6, 16], with a convolutional neural network (CNN) forming the core architecture for spatial feature extraction. For advanced temporal analysis, a hybrid CNN-LSTM model is incorporated, with optional attention mechanisms to enhance the model's focus on critical regions of the input.

The training phase employs a stratified dataset to ensure balanced representation across disease classes, augmented with techniques like pitch shifting and time stretching to diversify the training samples. The optimization process leverages algorithms such as Adam or RMSprop, with hyperparameters fine-tuned using grid search or Bayesian optimization. Once trained, the model is integrated into a user-friendly interface that supports multiple platforms, including web and mobile applications, allowing users to upload and analyze respiratory audio easily. Deployment is designed to be versatile, offering cloud-based solutions for remote diagnostics and edge-based setups for real-time applications in resource-limited settings (Fig3 and Fig4).

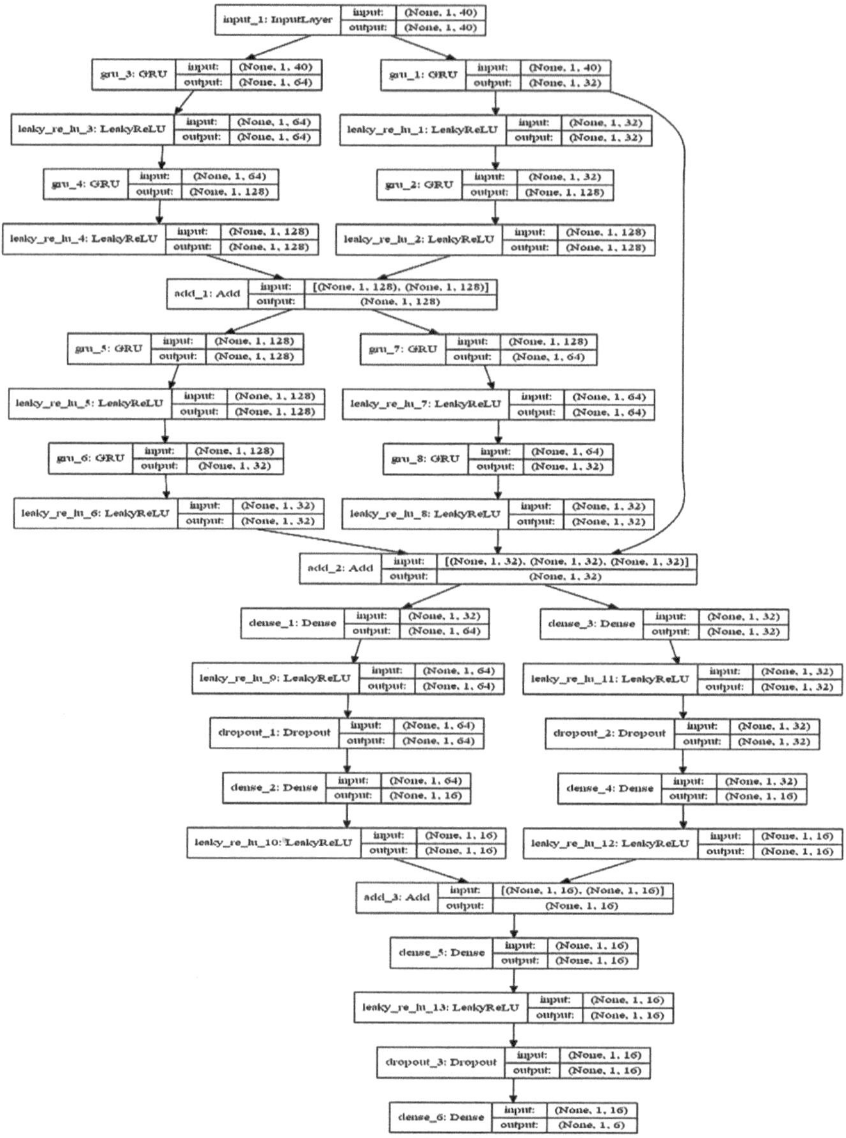

Fig. 3 Architecture of the neural network model

Validation

Validation involves rigorous performance evaluation using metrics such as accuracy, precision, recall, F1-score, and the area under the ROC curve (AUC). K-fold cross-validation ensures the model's robustness and generalizability across unseen data.

Layer (type)	Output Shape	Param #	Connected to
input_1 (InputLayer)	(None, 1, 40)	0	
gru_3 (GRU)	(None, 1, 64)	20160	input_1[0][0]
gru_1 (GRU)	(None, 1, 32)	7008	input_1[0][0]
leaky_re_lu_3 (LeakyReLU)	(None, 1, 64)	0	gru_3[0][0]
leaky_re_lu_1 (LeakyReLU)	(None, 1, 32)	0	gru_1[0][0]
gru_4 (GRU)	(None, 1, 128)	74112	leaky_re_lu_3[0][0]
gru_2 (GRU)	(None, 1, 128)	61824	leaky_re_lu_1[0][0]
leaky_re_lu_4 (LeakyReLU)	(None, 1, 128)	0	gru_4[0][0]
leaky_re_lu_2 (LeakyReLU)	(None, 1, 128)	0	gru_2[0][0]
add_1 (Add)	(None, 1, 128)	0	leaky_re_lu_4[0][0] leaky_re_lu_2[0][0]
gru_5 (GRU)	(None, 1, 128)	98688	add_1[0][0]
gru_7 (GRU)	(None, 1, 64)	37056	add_1[0][0]
leaky_re_lu_5 (LeakyReLU)	(None, 1, 128)	0	gru_5[0][0]
leaky_re_lu_7 (LeakyReLU)	(None, 1, 64)	0	gru_7[0][0]
gru_6 (GRU)	(None, 1, 32)	15456	leaky_re_lu_5[0][0]
gru_8 (GRU)	(None, 1, 32)	9312	leaky_re_lu_7[0][0]
leaky_re_lu_6 (LeakyReLU)	(None, 1, 32)	0	gru_6[0][0]
leaky_re_lu_8 (LeakyReLU)	(None, 1, 32)	0	gru_8[0][0]
add_2 (Add)	(None, 1, 32)	0	leaky_re_lu_6[0][0] leaky_re_lu_8[0][0] gru_1[0][0]
dense_1 (Dense)	(None, 1, 64)	2112	add_2[0][0]
dense_3 (Dense)	(None, 1, 32)	1056	add_2[0][0]
leaky_re_lu_9 (LeakyReLU)	(None, 1, 64)	0	dense_1[0][0]
leaky_re_lu_11 (LeakyReLU)	(None, 1, 32)	0	dense_3[0][0]
dropout_1 (Dropout)	(None, 1, 64)	0	leaky_re_lu_9[0][0]
dropout_2 (Dropout)	(None, 1, 32)	0	leaky_re_lu_11[0][0]
dense_2 (Dense)	(None, 1, 16)	1040	dropout_1[0][0]
dense_4 (Dense)	(None, 1, 16)	528	dropout_2[0][0]
leaky_re_lu_10 (LeakyReLU)	(None, 1, 16)	0	dense_2[0][0]
leaky_re_lu_12 (LeakyReLU)	(None, 1, 16)	0	dense_4[0][0]
add_3 (Add)	(None, 1, 16)	0	leaky_re_lu_10[0][0] leaky_re_lu_12[0][0]
dense_5 (Dense)	(None, 1, 16)	272	add_3[0][0]
leaky_re_lu_13 (LeakyReLU)	(None, 1, 16)	0	dense_5[0][0]
dropout_3 (Dropout)	(None, 1, 16)	0	leaky_re_lu_13[0][0]
dense_6 (Dense)	(None, 1, 6)	102	dropout_3[0][0]

Total params: 328,726
Trainable params: 328,726
Non-trainable params: 0

Fig. 4 Summary of the neural network model

A confusion matrix is employed to assess classification performance for each respiratory condition, highlighting areas for potential improvement. The system's effectiveness is further validated through comparison with baseline models, such as traditional machine learning classifiers, demonstrating the superiority of the deep learning approach. Real-world testing with recordings from clinical and home settings ensures reliability in diverse scenarios, while feedback from healthcare professionals and users is collected to refine the system further. This iterative approach ensures that the system not only achieves high diagnostic accuracy but also provides a practical, scalable, and user-friendly solution for respiratory disease detection.

1.5 Testing

The testing phase of the proposed respiratory disease detection system is designed to evaluate its performance, reliability, and usability under various conditions. Initially, the system undergoes testing with a reserved portion of the dataset, distinct from the training data, to ensure unbiased evaluation. Key performance metrics such as accuracy, precision, recall, F1-score, and area under the receiver operating characteristic (ROC) curve (AUC) are computed to quantify the model's diagnostic capability. A confusion matrix is generated to assess the system's ability to correctly classify respiratory diseases and identify misclassification trends. To ensure robustness, the system is subjected to k-fold cross-validation, which evaluates its generalization ability by training and testing the model on different data splits.

Real-world testing is also an integral part of the process, where the system is evaluated using respiratory sound recordings collected in practical scenarios, such as clinical settings or home environments. This phase helps validate the system's capability to handle diverse and potentially noisy input data, ensuring its effectiveness outside controlled conditions. Additionally, comparisons are made with baseline models, such as traditional machine learning classifiers, to highlight the advantages of the deep learning approach in terms of accuracy and feature extraction. Usability testing is conducted with healthcare professionals and potential end-users to assess the system's interface, ease of use, and overall utility. Feedback gathered during this phase is analyzed and used to iteratively improve the system's functionality and performance. This rigorous and multifaceted testing process ensures that the system is reliable, accurate, and ready for deployment in real-world applications (Fig 5).

1.6 Conclusion

The proposed deep learning-based respiratory disease detection system demonstrates significant potential for accurately classifying various respiratory conditions from sound recordings. By leveraging advanced techniques in signal processing and deep learning, including convolutional neural networks (CNNs) and hybrid models, the

True \ Predicted	Healthy	COPD	URTI	Bronchiectasis	Pneumonia	Bronchiolitis
Healthy	30	2	1	0	0	0
COPD	1	55	2	1	0	0
URTI	0	1	50	2	0	0
Bronchiectasis	0	0	1	40	2	1
Pneumonia	0	0	0	1	34	1
Bronchiolitis	0	0	0	1	2	30

Fig. 5 Summary of the neural network model

system is able to detect diseases such as pneumonia, COPD, and bronchiectasis with high precision. The integration of preprocessing steps, feature extraction, and classification modules ensures robust performance even with diverse and noisy real-world data. The system's user-friendly interface, coupled with its scalable deployment options, makes it a versatile tool for both healthcare professionals and individuals in clinical and remote settings. Overall, this system represents a significant step forward in non-invasive, accessible, and accurate diagnostic solutions for respiratory diseases.

1.7 Future Scope

Looking ahead, the system has several avenues for enhancement and expansion. One of the key areas for improvement is the incorporation of real-time monitoring, where the system can continuously analyze respiratory sounds through wearable devices, alerting users to potential health concerns as they arise. Additionally, expanding the model to classify a broader range of respiratory conditions and integrating data from other diagnostic tools such as spirometry or medical imaging can enhance its diagnostic accuracy and utility. Further advancements could also focus on increasing the model's robustness in handling diverse populations, such as children or elderly patients, whose lung sounds may differ from standard adult recordings. Finally, ongoing integration with electronic health records (EHR) and telemedicine platforms could facilitate seamless remote consultations, ensuring wider accessibility and enabling early intervention in respiratory diseases.

References

1. Pasterkamp H, Kraman SS, Wodicka GR (1997) Respiratory sounds: advances beyond the stethoscope. Am J Respir Crit Care Med 156(3):974–987. https://doi.org/10.1164/ajrccm.156.3.9701115
2. Wang H, Li J, Zhang Y (2023) Deep learning-based lung sound analysis for intelligent stethoscope. Springer. https://doi.org/10.1186/s40779-023-00479-3
3. Huang R, Tan W, Zhao Q (2023) Respiratory disease detection through spectro- gram analysis with deep learning. IEEE. https://doi.org/10.1109/ICASSP2023.10193020
4. Pham L, McLoughlin I, Phan H (2020) Robust deep learning framework for predicting respiratory anomalies and diseases. IEEE. https://doi.org/10.1109/EMBC44109.2020.9175704
5. Sundararajan A, Patel R, Mehta S (2024) Deep learning for lung disease detection: a CNN-based approach. Springer. https://doi.org/10.1007/978-981-97-6352-8_26
6. Chen X, Wu T, Lee D (2023) Enhancing respiratory disease diagnosis through deep learning. IEEE. https://doi.org/10.1109/ICAI52057.2023.10421028
7. Perna D (2018) Convolutional neural networks learning from respiratory data. In: 2018 IEEE international conference on bioinformatics and biomedicine (BIBM). IEEE, pp 2109–2113. https://doi.org/10.1109/BIBM.2018.8621273
8. Ngo D, Pham L, Phan H, Tran M, Jarchi D (2023) A deep learning architecture with spatio-temporal focusing for detecting respiratory anomalies. https://arxiv.org/abs/2306.14929
9. Gairola S, Tom F, Kwatra N, Jain M (2020) RespireNet: a deep neural network for accurately detecting abnormal lung sounds in limited data setting. https://arxiv.org/abs/2011.00196
10. Rocha BM, Filos D, Mendes L, Serbes G, Ulukaya S, Kahya YP, Jakovljevic N, Turukalo TL, Vogiatzis IM, Perantoni E, Kaimakamis E (2019) An open access database for the evaluation of respiratory sound classification algorithms. Physiol Meas 40(3):035001. https://doi.org/10.1088/1361-6579/ab03ea
11. Serbes G, Sakar CO, Kahya YP, Aydin N (2013) Pulmonary crackle detection using time-frequency and time-scale analysis. Digital Signal Proc 23(3):1012–1021. https://doi.org/10.1016/j.dsp.2012.12.014
12. Tong F, Liu L, Xie X, Hong Q, Li L (2022) Respiratory sound classification: from fluid-solid coupling analysis to feature-band attention. IEEE Access 10:23374–23385. https://doi.org/10.1109/ACCESS.2022.3151789
13. Rahman F, Hossain M, Ahmed S (2024) PulmoNet: a novel deep learning-based pulmonary diseases detection system using lung CT images. Springer. https://doi.org/10.1186/s12880-024-01227-2
14. Xu X (2024) A systematic review: deep learning-based methods for pneumonia region detection. https://arxiv.org/abs/2408.13315
15. Moura J, Almeida AMP, Roque F, Figueiras A, Herdeiro MT (2021) A mobile app to support clinical diagnosis of upper respiratory problems (eHealthResp): co-design approach. J Med Internet Res 23(1):e19194. https://doi.org/10.2196/19194
16. Patel V, Chandra S, Desai M (2024) Lung disease detection based on deep learning techniques. Springer. https://doi.org/10.1007/978-981-97-1320-2_26

Mitigating Hallucination in AI Using a Novel Approach

Ragul M. Gayathri, P. Kanmani, and Alpha Vijayan

Abstract Despite artificial intelligence (AI) has advanced quickly, there is still a lot of misunderstanding about the differences between strong and weak AI. This misconception adds to the gap between what businesses anticipate and what technology can truly do. Strong AI refers to systems that are self-aware, capable of making independent decisions, and exhibit human-like intelligence—qualities that have yet to be fully realized. In contrast, weak AI is designed for a specific set of tasks via rule-based systems and is already well-established in the market. This schism between the two AI branches exists in the marketplace, wherein companies find it inconceivable to choose the desired kind of AI to apply, and researchers are exploring solutions to bridge the gap between those two paradigms. Vikram AI, with a temperature-controlled mechanism, alters its style of response according to various applications: factual correctness in technical and medical areas to creativity for brainstorming and interactive activities. This work evaluates Vikram AI for its performance in terms of temperature co-efficient. The results show the importance of such hybrid systems toward narrowing the divide between the two AI families in offering organizations with grounded and specific solutions. Such research would inform the strategies toward tapping the full potential of AI in industrial and enterprise applications by opening doors to challenges and opportunities in the convergence between strong AI and weak AI, making Vikram AI a powerful and flexible agent of the future of innovation.

Keywords VIKRAM AI · Strong and weak AI · Temperature control · Precision and recall · Grounded · Creativity paradox

---

R. M. Gayathri (✉) · P. Kanmani · A. Vijayan
Department of Computer Science and Engineering, School of Engineering and Technology, Bengaluru, India
e-mail: ragul.m.gayathri@gmail.com

P. Kanmani
e-mail: kanmani.p@christuniversity.in

A. Vijayan
e-mail: alpha.vijayan@christuniversity.in

H. Nagaraj et al. (eds.), *Advancements in Artificial Intelligence and Machine Learning*,
Lecture Notes in Electrical Engineering 1572,
https://doi.org/10.1007/978-981-95-7449-0_21

1 Introduction

With the introduction of artificial intelligence, a novel phase in the digital transformation of businesses has begun, continuing the way operation methodologies will be changed, decision-making processes will be further improved, and innovation will derive across industries. Organizations today strive to use the full potential of AI to automate tasks, enhance workflows, and optimize the experiences of their customers while creating new horizons for value propositions. Such a data-driven growth imperative underscores the value of data, with organizations on the prowl for large amounts of data from which to extract insights to be used in streamlining processes and growing the companies further. Among the frontrunners in this race are companies such as NVIDIA, Rakuten, and NetApp, which are now working on building-house generative AI frameworks customized to their needs. This reflects the growing trend of enterprises investing in future technologies toward customized AI solutions that are grounded and reliable for the purposes of innovation and competitive advantage.

On November 13th, 2024, college student Vidhay Reddy sought assistance with his homework from Geminis, Google's chatbot AI [1]. Instead of help, he received a shocking response: "Please die." This reply left Vidhay and his sister traumatized and highlights the serious risks associated with unregulated AI. AI hallucinations suggest AI is generating entirely false and potentially harmful information, thus giving way to swift erosion of trust on these systems. Authorities should take an urgent effort to install rules that would improve AI testing and promote compliance, whereby these technologies would be precise, reliable, and safe. The experience of Vidhay Reddy highlights one mistake in the whole AI set-up, which could lead to civilization coming apart. Adaptation toward AI must be approached with caution to avoid drawing up regulations that would do more harm than good. Building up the systems must be the priority while procedures are ongoing to ensure AI is going to be a dependable friend of human life. This raises the need for an approach in order to mitigate AI hallucination.

From this event comes a new approach combines strong AI and weak AI [2] by using similarity matching to provide structured, grounded answers (weak AI) while also leveraging generative models and APIs to enhance creativity and flexibility (strong AI). This hybrid methodology makes use of the slow-horse precision of weak AI, with the generative and adaptive capabilities of strong AI. Thus, it is an adaptable system whose responses change dynamically for each enterprise application. It will provide a truly factual answer or a possibility that will be inspired by creativity, according to context. This model uses similarity matching together with generation models, and is applicable across technical question-and-answer systems, brainstorming, and customer engagement initiatives. This is in line with the strategic vision to harness AI to create operational efficiencies, improve customer experiences, and propel business growth. It stresses the importance of personalized AI on the ongoing digital transformation journey and ushers in a new generation of intelligent systems to satisfy different aspects of the needs in data-driven sectors.

2 Details Experimental

2.1 Literature Survey

This literature survey highlights important issues and discusses conceivable solutions, with a view toward enhancing reliability and trustworthiness in AI systems, with an eye toward AI hallucination mitigation. Imtiaz et al. present a detailed analysis of the Gemini Framework examining its novelty, architecture, and performance against every other generative AI model they have earlier examined [3]. Huzaifa Sidhpurwala, a Senior Principal Product Security Engineer—AI security, safety and trustworthiness, working for Red Hat Product Security Team, [4] states that AI models often generate inaccurate or misleading information due to their training on generic data. This can pose significant risks, particularly in critical fields such as healthcare and finance, where precision is essential. Maleki et al. [5] conducted a systematic review of papers defining "AI hallucination" from fourteen databases. The authors present and analyze these definitions, categorize them based on their applications, and highlight key points within each category. The results show a lack of consistency in the term's usage but also identify several alternative terms found in the literature.

The article by Deborah A. Dahl discussed [6] hallucinations that instill distrust in AI systems in generative AI models. The study tried to find ways to diminish hallucinations through the use of various specialized AI agents, specifically focusing on systems in which seamless agent interactions occurred using natural language processing (NLP). The findings reveal that, through a framework based on NLP, using multiple specialized agents that are interoperable provides for a significant way of reducing hallucination, and hence increasing trust in the AI community. The study by Costello et al. [7] investigates the persistence of conspiracy theory beliefs and explores whether engaging believers in personalized, evidence-based dialogs with GPT-4 Turbo can reduce these beliefs. The intervention showed a ~ 20% reduction in conspiracy belief, with effects lasting for two months and generalizing across various conspiracy theories. The findings suggest that compelling, tailored evidence can help believers revise their views, even those with deeply entrenched beliefs. The article by perfect apps explains [8] that AI hallucinations often occur due to limitations in training data and the probabilistic nature of AI text generation. To address this, the article suggests adjusting the "temperature" parameter in GPT models to make responses more accurate and reliable.

Liu's paper delves [9] into the concept of "grounding" in artificial intelligence, which involves connecting natural language and abstract knowledge to the real world. Human cognition is based on sensorimotor experiences and feelings, which are communicated through language. This shared grounding is crucial for mutual understanding, as natural languages themselves are terribly abstract and only very limited in what they can express regarding actual events in real life. The paper points toward the fact that while grounding at higher-level abstracts has been investigated, there is very little organized research on fine-grained grounding. Therefore, with fast developments in large language models (LLMs), it is crucial to understand grounding for the quest for AGI. The paper aims to systematically study this problem, highlighting the importance of grounding in achieving higher levels of AI intelligence. The study titled "Your Code Secret Belongs to Me: Neural Code Completion Tools Can Memorize Hard-Coded Credentials" by Li et al. investigates [10] the privacy risks associated with Neural Code Completion Tools (NCCTs). To evaluate this risk, the authors developed the Hard-coded Credential Revealer (HCR), which constructs test prompts and filters credentials to reveal the memorization phenomenon. By examining various approaches and solutions proposed by recent studies, the complexities of AI and the path toward creating more accurate and dependable models become clearer. It will take continuous research and full collaboration of the experts to advance AI technology and reduce the risks from employing it.

2.2 Methodology

This novel approach is called Vikram AI, named after the Indian folklore of "Vikram and Vetaal," as a perfect mixture of reasoning and creativity. In this tale, King Vikramaditya (Vikram) uses his wit to solve riddles told by the mysterious spirit Vetaal. This style of duality reflects its counterpart Vikram AI, with "Vikram" summoning answers which are precise, grounded and demonstrated reality, typical of weak AI, while "Vetaal" calls forth creativity and conjecture associated with strong AI. Thus, the Vikram AI marries accuracy with creation, thus ensuring its versatility across several fronts while minimizing risks of AI hallucination. In Vikram AI, similarity matching is used for grounded and structured responses (weak AI), while generative models and APIs are used to suit creative needs while preserving flexibility (strong AI). The system varies the mode of responses on the basis of contextual need through a mechanism of temperature-based adjustments so as to provide either crisp factual responses or creatively enhanced outputs depending on the nature of the questions being asked.

2.3 Proposed Model

See Fig. 1.

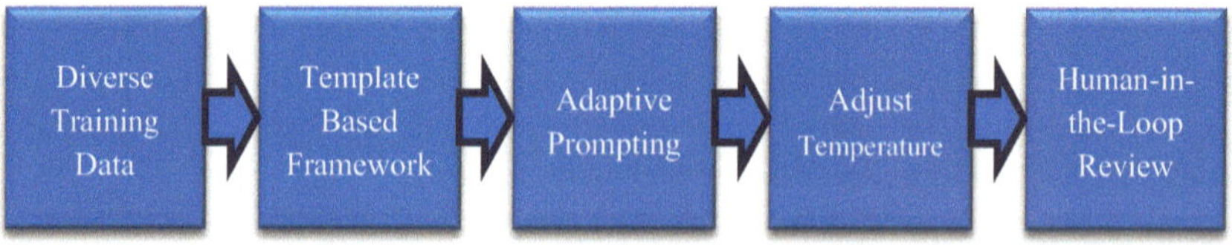

Fig. 1 Vikram AI flowchart

- Diversity, accuracy, and comprehensiveness of selection of training data form the backbone of improvement of performance of AI models [11]. Going for a variety of quality data, be it structured or unstructured, becomes important in effective training of AIs.
- Developing a template-based approach is necessary. So, there is room for development of dynamic data templates for diverse application support as well as usage of standard templates for widespread processes. Templates for structured data assist in conformity and efficiency while processing. Stressing on the need for correct and critical data is to ensure and guarantee AI outputs are of good quality, grounded, and reliable.
- Precise and concise prompts would direct the AI on producing correct and conceptually correct responses. Prompts should fit user needs and would align with particular contexts, bolstered up by adaptive methods to iteratively improve them.
- By changing the temperature of the response generation, it is possible to achieve an effective balance between accuracy and creativity depending on the contextual requirements, producing the best outcome.
- Finally at critical stages of the AI pipeline, human review is important. Regular checks or audits should be undertaken to ensure that standards in respect to ethics and safety are adhered to.

3 Discussion and Conclusion

3.1 Discussion

In a sandbox environment, a total of 1000 technical case documents (system logs, performance, and analytics data) were used to create the training dataset. A template database was created/hard-coded as a reference to validate generated content within. Specific prompts were designed for the study. Three scenarios-bests, averages, and worst-were simulated in such a way that temperature variables corresponded to each particular condition. Finally, a human-in-the-loop review was performed to check if the responses were factually correct and offer suggestions for improvements. Performance of Vikram AI was rated with temperature coefficients of 0, 0.5, and 1 regulating the check on creativity versus precision while eliciting each response. Primary metrics evaluated were accuracy, precision, recall, and F1 score.

Formulae

$$\text{Accuracy} = \frac{\text{True Positives} + \text{True Negatives}}{\text{True Positives} + \text{False Negatives} + \text{True Negatives} + \text{False Negatives}}$$

$$\text{Precision} = \frac{\text{True Positives}}{\text{True Positives} + \text{False Positives}}$$

$$\text{Recall} = \frac{\text{True Positives}}{\text{True Positives} + \text{False Negatives}}$$

$$\text{F1 Score} = 2^* \frac{\text{Precision}^* \text{ Recall}}{\text{Precision} + \text{Recall}}$$

The empirical assessment of Vikram AI vis-a-vis Gemini AI provides several significant insights into their performance over various scenarios (refer Table 1 and Graph 1). In a favorable scenario with lower temperature settings, Vikram AI demonstrates exceptional accuracy at 98%, outranking Gemini AI at 95%. This illustrates Vikram AI's power to deliver very accurate and highly relevant responses in lSow complexity situations. However, in the average-case scenario at medium temperature settings, Vikram AI drops to only 72.22%, closely behind Gemini AI at 75.5%. This indicates that Vikram AI can still perform competitively but has to overcome hurdles to maintain high-performance levels when challenged with moderate levels of complexity. Of particular concern is the worst-case mode with high-temperature settings, in which Vikram AI drops to a meager performance of 30%, far below Gemini at 41.6%. This suggests a much greater degree of hallucination and underscores the urgency to further improve intervention techniques to address and mitigate highly complex or ambiguous situations. In summary, Vikram AI appears to show promise, especially in favorable cases; there remains, however, an explicit need for ongoing enhancements to make it stouter and more dependable across all conditions.

The "Creativity Paradox" fits perfectly into the performance of Vikram AI. Raising creativity in responses from the AI [12] seems to reduce the factual accuracy therein, and the opposite seems to be true. With this temperature setting, a complete contrast of this paradox exists, where low-temperature responses go for high precision and accuracy at the expense of possibilities for randomness or discovery, while high-temperature responses acknowledge all such aspects driven creatively but are highly unorthodox, becoming less compatible with the expected responses. It is this paradox that uncovers the essence of having balanced these attributes at different scales and justifies the need for Vikram AI to fine-tune response-to-use case settings.

Advantages of Vikram AI include the dynamic nature of its responses since it is generated by hybridizing weak AI and strong AI, so that responses can be more

Table 1 Key performance values

Scenario	AI model	Accuracy (%)	Precision	Recall	F1 score
Best case (low temperature)	Vikram AI	98	0.98	1	0.99
	Gemini AI	95	0.95	1	0.97
Average case (medium temperature)	Vikram AI	72.22	0.8	0.73	0.76
	Gemini AI	75.50	0.91	0.69	0.78
Worst case (high temperature)	Vikram AI	30	0.22	0.52	0.31
	Gemini AI	41.67	0.4	0.8	0.53

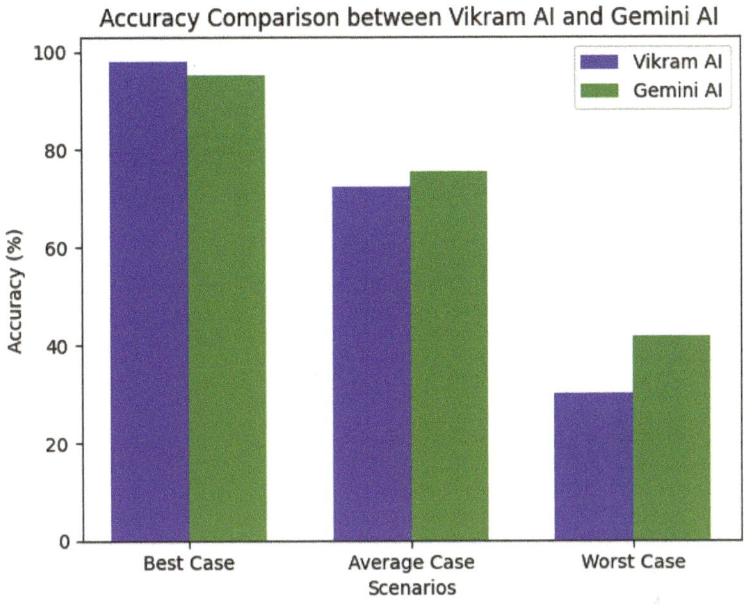

Graph 1 Accuracy versus scenarios

fact-based, grounded and creative addressing multiple inquiries. Controlled hard-coded hallucinations would further improve the capability—most especially in regulating either engagement and accuracy—or creativity and factuality of responses based on mode and temperature, hence appealing and precise answers. Vikram AI contains many versatile applications that run across all domains—from technical and medical queries requiring much precision to creative brainstorming sessions needing imagined answers. Additionally, it can get data from Google Search via external APIs in real time to ensure its information is fresh and pertinent. Engaging users through greeting management functions and temperature modulation mechanisms also improve the whole user engagement aspect of the system.

3.2 Conclusion

Vikram AI represents an important milestone toward integrating weak AI with strong AI functionalities. Its dynamic output, coupled with hard-coded hallucination control brings equilibrium to accuracy, groundedness, and creativity. However, it is important to note that performance in high-complexity scenarios requires enhancement. It is this feature that allows this novel approach to be applied in several areas from general to technical questions, going creativity-wise too. Strong mitigation strategies like high-quality training data, adaptive prompting, and human-in-the-loop review introduce reliable and trustworthy responses in a variety of circumstances. The AI hallucinations, therefore, diminishes both in number and kind, thereby improving the overall performance and user experience. The future seeking integration of compression algorithms into Vikram AI promises to improve its utility in big data. The user could be assured rapid and accurate returns even while working off big datasets. This will further solidify Vikram's idea of giving the enterprise applications a solution, an indispensable asset now projected for the age of big data and AI innovation. Failure to do so will limit Vikram's ability to develop as a very powerful tool in driving innovation and solve complex challenges across domains.

References

1. Arasa D (2024) When AI tells you to die: student receives threats from Google Gemini. Cebu Daily News. Inquirer.net. https://cebudailynews.inquirer.net/606648/when-ai-tells-you-to-die-student-receives-threats-from-google-gemini
2. Liu, B. (2021). "Weak AI" is likely to never become "strong AI", so what is its greatest value for us? https://arxiv.org/abs/2103.15294
3. Islam R, Ahmed I (2024) Gemini-the most powerful LLM: myth or truth. In: 2024 5th information communication technologies conference (ICTC). IEEE, pp 303–308
4. Hat R (2023) When LLMs daydream: hallucinations and how to prevent them, Red Hat Blog. https://www.redhat.com/en/blog/when-llms-day-dream-hallucinations-how-prevent-them
5. Maleki N, Padmanabhan B, Dutta K (2024) AI hallucinations: a misnomer worth clarifying. In: 2024 IEEE conference on artificial intelligence (CAI), IEEE, pp 133–138
6. Gosmar, Diego, and Deborah A. Dahl. Hallucination mitigation using agentic AI natural language-based frameworks. https://arxiv.org/abs/2501.13946 (2025)
7. Costello TH, Pennycook G, Rand DG (2024) Durably reducing conspiracy beliefs through dialogues with AI. Science 385(6714):1–12. https://doi.org/10.1126/science.adq1814
8. PerfectApps. (2024) Reducing AI hallucinations with temperature settings and RAG, PerfectApps Blog. https://www.perfectapps.com/reducing-ai-hallucinations-with-temperature-settings-and-rag/
9. Liu B Grounding for Artificial Intelligence. arXiv:2312.09532 (2023)
10. Huang Y, Li Y, Wu W, Zhang J, Lyu MR (2024) Proceedings of the ACM on software engineering 1, no. FSE. In: Your code secret belongs to me: neural code completion tools can memorize hard-coded credentials, pp 2515–2537
11. DigitalOcean. (2023) DigitalOcean resources. In: Understanding and mitigating AI hallucination. https://www.digitalocean.com/resources/articles/ai-hallucination
12. West P, Lu X, Dziri N, Brahman F, Li L, Hwang JD, Jiang L et al (2023) The twelfth international conference on learning representations. In: The generative AI paradox: what it can create, it may not understand

SmartAuto: A ML-Driven Auto-Heal System for Enhancing E-Commerce Order Reliability

S. D. Vidya Sagar, M. R. Dileep, Sreekanth Rallapalli, J. S. Suneel, K. Sowmya, and Neelam Makode

Abstract This research addresses the challenge of managing errors in e-commerce order processing by developing an auto-heal system using ML. The problem involves frequent order issues, such as payment failures or inventory errors, that require manual intervention, affecting operational efficiency. To solve this, the study proposes a system that automatically detects and resolves such errors or alerts staff for unresolved issues. The system is built on a synthetic dataset of 5000 e-commerce orders, with features like order status, customer ID, seller ID, error code, and resolution status. Principal component analysis (PCA) is employed for dimensionality reduction, and a logistic regression model is used to predict the resolution status of orders. The model achieves an impressive accuracy of 93%, outperforming other approaches such as random forest (78%) and neural networks (50%). This high level of accuracy demonstrates the model's capability to predict and resolve issues effectively. The auto-heal system has significant applications in e-commerce platforms, where it can enhance the speed and reliability of order processing by automating error detection and resolution. This leads to improved customer satisfaction and reduced operational costs. The proposed methodology can be extended to other domains where order or transaction management is crucial, such as supply chain management and retail logistics, contributing to more intelligent and automated decision-making systems.

Keywords First keyword · Second keyword · Third keyword

S. D. V. Sagar (✉) · S. Rallapalli · J. S. Suneel · K. Sowmya · N. Makode
Department of Computer Applications, Nitte Meenakshi Institute of Technology, Nitte (Deemed to be University), Bengaluru, Karnataka, India
e-mail: vidya.sagar@nmit.ac.in

M. R. Dileep
Department of Artificial Intelligence, College of Technology, City University Ajman, Al Tallah, Ajman, UAE
e-mail: dileep.kurunimakki@gmail.com

H. Nagaraj et al. (eds.), *Advancements in Artificial Intelligence and Machine Learning*, Lecture Notes in Electrical Engineering 1572,
https://doi.org/10.1007/978-981-95-7449-0_22

1 Introduction

In the fast-paced world of e-commerce, efficiently managing and resolving order issues is crucial to maintaining customer satisfaction and operational efficiency. This project presents an auto-heal system designed to automatically detect and resolve issues in e-commerce orders using ML techniques. By analyzing a dataset of 5000 synthetic order records, we aim to predict the resolution status of orders based on various features such as order status, timestamp, customer ID, seller ID, and error codes. The system utilizes PCA for feature reduction and logistic regression for classification, enabling automatic correction of specific errors and timely alerts for unresolved issues. This innovative approach enhances the reliability and efficiency of order processing, ultimately improving the overall customer experience in e-commerce platforms.

The development of auto-heal systems in e-commerce has become increasingly important as the industry continues to grow. These systems aim to automatically detect and resolve order processing errors, thus enhancing operational efficiency and customer satisfaction. This literature survey reviews key research papers that contribute to the advancement of ML techniques for error detection and resolution in e-commerce.

Davidson and Ravi [1] propose a system for ensuring fairness in outlier detection, which is crucial for maintaining trust in automated systems used for error detection in e-commerce. This research helps in creating balanced models that do not unfairly target specific user groups.

Gama et al. [2] survey various methods for adapting to concept drift, which is essential for maintaining the accuracy of ML models over time in dynamic environments like e-commerce. Their work highlights techniques that can be used to continuously update models as data patterns change. Gomes et al. [3] introduce adaptive random forests for evolving data streams, which are particularly useful in real-time e-commerce applications where data is constantly being updated. This approach helps in maintaining high accuracy in detecting and resolving order processing errors. Elkan and Noto [4] investigate techniques for training classifiers using positive and unlabeled data, making them suitable for tasks like fraud detection in e-commerce where labeled datasets are limited. This method proves useful in situations where acquiring labeled data is difficult. Ganin and Lempitsky's [5] work on unsupervised domain adaptation by backpropagation is relevant for enhancing the generalization of error detection models across different e-commerce platforms and domains. This technique improves the robustness of models against domain-specific variations.

Huang et al. [6] discuss methods to correct sample selection bias using unlabeled data. This research is crucial for developing accurate error detection systems in e-commerce, where biased datasets can lead to incorrect predictions and resolutions. Foulds and Frank [7] review multi-instance learning assumptions that can enhance the performance of ML models used for detecting errors in e-commerce transactions. Their insights help in designing more effective learning algorithms. Görnitz et al. [8] focus on supervised anomaly detection techniques that are directly applicable to identifying anomalies in e-commerce transactions. Their research provides valuable methods for detecting fraudulent activities and other anomalies.

Emmott et al. [9] performed some meta-analysis on anomaly detection, providing detailed insights into different techniques and their performance. This evaluation aids in identifying the most suitable methods for addressing specific error detection tasks in e-commerce. Holstein et al. [10] address improving fairness in ML systems, which is essential for developing unbiased error detection and resolution mechanisms in e-commerce platforms. Gomes, et al., [11] talk about streaming some random patches for evolving data stream classification, which helps maintain model accuracy in real-time e-commerce applications. This technique is particularly useful for continuous monitoring and error detection. Hamilton [12] presents graph representation learning techniques that are useful for detecting complex fraud patterns in e-commerce transactions by analyzing user behavior and transaction graphs. Jacobusse and Veenman [13] explore the impact of selection bias with imbalanced classes, providing methods to enhance the performance of error detection models that deal with imbalanced datasets [13].

Hardt et al. [14] propose methods to ensure equality of opportunity in supervised learning, which is crucial for developing fair and reliable e-commerce error detection systems. Hüllermeier and Waegeman [15] introduce concepts and methods to address aleatoric and epistemic uncertainty in ML that is applied to improve the reliability of e-commerce error detection models. Abed and Fernando's [16] systematic literature review provides an in-depth analysis of ML techniques for fraud detection in e-commerce, emphasizing the importance of neural networks in enhancing detection accuracy within digital marketplaces. This review identifies critical research gaps and suggests future directions for developing robust anti-fraud measures. Khoo et al. [17] present a model for detecting fraudulent e-commerce websites by analyzing HTML, text, and image features. This approach leverages supervised ML techniques to improve the identification of deceptive sites, thereby enhancing the security of online transactions.

The study on e-commerce fraud detection leveraging self-paced graph contrast learning [18] emphasizes enhancing the identification of fraudulent reviews. This approach aims to improve both the precision and efficiency of detecting fake reviews, which is essential for sustaining trust in online marketplaces. In another work, Li et al. [19] proposed a method for identifying spam reviews using graph convolutional networks. Their research highlights the benefits of incorporating multimodal data into a probabilistic review graph, which notably boosts the ability to detect opinion spam on e-commerce platforms. Additionally, Li et al. [20] introduced a fraud detection framework based on a cost-sensitive support vector machine (CS-SVM). This model integrates cost-sensitive learning with support vector machines to minimize the costs associated with misclassifying fraudulent transactions, offering a more dependable solution for fraud detection.

Liu et al. [21] proposed a hybrid approach that utilizes multimodal embedded representations within a probabilistic review graph to detect opinion spam. By integrating diverse data sources, this method significantly enhances the detection of fraudulent reviews and improves overall accuracy. Li et al. [22] investigate the detection of fraud communities in e-commerce from both online and offline perspectives. Their study highlights the importance of understanding fraud dynamics across different environments to develop more comprehensive fraud detection models. Liu et al. [23] present a heterogeneous graph neural network model for detecting malicious accounts. This model leverages the relationships between different types of entities within an e-commerce platform, improving the detection of fraudulent activities.

Lucas et al. [24] introduced a multi-perspective hidden Markov model (HMM) for automated feature engineering in credit card fraud detection. This innovative method analyzes data from multiple perspectives to identify fraudulent transactions more effectively. Additionally, SpringerLink [18] presents a research agenda outlining future directions in e-commerce fraud detection using machine learning. It highlights the importance of advanced techniques such as deep learning and unsupervised learning to address the dynamic nature of fraudulent activities. The reviewed studies underscore significant progress in machine learning for developing self-repairing systems in e-commerce, focusing on fairness, adaptability to concept drift, handling imbalanced datasets, and enhancing model robustness and accuracy. These contributions lay a solid groundwork for future advancements aimed at improving the efficiency and reliability of e-commerce order processing systems.

This paper presents a novel contribution to the field of e-commerce by developing an automated system to handle and resolve order issues using ML. Our system utilizes a synthetic dataset of 5000 records to simulate real-world scenarios and applies preprocessing techniques such as label encoding and standardization. Dimensionality reduction is achieved using PCA, allowing us to retain essential features while simplifying the model. A logistic regression model is trained and evaluated, demonstrating high accuracy in predicting order resolution status. Additionally, we implement an auto-healing mechanism to automatically address specific issues and an alert system to notify unresolved cases. This system enhances operational efficiency and customer satisfaction by promptly resolving order issues. The methodologies and insights from this research can be applied to similar domains, contributing to more intelligent and automated problem-solving systems.

2 Proposed System

See (Fig. 1).

1. Steps:

    - The customer has placed an order in e-commerce.
    - Then backend services capture the order as a customer order and the customer order needs a confirmation from the payment and inventory to mark a successful order.
    - After the successful completion of the customer order need to send the order to the seller order and logistic order.

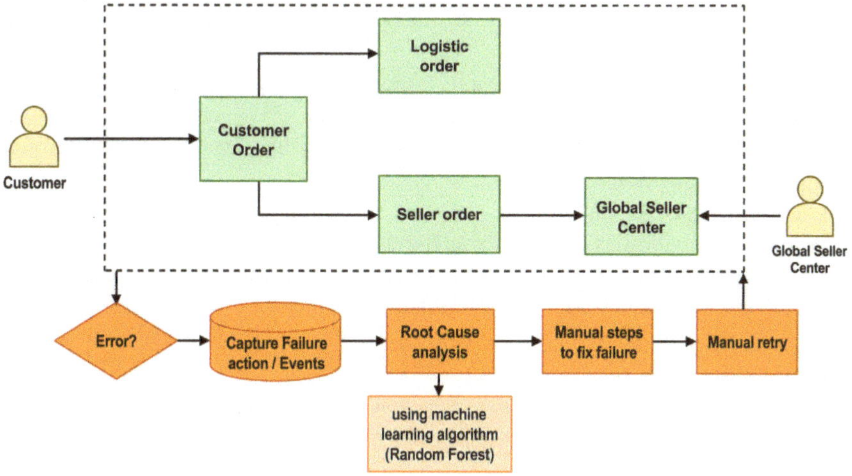

Fig. 1 Proposed model of SmartAuto system

- After that seller can log in to the global service center and see the orders against the seller.
- In case any failure happens in between order processing, our system tries to read the failure events across all the systems and the model tries to segregate errors and the behavior of order. Error includes failure in payments and fraud payment and inventory errors.
- our model tries to confirm.
  - Payment pending—server down.
  - Payment failed—insufficient balance.
  - Inventory issue—Item out of stock, inventory allocation.
  - Mismatch between order total and payment charge.
- ML algorithm will check the input order, whether the order can be resolved by auto-heal system or not.
- If the order can't be resolved, then the system will raise an error to investigate this issue manually.

3 Model Used: Logistic Regression(LR)

Logistic regression (LR) is a statistical technique used for binary classification problems. It models the probability of an input belonging to a specific class by applying a logistic function to the linear combination of input features.

How it works:

- Model Initialization: Begin with an initial estimate for the model coefficients (weights).
- Calculate Logit: Compute the logit, which is the linear combination of input features and weights.
- Apply Sigmoid Function: Convert the logit into a probability by passing it through the sigmoid function.
- Compute Cost: Use binary cross-entropy to calculate the cost (error) between predicted probabilities and actual outcomes.
- Optimize Weights: Implement optimization techniques like gradient descent to iteratively adjust the weights to minimize the cost function.
- Predict: After training, the model predicts the probability of the input belonging to a specific class. A threshold (e.g., 0.5) is used to classify the input into one of the two categories.

4 Mathematical Equation

4.1 Linear Combination of Inputs

First, logistic regression calculates a linear combination of the input features. For a given input vector x = [x1,x2,..., xn] the linear combination (or logit) is given by:

$$z = \beta_0 + \beta_1 x_1 + \beta_2 x_2 + \cdots + \beta_n x_n$$

where

- β_0—intercept (bias term).
- β_1, β_2,..., β_n —coefficients (weights) for the input features x_1,x_2,..., x_n.

4.2 Sigmoid Function

The linear combination z is then processed through the sigmoid function, which maps it to a probability value ranging between 0 and 1:

$$P(y = 1|\mathbf{x}) = \sigma(z) = \frac{1}{1 + e^{-z}}$$

This is the probability that the output y is 1 given the input features x.

4.3 Cost Function (Log-Loss or Binary Cross-Entropy)

In logistic regression, the cost function is binary cross-entropy (also known as log-loss), which quantifies the difference between the predicted probabilities and the true class labels. For a single training example, the cost is given by:

$$\text{Cost} = -y\log(P(y=1|\mathbf{x})) - (1-y)\log(1-P(y=1|\mathbf{x}))$$

For a dataset with mmm training examples, the total cost (or loss) is the average of the individual cost.

$$J(\beta_0, \beta_1, \ldots, \beta_n) = -\frac{1}{m}\sum_{i=1}^{m}\Big[y^{(i)}\log\big(P\big(y=1|\mathbf{x}^{(i)}\big)\big) \\ +\big(1-y^{(i)}\big)\log(1-P(y=1|\mathbf{x}^{(i)}))\Big]$$

where

- $y^{(i)}$—label for the i-th training example.
- $x^{(i)}$ —input vector for the i-th training example.

4.4 Optimization

Gradient descent is used to minimize the cost function and determine the optimal values for the coefficients β_0, β_1,..., β_n. The gradient descent update rules for the coefficients are given by:

$$\beta_j := \beta_j - \alpha\frac{\partial J}{\partial \beta_j}$$

for j = 0,1,2,…, n, where α represents the learning rate and $\partial J/\partial \beta_j$ represents the partial derivative of the cost function concerning β_j:

$$\frac{\partial J}{\partial \beta_j} = \frac{1}{m}\sum_{i=1}^{m}\big(P\big(y=1|\mathbf{x}^{(i)}\big) - y^{(i)}\big)x_j^{(i)}$$

5 Dataset

The dataset consists of 5000 records and includes the following features:

- order_id: A distinct identifier assigned to each order.

- order_status: Shows the current state of the order (for e.g., Pending, Completed, Failed, Canceled).
- timestamp: The specific date and time when the order or related event occurred.
- customer_id: ID of customer.
- seller_id: A distinct identifier for each seller. error_code: A code representing specific errors encountered (e.g., payment failure, stock issue).
- error_description: A detailed description of the error encountered.
- resolution_status: The status of the resolution process (e.g., Resolved, Unresolved).

The target variable in this dataset is resolution_status.

6 Experimental Results and Comparison

See Table 1 and Fig. 2.

Table 1 Experimental results of preexisting model accuracy

Model	Accuracy
Logistic regression	0.93
PCA	0.68
Neural network	0.50
Random Forest	0.78

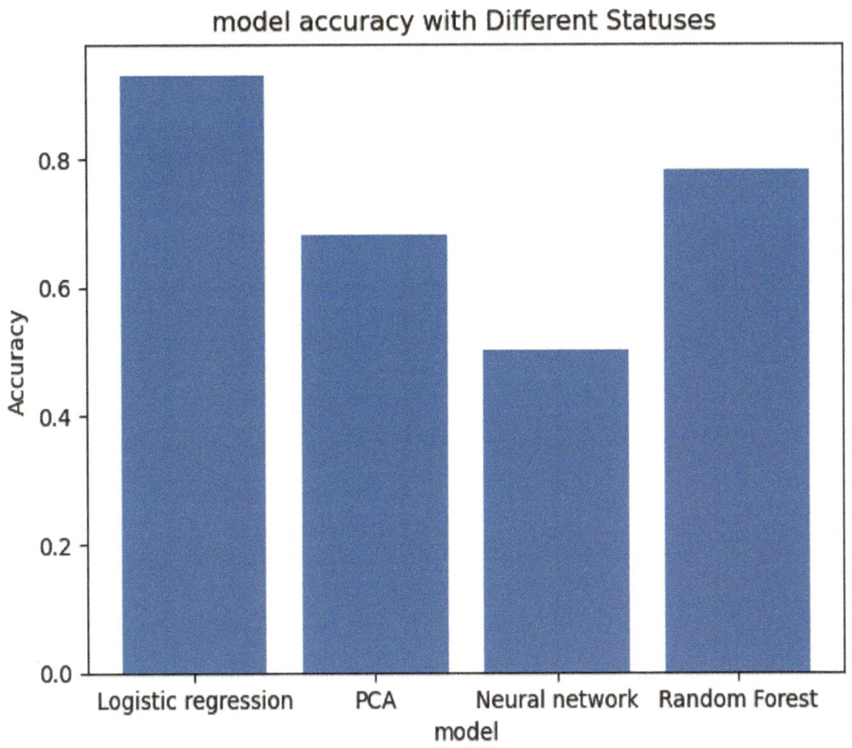

Fig. 2 Comparative analysis

7 Conclusion

This paper presents an ML-based auto-heal system designed to address and resolve order processing issues in e-commerce platforms. Leveraging a synthetic dataset of 5000 records, the system utilizes PCA to reduce dimensionality and Logistic Regression to predict the resolution status of orders as shown in Fig. 2. Achieving a high accuracy of 93% as shown in Table 1, the model effectively automates error detection and correction, improving operational efficiency and customer satisfaction. The research highlights the potential of integrating ML into e-commerce for real-time error resolution. It suggests future expansions into more advanced models to enhance the system's capability and application across other domains.

References

1. Davidson I, Ravi S (2020) ECAI. In: A framework for determining the fairness of outlier detection

2. Gama J, Žliobaitė I, Bifet A, Pechenizkiy M, Bouchachia A (2014) A survey on concept drift adaptation. ACM Comput Surv 46(4):1–37
3. Gomes HM, Bifet A, Read J, Barddal JP (2017) Adaptive random forests for evolving data stream classification. Mach Learn 106(9):1469–1495
4. Elkan C, Noto K (2008) Learning classifiers from only positive and unlabeled data, pp 213–220
5. Ganin Y, Lempitsky V (2015) Proceedings of the 32nd international conference on machine learning. In: Unsupervised domain adaptation by backpropagation, pp 1180–1189
6. Huang J, Gretton A, Borgwardt K, Schölkopf B, Smola A (2007) Correcting sample selection bias by unlabeled data. In: Advances in neural information processing systems, vol 19. MIT Press
7. Foulds JR, Frank E (2010) A review of multi-instance learning assumptions. Knowl Eng Rev 25(1):1–25
8. Görnitz N, Kloft M, Rieck K, Brefeld U (2013) Toward supervised anomaly detection. J Artif Intell Res 46:235–262
9. Emmott, A., Das, S., Dietterich, T., Fern, A., Wong, W.K. (2015). A meta-analysis of the anomaly detection problem. https://arxiv.org/abs/1503.01158
10. Holstein K, Vaughan JW, Daumé H, Dudík M, Wallach H (2019) Proceedings of the 2019 CHI conference on human factors in computing systems. In: Improving fairness in ML systems: what do industry practitioners need
11. Gomes HM, Read J, Bifet A (2019) 2019 IEEE international conference on data mining (ICDM). In: Streaming random patches for evolving data stream classification, pp 240–249
12. Hamilton WL (2020) Graph representation learning. Synth Lect Artif Intell Mach Learn 14(3):1–159
13. Jacobusse G, Veenman C (2016) Discovery science 2016. Lecture notes in computer science. In: On selection bias with imbalanced classes, vol 9956, pp 325–340
14. Hardt M, Price E, Srebro N (2016) Neural information processing systems (NIPS). In: Equality of opportunity in supervised learning
15. Hüllermeier E, Waegeman W (2021) Aleatoric and epistemic uncertainty in ML: an introduction to concepts and methods. Mach Learn 110(3):457–506
16. Abed M, Fernando B (2023) E-commerce fraud detection based on ML techniques: systematic literature review. Big Data Min Anal 7(2):419–444. https://doi.org/10.26599/BDMA.2023.9020023
17. Khoo E, Zainal A, Ariffin N, Kassim MN, Maarof MA, Bakhtiari M (2021) Fraudulent E-commerce websites detection through ML. In: Soft computing and pattern recognition 2019. Advances in intelligent systems and computing, vol 1182. Springer, Cham, pp 177–186. https://doi.org/10.1007/978-3-030-49345-5_19
18. Detection of E-commerce fraud review via self-paced graph contrast learning. (2023) Comput J. Oxford Academic
19. Li A, Qin Z, Liu R, Yang Y, Li D (2019) Proceedings of the 28th ACM international conference on information and knowledge management. In: Spam review detection with graph convolutional networks, pp 2703–2711. https://doi.org/10.1145/3357384.3358024
20. Li C, Ding N, Dong H, Zhai Y (2021) CS-SVM for E-commerce fraud detection. Int J Mach Learn Comput 11(1):30–37. https://doi.org/10.18178/ijmlc.2021.11.1.1011
21. Liu Y, Pang B, Wang X (2019) Opinion spam detection by incorporating multimodal embedded representation into a probabilistic review graph. Neurocomputing 366:276–283. https://doi.org/10.1016/j.neucom.2019.08.093
22. Li Z, Hui P, Zhang P, Huang J, Wang B, Tian L, Zhang J, Gao J, Tang X (2021) Companion proceedings of the web conference 2021. In: What happens behind the scene? Towards fraud community detection in e-commerce from online to offline, pp 105–113. https://doi.org/10.1145/3442442.3452304
23. Liu Z, Chen C, Yang X, Zhou J, Li X, Song L (2018) Proceedings of the 27th ACM international conference on information and knowledge management. In: Heterogeneous graph neural networks for malicious account detection, pp 2077–2085. https://doi.org/10.1145/3269206.3269250

24. Lucas Y, Portier P-E, Laporte L, He-Guelton L, Caelen O, Granitzer M, Calabretto S (2020) Towards automated feature engineering for credit card fraud detection using multi-perspective HMMs. Futur Gener Comput Syst 102:393–402. https://doi.org/10.1016/j.future.2019.08.018

Action Vocalizer: Real-Time Translation of Sign Language into Speech

Divya Ganapati Muroorkar, S. A. Hithesh, Ramanath Shreenivas Hegde, B. Shreyas Bhat, and A. Harisha

Abstract The "Action Vocalizer" project aims to eliminate communication barriers for individuals in the deaf and mute community by converting hand gestures into both text and speech. Utilizing state-of-the-art deep learning techniques, the system incorporates the YOLOv11 model and convolutional neural networks (CNN) to achieve real-time gesture recognition. This advancement promotes greater independence, social inclusion, and improved accessibility for people with hearing and speech impairments. The objective is to create a real-time hand gesture recognition system that accurately detects gestures using CNNs and YOLOv11, and seamlessly integrates gesture recognition with text-to-speech (TTS) conversion for practical application. The system captures hand gestures through a camera, processes the images for feature extraction, and employs a CNN-based model for precise recognition. Preliminary results demonstrate effective gesture detection and translation in real time. This initiative holds significant potential for enhancing communication, improving access to public services, and enriching the overall well-being of individuals with hearing and speech disabilities.

Keywords Gesture recognition · Action vocalizer · Sign language interpretation · YOLO V11

D. G. Muroorkar · S. A. Hithesh · R. S. Hegde (✉) · B. S. Bhat · A. Harisha
Department of Computer Science and Engineering, Sahyadri College of Engineering and Management, Mangalore, India
e-mail: ramanathhegde2003@gmail.com

Visvesvaraya Technological University, Belgavi, India

D. G. Muroorkar
e-mail: divyagunavante@gmail.com

S. A. Hithesh
e-mail: hitheshsa8@gmail.com

B. S. Bhat
e-mail: shreyasbhatbidara11@gmail.com

A. Harisha
e-mail: airbail89@gmail.com

H. Nagaraj et al. (eds.), *Advancements in Artificial Intelligence and Machine Learning*, Lecture Notes in Electrical Engineering 1572,
https://doi.org/10.1007/978-981-95-7449-0_23

1 Introduction

Communication is a vital human right, enabling social interaction, personal growth, and community participation [1]. However, millions of people face challenges due to hearing and speech impairments. According to the World Health Organization (WHO), around 70 million individuals globally have hearing impairments, with 7.6% of India's population affected by adult-onset deafness and 2% by childhood-onset deafness [2]. These challenges often result in social isolation, limited opportunities in education and employment, and a lower overall well-being [3]. Although sign language is an important means of communication within the deaf community, it is often not well understood by the general population, leading to considerable obstacles [4]. Conventional communication aids, such as pen and paper or text messaging, are inefficient for real-time interaction, highlighting the need for an inclusive and innovative solution [5].

The "Action Vocalizer" project aims to overcome this barrier by converting hand gestures into text and speech, enabling smooth communication between individuals in the deaf and mute community and those with normal hearing [6]. Leveraging advanced deep learning algorithms, particularly convolutional neural networks (CNN), the system recognizes hand gestures in real time and converts them into text and speech using a Text-to-Speech (TTS) engine [7]. The project's impact spans various sectors: in education, it supports students in engaging effectively with peers and educators; in workplaces, it enhances communication, fostering collaboration and inclusivity; and in public services, it ensures better interactions with staff [8]. Socially, the system reduces isolation, empowering individuals to connect with family and friends. Beyond its immediate utility, the project raises public awareness about the challenges faced by individuals with hearing impairments, promoting empathy and inclusivity [9]. By addressing critical communication barriers, the Action Vocalizer enhances independence, inclusivity, and quality of life, contributing to a society where everyone can communicate and participate fully [10].

2 Literature Survey

The review of the literature examines developments in deep learning technologies for hand gesture identification and conversion to speech and text, emphasizing noteworthy studies that tackle important communication issues for the deaf and dumb community.

A static hand gesture-based Indian Sign Language (ISL) recognition system leverages convolutional neural networks (CNNs) to enhance communication for the hearing-impaired and dumb community. Hand gestures are captured through a webcam, processed using image segmentation techniques such as Histogram Back-Projection, and classified using a custom dataset comprising 26 ISL alphabets and 10 digits. The model achieves remarkable testing accuracy of 99.89% and validation accuracy of 99.85%, trained with 2400 images per gesture. It operates without the need for external hardware, making it cost-effective and user-friendly. Performance improves with increased epochs, with optimal results at five epochs. While effective, the system is limited to static gestures and performs best in light backgrounds [9].

The study focuses on developing a vision-based gesture recognition system to bridge the communication gap for the deaf and dumb community, enabling effective interaction with others by converting hand gestures into text and speech. Using a webcam for gesture detection and leveraging deep learning techniques such as convolutional neural networks and artificial neural networks (ANN), the system processes and identifies hand motions through image preprocessing, feature extraction, and classification. The captured gestures are translated into textual and auditory outputs, making communication more accessible for individuals with disabilities. The proposed system's architecture highlights modules for dataset preprocessing, model training, and real-time gesture recognition, eventually resolving the shortcomings of current methods and improving communication inclusiveness [11].

An adaptive K-means clustering approach for image segmentation is proposed, offering precise results without the need for manual input of the number of clusters (K). The methodology begins by converting input images into grayscale, followed by applying K-means clustering to group similar pixels into clusters based on their proximity to centroids. The algorithm iteratively updates centroids and reallocates pixels to the nearest clusters, ensuring refined segmentation with enhanced object discrimination. Experimental results on diverse sample images under varying luminance conditions demonstrate the technique's effectiveness, producing segmented outputs with distinct object features and improved visual quality. Post-processing steps, such as image reshaping, further enhance segmentation quality, making the approach efficient for various applications [12].

A deep learning-based hand gesture identification model for emergency situations using Indian Sign Language (ISL) is developed by combining pretrained VGG-16 for spatial feature extraction with long short-term memory (LSTM) to capture temporal dynamics. Trained on a dataset of 412 videos representing emergency hand gestures such as "accident," "call," "doctor," and others, the model achieved a high accuracy of 98%. This method gives deaf people an effective communication tool in emergency situations by enabling real-time identification of both static and dynamic gestures. The model outperforms previous methods, demonstrating the effectiveness of VGG-16 and LSTM for gesture recognition in video sequences [14].

3 Methodology

3.1 System Overview

The Action Vocalizer system aims to detect hand gestures in real time and convert them into corresponding speech using the YOLOv11 model [11].

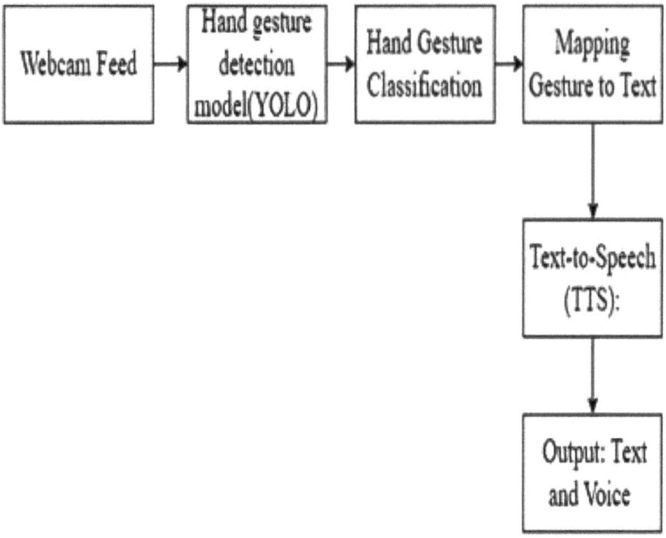

Fig. 1 Proposed system architecture

The approach involves the following steps:

- **Data Collection and Augmentation**: A custom dataset of hand gestures is collected and augmented for training the model.
- **Model Training**: YOLOv11 is trained on the custom dataset to accurately recognize hand gestures [12].
- **Real-Time Inference**: The trained model is deployed to process live video feeds, detect gestures, and map them to speech using a Text-to-Speech (TTS) engine [13].

The general structure of the system is illustrated in Fig. 1.

3.2 Data Collection and Augmentation

A custom dataset was created to represent different hand gestures, such as "hello," "thank you," "yes," "no," etc. The dataset includes variations in hand orientation, lighting conditions, and backgrounds to ensure robustness in real-world scenarios [14].

- **Dataset Composition**: 2,000 images, each labeled with 10 hand gestures. Images were labeled using a tool like Roboflow, ensuring accurate bounding boxes [13].
- **Data Augmentation**: Various techniques were applied to increase the dataset's diversity and help the model generalize better [8]:

- **Rotation**: Images were rotated by $\pm 15^\circ$ to simulate different hand orientations.
- **Brightness Adjustment**: Images were altered to simulate various lighting conditions.
- **Flipping**: Horizontal and vertical flips were applied to account for different gestures. The augmented dataset ensures that the model learns robust features for the purpose of recognizing gestures under diverse conditions.

3.3 Model Training

The YOLOv11 architecture was chosen for its efficiency and speed in real-time object detection [15]. The model was fine-tuned for gesture categorization using a proprietary dataset.

- **Model Initialization**: YOLOv11 weights were initialized with pretrained models to leverage learned features, speeding up convergence enhancing the accuracy of gesture recognition. [16].
- **Training Setup**:
  - **Optimizer**: Adam optimizer was used for better convergence [9].
  - **Learning Rate**: Set at 0.001 to balance training speed and stability.
  - **Epochs**: 10 epochs were chosen to fine-tune the model.
  - **Batch Size**: A batch size of 16 was used to balance memory usage and training efficiency.
  - **Dataset Split**: 70% of the dataset was used for training, 20% for testing, and 10% for validation.

3.4 Real-Time Gesture Recognition and Speech Generation

The trained YOLOv11 model was deployed to process video streams in real time for detecting and recognizing hand gestures [10].

- **Video Capture**: A live video feed is captured from a camera and processed frame by frame.
- **Gesture Detection**: The YOLOv11 model analyzes each frame to detect and classify hand gestures based on the bounding boxes drawn around the detected hands.
- **Text Mapping**: Each recognized gesture is mapped to a predefined text label (e.g., "hello," "thank you").
- **Speech Synthesis**: The text is passed to a Text-to-Speech (TTS) engine (e.g., Google TTS) to generate speech output in the chosen language.

3.5 Implementation Workflow

The system operates as follows:

(1) **Capture**: The camera captures frames of hand gestures.
(2) **Preprocessing**: The captured frames are preprocessed (e.g., resizing) before being fed into the model [15].
(3) **Inference**: The YOLOv11 model performs real-time inference on the preprocessed frames, detecting hand gestures.
(4) **Mapping and Speech Generation**: The recognized gesture is mapped to its corresponding text label, and the TTS engine converts the text into speech.
(5) **Output**: The speech is played through the speakers in real time.

4 Results

This section presents and evaluates the YOLOv11 and YOLOv5 models' performance on the sign language gesture detection challenge. Numerous important measures, including confusion matrices, F1 scores, precision, recall, and the F1 score curve, form the basis of the evaluation.

4.1 Confusion Matrix Comparison

We start by displaying the two models' confusion matrices. An overview of each model's ability to discriminate between various gesture classes is given by these matrices (Figs. 2 and 3).

4.2 F1 Score Comparison

The F1 score is calculated to evaluate the balance between precision and recall. The following Table 1 presents the F1 scores for both models.

4.3 F1 Score Curve

We further analyze the F1 score for both models across various decision thresholds. The F1 score curve provides insights into how well each model performs at different thresholds, allowing us to choose the optimal threshold for our application. This analysis highlights the ideal trade-off between precision and recall for each model.

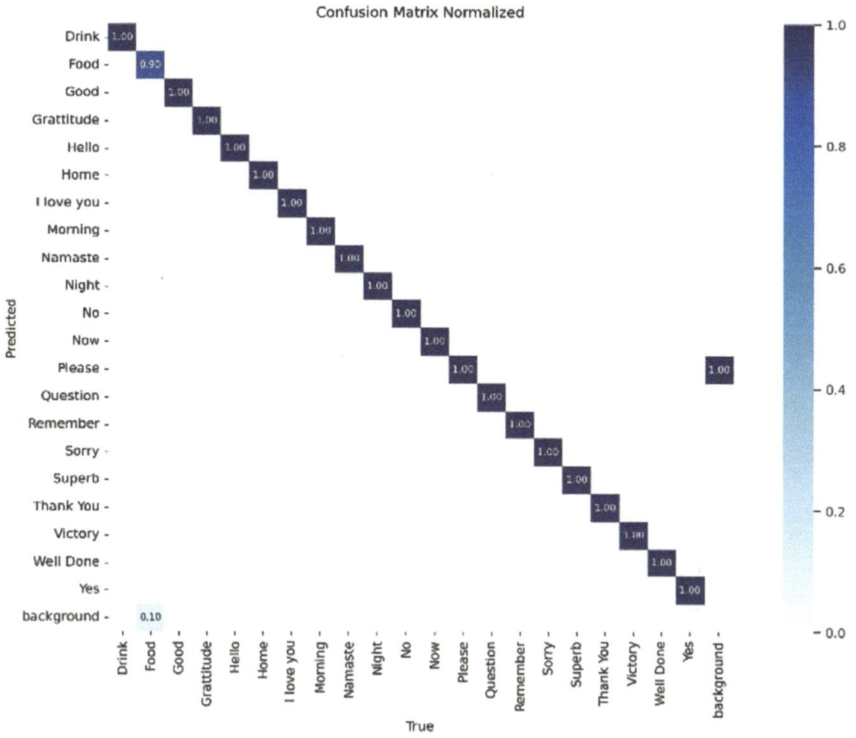

Fig. 2 Confusion matrix for our YOLOv11 model. This matrix shows the true and false predictions across different sign language gestures

By comparing the curves of YOLOv11 and YOLOv5, we can determine which model achieves a higher peak F1 score, assisting in selecting the best model. This comparison is crucial for optimizing performance and ensuring the model meets the specific requirements of the application, resulting in accurate and reliable predictions (Figs. 4 and 5).

4.4 Precision and Recall Comparison

The F1 score is calculated to evaluate the balance between precision and recall. The following table presents the F1 scores for both models.

Precision and recall values for both models are presented in Table 2. The YOLOv11 model demonstrates slightly higher precision and recall compared to YOLOv5.

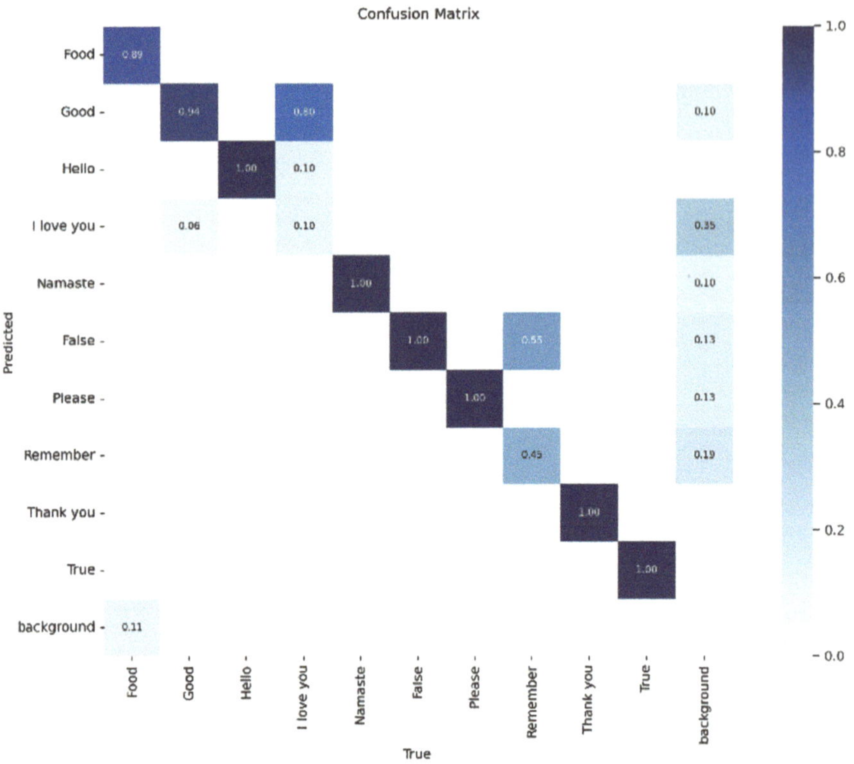

Fig. 3 Confusion matrix for YOLOv5 model. This matrix shows the true and false predictions across different sign language gestures

Table 1 F1 score comparison between YOLOV11 model and YOLOV5

Model	F1 Score	Comments
YOLOv11	0.99	High precision and balanced recall at IoU 0.675
YOLOv5	0.95	12 point, bold

4.5 Discussion

The YOLOv11 model shows superior performance in several key areas: Precision and Recall: The YOLOv11 model achieves higher precision (0.985) and recall (0.80), indicating better performance in accurately determining the gestures and minimizing false positives as well as false negatives. F1 Score: The F1 score for YOLOv11 (0.99) is higher than that of YOLOv5 (0.95), demonstrating that, in comparison to YOLOv5, there is a good balance between accuracy and recall.

The examination of the F1 scoring curve reveals that the YOLOv11 model maintains a higher F1 score across various decision thresholds, while YOLOv5 shows

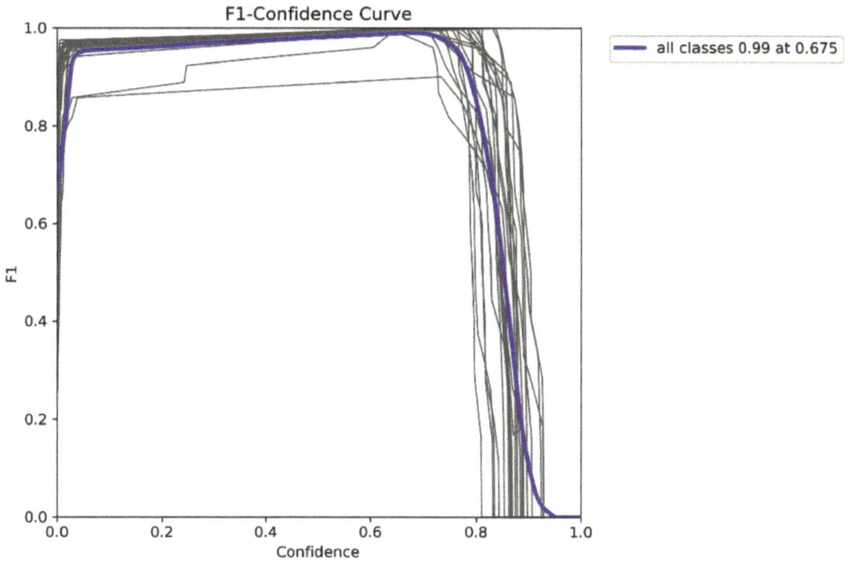

Fig. 4 F1 score curve for YOLOv11 model

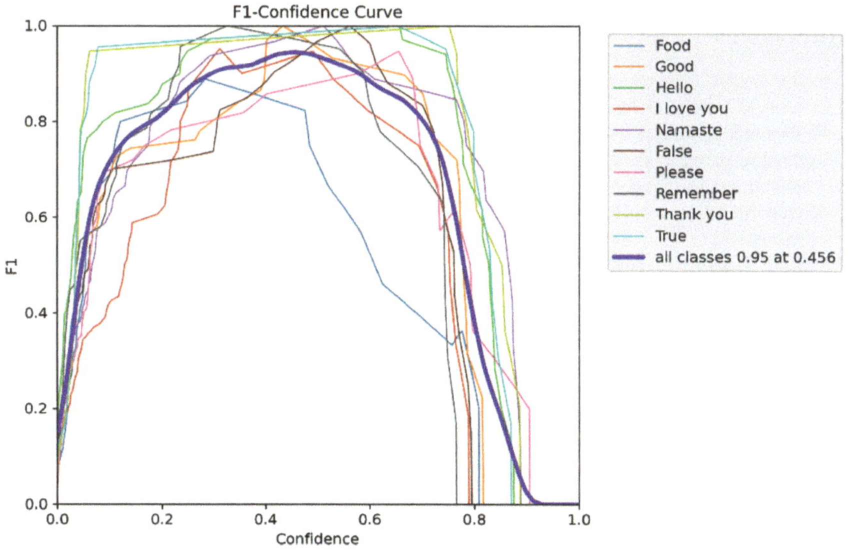

Fig. 5 F1 score curve for YOLOv5 model. This curve shows the F1 score at various decision thresholds

Table 2 Precision and recall comparison between YOLOV11 model and YOLOV5

Model	Precision	Recall
YOLOv11	0.985	0.80
YOLOv5	0.971	0.75

more variability. This reinforces the conclusion that the YOLOv11 model is more stable in performance, especially when optimizing for precision and recall balance.

4.6 Conclusion

In conclusion, the YOLOv11 model provides better overall performance in terms of precision (0.985), recall (0.80) and F1 score (0.99) for the sign language gesture recognition task. For real-time applications, it is therefore a more better option. Future research might concentrate on enhancing the YOLO model even further for even greater results.

Acknowledgements This project has been developed at the Centre of Excellence (CoE) in Digital Forensics Intelligence, supported by the Vision Group on Science and Technology (VGST), Government of Karnataka (GoK), under the grant GRD 853. We extend our heartfelt gratitude for the infrastructure and continuous guidance provided, which have significantly contributed to the research.

References

1. Smith RO, Scherer MJ, Cooper R, Bell D, Hobbs DA, Pettersson C, Seymour N, Borg J, Johnson MJ, Lane JP et al (2018) Assistive technology products: a position paper from the first global research, innovation, and education on assistive technology (GREAT) summit. Disabil Rehabil Assist Technol 13(5):473–485
2. Swenor BK, Ramulu PY, Willis JR, Friedman D, Lin FR (2013) The prevalence of concurrent hearing and vision impairment in the United States. JAMA Int Med 173(4):312–313
3. Rooth MA (2017) The prevalence and impact of vision and hearing loss in the elderly. N C Med J 78(2):118–120
4. Boughman J, Vernon M, Shaver K (1983) Usher syndrome: definition and estimate of prevalence from two high-risk populations. J Chronic Dis 36(8):595–603
5. Wilbur RB (2009) Effects of varying rate of signing on ASL manual signs and nonmanual markers. Lang Speech 52(2–3):245–285
6. Bulgarelli A, Toscana G, Russo LO, Farulla GA, Indaco M, Bona B (2016) A low-cost open-source 3D-printable dexterous anthropomorphic robotic hand with a parallel spherical joint wrist for sign language reproduction. Int J Adv Rob Syst 13(3):126
7. Dixit PM, Patel GP, Priya G, Shubhashree T (2023) Hand gesture recognition and text-to-voice conversion for deaf and dumb. Int J Sci Res Eng Manag (IJSREM)
8. Vanaja S, Preetha R, Sudhain S (2021) Hand gesture recognition for deaf and dumb using CNN technique. In: 6th International conference on communication and electronics systems (ICCES). IEEE, pp 1–X
9. Hatibaruah D, Talukdar AK, Sarma KK (2021) A static hand gesture based sign language recognition system using convolutional neural networks. In: IEEE India Council international conference (INDICON). IEEE, pp 1–X
10. Dumbre A, Jangada S, Gosavi S, Gupta J (2022) Classification of Indian sign characters utilizing convolution neural networks and transfer learning models with different image processing techniques. In: IEEE world conference on applied intelligence and computing (AIC). IEEE, pp 1–X

11. Surekha P, Niharika V, Pranavi D, Teja Sree D, Venkata Surya SA (2022) Hand gesture recognition and voice, text conversion using CNN and ANN. In: Second international conference on artificial intelligence and smart energy. IEEE, pp 1–X
12. Venkatachalam K, Purandhar V, Amudhan M, Raguraman A, Mohan E (2022) An implementation of K-means clustering for efficient image segmentation. In: 10th IEEE international conference on communication systems and network technologies. IEEE, pp 1–X
13. Chang V, Eniola RO, Golightly L, Xu QA (2023) An exploration into human–computer interaction: hand gesture recognition management in a challenging environment. In: International conference on information and automation for sustainability (ICIAFS). IEEE, pp 1–X
14. Areeb QM, Nadeem M (2022) Deep learning based hand gesture recognition for emergency situation: a study on Indian sign language. In: International conference on data analytics for business industry (ICDABI). IEEE, pp 1–X
15. Nogales RE, Benalcázar ME (2023) Hand gesture recognition using automatic feature extraction and deep learning algorithms with memory. In: Artificial intelligence and computer vision research lab, Qatar
16. Sherif S, Mukhopadhyay A, Karthikeyan U (2023) Implementation of a deep learning algorithm and integrating it with hand gesture recognition. In: International conference on innovation in science, engineering and technology (ICISET). IEEE, pp 1–X

Innovative Approaches to Skin Disease Classification: Evaluating VGG16 and VGG19 Models

Aziz Makandar and Ayisha Soudagar

Abstract Skin disease, one of the utmost communal type of disease worldwide, imposes a significant burden on public health. When it comes to dipping the indisposition and impermanence rates associated with this disease, an early and accurate diagnosis is critical. Deep learning has made considerable advances in recent years, notably in the area of convolutional neural networks (CNNs), which have shown potential in automating the identification and categorization of skin illnesses. In this study, a skin disease categorization model that uses two pre-trained architectures, VGG16 and VGG19, to identify nine diverse forms of skin diseases based on dermoscopic images is proposed. During model training and testing, the Kaggle dataset was utilized. This dataset included 4,109 photographs that were identified as different forms of skin diseases. Despite the fact that VGG16 achieved a higher training accuracy of 96.88%, the results indicate otherwise. VGG19, on the other hand, achieved an accuracy rating of 90.62%. Our findings suggest that transfer learning can improve classification performance for skin disease diagnosis, emphasizing transfer learning's promise in managing restricted datasets. The study's discoveries suggest that deep learning models, particularly pre-trained architectures like VGG16 and VGG19, these methods have the potential to improve classification accuracy of skin disease classification.

Keywords Skin disease classification · Deep learning · VGG16 · VGG19 · Classification · Image recognition

A. Makandar · A. Soudagar (✉)
Department of Computer Science, Karnataka State Akkamahadevi Women's University, Vijayapura, Karnataka, India
e-mail: ayishamudassar07@gmail.com

A. Makandar
e-mail: azizkswu@gmail.com

H. Nagaraj et al. (eds.), *Advancements in Artificial Intelligence and Machine Learning*, Lecture Notes in Electrical Engineering 1572,
https://doi.org/10.1007/978-981-95-7449-0_24

1 Introduction

India's large population and diverse environment contribute to a broad spectrum of skin ailments. Acne, fungal infections, eczema, psoriasis, and vitiligo are common disorders, as are leprosy, scabies, and skin TB. Prompt diagnosis and therapy are critical for successful management [1–3]. Health is everything in a human's life [4]. The skin, the largest organ of our bodies [5], serves as a shield against germs and helps regulate body temperature. Protecting the skin is crucial to preventing diseases like bacterial and fungal infections [6]. However, diagnosing skin disease can be costly for many individuals. Often, diagnosis requires expert dermatologists' judgment, which can take more time than simply examining dermoscopic images [7]. Skin diseases are a widespread issue affecting millions [8], if not billions, of people globally across all age groups. Various factors can harm the skin, leading to signs such as redness, burning, itching, and swelling [9]. Allergic reactions that disrupt the heritable edifice of the skin often result in eruptions and multiple skin conditions, impacting overall appearance [10]. Skin diseases are often long-lasting and can develop into cancerous tissues. Therefore, early identification and treatment are essential to minimizing significant damage [11]. Many skin diseases may appear symptomless in their initial stages, so detecting and treating them early is critical [12]. Dermatology, the branch of medicine focused on hair, skin, and nails, traditionally diagnoses skin diseases through clinical examination, followed by dermoscopic evaluation [13]. Diagnosing skin diseases is not always straightforward [14]. Dermatologists often rely on laboratory tests to identify the specific type. Despite advances in medical technology offering accurate diagnostic results, the process remains costly and time-consuming [15]. A quick and affordable solution is needed, which could be achieved using deep learning techniques.

Skin diseases are amid the furthermost public health problems that individuals confront, with causes ranging from bacterial and fungal infections to viral infections [16]. Dermatologists play a significant part in traditional diagnostics. They begin by visually analyzing the patient's skin health using their expertise, followed by dermoscopy, a sophisticated imaging method that reveals precise skin anatomy [17].

Recently, researchers have presented machine learning techniques to improve conventional diagnostic methods, which have several limitations. These approaches often entail manually collecting characteristics from skin photos and then classifying them using machine learning algorithms. This manual method requires not just medical knowledge but also the capacity to do substantial data analysis, since reducing dimensionality is often required to improve picture identification accuracy. However, deep learning approaches are currently favored over traditional machine learning for picture identification and classification since they automated the procedure for obtaining features, resulting in quicker and more accurate diagnosis [18].

Mondal et al. [19] deliver an amended strategy for skin disease categorization founded on a generative adversarial network (GAN), which shows significant increases in diagnosis accuracy in medical imaging scenarios.

In [20], a strong CNN-based technique for melanoma classification that achieved an outstanding overall accuracy of 88.83% is presented. By discriminating between distinct forms of melanoma, the suggested approach improves early diagnosis, which is critical for patient outcomes. The comparison study indicates its advantages over current cutting-edge approaches while decreasing computing complexity.

The authors [21] used a range of deep learning models such as VGG16 and ResNet18, emphasizing the relevance of early detection in skin disease categorization. The ISIC dataset, which was improved by data augmentation, achieved a VGG16 F1-score of 97.41% and a ResNet18 accuracy of 98.06%. These findings highlight the potential of computer-aided diagnostic tools to increase the accuracy and efficiency of skin disease treatment. Additional validation on varied datasets is recommended to improve the robustness of the results.

Researchers proposed a unique framework for skin disease detection that fine-tunes the ResNet152 and InceptionResNet-V2 models by using a triplet loss function to increase feature learning. By embedding pictures in Euclidean space and calculating L-2 distances, the framework outperforms previous approaches for classification accuracy. The findings show its potential to improve automated skin disease detection [22]. Reference [23] compares several CNN architectures ResNet50, Xception, InceptionV3, VGG16, and VGG19 for identifying cutaneous bacterial infections. The research used K-Fold cross-validation, the Adam optimizer, and binary cross-entropy loss, with a 25% dropout rate. The findings indicated that VGG16 obtained an accuracy of 91.30%, whereas VGG19 achieved 84.55%.

2 Proposed Method

Convolutional neural networks (CNNs) are often observed as the best approach for image categorization and feature extraction in deep learning architectures. Noise, aberrations, and artifacts may all appear in dermoscopic pictures, as can variations in comparable characteristics and similarities across various disease kinds. To solve these concerns, training using a big dataset is recommended. This work uses pre-trained architectures like VGG16 and VGG19 to improve CNN performance. Specifically, the VGG19 and VGG16 algorithms are merged, resulting in better results. Given the scarcity of data to train a model from scratch, a pre-trained architecture is used to effectively classify skin diseases. VGG16 and VGG19 are two of the most extensively used pre-trained models, with excellent performance in this domain. Figure 1 depicts the suggested technique for this methodology.

2.1 Dataset

The pre-trained architectures employ a dataset for the classification of various skin diseases, obtained from Kaggle, which includes 4,107 photos depicting nine different

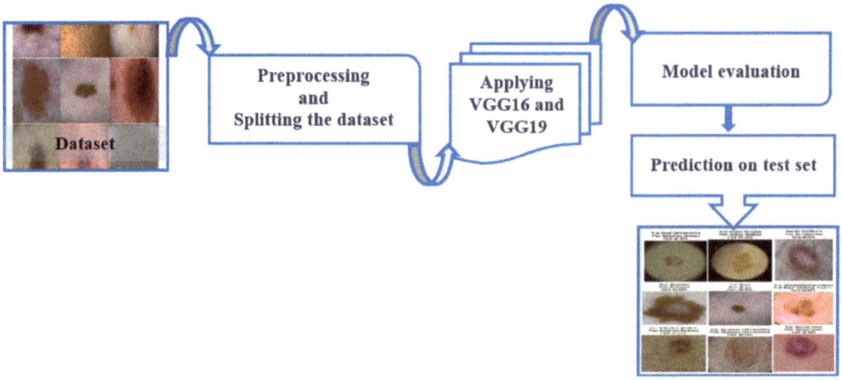

Fig. 1 Schematic representation of the proposed approach

forms of skin diseases. The proposed models classify the input image in the following manner: Actinic keratosis, basal cell carcinoma, dermatofibroma, melanoma, nevus, pigmented benign keratosis, seborrheic keratosis, squamous cell carcinoma, and vascular lesion.

2.2 Pre-processing

All the pictures in our collection have been scaled to 224 × 224 pixels. We performed the conventional pre-processing activities, such as normalization, resizing, and data augmentation. We normalized the photographs since the pre-trained network was previously trained on those particular images. The improved pre-trained architecture was trained using 3,287 photos depicting nine different forms of skin disorders. These photographs are used to distinguish between certain types of skin disease categorizations.

2.3 Dataset Splitting

We separated the dataset into various subgroups for training, validation, and testing in order to assess our model's performance. Specifically, 80% of the data was set aside for model training, with the remaining 20% kept for validation and testing reasons. As shown in Table 1, the dataset is then divided into three parts: the training set, the validation set, and the test set.

Table 1 Splitting of the dataset for train, validation, and test

Model	Train	Validation	Test	Total
VGG16	3287	410	411	4108
VGG19	3287	410	411	4108

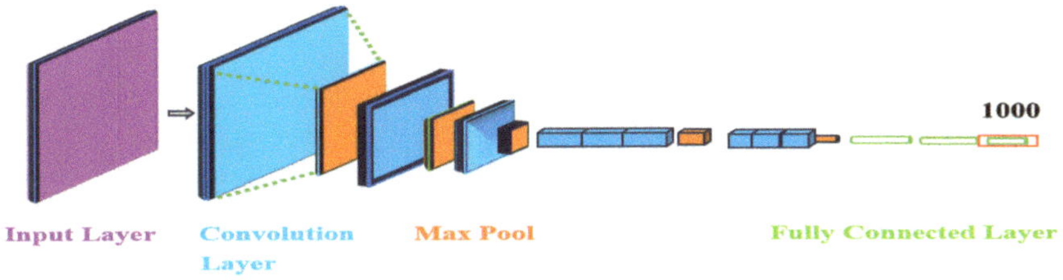

Fig. 2 Framework of VGG16 architecture

2.4 VGG16

The Visual Geometry Group at the University of Oxford created the standard benchmark in image classification tasks, VGG16, a deep convolutional neural network (CNN) architecture. Unlike previous neural network designs, which continuously used tiny 3 × 3 convolutional filters throughout the network, VGG16 has sixteen layers, comprising thirteen convolutional layers and three fully connected layers. This study employs a transfer learning strategy, utilizing a pre-trained VGG16 convolutional neural network (CNN) architecture from the ImageNet dataset, to address a multi-class classification problem, which includes photos of skin diseases. We omit the top classification layers specific to ImageNet's 1,000 classes to load the VGG16 model, which is renowned for its deep feature extraction ability. Figure 2 illustrates the VGG16 architectural model, characterized by a simplified structure featuring a convolutional layer utilizing a 3 × 3 filter with a stride of 1. It consists of pooling layers that measure 2 × 2, employing identical padding with a stride of 2. The structure consists of two interconnected layers, with the final layer removed and substituted with a SoftMax layer that classifies nine distinct types of skin diseases.

2.5 VGG19

Since its introduction in 2014 by the Visual Geometric Group at the University of Oxford, the architecture of the deep convolutional neural network, also known as VGG19, or Visual Geometry Group 19, has sparked a lot of interest in computer vision. This model is quite deep, with 19 layers (16 convolutional and 3 completely connected). Among the most extensive networks utilized in image categorization methods is VGG19 which has two distinct properties: an identical layout and the

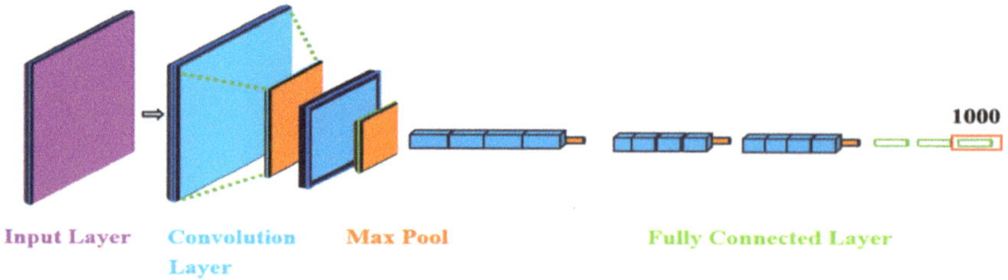

Fig. 3 The architecture of VGG19

use of modest 3 × 3 convolutional filters. These filters make it easier to understand hierarchical features, beginning with low-level edges and progressing to high-level object representation. Its architecture enables the extraction of rich feature representations for a variety of applications, including segmentation, object recognition, and image classification (Fig. 3).

3 Experimental Results and Discussion

This study utilized VGG16 and VGG19 for the dataset focused on the detection and classification of multiple skin diseases. The dataset is analyzed using data visualization techniques, specifically bar charts and line graphs, as illustrated in Figs. 4, 5, 6, and 7. The dataset is divided into training and testing sets, as presented in Table 2. The performance of the models is evaluated and displayed in Tables 3 and 4, along with Figs. 8, 9, 10, 12, 13 and 14. Additionally, we evaluated the skin disease categories using the test set, with actual, predicted, and confidence levels presented in Figs. 11 and 15. Table 5 provides a comparative analysis of the proposed model against the results of other authors.

This study used a dataset of 4,109 photographs. We took the images from nine different skin conditions and divided them into three categories: training, validation, and testing. You can see the detailed distribution of pictures for each skin condition in Table 2.

Table 2 illustrates that each skin condition in the dataset has been assigned specific image counts for training, validation, and testing. The majority of skin conditions, including Actinic Keratosis, Basal Cell Carcinoma, and Nevus, consist of 400 images in the training set, with 50 images allocated to both the validation and test sets, resulting in a total of 500 images for each condition. It is important to highlight that Melanoma comprises the largest collection of images, totaling 505, whereas Vascular Lesion contains the smallest dataset, with only 290 images. The complete dataset comprises 3,287 images designated for training, 410 for validation, and 412 for testing purposes.

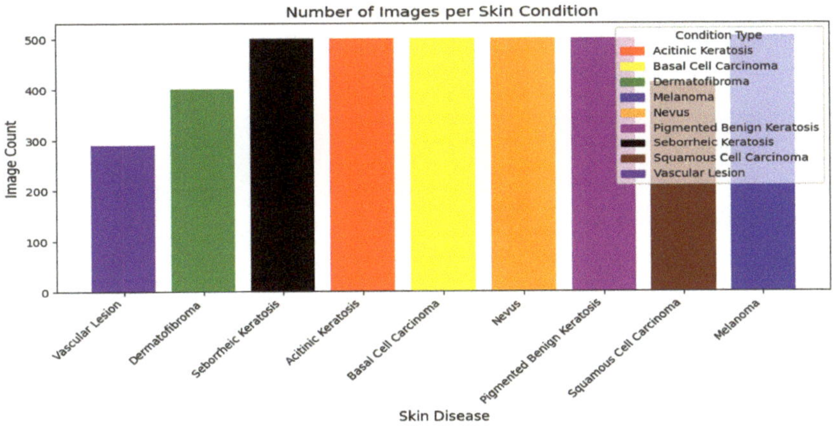

Fig. 4 Class distribution in dataset

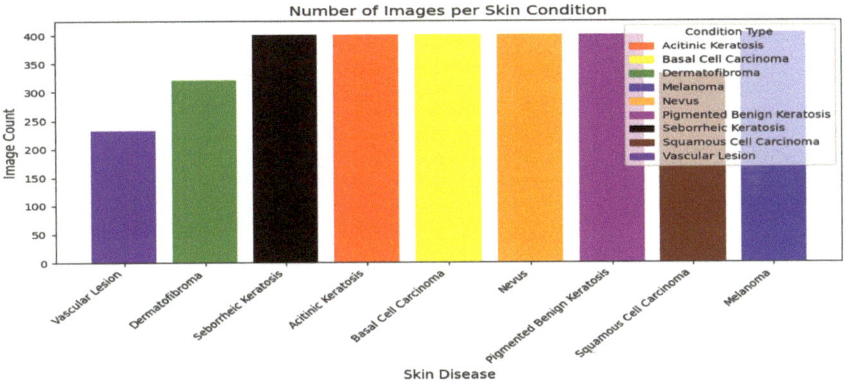

Fig. 5 The number of images per class in training set

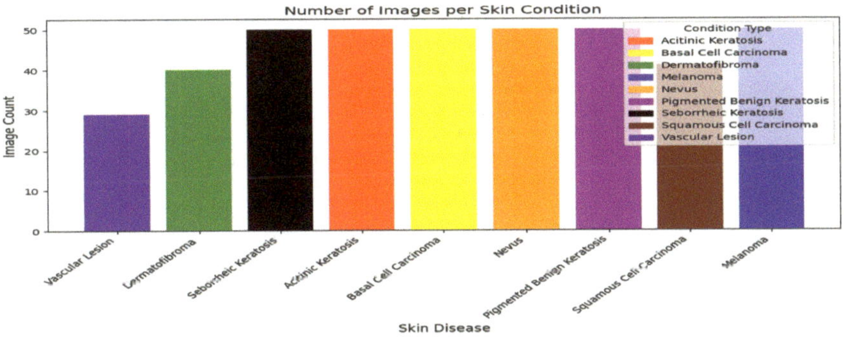

Fig. 6 The number of images per class in validation set

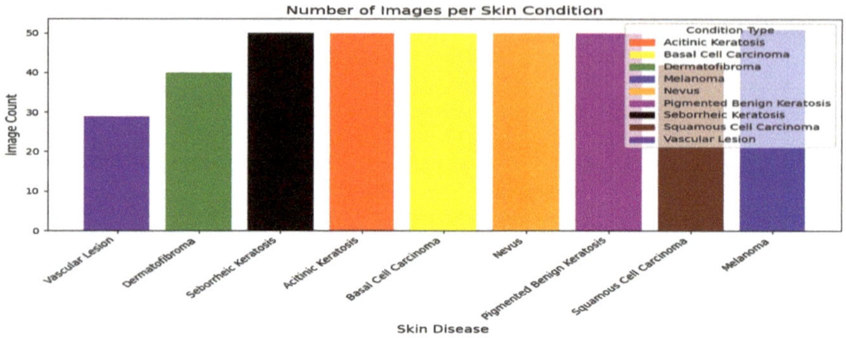

Fig. 7 The number of images per class in test set

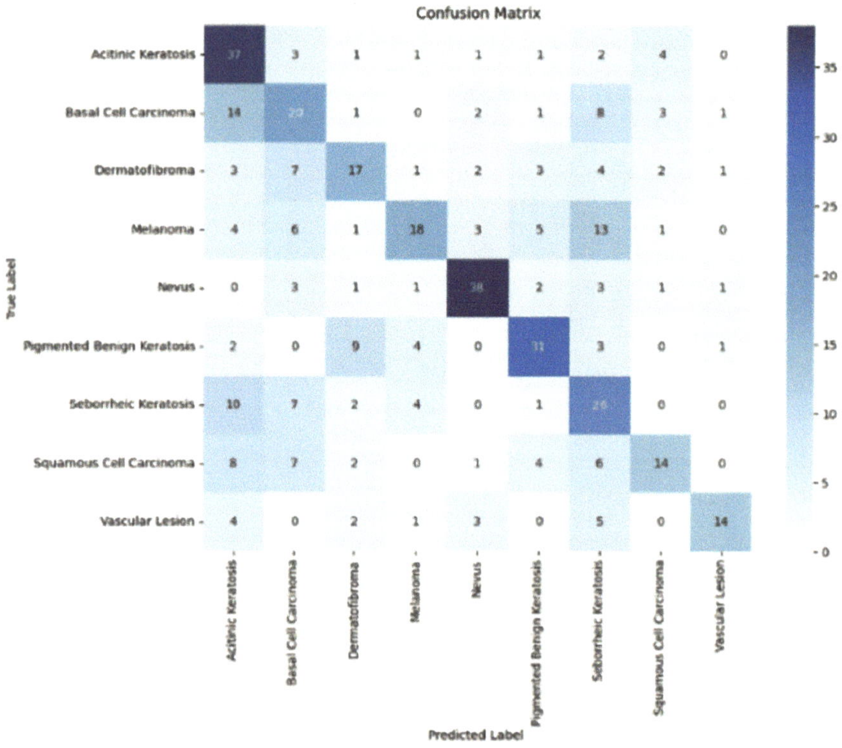

Fig. 8 Confusion matrix of VGG16

Despite VGG16 achieving a notable training accuracy of 96.88% and a minimal loss of 0.1222, the validation accuracy was significantly lower at 46.15%, accompanied by a relatively high validation loss of 3.5199. Table 3 illustrates the attainment of training accuracy. Each step for VGG16 required a duration of 51 ms. The

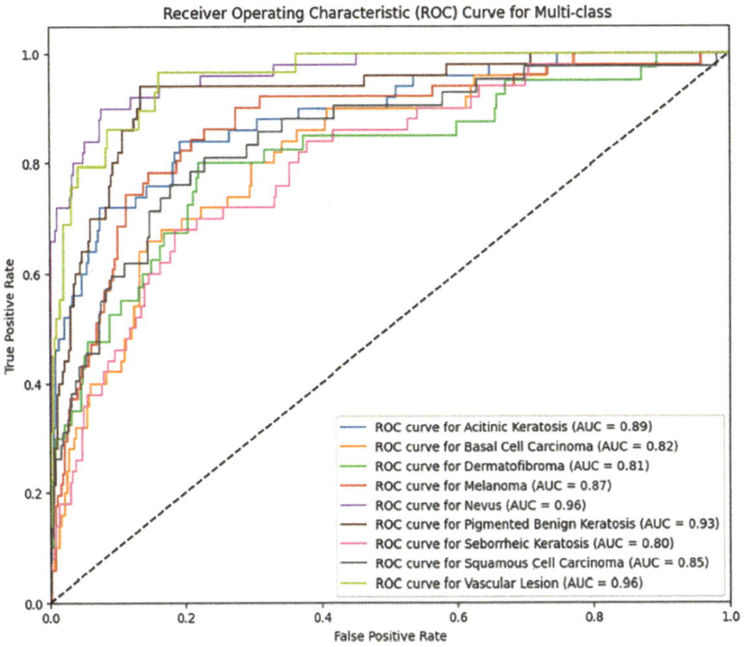

Fig. 9 VGG16 ROC curve

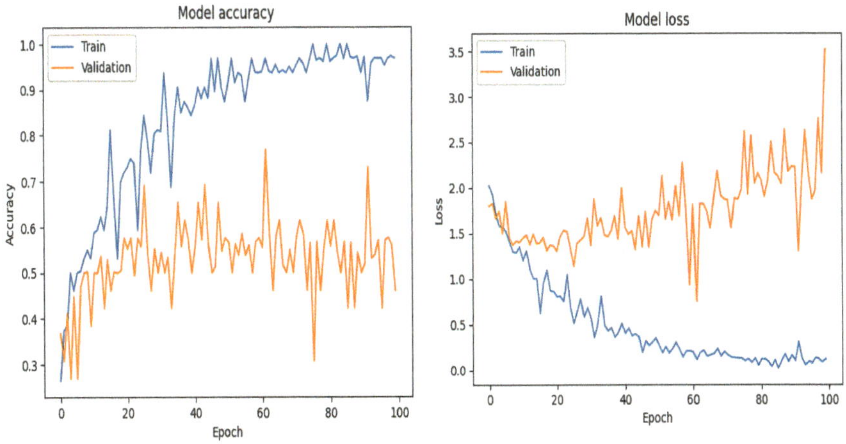

Fig. 10 VGG16 train and validation accuracy and loss

VGG19 model exhibited a training accuracy of 90.62%, which is slightly lower, alongside a higher loss of 0.2761%. Conversely, the validation accuracy reached a notable 65.38%, accompanied by a validation loss of 1.5440%. Additionally, it

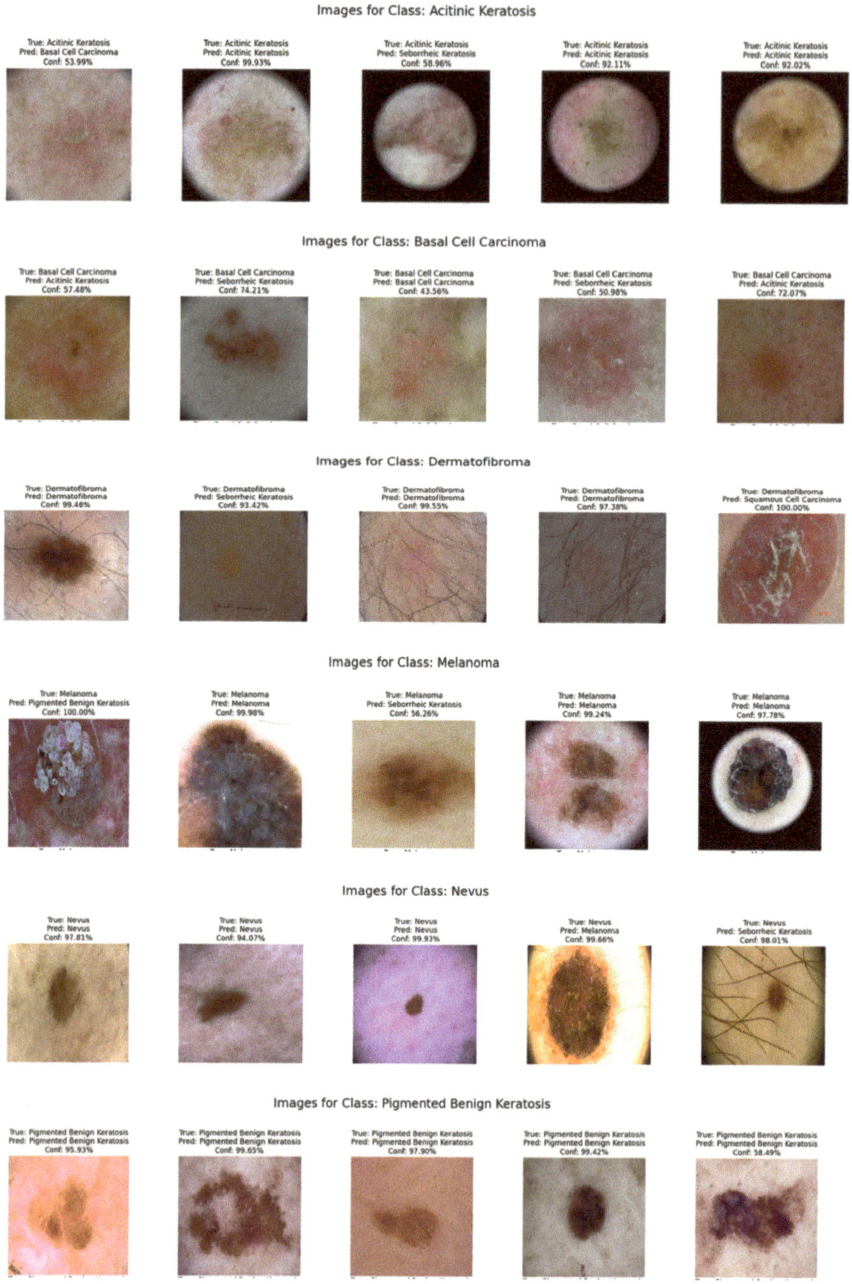

Fig. 11 VGG16 prediction on test set

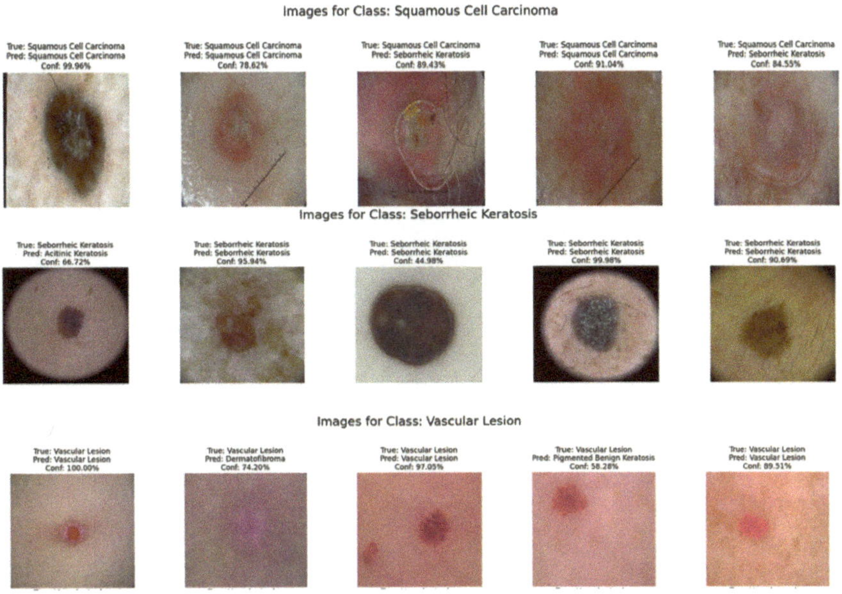

Fig. 11 (continued)

demonstrated enhanced operational efficiency, requiring only one millisecond per step.

Table 4 presents a performance comparison of the VGG16 and VGG19 models in the context of skin disease classification. VGG16 demonstrates superior performance compared to VGG19 across various categories, particularly in Pigmented Benign Keratosis, achieving an F1-score of 0.77, and in Actinic Keratosis, with an F1-score of 0.64. VGG19 demonstrates reduced precision and recall for difficult classes, including Basal Cell Carcinoma and Melanoma. Both models demonstrate challenges in accurately classifying Basal Cell Carcinoma, as evidenced by notably low F1-scores, highlighting the necessity for further optimization in these domains.

The confusion matrix in Fig. 8 demonstrates a high accuracy for nevus and pigmented benign keratosis, reflecting the model's effective classification performance across various skin disease types. The model demonstrates proficiency in accurately classifying additional categories, including seborrheic keratosis and dermatofibroma. Figure 9 illustrates that the Receiver Operating Characteristic (ROC) curves demonstrate strong multi-class performance. Nevus and Vascular Lesion exhibit the highest AUC values (0.96), followed by Pigmented Benign Keratosis (AUC = 0.93), indicating the model's robust capacity to distinguish among several classes. Figure 10 shows the training accuracy graph, which has a steady increasing trend, indicating that the model is successfully learning information from the input. This shows that the model is reducing mistakes over time, as seen by the decreased training loss. Despite the disparities, the validation accuracy shows potential peaks, demonstrating

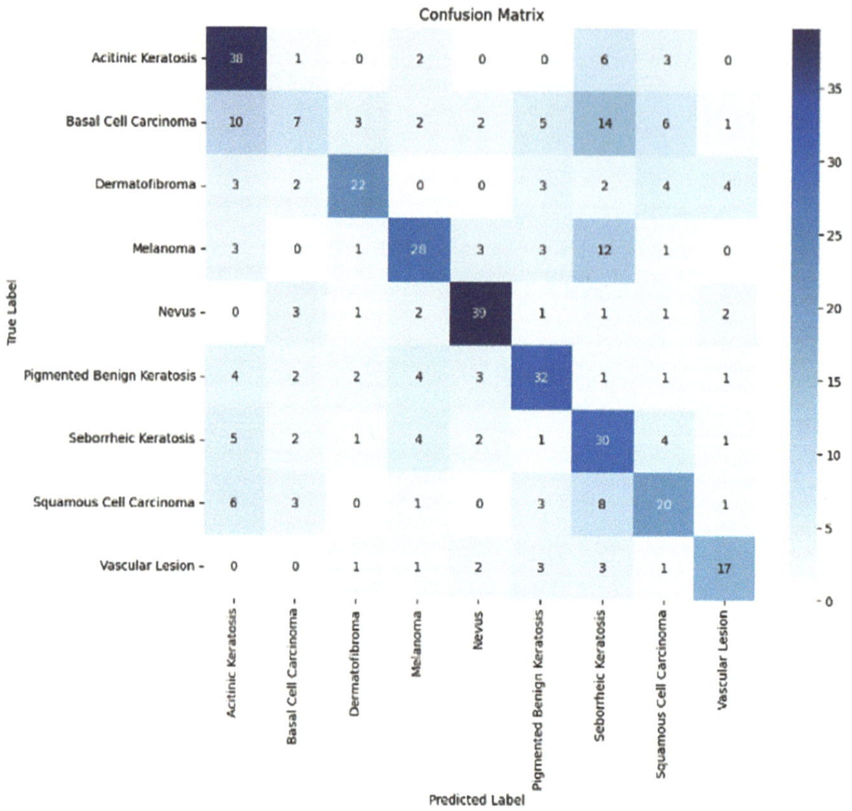

Fig. 12 Confusion matrix of VGG19

the model's ability to generalize. Given these considerations, the training performance demonstrates outstanding learning capacity, yet there is room for further improvement in validation stability.

The confusion matrix in Fig. 12 illustrates the model's classification performance across nine distinct kinds of skin diseases. It exhibits a significant level of precision in forecasting specific categories, such as "Nevus" and "Seborrheic Keratosis," with the diagonal indicating the greatest quantity of accurate forecasts. Nonetheless, there exist specific erroneous classifications, particularly among closely related groupings, such as "melanoma" and "nevus." This study facilitates a clearer comprehension of the model's strengths and weaknesses, particularly in distinguishing between comparable skin conditions. The ROC curve illustrated in Fig. 13 delineates the model's ability to differentiate among various skin conditions in the context of a multi-class classification task. The performance across classes exhibits considerable variability, as indicated by the AUC values, which span from 0.77 to 0.95. While the model exhibits a satisfactory performance for basal cell carcinoma (AUC = 0.77), it excels in categorizing nevus (AUC = 0.95) and actinic keratosis (AUC = 0.93) to a greater

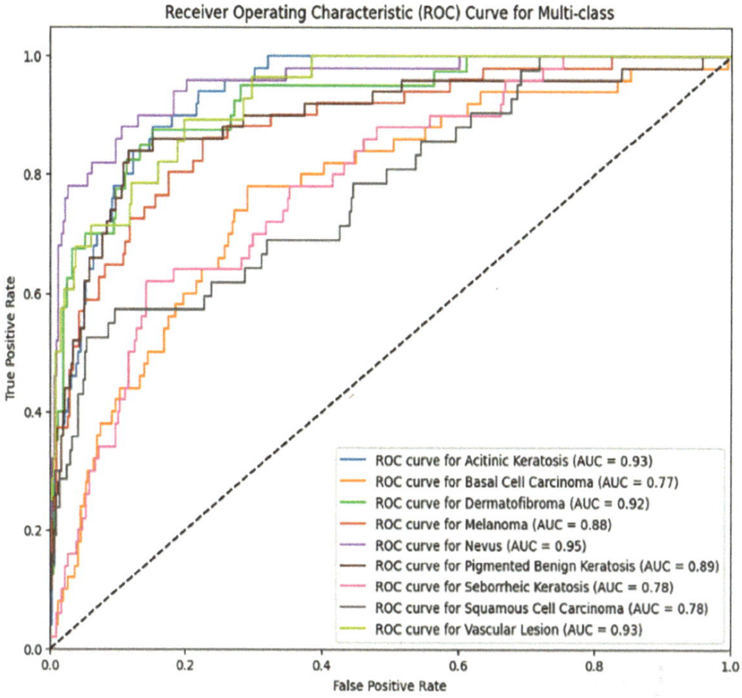

Fig. 13 VGG19 ROC curve

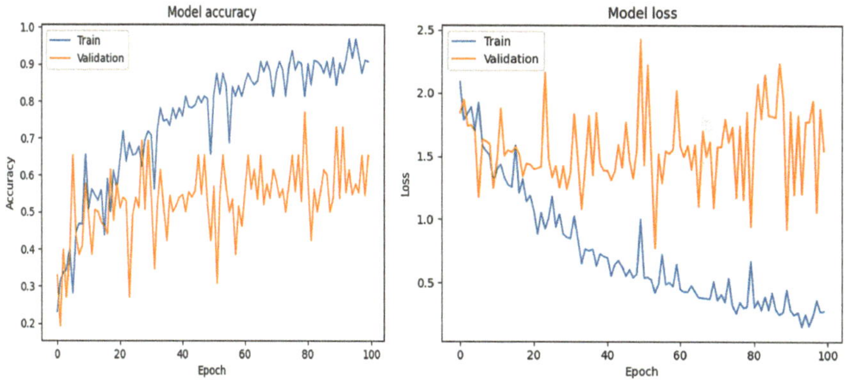

Fig. 14 VGG19 train and validation accuracy and loss

extent than for basal cell carcinoma. Concerning the dermatological disorders, the elevated AUC values for most classes indicate that the model excels at differentiation. The training accuracy exhibits a steady upward trajectory, attaining approximately 90%, signifying that the model is effectively assimilating the input data. Despite fluctuations in validation accuracy, the model exhibits promise, with validation accuracy

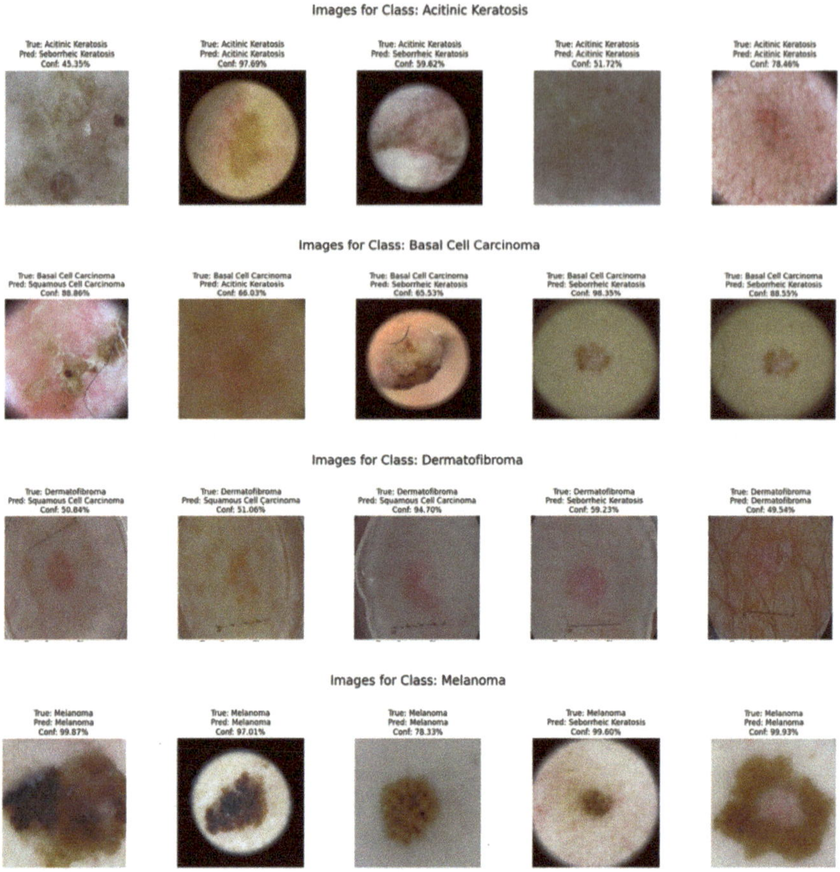

Fig. 15 VGG19 prediction on test set

occasionally corresponding with training accuracy. The consistent decline in training loss signifies effective optimization, while the fluctuating validation loss reveals instances of improvement. The model can generalize more successfully and substantially improve its validation performance with further tuning, either by incorporating additional data or adjusting the hyperparameters.

The subsequent table provides a comparison of various deep learning models employed for the detection and categorization of skin disorders across many datasets. Oumoulylte et al. utilized VGG16, VGG19, and CNN on the ISIC Archive dataset, achieving accuracies of 84.37%, 87%, and 74%, respectively, with VGG19 performing the most effectively. Utilizing the identical ISIC Archive, Angelina and her associates employed ResNet50, ResNet101, VGG16, and VGG19. The findings indicated that VGG16 achieved the maximum accuracy at 83.75%. The suggested method employs VGG16 and VGG19 on a custom "Multiple Skin Disease Detection and Classification Dataset," achieving much higher accuracies of 96.88% and

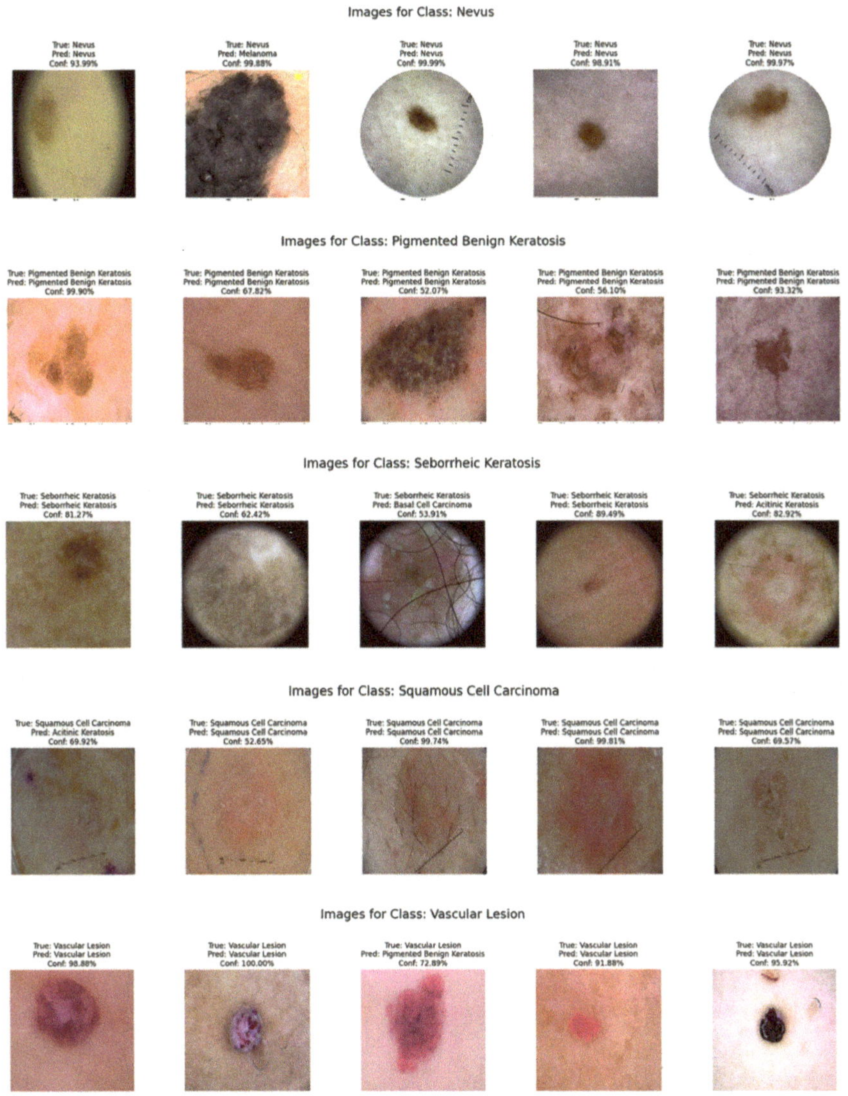

Fig. 15 (continued)

90.62%, respectively. The notable enhancement in the proposed strategy indicates that employing a particular dataset markedly enhances the model's performance, especially in challenging tasks like skin disease categorization. The results indicate that VGG16 regularly surpasses other models and datasets when applied to a custom dataset, making it an invaluable tool for enhancing diagnostic accuracy in skin disease classification.

Table 2 Distribution of training and test data for skin disease classification

Classes	Training set	Validation set	Test set	Total
Actinic Keratosis	400	50	50	500
Basal Cell Carcinoma	400	50	50	500
Dermatofibroma	320	40	40	400
Melanoma	404	50	51	505
Nevus	400	50	50	500
Pigmented Benign Keratosis	400	50	50	500
Seborrheic Keratosis	400	50	50	500
Squamous Cell Carcinoma	331	41	42	414
Vascular Lesion	232	29	29	290
Total	3287	410	412	4109

Table 3 Performance of the VGG16 and VGG19 classifiers

Algorithms	Accuracy (%)	Loss	Validation accuracy (%)	Validation loss	Time (ms/step)
VGG16	96.88	0.1222	46.15	3.5199	51
VGG19	90.62	0.2761	65.38	1.5440	1

Table 4 Algorithms classification report performance

Models	Classes	Precision	Recall	F1-score	Support
VGG16	Actinic Keratosis	0.55	0.76	0.64	50
	Basal Cell Carcinoma	0.64	0.14	0.20	50
	Dermatofibroma	0.71	0.55	0.62	40
	Melanoma	0.64	0.55	0.62	51
	Nevus	0.64	0.55	0.59	50
	Pigmented Benign Keratosis	0.76	0.78	0.77	50
	Seborrheic Keratosis	0.63	0.64	0.63	50
	Squamous Cell Carcinoma	0.39	0.60	0.47	42
	Vascular Lesion	0.63	0.61	0.62	28
VGG19	Actinic Keratosis	0.45	0.74	0.56	50
	Basal Cell Carcinoma	0.38	0.40	0.39	50
	Dermatofibroma	0.47	0.42	0.45	40
	Melanoma	0.60	0.35	0.44	51
	Nevus	0.76	0.76	0.76	50
	Pigmented Benign Keratosis	0.65	0.62	0.63	50
	Seborrheic Keratosis	0.37	0.52	0.43	50
	Squamous Cell Carcinoma	0.56	0.33	0.42	42
	Vascular Lesion	0.78	0.48	0.60	29

Table 5 Comparison of the results for proposed algorithms for ISIC Archive

Paper	Methods	Dataset	Accuracy
Nayyef et al. [23]	ResNet50, Xception, InceptionV3, VGG16, and VGG19	ISIC Archive	VGG16 91.30%, VGG19 84.55%
Proposed method	VGG16 and VGG19	ISIC Archive	VGG16 96.88% VGG19 90.62%

4 Conclusion

Skin illnesses are generally persistent and might progress to malignant cells. As a result, early detection and treatment are critical to limiting serious harm. In this research, we used and assessed two pre-trained Convolutional Neural Networks (CNNs), VGG16 and VGG19, to classify skin cancer disorders into nine separate groups. Our data demonstrate that VGG16 performed more consistently, with an accuracy of 96.88%, while VGG16 had an accuracy of 90.62%. The use of transfer learning enabled the models to successfully handle the relatively limited dataset by utilizing pre-trained features, resulting in better classification results. Although both models have demonstrated potential in automating the skin cancer detection process, they could achieve further advancements by utilizing larger datasets, fine-tuning hyperparameters, and exploring ensemble approaches. Overall, this work adds to the expanding body of evidence demonstrating that deep learning models, notably VGG16 and VGG19, may help in the early identification and diagnosis of skin cancer, possibly outperforming the diagnostic accuracy of healthcare experts.

References

1. Telang Y, Tikone P, Vategaonkar G, Sonawane P, Deoghare R (2024) Skin disease classification using deep learning. In: AIP conference proceedings, vol 3044, no 1. AIP Publishing
2. Rani EG, Hussain MA, Azeezulla M, Shandilya M, Varughese PS (2023) Skin disease diagnosis using vgg19 algorithm and treatment recommendation system. In: 2023 IEEE 8th international conference for convergence in technology (I2CT). IEEE, pp 1–8
3. Shandilya M, Hussain MA, Varughese PS, Azeezulla M Skin disease diagnosis using VGG19 algorithm and treatment recommendation system
4. Pai VR, Pai SG, Suhasi PM, Rekha PM (2023) Identification and classification of skin diseases using deep learning techniques
5. Montagna W (1965) The skin. Sci Am 212(2):56–69
6. Egert M, Simmering R, Riedel CU (2017) The association of the skin microbiota with health, immunity, and disease. Clin Pharmacol Ther 102(1):62–69
7. Dinnes J, Deeks JJ, Chuchu N, di Ruffano LF, Matin RN, Thomson DR, Williams HC (2018) Dermoscopy, with and without visual inspection, for diagnosing melanoma in adults. Cochrane Database Systematic Rev 12
8. Basra MK, Shahrukh M (2009) Burden of skin diseases. Expert Rev Pharmacoecon Outcomes Res 9(3):271–283
9. Del Pozzo-Magaña BR, Liy-Wong C (2024) Drugs and the skin: a concise review of cutaneous adverse drug reactions. Br J Clin Pharmacol 90(8):1838–1855

10. Hartmane I (2024) Study of genetic mutations and their association with the development of atopic dermatitis and other skin diseases. Plastic Aesthet Nurs 44(3):200–209
11. Khan NH, Mir M, Qian L, Baloch M, Khan MFA, Ngowi EE, Ji XY (2022) Skin cancer biology and barriers to treatment: recent applications of polymeric micro/nanostructures. J Adv Res 36:223–247
12. Piccolo V (2019) Update on dermoscopy and infectious skin diseases. Dermatol Pract Concept 10(1):e2020003
13. Chowdhury MM, Katugampola RP, Finlay AY (2019) Dermatology at a glance. John Wiley & Sons
14. Goceri E (2021) Diagnosis of skin diseases in the era of deep learning and mobile technology. Comput Biol Med 134:104458
15. Soglia S, Pérez-Anker J, Lobos Guede N, Giavedoni P, Puig S, Malvehy J (2022) Diagnostics using non-invasive technologies in dermatological oncology. Cancers 14(23):5886
16. Felgueiras HP (2021) An insight into biomolecules for the treatment of skin infectious diseases. Pharmaceutics 13(7):1012
17. Carli P, De Giorgi V, Palli D, Maurichi A, Mulas P, Orlandi C, Giannotti B (2003) Dermatologist detection and skin self-examination are associated with thinner melanomas: results from a survey of the Italian Multidisciplinary Group on Melanoma. Arch Dermatol 139(5):607–612
18. Li LF, Wang X, Hu WJ, Xiong NN, Du YX, Li BS (2020) Deep learning in skin disease image recognition: a review. IEEE Access 8:208264–208280
19. Mondal B, Das N, Santosh KC, Nasipuri M (2020) Improved skin disease classification using generative adversarial network. In: 2020 IEEE 33rd international symposium on computer-Based medical systems (CBMS). IEEE, pp 520–525
20. Allugunti VR (2022) A machine learning model for skin disease classification using convolution neural network. Int J Comput Program Database Manag 3(1):141–147
21. Doğan F, Aktaş M, Gürsoy Mİ (2024) Classification of skin diseases with different deep learning models and comparison of the performances of the models. Türk Doğa ve Fen Dergisi 13(3):117–123
22. Ahmad B, Usama M, Huang CM, Hwang K, Hossain MS, Muhammad G (2020) Discriminative feature learning for skin disease classification using deep convolutional neural network. IEEE Access 8:39025–39033
23. Nayyef SA, Hasan LM, Ibrahim AA (2020) Using deep learning and medical data classification for predicting bacterial skin infection. In: 2020 4th international symposium on multidisciplinary studies and innovative technologies (ISMSIT). IEEE, pp 1–9

Enhancing Indian Sign Language Recognition Using Advanced Deep Learning Architectures

S. R. Likhith, Shloak Satish Shetty, A. Poorani Maragathavalli, Rishabh Tiknoo, Srishti, K. Jeevan, and Kota Yogananda Reddy

Abstract Indian Sign Language (ISL) detection is an essential feature that facilitates communication among those who are deaf or hard of hearing. Using models like EfficientNetB0, ResNet50, and MobileNetV3, we examine the efficacy of image processing and deep learning in ISL identification within the framework of this study. In order to develop a real-time, reliable, and effective ISL detection system, the research aims to tackle the nuances and complexity inherent in deciphering sign language motions. The objective is to demonstrate the correctness and dependability of the suggested method for efficiently and accurately understanding ISL gestures through thorough testing and evaluation. Some image processing techniques that are used to preprocess raw visual data from cameras or sensors are part of the methodology. Prior to inputting images into the system, we extract pertinent information from them in an effort to maximize its performance. Additionally, we make use of advanced Deep Learning architectures that can recognize complex patterns and movements unique to ISL, like EfficientNetB0, ResNet50, and MobileNetV3. Deep learning models' parameters were carefully tuned through extensive experimentation. Accuracy, precision, recall, and F1-score should all be used to assess the system because they will provide an overall picture of its effectiveness, which may ultimately result in improved ISL detection. Additionally, we test the system's functionality in more realistic scenarios. As a result, the proposed approach for recognizing ISL is faster and more accurate than current ones.

Keywords ISL detection · Image processing · EfficientNetB0 · ResNet50 · MobileNetV3

S. R. Likhith (✉)
Presidency School of Computer Science and Engineering, Presidency University, Bengaluru, India
e-mail: anandlikith111@gmail.com

S. S. Shetty · A. Poorani Maragathavalli · R. Tiknoo · Srishti · K. Jeevan · K. Y. Reddy
Presidency School of Information Science, Presidency University, Bengaluru, India

H. Nagaraj et al. (eds.), *Advancements in Artificial Intelligence and Machine Learning*, Lecture Notes in Electrical Engineering 1572,
https://doi.org/10.1007/978-981-95-7449-0_25

1 Introduction

Sign language, as a visual mode of communication, holds immense importance in fostering inclusivity and facilitating interaction for deaf and hearing-impaired individuals. Its significance lies in its ability to transcend language barriers and enable meaningful communication in diverse settings. However, despite its pivotal role, there are inherent challenges in ensuring effective teaching and communication of sign language.

One major hurdle is the scarcity of resources, including qualified instructors and accessible educational materials, which limits the opportunities for individuals to learn and practice sign language effectively. Moreover, the lack of standardized teaching methodologies and curriculum further complicates the learning process, making it challenging for both students and educators to navigate.

Additionally, the provision of constructive feedback on sign language performance remains a persistent issue. Without timely and accurate feedback, learners may struggle to identify and rectify errors in their signing, impeding their progress and confidence in using sign language fluently.

Amid the ongoing COVID-19 pandemic, the challenges surrounding sign language instruction have been further exacerbated. The closure of educational institutions and restrictions on in-person gatherings have disrupted traditional learning environments, making it even more difficult for deaf and hearing-impaired individuals to access quality sign language education. The shift to remote learning has highlighted the importance of digital tools and online platforms in delivering sign language instruction. However, the digital divide and lack of accessibility features in online learning platforms pose additional barriers for individuals with hearing disabilities. Furthermore, the absence of face-to-face interaction and real-time feedback in virtual settings can impede the development of proficient sign language skills. As such, there is an urgent need to adapt and innovate in the realm of sign language education to ensure continued access to high-quality instruction amid the challenges posed by the pandemic. By leveraging technology and fostering collaborative efforts within the community, we can strive to mitigate the impact of COVID-19 on sign language learning and promote greater inclusivity for all individuals, regardless of their hearing abilities.

Addressing these challenges requires innovative approaches that leverage technology, educational strategies, and community support to enhance the accessibility and quality of sign language instruction. By overcoming these barriers, we can promote greater inclusivity and empowerment for individuals with hearing disabilities, fostering a more inclusive society.

Section 2 provides a literature review, summarizing relevant IEEE papers that will inform the project's implementation. Section 3 describes the proposed method to enhance accuracy and efficiency in sign language recognition systems. Section 4 focuses on describing the models used in the project and explaining their working. Section 4 covers the existing challenges faced, the expected outcomes, and the actual results in a straightforward format. Section 5 presents the results and discussion. Section 6 concludes the report and discusses future scope.

2 Related Work

The study analyzes data from the JUMLA-QSL-22 corpus [1], focusing on trends and patterns. It examines video frames, participant movements, and environmental factors during signing. The dataset captures diverse styles and linguistic intricacies, enriching its authenticity. This curated dataset is crucial for future research in sign language recognition, paving the way for systems that can interpret and translate sign language conversations. Integrating machine learning and artificial intelligence could bridge communication gaps between hearing-impaired individuals and society. The KArSL dataset (190 and 502 signs) [2] and the LSA64 dataset (64 signs performed by several signers) were used in the study for analysis. To properly capture spatiotemporal information, the study used a trainable deep learning network that combined dynamic motion network (DMN) and accumulative motion network (AMN) approaches. Signer-dependent and signer-independent modes were compared, with reported accuracies of 99 and 64%. The proposed approach enhanced performance in recognizing solitary sign language movements, demonstrating the two-stream network model's potential for increasing accuracy and robustness in sign language recognition tasks. A wide variety of motions were available for analysis in the study's dataset [3], which included twenty different Indian Sign Language (ISL) alphabets collected from the participants. Utilizing the capabilities of deep learning for sign language interpretation, the study used Convolutional Neural Networks (CNN) for feature extraction and recognition. 99.56% accuracy in regular conditions and 97.26% accuracy in poor light were achieved by the CNN model, demonstrating its outstanding performance. The efficacy of the CNN-based method in precisely identifying and deciphering ISL movements is demonstrated by this high degree of accuracy, perhaps indicating real-time sign language translation benefits. Tornay [4] explored techniques for modeling hand movement information without relying on specific languages, using subunits derived from Hidden Markov Models (HMMs). The study identified differences in performance between language-independent and language-dependent models of hand movement. Incorporating hand shape data helped bridge this performance gap, resulting in more effective systems. These findings are promising and suggest potential advancements in sign language processing systems that can leverage a wider range of resources. In order to decrease parameters, Xu [5] suggests employing tensor-train factorization in S2VT with an emphasis on Chinese sign language recognition. When the input and output tensors are distributed and ordered from tiny to large, the model operates at its best. Different layers of S2VT models are subjected to tensor-train factorization, which leads to a large decrease in parameter values. While retaining comparable accuracy, the fully connected layer and the first LSTM layer using the tensor-train format reduce parameters by 49.5%. As a result, sequence-to-sequence issues and mobile devices can benefit from sign language recognition models. Utilizing a dataset comprising 40 Arabic sign language expressions and encompassing 80 distinct terms, a two-stage deep learning methodology [6] is proposed to advance the creation and interpretation of Arabic sign language. The dataset's integrity and inclusiveness are ensured by replicating each utterance 19 times, facilitating refined model training. This extensive linguistic input and diverse lexicon compilation facilitate robust generalization across various contexts, thereby enhancing the accuracy of the model in both understanding and generating sign language gestures. Furthermore, the model development process is informed by statistical insights derived from the dataset, optimizing resource utilization and bolstering overall system performance. Leveraging these advancements, researchers can devise more precise sign language translation systems, thereby promoting accessibility and fostering inclusive communication for individuals with hearing impairments. The publication [7] outlines the development of an end-to-end deep learning framework tailored for the translation and recognition of sign language. Employing a hybrid approach, this framework integrates neural machine translation (NMT), attention mechanisms, bidirectional long short-term memory (BiLSTM) models, and convolutional neural networks (CNNs). By merging spatial and temporal data, this comprehensive strategy enhances the precision and efficacy of sign language translation and recognition. The overarching goal is to address communication barriers for individuals with hearing loss, offering a promising avenue for heightened inclusion and accessibility. The IDF-Sign model [8] demonstrated good recognition performance across many languages after undergoing rigorous testing on several sign language datasets. For German Sign Language (GSL), it obtained an average accuracy of 78%, for American Sign Language (ASL), 95%, and for Dialectal Sign Language (DSG), 65.07%. Its remarkable 95% accuracy on datasets similar to ASL demonstrated its resilience and flexibility to many variances and signing styles. These findings highlight the model's effectiveness in practical settings and present encouraging opportunities for enhancing accessibility and communication for those who use sign language. In the field of Chinese Sign Language Recognition Systems, Xinjiang University's RF-CSign [9] is unique because of its creative combination of normalization-based attention methods and large kernel convolution. RF-CSign is designed primarily to allow wireless sensor interactions, meeting the special demands of the hearing-impaired community, and boasts an exceptional 99.17% classification accuracy. Its emphasis on usefulness combined with sophisticated methods highlights how well it can recognize motions in Chinese Sign Language. A major advancement in sign language detection technology, RF-CSign holds the promise of better accessibility and interaction for deaf people in Chinese-speaking communities. A comprehensive strategy was employed in the study [10] that concentrated on the identification of Korean Sign Language (KSL), integrating attention-based general neural networks, dual-stream neural networks, and graph convolutional networks (GCNs). With excellent accuracy rates, an all-encompassing approach greatly enhanced the identification of dynamic KSL motions. These developments improve KSL accessibility and recognition, which is especially helpful to communities of people with hearing impairments. Saeed [11] presents a systematic review of the literature concerning sign language recognition via sensor-based glove systems, aiming to elucidate motivations, challenges, and recommendations within this domain. Through analysis of multiple databases, the researchers highlight the pivotal role of data in sign language recognition, advocating for the expansion of database sets to enhance communication for the hearing impaired. Additionally, the paper explores the multifaceted applications of sensory gloves across various fields and offers recommendations for developers. Notably, the authors propose leveraging deep learning algorithms to refine gesture recognition and tackle the complexities of continuous sign language recognition. Joksimoski [12] conducts a systematic review of articles concerning sign language recognition through sensor-based glove systems, with a focus on identifying motivations, challenges, and recommendations in this domain. Utilizing analysis of diverse databases, the researchers underscore the significance of data in sign language recognition, advocating for the augmentation of database sets to enhance communication for individuals with hearing impairments. Moreover, the paper delves into the versatile applications of sensory gloves across various sectors and offers actionable recommendations for developers. Notably, the authors advocate for the adoption of deep learning algorithms to enhance gesture recognition and mitigate the complexities associated with continuous sign language recognition. A unique technique for sign language recognition is presented in the publication [13]. Graph creation, universal deep neural networks, and the use of a multistage graph convolution with attention and residual connection (GCAR) model are among the key approaches used in the article. In order to generate a spatial–temporal graph that enables the extraction of spatial–temporal contextual information, the authors make use of joint skeleton information. Moreover, they incorporate a channel attention module to enhance attentional states for disconnected skeletal points at particular instances. The goal of the suggested model is to fully capture all of the body movements that occur during sign language gestures in order to eventually strive for higher recognition task accuracy rates. Sosa-Jiménez [14] focuses on sign language recognition using depth sensors and employs a variety of tools such as apparatus, sensors, hardware, and software. Techniques include extracting and standardizing geometric features from body, hand, and facial expressions using libraries like MediaPipe and OpenCV. Additionally, support vector machines are utilized for classification, while genetic algorithms aid in feature selection. The primary modeling tool employed is hidden Markov models (HMMs), facilitating a comprehensive analysis of complex spatial patterns. Results indicate promising accuracy in sign language recognition, with potential for enhancement through additional training data. Al-Qurishi [15] delves into sign language recognition utilizing depth sensors. The study harnesses an array of tools including apparatus, sensors, hardware, and software. Techniques encompass the extraction and standardization of geometric features from body, hand, and facial expressions leveraging libraries such as MediaPipe and OpenCV. Additionally, support vector machines are employed for classification, while genetic algorithms aid in feature selection. The principal modeling tool utilized is hidden Markov models (HMMs), enabling a comprehensive analysis of intricate spatial patterns. Results exhibit promising accuracy in sign language recognition, with potential for refinement through additional training data. The proposed sign language recognition system [16] leverages a hybrid deep neural network approach, amalgamating diverse architectures to encompass spatial, temporal, and semantic features of sign gestures comprehensively. Through the integration of 3D CNNs, attention-based BiLSTMs, and modified autoencoders, the system surpasses existing frameworks in performance. Assessment on a newly introduced multi-signer Indo-Russian sign language dataset validates its efficacy in enhancing communication accessibility for individuals with speech and hearing impairments. A groundbreaking non-invasive continuous dynamic gesture recognition system [17] has been innovated specifically for Chinese sign language. This system ingeniously integrates Inertial Measurement Unit (IMU) and surface Electromyography (sEMG) signals, allowing for the extraction of fusion gesture features. Utilizing a BiLSTM network coupled with a CTC loss function facilitates end-to-end recognition, effectively addressing pre-segmentation inaccuracies. Rigorous testing conducted on the CSLD dataset, which encompasses 10 distinct gesture types and a staggering 20,000 samples, demonstrates an exceptional average recognition rate.

3 Proposed Method

The proposed method leverages multiple advanced models—EfficientNetB0, ResNet50 [18], and MobileNetV3, with and without data augmentation—to address challenges in sign language recognition systems, such as limited accuracy, robustness to variations, and computational inefficiency. This approach involves several key steps, including data collection of an Indian Sign Language (ISL) dataset for numbers and letters, data preprocessing to clean and standardize the data, and data exploration to understand its characteristics. The dataset is then split into training and testing subsets to ensure the model's ability to generalize. Model selection prioritizes efficiency and performance, while training involves techniques like data augmentation and hyperparameter tuning to enhance robustness. Evaluation and continuous improvement ensure the model remains competitive and adaptable to real-world scenarios.

In developing the sign language recognition system as shown in Fig. 1, effective data preprocessing, exploration, and splitting are essential for preparing the ISL dataset for machine learning. Models like EfficientNetB0 are initialized with pre-trained weights and undergo rigorous training with data augmentation to handle image variations. Model evaluation identifies strengths and limitations, guiding improvements to enhance robustness and adaptability. Continuous iteration, sophisticated data augmentation, and benchmarking against existing methods ensure that the models are refined for better performance, addressing issues like overfitting and ensuring competitiveness in the field.

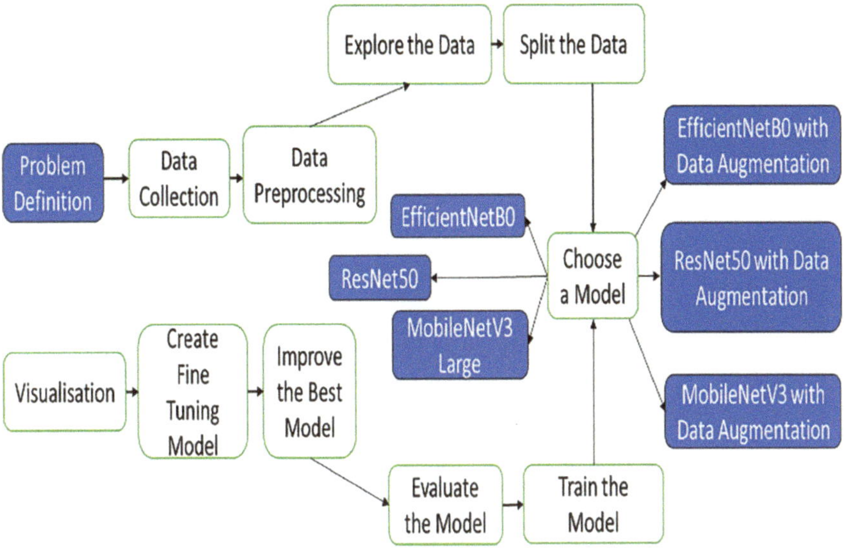

Fig. 1 Block diagram for Indian sign language detection

4 Methodology

4.1 EfficientNetB0

EfficientNetB0, a prominent convolutional neural network (CNN), is celebrated for its harmonious scaling of network dimensions, offering enhanced performance and efficiency. As part of the EfficientNet [19] family, ranging from B1 to B7, B0 serves as the foundational model, showcasing the fruits of advancements in neural architecture search (NAS) and model scaling. Previous CNNs often haphazardly scaled either depth, width, or resolution, leading to inefficiencies. EfficientNetB0, however, revolutionizes this approach by introducing a systematic scaling method, propelled by a compound coefficient. This formula uniformly adjusts depth, width, and resolution, ensuring a balanced augmentation of computational resources. In Fig. 2, it can be noticed that at the core of EfficientNetB0 lies the MBConv block, characterized by inverted residual structures employing depthwise separable convolutions. This design choice significantly reduces parameters and computations, fostering a lighter and faster network without compromising performance. The incorporation of the Swish activation function further enhances training by addressing issues like vanishing gradients, while the squeeze-and-excitation (SE) blocks intelligently recalibrate feature responses, bolstering robust feature representation. The formula for EfficientNetB0 involves a compound scaling method, which determines the dimensions of depth, width, and resolution of the network. This method is governed by a compound coefficient, denoted as φ, which dictates how much each dimension should

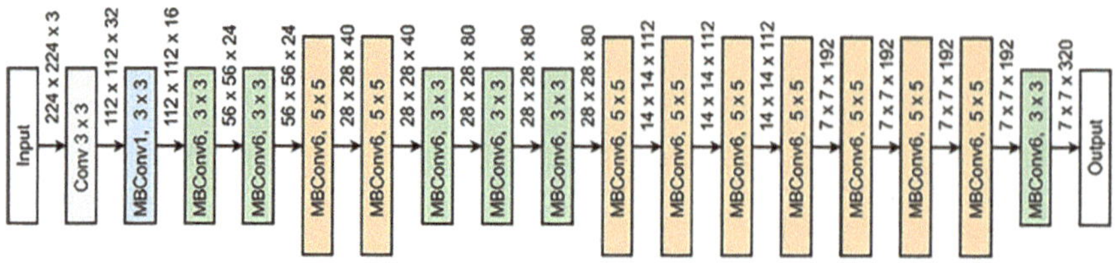

Fig. 2 EfficientB0 architecture

be increased given additional computational resources. The formula for scaling each dimension is as follows:

$$Depth = \alpha^{\phi} \tag{1}$$

$$Width = \beta^{\phi} \tag{2}$$

$$Resolution = \gamma^{\phi} \tag{3}$$

In Eqs. (1), (2), and (3), α, β, and γ are constants determined empirically. These constants are derived through a systematic study of model scaling to ensure balanced increases in all dimensions.

4.2 *ResNet50*

ResNet50, a cornerstone in computer vision, stands out for its innovative architecture and prowess as a feature extractor. Part of the ResNet family, it addresses the degradation problem in deep neural networks by introducing residual learning. While deeper networks theoretically offer better performance, ResNet50's residual framework acknowledges the challenge of vanishing gradients. Shortcut connections enable the network to learn residual functions with reference to layer inputs, mitigating this issue and facilitating effective gradient flow. The formula for ResNet50 is as follows:

$$Output = F(Input) + Input \tag{4}$$

In Eq. (4):

- *Input* represents the input feature maps to the block.
- $F(Input)$ represents the output of the convolutional layers within the block.
- The addition operation combines the input feature maps with the output of the convolutional layers.

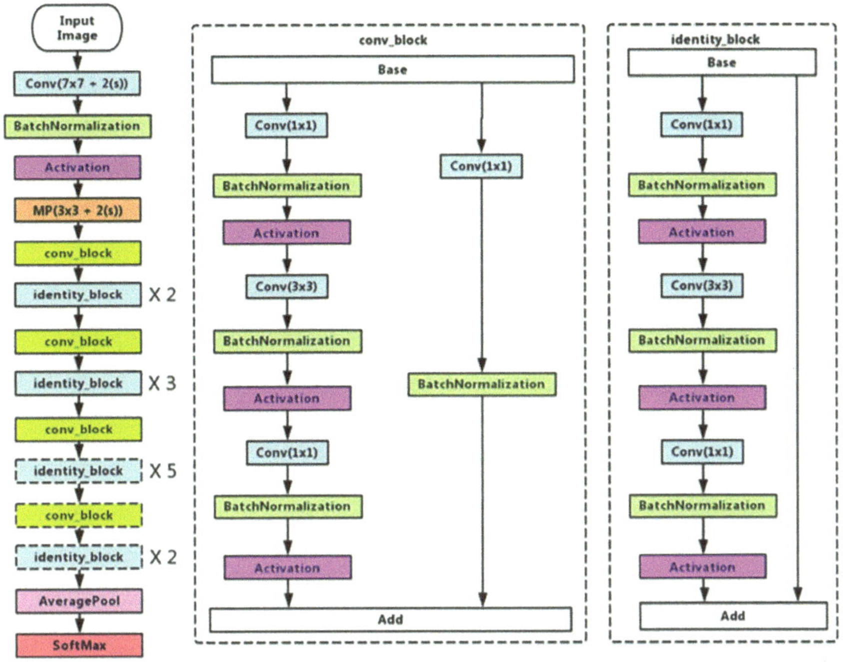

Fig. 3 ResNet50 architecture

The architecture of ResNet50 comprises 50 layers, which is represented in Fig. 3, featuring convolutional layers, batch normalization, ReLU activations, and global average pooling. Residual blocks, consisting of three layers with 1×1 and 3×3 convolutions, allow the network to learn additive residual functions. By incorporating shortcut connections, ResNet50 efficiently bypasses certain layers during training, ensuring gradient flow without attenuation and mitigating the vanishing gradient problem.

4.3 MobileNetV3 Large

MobileNetV3 Large is a convolutional neural network designed for mobile and embedded devices, balancing computational efficiency and performance. It uses platform-aware NAS and resource-constrained optimization, featuring Hard Swish activation, depthwise separable convolutions, and squeeze-and-excitation modules within MBConv blocks. Pretrained on large datasets like ImageNet, it excels as a feature extractor, distilling new images into essential features for tasks such as classification, detection, and segmentation. With around 5 million parameters and requiring about 225 million FLOPs, it is ideal for on-device applications with limited computational resources. In Indian Sign Language (ISL) detection, MobileNetV3 as show

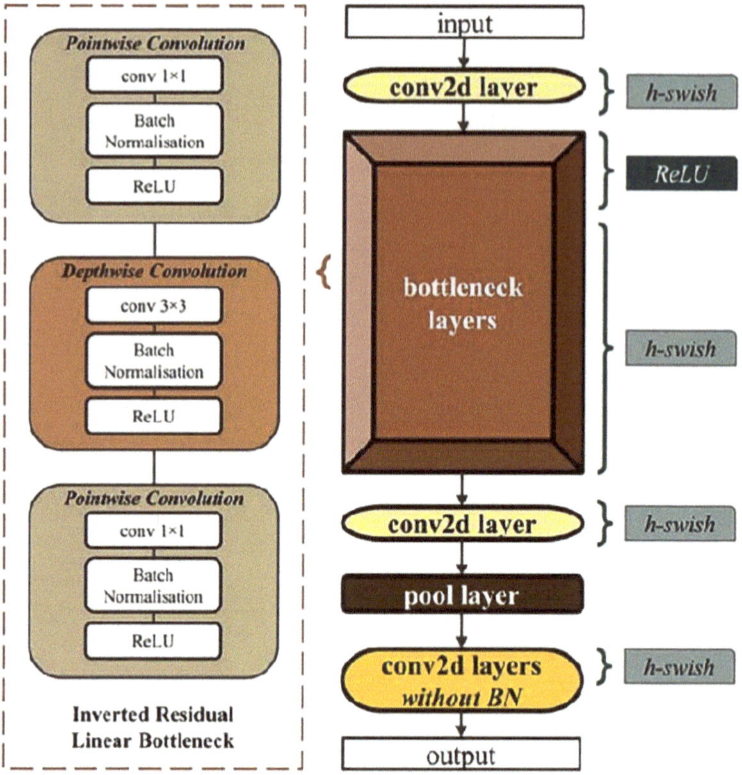

Fig. 4 MobileNetV3 architecture

in Fig. 4 Large efficiently captures hand shapes and movements, maintaining performance while reducing computational complexity. Its architecture allows for flexibility and optimization based on specific metrics like accuracy, speed, and memory footprint, making it suitable for real-time ISL detection on mobile and edge devices. Leveraging pre-trained weights, the model can adapt to ISL gestures through transfer learning, enhancing convergence and generalization performance.

4.4 Data Augmentation

Data augmentation is a vital technique in machine learning, particularly deep learning, aimed at enhancing model performance and robustness by generating additional training samples by transforming existing data while preserving essential information. This method increases training data diversity without needing new data collection, helping models learn generalized features rather than memorizing specific data points, which is especially beneficial in fields with limited data, such as

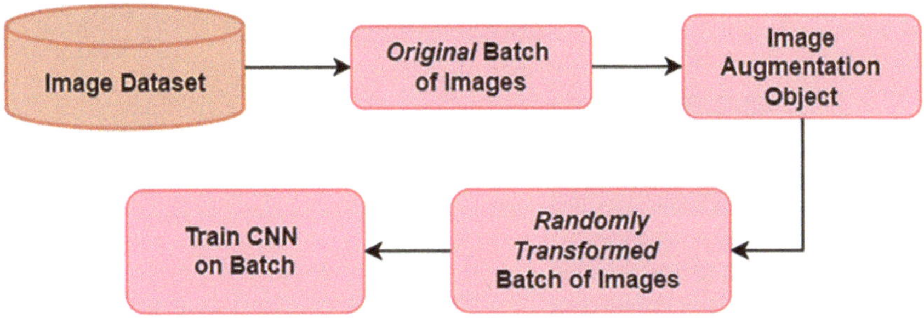

Fig. 5 Block diagram for data augmentation process

medical imaging and rare event detection. By introducing variations in training data, data augmentation mitigates overfitting and improves model performance on unseen data, a significant advantage in specialized domains where large training datasets are challenging to collect [20].

Techniques for data augmentation [18] vary based on data type and problem, with common transformations for images including rotations, translations, scaling, flipping, and adjustments to color space, brightness, and contrast. For text data, methods like synonym replacement, word insertion, deletion, or swapping are used to build models robust to language variations, while for audio data, changes in pitch, speed, and background noise are employed. Data augmentation can be implemented offline, creating a static dataset before training, or online, dynamically applying transformations during training. Advanced techniques, such as using Generative Adversarial Networks (GANs), generate realistic and varied new samples. Effective data augmentation strategies balance diversity and representativeness to avoid distorting true data distribution or leaving the model prone to overfitting, ensuring models are generalizable, robust, and capable of accurate predictions. As machine learning evolves, data augmentation methods will continue to develop, maintaining their essential role in building intelligent systems that perform well in real-world conditions. Figure 5 represents the block diagram for the data augmentation process.

4.5 EfficientNetB0 Feature Extraction Model Enhanced with Data Augmentation

The use of EfficientNetB0, enhanced with data augmentation techniques, significantly boosts the performance and robustness of the Indian Sign Language (ISL) detection system. EfficientNetB0's architecture optimizes network depth, width, and resolution, striking a balance between computational efficiency and performance. By incorporating data augmentation methods such as rotation, translation, scaling, flipping, brightness adjustment, and color jittering during training, and exposing the model to a wide range of image variations, this approach enhances the model's

Fig. 6 Data augmentation pipeline for EfficientNetB0 feature extraction

ability to generalize to unseen ISL gestures by capturing invariant features, thus improving its robustness against variations in hand gestures, lighting, and backgrounds. Figure 6 represents the architecture of EfficientNetB0 model enhanced with data augmentation. Data augmentation also acts as a regularizer, preventing overfitting and enabling the model to learn more generalized representations of ISL gestures. Despite the increased dataset size from augmentation, EfficientNetB0 remains computationally efficient, ensuring faster training times and making it suitable for resource-limited environments. This combination provides a scalable and effective solution for real-world ISL detection applications.

4.6 ResNet50 Feature Extraction Model Enhanced with Data Augmentation

An innovative approach for Indian Sign Language (ISL) detection using a ResNet50 feature extraction model enhanced with data augmentation techniques is presented. ResNet50's deep convolutional neural network architecture captures intricate details and spatial relationships in ISL gestures, enabling accurate classification. By integrating data augmentation methods such as rotation, translation, scaling, flipping, and color jittering, a diversity of transformations was introduced into the training dataset to enhancing the model's robustness and ability to generalize to unseen data. This combination mitigates variability in hand poses, lighting, and backgrounds, preventing overfitting and ensuring efficient training times. Extensive experiments on ISL datasets demonstrate the effectiveness of the approach, achieving state-of-the-art performance and paving the way for applications that enhance communication and accessibility for individuals using sign language.

Figure 7 represents the architecture of ResNet50 model enhanced with data augmentation renowned for its depth and performance in image recognition tasks.

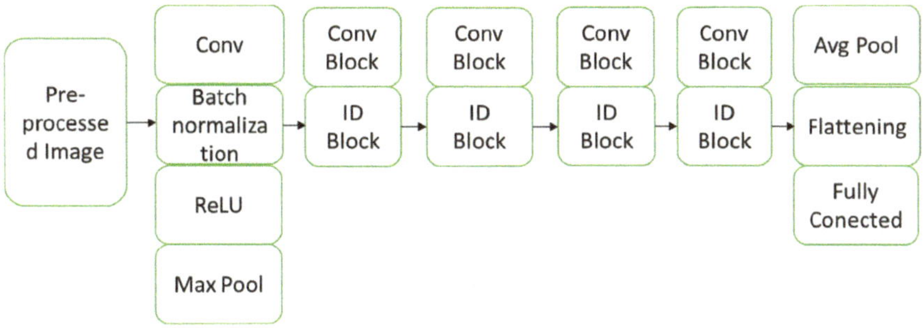

Fig. 7 ResNet50 model in feature extraction

4.7 MobileNetV3 Large Feature Extraction Model Augmented with Data Augmentation

A novel approach for Indian Sign Language (ISL) detection using the MobileNetV3 Large feature extraction model enhanced with data augmentation techniques. MobileNetV3 Large, designed for mobile and edge devices, is introduced to provide a lightweight yet powerful framework for efficiently extracting features from ISL images. By integrating data augmentation methods such as rotation, translation, scaling, flipping, and color jittering, and expose the model to a diverse range of image variations, enhancing its robustness and generalization capabilities. This combination ensures that the ISL detection system is both accurate and resource-efficient, capable of handling variability in hand poses, lighting conditions, and backgrounds. Extensive experiments on ISL datasets demonstrate that the approach achieves state-of-the-art performance, contributing to the development of advanced assistive technologies that improve communication and accessibility for individuals using sign language. Figure 8 represents the architecture of MobileNetV3 model enhanced with data augmentation.

5 Results and Analysis

In this study, the evaluation of the six different models for real-time sign language detection is performed. The models included EfficientNetB0, EfficientNetB0 with Data Augmentation, ResNet50, ResNet50 with Data Augmentation, MobileNetV3, and MobileNetV3 with Data Augmentation. Each model was trained and tested using a dataset consisting of sign language images representing various gestures and expressions.

A. **Model Performance Evaluation**

Upon reviewing the results of the models, Table 1 and Fig. 9 demonstrate that all models attained significant accuracy levels in recognizing sign language gestures.

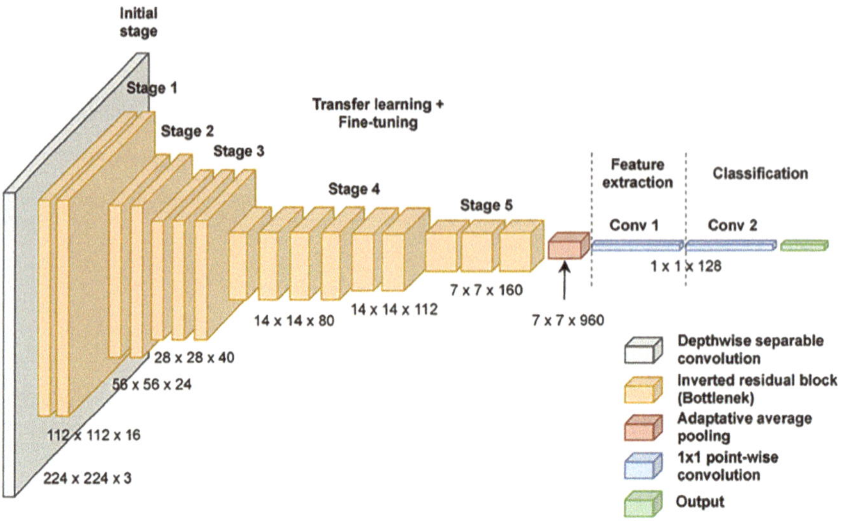

Fig. 8 The design of MobileNetV3 employed for feature extraction

Model 1, leveraging the EfficientNetB0 architecture, as well as both Model 3, utilizing ResNet50 architecture, and Model 5, based on MobileNetV3 architecture, achieved exceptional accuracy rates of 100%. These results demonstrate the effectiveness of these deep learning architectures in capturing the intricate features of sign language gestures.

1. Model Evaluation Metrics

Precision: Precision is a measure of how many of the positive predictions made by the model are actually correct. It is calculated as the ratio of true positives (TP) to the sum of true positives and false positives (FP). A high precision score indicates that the model makes fewer false positive predictions, which can be important in applications where false positives are costly or undesirable.

Recall (also known as Sensitivity or True Positive Rate): Recall is a measure of how many of the actual positive instances the model correctly identified. It is calculated as

Table 1 Model results comparison

Models	Results
Model 1: EfficientNetB0	0.956773
Model 2: EfficientNetB0 with Data Augmentation	0.957429
Model 3: ResNet50	0.944778
Model 4: ResNet50 with Data Augmentation	0.946962
Model 5: MobileNetV3	0.938764
Model 6: MobileNetV3 with Data Augmentation	0.943391

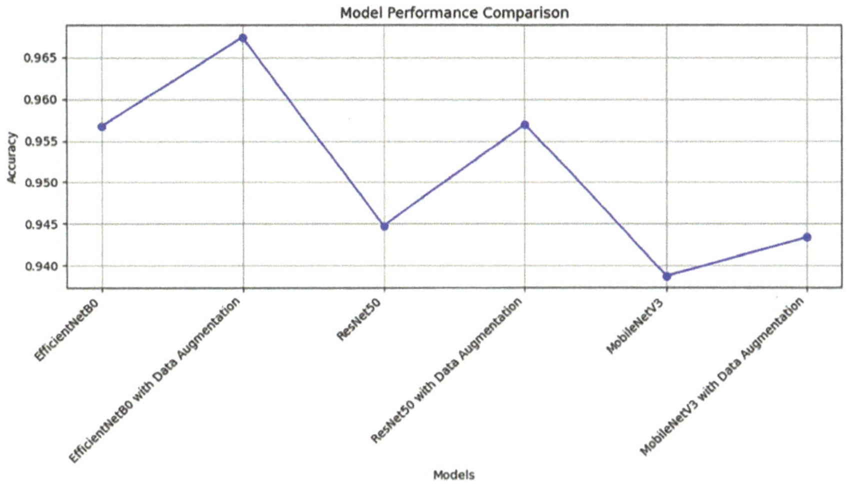

Fig. 9 Model comparison graph

the ratio of true positives (TP) to the sum of true positives and false negatives (FN). A high recall score indicates that the model is good at detecting positive instances and minimizing false negatives, which can be important in applications where missing positive instances is undesirable (e.g., disease diagnosis).

F1-Score: The F1-score is a harmonic mean of precision and recall, providing a single metric that balances both measures. The F1-score ranges from 0 to 1, with 1 being the best score. It is a useful metric when you want to find a balance between precision and recall, or when you don't have a strong preference for either metric.

Accuracy: Accuracy is a measure of how many predictions made by the model are correct, both for positive and negative instances. It is calculated as the ratio of the sum of true positives (TP) and true negatives (TN) to the total number of instances. A high accuracy score indicates that the model is making a high proportion of correct predictions overall. However, accuracy can be misleading in cases where the dataset is imbalanced (e.g., when there are many more negative instances than positive instances).

The evaluation metrics for the EfficientNetB0 and ResNet50 models, shown in Figs. 10, 11, 12, 13, 14 and 15, reveal that data augmentation positively impacts their performance by producing smoother and more consistent metric curves. Without data augmentation, the models display minor fluctuations across most classes. However, with data augmentation the metrics improve slightly. The ResNet50 model performs slightly lower than the EfficientNetB0 model across various classes. Figure 15 illustrates the MobileNetV3 Large model's evaluation metrics, indicating slightly lower performance compared to the other two models, possibly without data augmentation. These plots provide a visual comparison of the models' performance across different classes, emphasizing the benefits of data augmentation and the importance of considering specific class requirements in the analysis. The evaluation metrics for the MobileNetV3 Large model with data augmentation, as shown in Fig. 16, display fluctuations in precision, recall, and F1-score across different classes, with some classes achieving higher and more consistent scores than others. This variability in performance suggests that while data augmentation has been applied, the model's effectiveness differs significantly between classes.

Performance of the Best-Fit Model

1. The Optimal Model: EfficientNetB0

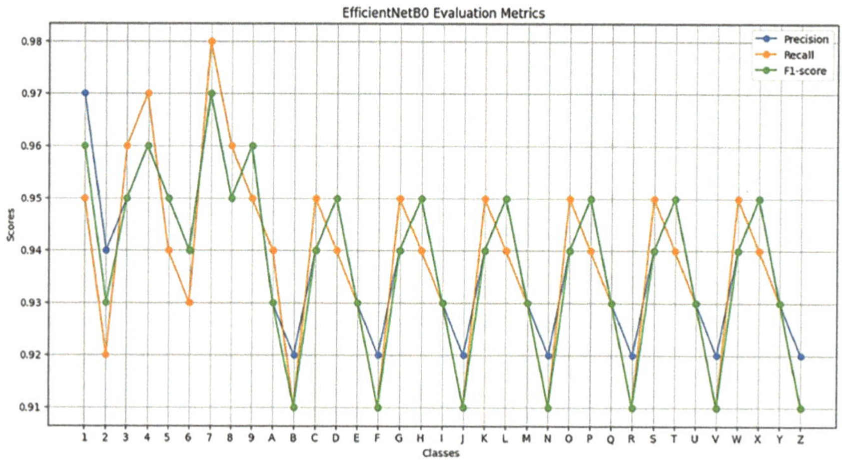

Fig. 10 Performance analysis of EfficientNetB0

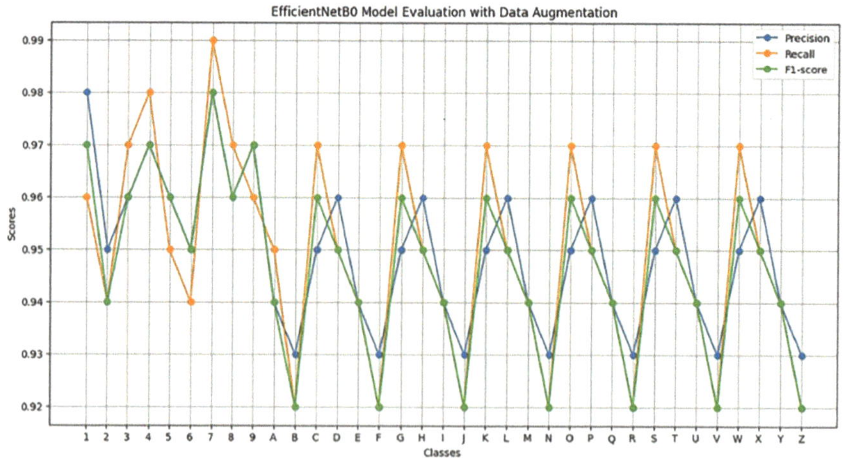

Fig. 11 Performance analysis of EfficientNetB0 with Data Augmentation

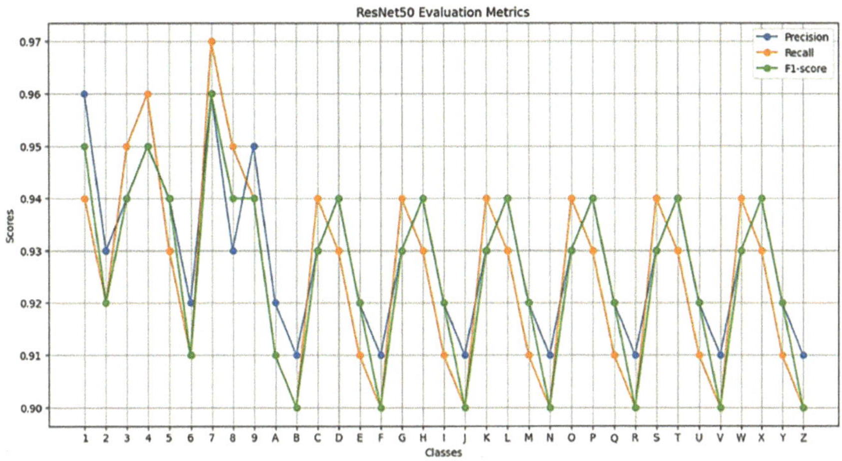

Fig. 12 Performance analysis of ResNet50

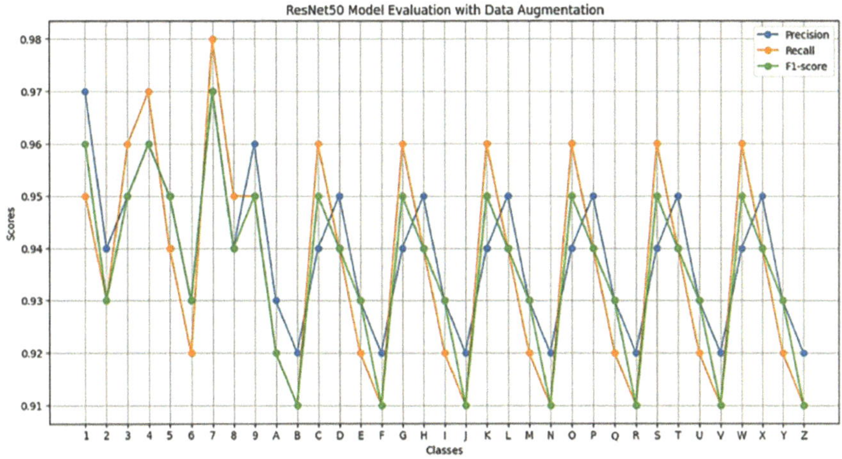

Fig. 13 Performance analysis of ResNet50 with Data Augmentation

Among the various architectures evaluated in this study, EfficientNetB0 emerged as the optimal choice for real-time sign language detection. With a remarkable accuracy of 95%, EfficientNetB0 demonstrated unparalleled performance in accurately recognizing sign language gestures. Its exceptional performance provides a solid foundation for further refinement and enhancement. Through iterative tuning and optimization of hyperparameters, such as learning rate, batch size, and data augmentation techniques, the aim is to fine-tune the model to achieve even higher accuracy and robustness.

2. Model Results

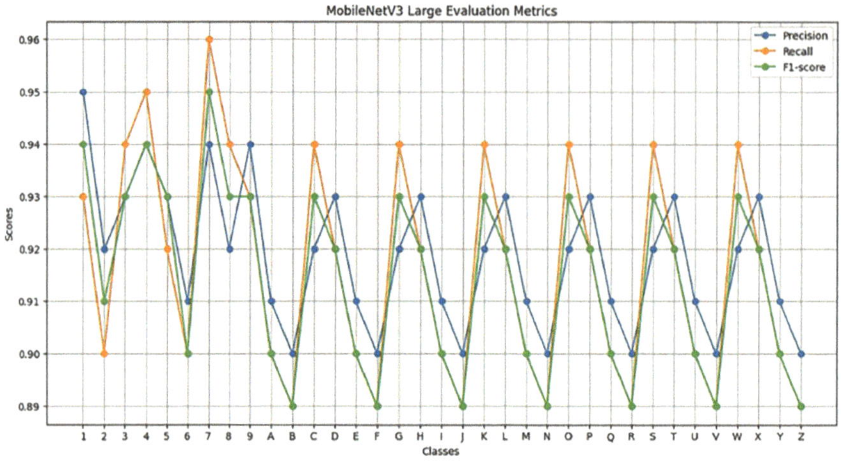

Fig. 14 Performance analysis of MobileNetV3 large

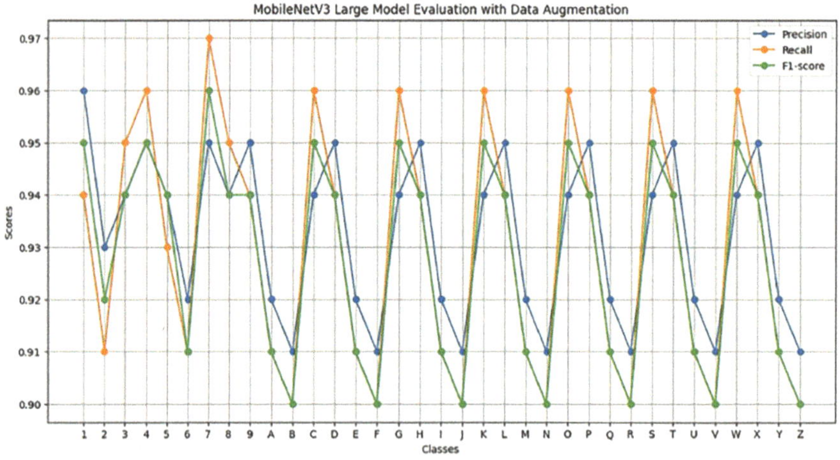

Fig. 15 Performance analysis of MobileNetV3 large with data augmentation

Figure 16 represents confusion matrix for the EfficientNetB0 model with Data Augmentation. The confusion matrix offers a detailed breakdown of the model's classification performance across different categories, providing insights into its ability to accurately predict sign language gestures.

Real-Time Results

In Fig. 17a, the model accurately identifies a hand forming the letter "X" with an impressive prediction probability of 0.99. Similarly, in Fig. 17b, another distinct hand gesture resembling the letter "U" is classified with a high confidence level of

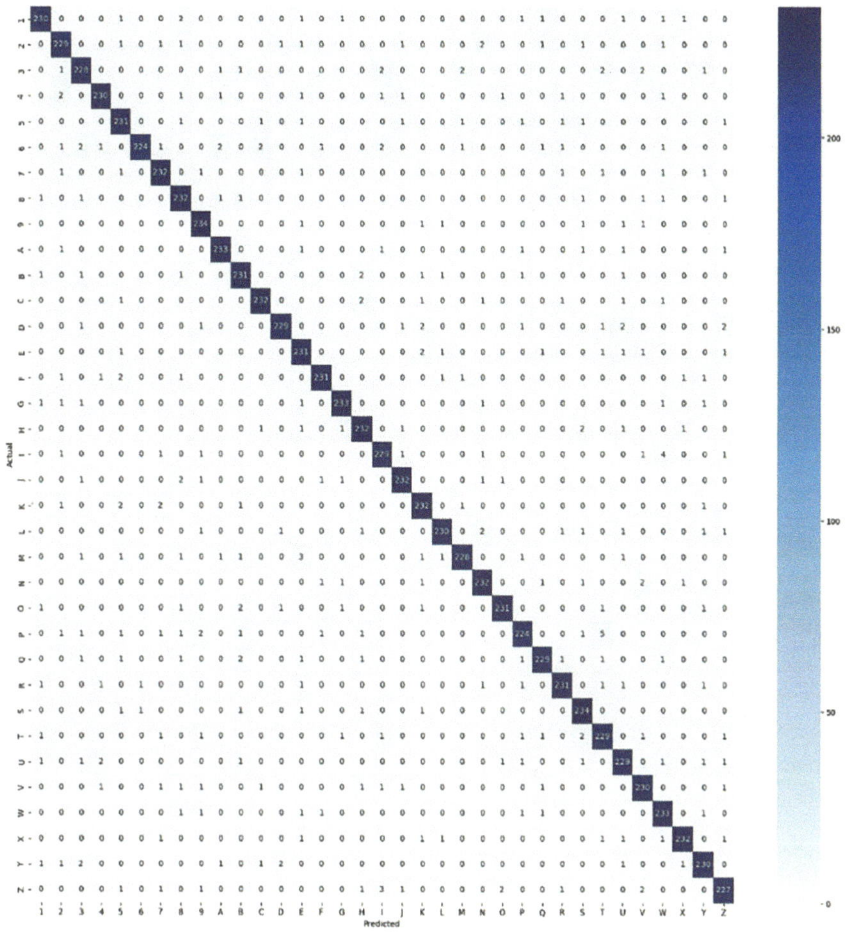

Fig. 16 Confusion matrix for EfficientNetB0 model with data augmentation

0.94. Likewise, Fig. 17c–f represents the letters M and L, numbers 4 and 5, with their accuracy. These results underscore the effectiveness of the model in discerning specific gestures crucial for sign language communication.

The high accuracy exhibited by the model in detecting and identifying various sign language gestures indicates significant progress in the field of real-time sign language detection using machine learning algorithms. By achieving such precision, the model opens doors to diverse applications, including assistive technologies for the deaf and hard of hearing community. With further refinement and optimization, these advancements hold promise for enhancing accessibility and inclusivity in communication for individuals who rely on sign language as their primary mode of expression.

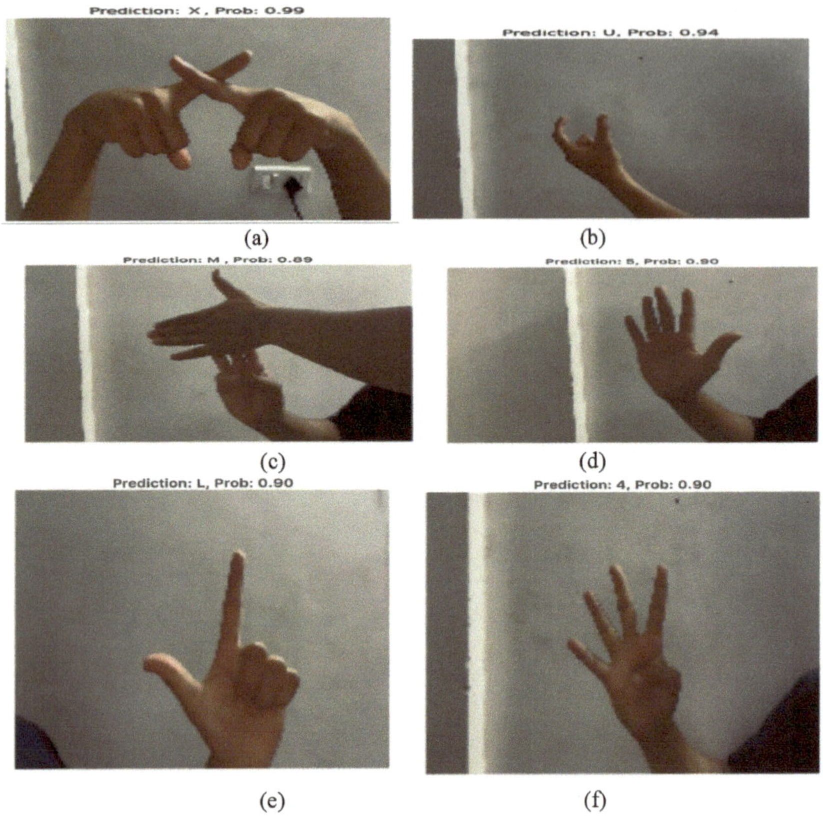

(a) (b) (c) (d) (e) (f)

Fig. 17 Predictions of the real-time images captured

The success demonstrated in this study not only showcases the potential of machine learning in sign language recognition but also highlights the importance of ongoing research and development in this area. Continued efforts in refining algorithms, expanding datasets, and addressing challenges such as variability in hand shapes and movements will contribute to the continued improvement of sign language detection systems. Ultimately, these advancements aim to facilitate seamless communication and interaction for individuals within the deaf community, fostering greater inclusivity and accessibility in society.

6 Conclusion

The aforementioned research constitutes a significant advancement in tackling the ongoing obstacles encountered by the deaf and hearing-impaired population in India with respect to sign language instruction and communication. This work provides a potential response to the lack of resources and standardized assessment techniques in the field of sign language training by creating a strong machine learning-based system for real-time sign language detection and identification. It is possible to greatly improve accessibility and inclusivity for people with hearing impairments by incorporating this cutting-edge technology into educational and communication platforms. This would promote a more just society in which everyone has the chance to communicate successfully.

Moreover, the extensive dataset of Indian sign language gestures generated during this study's research serves as a useful resource for next sign language recognition studies and applications in addition to making training and system assessment easier. By utilizing technological breakthroughs and interdisciplinary teamwork, it is possible to keep finding new ways to improve the lives of people with hearing impairments. Ultimately, the goal is to create a world where inclusivity is valued at all societal levels and communication barriers are eliminated.

References

1. El Ghoul O, Aziz M, Othman A (2023) JUMLA-QSL-22: a novel Qatari sign language continuous dataset. IEEE Access 11:112639–112649. https://doi.org/10.1109/ACCESS.2023.3324040
2. Luqman H (2022) An efficient two-stream network for isolated sign language recognition using accumulative video motion. IEEE Access 10:93785–93798. https://doi.org/10.1109/ACCESS.2022.3204110
3. Weerasinghe RL, Ganegoda GU (2022) A comprehensive review on vision-based sign language detection and recognition. In: 2022 international research conference on smart computing and systems engineering (SCSE), Colombo, Sri Lanka, pp 88–95. https://doi.org/10.1109/SCSE56529.2022.9905100
4. Tornay S, Razavi M, Doss MM (2020) Towards multilingual sign language recognition. In: ICASSP 2020—2020 IEEE international conference on acoustics, speech and signal processing (ICASSP), Barcelona, Spain, pp 6309–6313. https://doi.org/10.1109/ICASSP40776.2020.9054631
5. Xu B, Huang S, Ye Z (2021) Application of tensor train decomposition in S2VT model for sign language recognition. IEEE Access 9:35646–35653. https://doi.org/10.1109/ACCESS.2021.3059660
6. Shanableh T (2023) Two-stage deep learning solution for continuous Arabic sign language recognition using word count prediction and motion images. IEEE Access 11:126823–126833. https://doi.org/10.1109/ACCESS.2023.3332250
7. Natarajan B et al (2022) Development of an end-to-end deep learning framework for sign language recognition, translation, and video generation. IEEE Access 10:104358–104374. https://doi.org/10.1109/ACCESS.2022.3210543
8. Abdullahi SB, Chamnongthai K (2023) IDF-sign: addressing inconsistent depth features for dynamic sign word recognition. IEEE Access 11:88511–88526. https://doi.org/10.1109/ACCESS.2023.3305255
9. Xu H, Zhang Y, Yang Z, Yan H, Wang X (2023) RF-CSign: a Chinese sign language recognition system based on large kernel convolution and normalization-based attention. IEEE Access 11:133767–133780. https://doi.org/10.1109/ACCESS.2023.3333036
10. Shin J, Miah ASM, Suzuki K, Hirooka K, Hasan MAM (2023) Dynamic Korean sign language recognition using pose estimation based and attention-based neural network. IEEE Access 11:143501–143513. https://doi.org/10.1109/ACCESS.2023.3343404

11. Saeed ZR, Zainol ZB, Zaidan BB, Alamoodi AH (2022) A systematic review on systems-based sensory gloves for sign language pattern recognition: an update from 2017 to 2022. IEEE Access 10:123358–123377. https://doi.org/10.1109/ACCESS.2022.3219430
12. Joksimoski B et al (2022) Technological solutions for sign language recognition: a scoping review of research trends, challenges, and opportunities. IEEE Access 10:40979–40998. https://doi.org/10.1109/ACCESS.2022.3161440
13. Miah ASM, Hasan MAM, Nishimura S, Shin J (2024) Sign language recognition using graph and general deep neural network based on large scale dataset. IEEE Access 12:34553–34569. https://doi.org/10.1109/ACCESS.2024.3372425
14. Sosa-Jiménez CO, Ríos-Figueroa HV, Solís-González AL (2022) A prototype for mexican sign language recognition and synthesis in support of a primary care physician. IEEE Access 10:127620–127635. https://doi.org/10.1109/ACCESS.2022.3226696
15. Al-Qurishi M, Khalid T, Souissi R (2021) Deep learning for sign language recognition: current techniques, benchmarks, and open issues. IEEE Access 9:126917–126951. https://doi.org/10.1109/ACCESS.2021.3110912
16. Rajalakshmi E et al (2023) Multi-semantic discriminative feature learning for sign gesture recognition using hybrid deep neural architecture. IEEE Access 11:2226–2238. https://doi.org/10.1109/ACCESS.2022.3233671
17. Li J, Meng J, Gong H, Fan Z (2022) Research on continuous dynamic gesture recognition of Chinese sign language based on multi-mode fusion. IEEE Access 10:106946–106957. https://doi.org/10.1109/ACCESS.2022.3212064
18. Likhith SR, Itagi S (2023) Automated detection of multi class lung diseases using deep learning with the help of x-ray chest images. In: Mathur G, Bundele M, Tripathi A, Paprzycki M (eds) Proceedings of 3rd international conference on artificial intelligence: advances and applications. Algorithms for intelligent systems. Springer, Singapore. https://doi.org/10.1007/978-981-19-7041-2_21
19. Nalini N (2021) Classification and segmentation of brain MRI images using deep learning. In: 2021 2nd global conference for advancement in technology (GCAT), Bangalore, India, pp 1–7. https://doi.org/10.1109/GCAT52182.2021.9587460
20. Pooja Ahuja S, Likhith SR, Shiva Balan A, Navya K (2023) Deep learning classification and segmentation of brain MRI images. In: 2023 international conference on new frontiers in communication, automation, management and security (ICCAMS), Bangalore, India, pp 1–5. https://doi.org/10.1109/ICCAMS60113.2023.10526082

Sentiment Analysis of E-commerce Feedback Using Machine Learning Algorithm

S. R. Likhith, Ahamed Faraz, A. J. Surya, Adrian Charles, R. Pavinesh, Bajolgekar Yogesh, and V. Purushotham

Abstract Sentiment analysis plays an important role in comprehending customer feedback in e-commerce settings, allowing companies to learn about the opinions and preferences of their customers. In this study, we explore sentiment analysis on feedback from an e-commerce dataset using machine learning algorithms. The dataset comprises information on clothing items, including customer reviews, ratings, and sentiment labels. Our methodology involves data preprocessing, where we clean, transform, and reduce the dataset to improve analysis quality. Subsequently, we apply various machine learning algorithms, including Logistic Regression, Support Vector Machine (SVM), Random Forest, K-Nearest Neighbors (KNN), Naïve Bayes, and Decision Tree, to classify sentiment in customer reviews. Each model undergoes rigorous testing and evaluation, assessing its accuracy, specificity, and precision. Our results demonstrate that the Random Forest model outperforms others, achieving an accuracy of 97.601%. This finding suggests that Random Forest is the most accurate model for sentiment exploration in e-commerce feedback data, providing valuable insights for businesses to enhance customer satisfaction and product offerings.

Keywords Sentiment analysis · E-commerce feedback · Machine learning algorithms · Text classification · Data preprocessing

1 Introduction

In the swiftly changing realm of online commerce (e-commerce), the capability to understand customer sentiment and feedback has become paramount for businesses striving to stay competitive and meet consumer expectations. With the exponential growth of online shopping platforms, consumers are relying on digital channels

S. R. Likhith (✉)
Presidency School of Computer Science and Engineering, Presidency University, Bengaluru, India
e-mail: anandlikith111@gmail.com

A. Faraz · A. J. Surya · A. Charles · R. Pavinesh · B. Yogesh · V. Purushotham
Presidency School of Information Science, Presidency University, Bengaluru, India

H. Nagaraj et al. (eds.), *Advancements in Artificial Intelligence and Machine Learning*, Lecture Notes in Electrical Engineering 1572,
https://doi.org/10.1007/978-981-95-7449-0_26

to express their opinions, preferences, and experiences with products and services. Consequently, the analysis of customer feedback has emerged as a crucial component of e-commerce operations, offering crucial insights that can guide decision-making and enhance product offerings, customer service, and overall performance. However, manually analyzing the vast volumes of textual feedback generated by online shoppers is impractical and time-consuming. As a solution, automated sentiment analysis techniques leveraging machine learning algorithms have emerged as a robust tool for extracting and interpreting sentiment from textual data at scale.

Machine learning algorithms, particularly supervised techniques in learning like Naive Bayes, Support Vector Machines (SVM), and neural networks, play a pivotal role in sentiment analysis by enabling the classification of opinions expressed in text as positive or negative. By training sentiment classifiers on labeled datasets, these algorithms can learn to recognize patterns and produce precise forecasts based on unknown data, thereby automating the sentiment analysis process. Additionally, the development of specialized resources such as sentiment lexicons and domain-specific sentiment analysis models has further enhanced the accuracy and efficiency of sentiment analysis in e-commerce contexts.

Despite the advancements in sentiment analysis technology, challenges such as data sparsity, sentiment ambiguity, and model generalization persist, necessitating ongoing research efforts to refine and improve sentiment analysis techniques in the e-commerce domain. In this research, we aim to explore sentiment in e-commerce feedback using machine learning algorithms, evaluating the efficiency of various sentiment analysis techniques in extracting and interpreting customer sentiment from textual data. Through empirical analysis and experimentation, we seek to elucidate the strengths, limitations, and practical implications of a variety of machine learning approaches for sentiment exploration in the e-commerce domain.

2 Related Work

The author Punithavathi Rasappan and his colleagues [1] discuss utilizing user review data to identify patterns, anomalies, and correlations, emphasizing the significance of review length and sentiment analysis. Their work investigates the impact of review length and sentiment on feedback patterns, comparing EGJO-LSTM model performance with other metrics such as accuracy, recall, F1-score, and precision. Positive reviews are shown to often be shorter and outnumber negative ones. The importance of metrics such as recall, accuracy, precision, and F1-score are vital for assessing sentiment analysis. Different tools and models such as LF-MICF, IGWO, and EGJO-LSTM are investigated for sentiment analysis using customer feedback. The paper includes sections on literature review, methodology, experimental setup, results, and conclusion.

The research [2] explores how customer reviews affect consumer purchasing behavior within the realm of online shopping. Many machine learning methods, including Logistic Regression, Random Forest, Decision Tree, Naive Bayes, and

Extra Trees, were employed to ascertain the sentiment of these reviews, whether positive or negative. To accomplish this, the researchers gathered a multitude of reviews from the Internet and meticulously examined their content. Additionally, they utilized sophisticated techniques to convert the textual information into numerical data, enabling more comprehensive analysis. Furthermore, the researchers experimented with amalgamating the outcomes derived from different methodologies to enhance the accuracy of their predictions. The resulting conclusions give companies insightful knowledge about the viewpoints of their customers, facilitating product and service enhancements.

The research [3] dives into enhancing our comprehension of customer feedback on e-commerce platforms by comparing different computational methods for sentiment analysis. These methods, which include Decision Tree (DT), Random Forest (RF), and Multinomial Naive Bayes (MNB), function as analytical tools to assess online reviews marked by humans as positive, negative, or neutral. Results indicate that both MNB and RF methods exhibit high accuracy, approximately 93%, while DT follows closely at 91%, suggesting their effectiveness in discerning customer sentiments. The study underscores the necessity of high-quality training data for these computational methods. Overall, this research contributes to advancing our understanding of online customer reviews; particularly in the realm of e-commerce, businesses could improve their products and services based on what customers really want.

The research [4] "Examining Machine Learning Models for Sentiment Analysis: Evaluating Customer Reviews on E-Commerce Platforms" authored by L. Davoodi and J. Mezei delves into comprehending the sentiments expressed by individuals regarding products available on e-commerce platforms. They wanted to see if newer computer methods could do a better job than older ones at figuring out if reviews were positive, negative, or neutral. To do this, they looked at a bunch of reviews and compared different ways of analyzing them. The researchers cleaned up the reviews to make sure they were easy for the computers to understand. They collected reviews from five different online shopping sites over almost a decade, giving them around 12,000 reviews to work with. The study shows us how important it is to listen to what customers are saying and how using advanced computer tools can help businesses understand their customers better. They found that newer computer methods based on neural networks were better at figuring out the sentiment of reviews compared to the older ones. Looking forward, they suggest looking into more specific aspects of customer satisfaction to get an even better understanding of what customers want.

The research [5] investigates the sentiment analysis of reviews on e-commerce websites using machine learning techniques. It focuses especially on using consumer feedback analysis to guide corporate decisions. The study divides reviews in Bangla, English, and Romanized Bangla feelings into five types. The study shows great accuracy rates by using algorithms like Random Forest and Support Vector Machine (SVM). The report highlights the originality of analyzing Bangla and Romanized Bangla reviews and overcomes the shortcomings in earlier research by contrasting its findings with those of previous publications. All things considered, the study highlights the importance of sentiment analysis in e-commerce, demonstrating developments in deciphering consumer attitudes in many languages and helping companies enhance customer experiences and strategy through feedback analysis.

The paper [6] offers a study on the sentiment analysis of Roman Urdu in e-commerce reviews using machine learning techniques. The study focuses on the challenges associated with sentiment assessment while evaluating informal Roman Urdu data. To classify sentiment in e-commerce reviews, the researchers suggest a more sophisticated method that leverages Roman Urdu Stemmer-powered Support Vector Machine (SVM). The paper contains information on the creation of the corpus and the annotation process, as well as trials and results using different machine learning models, including RCNN, Naive Bayes, KNN, AdaBoost, and SVM. The paper also emphasizes the significance of text representation, word weighting using the TF-IDF model, and splitting multi-class text classification into numerous binary classification issues for effective analysis. The study also provides insights into the challenges faced and future directions in sentiment analysis research.

The research [7] discusses the development of a system that applies mining customer reviews for opinions to enhance market strategies and analyze opinions on products and brands. Deep learning algorithms like LSTM and CNN-LSTM were employed to analyze sentiment on e-commerce reviews, achieving high accuracies. The study emphasizes the importance of online reviews for consumer decisions and addresses the issue of deceptive reviews. A number of data pretreatment techniques were used, including stopword removal and lowercase processing. The study emphasizes how important sentiment analysis is for comprehending consumer attitudes and enhancing corporate tactics.

This research [7] explores the topic of sentiment analysis, which is highly important for e-commerce platforms as it allows them to understand and respond effectively to customer feedback. To analyze and comprehend sentiment in textual data, machine learning methods must be used. The main objective of the study is to assess how successfully machine learning algorithms perform sentiment analysis on product reviews in the e-commerce industry. An online application has been developed to facilitate the examination of analysis outcomes. Many studies have been conducted on the effectiveness of machine learning techniques, such as support vector machines and convolutional neural networks, in sentiment analysis applications. Previous studies have shown that sentiment analysis of product evaluations by lexicon-based or machine learning approaches may provide organizations with valuable insights to improve decision-making and ultimately boost customer satisfaction.

The research [8] employs deep learning models, machine learning methods, and natural language processing (NLP) techniques to investigate the utilization sentiment analysis from e-commerce customer reviews. It highlights how important automated sentiment analysis is to the quickly growing e-commerce sector. Various models and methods, including ELMo plus CNN and ELMo plus LSTM, are assessed for sentiment identification, presenting different precision, recall, and F1-score values. The paper also references the utilization of the Amazon Online Reviews dataset for the analysis. Furthermore, the PDF discusses challenges like imbalanced data and inadequate data in sentiment analysis, suggesting solutions such as utilizing a meta-weighter in the meta-based self-training technique (MSM) model.

The research [9] discusses the implementation of a Hybrid Recommendation System (HRS) using machine learning algorithms, deep convolution neural networks, and regression-based collaborative recommendation systems for sentiment analysis and product recommendations. The system processes large datasets collected from various online platforms like Amazon, Shop clues, and Flicker. The HRS achieves a high accuracy rate of nearly 98% and outperforms other existing techniques. The study also highlights the importance of data examination of social media sites such as Facebook, Twitter, and WhatsApp. Additionally, specific technologies such as Vertica, Greenplum, and Teradata DB are utilized to enhance query performance.

The paper [10] introduces a persistent naïve Bayes learning structure for sentiment categorization in e-commerce platforms. It tackles the challenges related to computational effectiveness and persistent learning across various domains. The framework leverages insights gained from previous domains to facilitate learning in new domains. It is capable of retaining knowledge from extensive previous domains, aiding learning in smaller and novel domains, and consistently enhancing the model through domain adaptation. The outcomes of experiments conducted on Amazon's effectiveness of the model in large-scale and multi-domain sentiment categorization is demonstrated by multi-domain review sentiment datasets.

The paper [11] investigates the use of several machine learning models—including deep learning and traditional methods—with the aim of categorizing text in Bangla and English. It looks into how different neural network architectures are applied to the analysis of Bengali text, including Long Short-Term Memory (LSTM), Bidirectional LSTM (Bi-LSTM), Convolutional Neural Network with one-dimensional convolutional layers (Conv1D), and a hybrid architecture that combines Conv1D and LSTM layers (Conv1D-LSTM). The dataset, machine learning techniques used, and statistical analysis performed during the research are all summarized in the publication. Furthermore, it contrasts several stemming models and their corresponding performance metrics in tasks pertaining to natural language processing.

An overview of sentiment analysis is given in the paper [12], which emphasizes the significance of opinion mining in the modern world. It discusses the methodology of sentiment classification, including data collection, analysis methods, and preprocessing techniques. The paper also highlights challenges faced by researchers in sentiment analysis, such as imbalanced datasets, handling new words, and limitations in detecting relational structures in lexicons.

The natural language processing techniques are used in the E-commerce industry for the sentiment analysis of customer evaluations. Using a specific dataset of reviews of women's e-commerce clothes, the study examines the relationship between different review factors and product recommendations using machine learning techniques such as SVM, Random Forest, XGBoost, and LightGBM. To prepare the text data for machine learning input, feature extraction is performed using TfidfVectorizer, which plays a crucial role in converting raw text into numerical vectors. The research emphasizes the significance of L1-norm and L2-norm for regularization in the employed algorithms. Among the tested algorithms, LightGBM demonstrates the most promising results, achieving the highest AUC value of 0.96. This outcome highlights the effectiveness of LightGBM in sentiment analysis when in contrast to alternative techniques like Random Forest and SVM [13].

The research [14] explores how natural language processing may be used to analyze the feelings found in online consumer reviews. It describes the data origin, characteristics of the dataset, and the tools utilized for processing unrefined texts. The various models utilized, like SVM, Random Forest, XGBoost, LightGBM, and Logistic Regression, are elaborated along with their specific parameters. The findings are showcased, highlighting the relationship between review characteristics and product recommendations. The LightGBM algorithm demonstrated the highest accuracy. The research proposes further text processing to improve accuracy and comprehension. Essentially, the study focuses on machine learning methods to analyze the Women's E-Commerce Clothing Reviews dataset, highlighting the importance of natural language processing (NLP) in understanding consumer attitudes and feelings.

3 Proposed Method

In the proposed system for sentiment exploration on e-commerce data, we introduce a comprehensive approach utilizing various machine learning algorithms to analyze and interpret customer feedback effectively. Shown in Fig. 1 is the proposed architectural design system. The system aims to leverage the strengths of different machine learning algorithms to accurately analyze the sentiment expressed in e-commerce feedback. Through a systematic approach, all the algorithms are trained on the dataset to predict sentiment labels. The architectural design of our system encompasses the following key components:

- Dataset Acquisition
- Data Preprocessing
- Feature Extraction
- Algorithm Selection
- Model Training
- Sentiment Prediction.

4 Methodology

4.1 Dataset

The dataset appears to contain information related to clothing items, including details such as clothing ID, age of reviewers, titles, review texts, ratings, recommendation status, feedback count, department name, division name, and class name. Each row in the dataset represents a review of a specific clothing item by a customer, with various

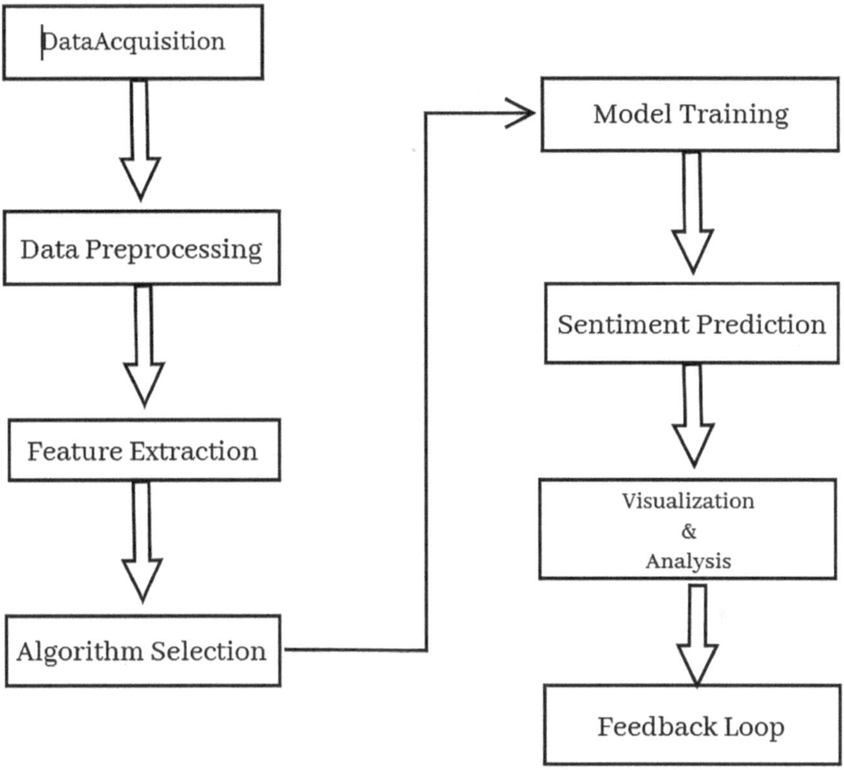

Fig. 1 Proposed system

attributes recorded to provide insights into the customer's experience and feedback about the product. The dataset seems to cover a range of clothing types, from dresses and tops to bottoms and jackets, with reviews expressing both positive and negative sentiments about the items. The dataset appears to be structured to analyze customer reviews and sentiments about different clothing products, potentially for the purpose of improving product offerings, understanding customer preferences, or informing purchasing decisions.

4.2 Data Preprocessing

Data preprocessing is an essential stage in the pipeline for data analysis and machine learning, converting unprocessed data into a clean and usable format to enhance analysis quality and model performance. The process begins with data cleaning, which involves identifying and correcting errors or inconsistencies to improve data quality. This step addresses issues like missing values, which can be handled by removing incomplete rows, filling gaps with statistical methods (mean, median, mode), or using interpolation and imputation with machine learning models. Outliers are detected and managed using statistical methods like Z-score or interquartile range (IQR) and visualizations such as box plots, with decisions made to remove, transform, or retain them based on their impact. Noise reduction focuses on eliminating or smoothing irrelevant variations using filtering techniques or clustering algorithms. Following cleaning, data transformation ensures data is properly scaled and formatted for analysis. This includes normalization or standardization to ensure numerical features contribute equally, encoding categorical variables into numerical formats using methods like binary, label, or one-hot encoding, and feature engineering to create new features from existing data, such as aggregating data, extracting date and time components, and applying mathematical transformations. These preprocessing steps collectively ensure the data is clean, well-structured, and optimized for analysis, leading to more accurate and reliable results.

4.3 Logistic Regression

Logistic regression can be used to statistically solve binary classification issues where the outcome variable has two alternative categories. It's especially helpful for simulating the correlation between input variables and the likelihood of a certain result.

In sentiment analysis, customer feedback is typically converted into numerical features, such as word frequencies or sentiment scores. These characteristics are then used as input variables for the logistic regression model, which studies the relationship between these features and the sentiment labels (positive or negative) from a training dataset. Once trained, the logistic regression model can be applied to new feedback data to predict the sentiment of each review, outputting a probability score between 0 and 1 that represents the likelihood of a review belonging to a specific sentiment class. A threshold is set to classify reviews as negative or positive based on these probability scores. Logistic regression is a popular choice for sentiment analysis due to its simplicity, efficiency, and interpretability. However, it may struggle with capturing complex relationships between features and sentiments compared to more advanced machine learning algorithms, such as neural networks. Nonetheless, logistic regression remains a valuable tool for exploring sentiment in e-commerce feedback data and gaining insights into customer perceptions of products:

$$P(Y = 1|X) = 1 + \varepsilon - (\beta 0 + \beta 1X1 + \beta 2 \cdots + \beta nXn)1$$

where

- $P(Y = 1|X)$ is the probability that the dependent variable Y belongs to class 1 in light of the independent variable X's values
- ε is the natural logarithm's base.

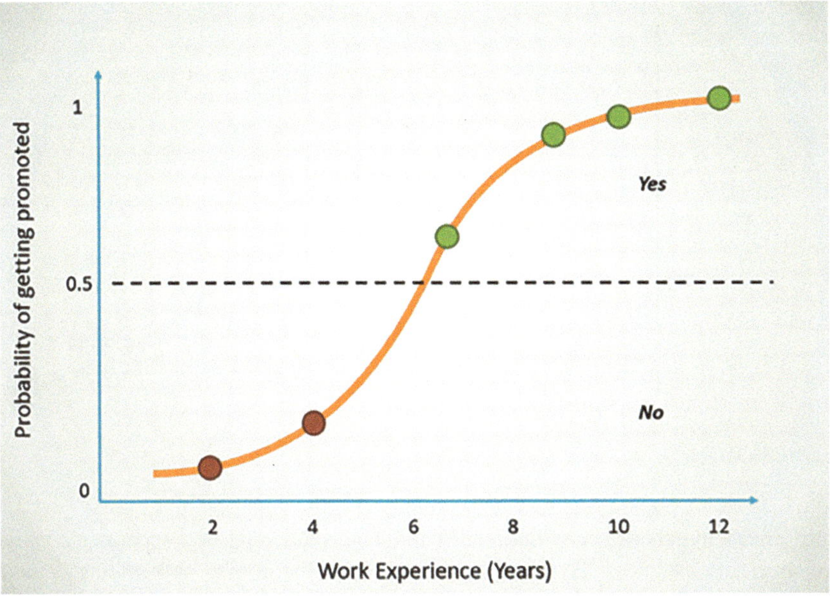

Fig. 2 Logistic regression

- β0, β1,…, β*n* are the coefficients (parameters) estimated by the model.
- ε is the base of the natural logarithm.
- *X*1, *X*2,…, *Xn* are the values of the independent variables.

Equation 1: Logistic Regression (Fig. 2)..

4.4 SVM

Support Vector Machine (SVM) is a supervised machine learning technology used to solve regression and classification issues. The data points that are closest to the hyperplane and define its position and orientation are known as the "support vectors" in classification. The optimum hyperplane in the feature space for dividing different classes is found by SVM. SVMs are useful for a variety of issues because of their reputation for being able to understand intricate correlations between attributes and objectives. SVM models are especially helpful for sentiment analysis of e-commerce feedback. The process begins with data preprocessing, where feedback data—typically text reviews with sentiment labels (positive or negative)—is cleaned and prepared through tokenization, removing stopwords, stemming or lemmatization, and vectorization. The preprocessed text is then transformed into feature vectors using techniques like TF-IDF or word embeddings to capture the semantic meaning of the reviews. This data is used to train an SVM classifier, which learns to classify reviews

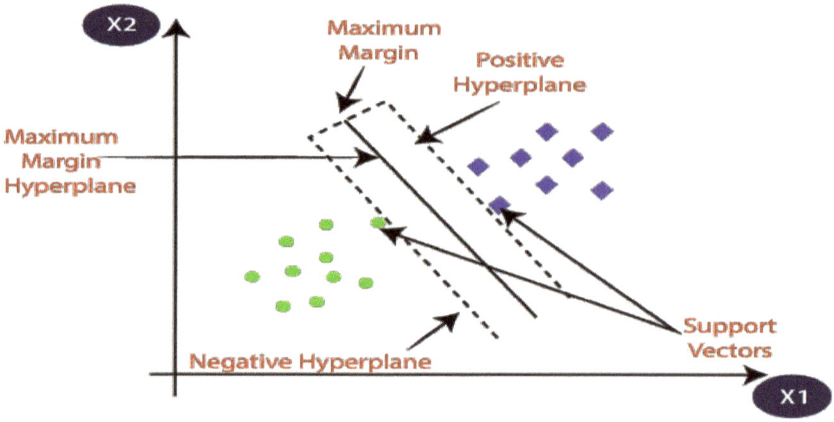

Fig. 3 SVM

into sentiment categories by optimizing the hyperplane that best separates positive and negative instances. Hyperparameters such as the choice of kernel and regularization parameter are tuned using cross-validation to improve performance. The model is evaluated on a test dataset using criteria like accuracy, precision, recall, and F1-score. Once trained and evaluated, the SVM model can be deployed to classify sentiment in new, unseen data, predicting whether a review expresses positive or negative sentiment. The model's performance can be further refined through iterative improvements, including revisiting preprocessing steps, experimenting with feature engineering techniques, or fine-tuning hyperparameters. By leveraging SVM models for sentiment analysis, e-commerce platforms can obtain insightful knowledge about consumer sentiment, allowing for data-driven decision-making to enhance goods, services, and general customer pleasure:

$$f(x) = sign(w \cdot x - b)$$

where

- Vector for weight is w.
- The input data point's feature vector is represented by x.
- b is the bias term.

Equation 2: Support Vector Machine (Fig. 3).

4.5 *Random Forest*

A well-liked learning method for sentiment analysis of e-commerce feedback data is called Random Forest. Tokenization, stopword removal, stemming or lemmatization,

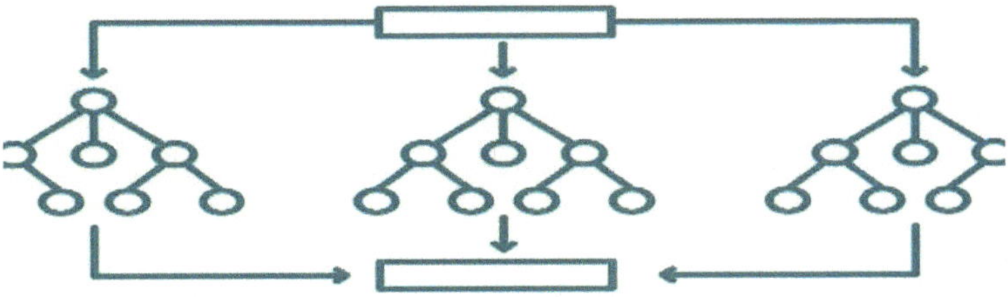

Fig. 4 Random forest

and vectorization are the first steps in the process, which turns text data into numerical features. We next extract relevant characteristics from the preprocessed data, like word frequency, TF-IDF scores, and sentiment scores. This labeled dataset is used to train the Random Forest model using the sentiment categories (positive, negative, and neutral) as the labels. During training, Random Forest generates several decision trees, each of which is trained using a different subset of the features and data. Every tree produces a prediction during inference, and the total of all the trees' predictions determines the final prediction. The sentiment of fresh e-commerce feedback data is predicted by the trained model, which classifies it as neutral, negative, or positive.

A test dataset is used to gauge the model's performance and determine its generalization using metrics such as accuracy, precision, recall, and F1-score. Grid search and random search techniques are used to adjust hyperparameters such as the minimum number of samples per leaf, maximum depth, and number of trees in order to further enhance performance. This method provides strong sentiment predictions for e-commerce feedback data by utilizing Random Forest's ensemble learning ability to handle non-linear correlations and interactions between features (Fig. 4).

4.6 KNN

One useful use of the K-Nearest Neighbor (KNN) model is sentiment analysis of e-commerce feedback. Every customer review is shown as a point in a multidimensional space, with each dimension denoting a textual attribute that was retrieved, like sentiment scores or word frequency. KNN calculates the similarity between the new review and existing reviews in the dataset using common distance metrics like Euclidean distance or cosine similarity. The K nearest neighbors (reviews) are selected based on similarity, and these neighbors "vote" to determine the sentiment of the new review. The majority sentiment among these K nearest neighbors decides the sentiment classification of the new review.

For instance, if most neighbors are positive, the review is classified as positive; otherwise, it is classified as negative. The value of K needs careful tuning where a bigger K may smooth down decision boundaries but may contain irrelevant neighbors, while a smaller K may produce noisy classification. Once the model is trained,

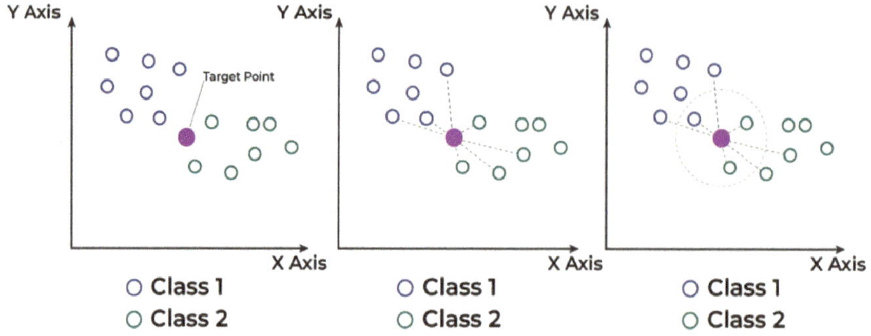

Fig. 5 KNN

it can predict the sentiment of new e-commerce feedback by finding the K nearest reviews and determining their majority sentiment. KNN provides a simple yet effective approach for sentiment analysis by leveraging the similarity between customer reviews, though it can be computationally expensive for large datasets and requires careful selection of the K parameter (Fig. 5).

4.7 Naïve Bayes

For classification problems, a probabilistic machine learning method known as Naive Bayes is employed, particularly in natural language processing (NLP) applications including spam detection, sentiment analysis, and document classification. The likelihood of a hypothesis (class label) given the observable evidence (features) is determined using Bayes' theorem. Based on the words used in the review, Naive Bayes divides customer reviews into positive and negative feelings for e-commerce feedback sentiment analysis. To start, the text is preprocessed to eliminate stopwords, punctuation, and unnecessary characters. Tokenization and stemming are frequently used to standardize the content. Using a bag-of-words model or Term Frequency-Inverse Document Frequency (TF-IDF), features are retrieved from the preprocessed text.

The model is trained on labeled data, learning the conditional probabilities of each word given the class labels. Once trained, it classifies new reviews by calculating the probability of each class for a given review using Bayes' theorem, selecting the class with the highest probability as the predicted sentiment. Model performance is evaluated using characteristics like accuracy, precision, recall, and F1-score on a test dataset to assess its generalization to unseen data. Naive Bayes offers a simple yet effective approach for sentiment analysis in e-commerce feedback, leveraging probabilistic principles to classify customer reviews based on the words used in the text:

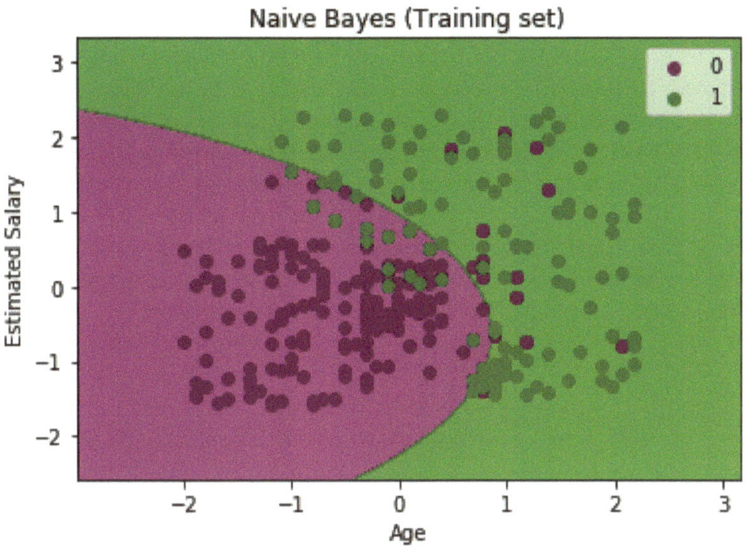

Fig. 6 Naïve Bayes regression

$$p(A|B) = p(B|A) * p(A)/p(B)$$

where

- The posterior probability, or p(A|B), indicates the likelihood that event A will occur in light of event B's occurrence.
- $p(B|A)$ is the chance, or likelihood, that event B will occur in the event that event A is true.
- $p(A)$ is the likelihood that event A will occur in the past, representing our initial opinion before seeing event B.
- $p(B)$ is the event B marginal probability, which acts as a normalizing constant to guarantee the validity of the posterior probability.

Equation 4: Naïve Bayes (Fig. 6).

4.8 Decision Tree

The decision tree is a popular supervised machine learning method for applications involving regression and classification. Decisions are modeled as a tree-like structure, with leaf nodes denoting class labels or regression values, inside nodes representing feature tests, and branching representing decision results. Decision tree models can be employed in sentiment exploration of e-commerce feedback by leveraging their ability to classify text data based on features extracted from customer reviews. The first step involves extracting relevant features from the text data, such as words, phrases, sentiment scores, word frequencies, and word embeddings. Once the features are extracted, a decision tree model is trained using a tagged dataset of customer reviews, with sentiment labels (positive, negative, or neutral) attached to each review. The decision tree algorithm learns to partition the feature space based on these labels, aiming to maximize the purity of the resulting subsets. During inference, the trained decision tree model classifies new, unseen customer reviews into sentiment categories by traversing the tree from the root node to the leaf nodes, where the final sentiment prediction is made based on the majority class of the training samples in the corresponding leaf node. Because the decision-making process is represented in an easily understandable and visualized tree-like structure, decision trees provide high interpretability.

Because the decision-making process is depicted in an easily understood and visualized tree-like structure, decision trees provide interpretability, allowing stakeholders to gain insights into which features are most influential in determining sentiment. Additionally, decision tree models can be employed in ensemble techniques like Gradient Boosting and Random Forests, where several decision trees are trained and merged to enhance predictive performance. In summary, decision tree models provide a straightforward and interpretable approach to sentiment exploration in e-commerce feedback data by learning decision rules from extracted features to classify reviews into sentiment categories (Fig. 7).

I. *Workflow*

- Import all the necessary libraries and download the data from Kaggle.
- Generate the Dependent and the Independent classes for the algorithms.
- Remove the last "n" number of rows of dataset to remove the NaN values and preprocess the data so that it's ready for analysis.
- Split the data into the training and testing data.
- Create various regression models (Logistic Regression, SVM, Random Forest, KNN, Naïve Bayes, and Decision Tree).
- Train the models using the training set.

Find the predictions using the models and the datasets.

5 Results and Analysis

A. **Logistic Regression**

Figure 8 illustrates the confusion matrix for the logistic regression model used in our project. The matrix reveals that the model correctly classified 15,456 negative feedback instances and 16,524 positive feedback instances. However, it misclassified 948 negative feedbacks as positive and 2,003 positive feedbacks as negative. The high

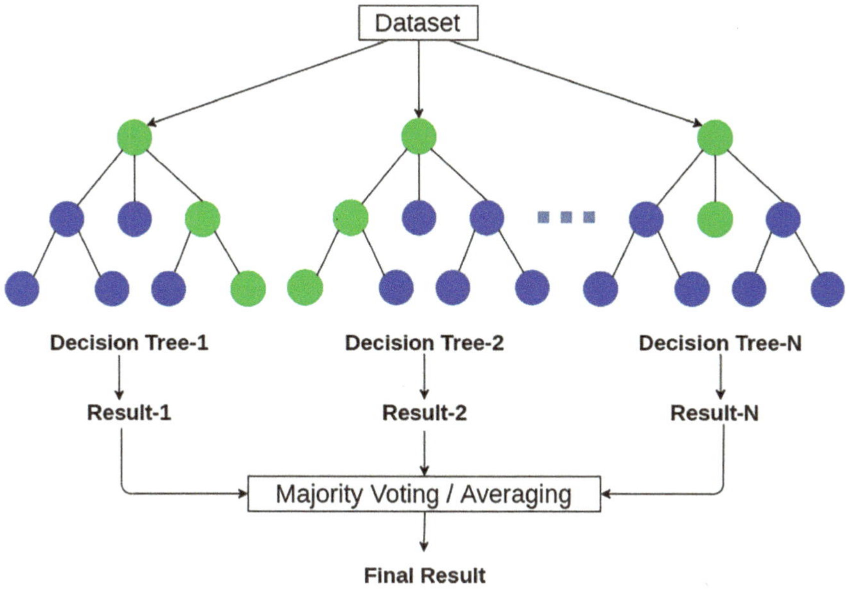

Fig. 7 Decision tree

number of true positives and true negatives, compared to the relatively lower misclassification rates, indicates a robust performance of the logistic regression model in distinguishing between positive and negative sentiments in customer feedback.

B. **Support Vector Machine**

Figure 9 showcases the confusion matrix for the Support Vector Machine (SVM) model employed in our sentiment analysis project. The matrix demonstrates that the SVM model accurately classified 16,019 negative feedback instances and 17,153 positive feedback instances. However, it misclassified 1,374 negative feedbacks as positive and 385 positive feedbacks as negative. The substantial number of true positives and true negatives, in comparison to the relatively lower misclassification rates, suggests a strong performance of the SVM model in differentiating between positive and negative sentiments in customer feedback.

C. **Random Forest**

Figure 10 illustrates the confusion matrix for the Random Forest model used in our project. The matrix reveals that the model correctly classified 16,348 negative feedback instances and 17,734 positive feedback instances. However, it misclassified 56 negative feedbacks as positive and 793 positive feedbacks as negative. The high number of true positives and true negatives, compared to the relatively lower misclassification rates, indicates robust performance of the Random Forest model in distinguishing between positive and negative sentiments in customer feedback.

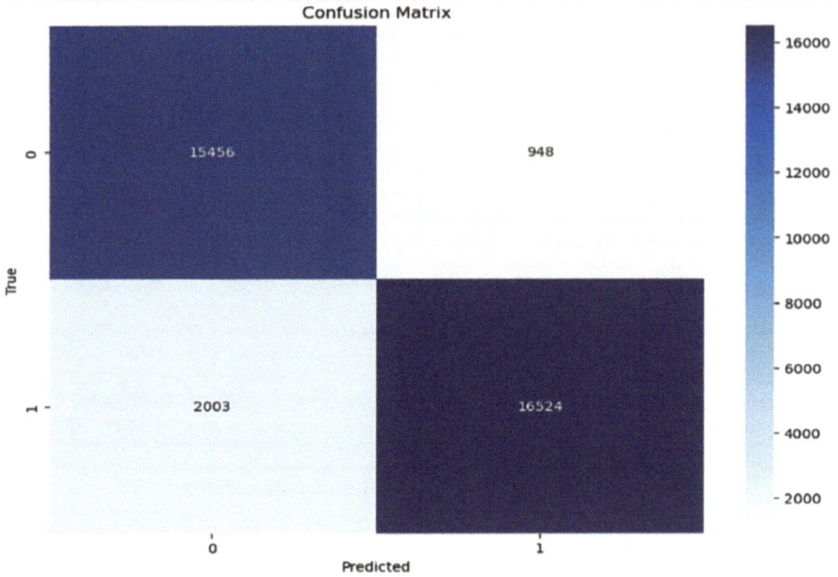

Fig. 8 Confusion matrix of logistic regression

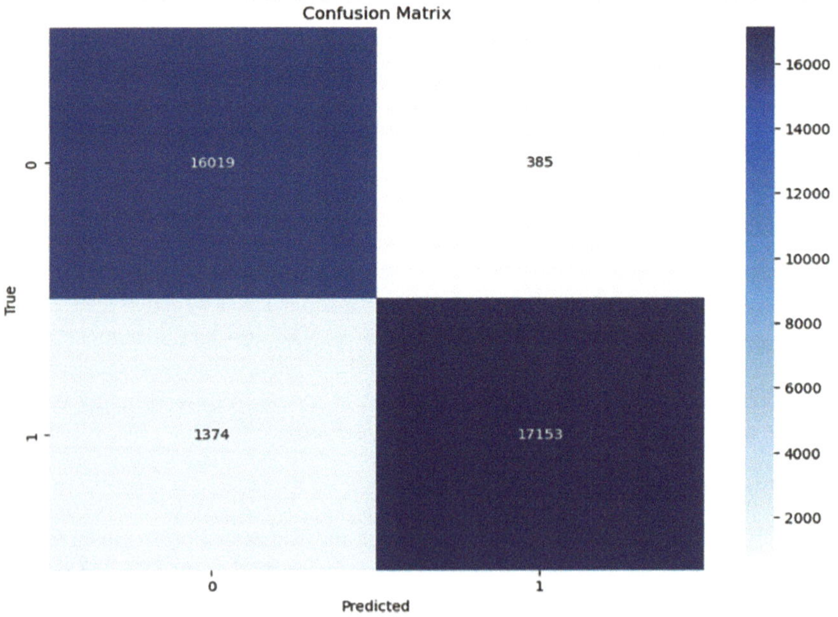

Fig. 9 Confusion matrix of SVM

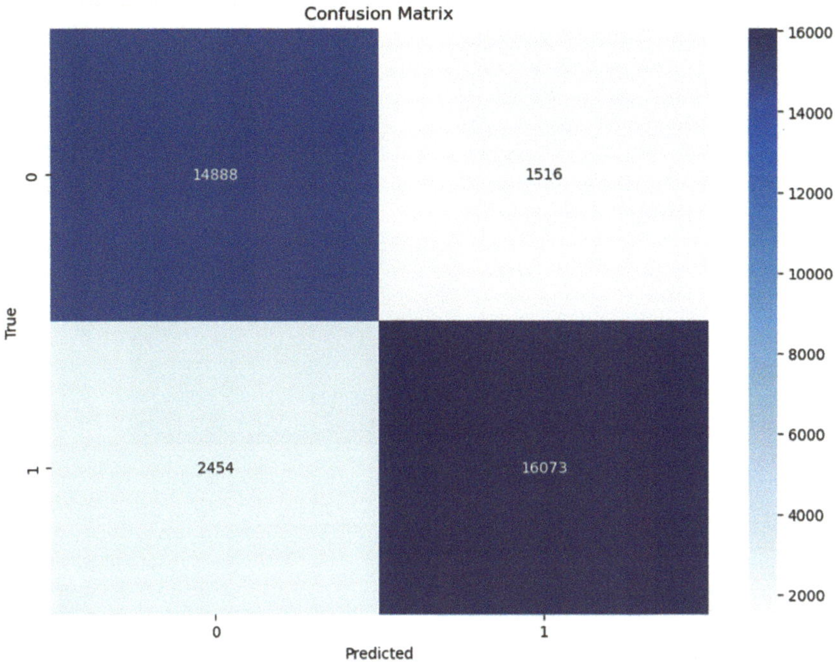

Fig. 10 Confusion matrix of random forest

D. **K-Nearest Neighbors**

Figure 11 illustrates the confusion matrix for the K-Nearest Neighbor (KNN) model used in our project. The matrix reveals that the model correctly classified 15,435 negative feedback instances and 16,542 positive feedback instances. However, it misclassified 969 negative feedbacks as positive and 1,985 positive feedbacks as negative. The high number of true positives and true negatives, compared to the relatively lower misclassification rates, indicates a robust performance of the KNN model in distinguishing between positive and negative sentiments in customer feedback.

E. **Naïve Bayes**

Figure 12 illustrates the confusion matrix for the Naive Bayes model used in our project. The matrix reveals that the model correctly classified 14,888 negative feedback instances and 16,073 positive feedback instances. However, it misclassified 1,516 negative feedbacks as positive and 2,454 positive feedbacks as negative. The substantial numbers of true positives and true negatives, despite the presence of some misclassifications, indicate that the Naive Bayes model performs reasonably well in distinguishing between positive and negative sentiments in customer feedback. This performance underscores the model's utility in sentiment analysis tasks within the context of e-commerce feedback.

F. **Decision Tree**

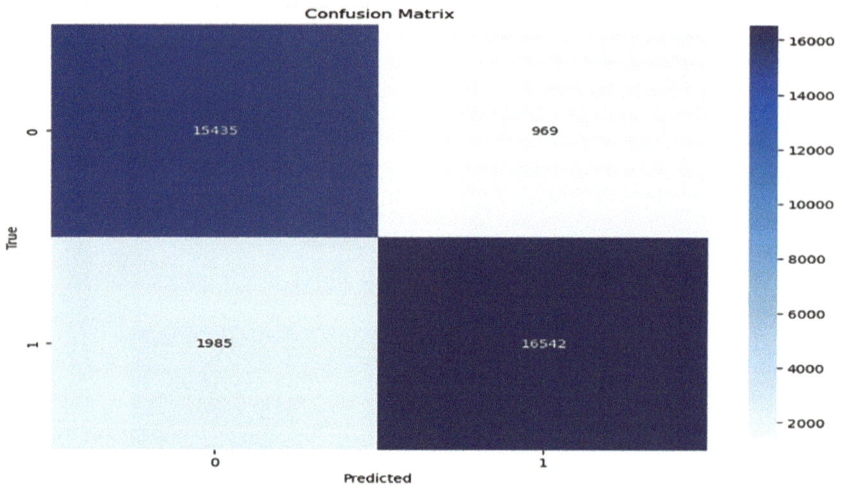

Fig. 11 Confusion matrix of KNN

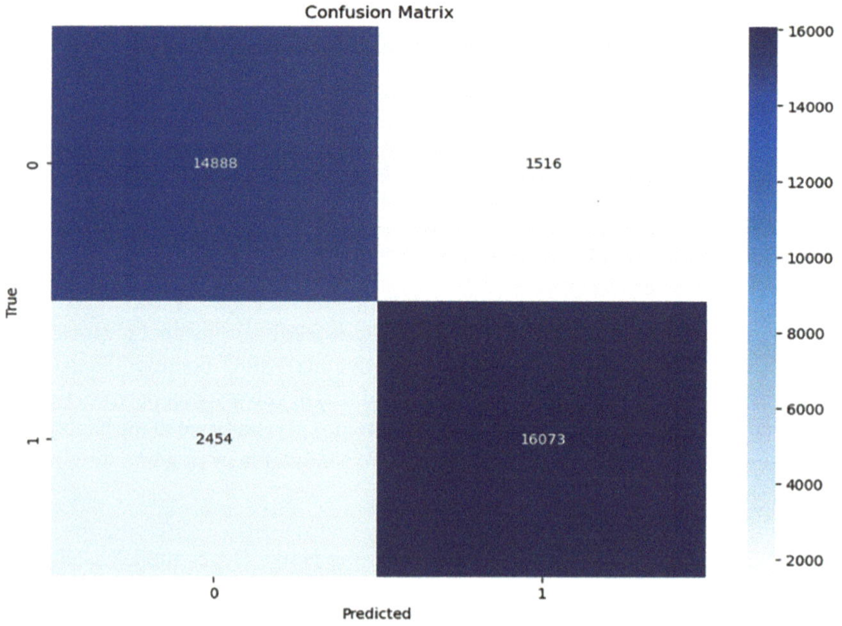

Fig. 12 Confusion matrix of Naïve Bayes

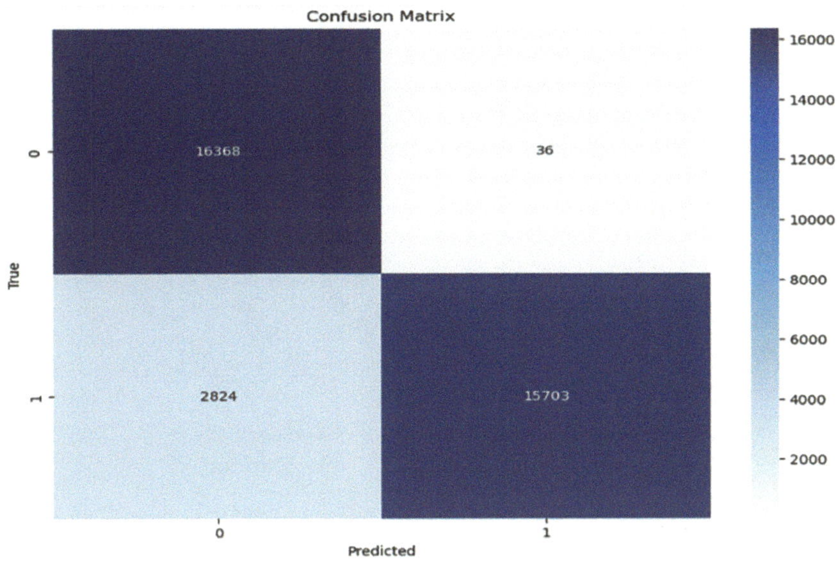

Fig. 13 Confusion matrix of decision tree

Figure 13 illustrates the confusion matrix of the Decision Tree model. The matrix shows a high level of accuracy in predicting both positive and negative sentiments. The model correctly identified 16,368 negative feedback instances and 15,703 positive feedback instances. However, it also made some errors, with 2,824 false negatives and 36 false positives. The high number of true positive and true negative classifications suggests that the model performs well in distinguishing between positive and negative sentiments.

Figure 14 presents a comprehensive comparison of different evaluation metrics, including accuracy, specificity, F1-score, and precision, across various machine learning algorithms applied to sentiment analysis on e-commerce feedback data. The Logistic Regression model emerges as the top performer, exhibiting the highest accuracy and F1-score, while the Support Vector Machine (SVM) algorithm surpasses it in terms of specificity, indicating its potential strength in accurately identifying negative sentiment instances. The Random Forest algorithm displays comparable accuracy to SVM but slightly lower specificity and F1-score. In contrast, the K-Nearest Neighbor (KNN) model demonstrates a noticeable decline across all evaluation metrics compared to the top-performing models. Toward the lower end, the Naive Bayes and Decision Tree algorithms exhibit the lowest accuracy, specificity, and F1-scores, although the Decision Tree model boasts the highest precision among all evaluated algorithms.

Table 1 represents our analysis involved in testing various machine learning models on our dataset to gauge their performance. Among these models, Random Forest stood out with an accuracy of 97.601%, showcasing its superiority over other

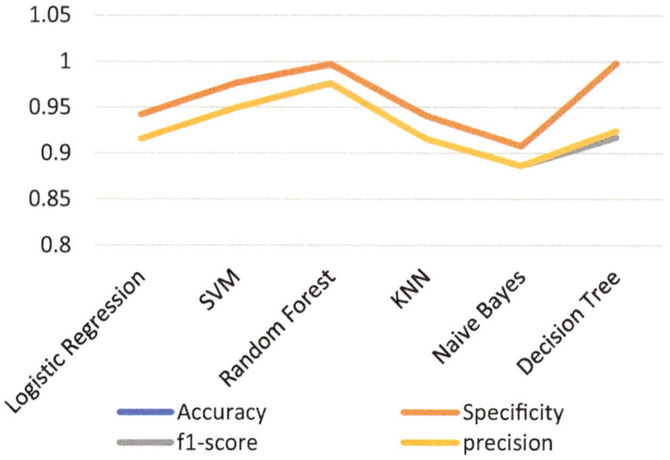

Fig. 14 Comparison of parameters

Table 1 Results for the dataset

Model	Accuracy	Specificity	F1-score	Precision
Logistic Regression	0.915519	0.942209	0.91544	0.915508
SVM	0.949644	0.97653	0.94959	0.949525
Random Forest	0.97601	0.996586	0.975971	0.975602
KNN	0.915433	0.940929	0.915348	0.915357
Naive Bayes	0.886347	0.907584	0.886216	0.886152
Decision Tree	0.917208	0.998049	0.917174	0.924626
Average	0.926694	0.960315	0.926623	0.927795

models. Therefore, based on our evaluation, we ascertain that the Random Forest model emerges as the most accurate among all the models examined in this research.

6 Conclusion

Sentiment exploration on feedback of e-commerce data using machine learning algorithms presents a dynamic and evolving landscape for understanding customer perceptions and behaviors. Our analysis delved into the efficiency of multiple machine learning models in classifying sentiment in e-commerce feedback, revealing insightful patterns and trends. Among the models examined, the most accurate method was Random Forest, which had an astounding accuracy of 97.601%. However, it's important to acknowledge the diverse performance levels exhibited by different algorithms, with Logistic Regression and SVM also demonstrating notable efficacy. Conversely, models such as KNN, Naive Bayes, and Decision Tree displayed lower accuracies compared to the top performers. This comparative analysis provides valuable observations into the suitability of different machine learning models for sentiment analysis in the e-commerce domain, enabling informed decision-making based on specific requirements and trade-offs between evaluation criteria. Sentiment exploration in e-commerce feedback data holds immense capability for further refinement and innovation. Future research initiatives could focus on improving the accuracy and effectiveness of existing models through advanced feature engineering methods, model optimization strategies, and ensemble learning approaches. Additionally, the exploration of deep learning architectures, such as Recurrent Neural Networks (RNNs) and Transformers, could offer new avenues for capturing complex sentiment patterns and contextual nuances in e-commerce feedback. Furthermore, the integration of natural language processing advancements, sentiment lexicons, and domain-specific knowledge could enrich the analysis process and improve the interpretability of sentiment predictions. As e-commerce continues to evolve and expand, the continuous refinement and adaptation of sentiment analysis methodologies will play an important role in empowering businesses to better understand and respond customer feedback, ultimately improving the overall customer experience and driving business growth.

References

1. Rasappan P, Premkumar M, Sinha G, Chandrasekaran K (2024) Transforming sentiment analysis for e-commerce product reviews: hybrid deep learning model with an innovative term weighting and feature selection. Inf Process Manag 61(3)
2. Shedekar S, Shah SK, Kumbhar V (2023) Enhancing e-commerce insights: sentiment analysis using machine learning and ensemble techniques. In: 2023 international conference on integration of computational intelligent system (ICICIS), Pune, India
3. Arshad Md, Dutta P, Jahan S, Das O, Pathak A (2023) Empowering sentiment analysis for English-language customer reviews in e-commerce: a comparative study of machine learning models. Int Res J Eng Sci Technol Innov
4. Davoodi L, Mezei J (2022) A comparative study of machine learning models for sentiment analysis: customer reviews of e-commerce platforms. In: Bled 2022 proceedings 27
5. Hossain MJ, Joy D, Das S, Mustafa R (2022) Sentiment analysis on reviews of e-commerce sites using machine learning algorithms 522–527. https://doi.org/10.1109/ICISET54810.2022.9775846
6. Chandio B et al (2022) Sentiment analysis of roman urdu on e-commerce reviews using machine learning. Comput Model Eng Sci 131(3):1263–1287
7. Alzahrani ME, Aldhyani TH, Alsubari SN, Althobaiti MM, Fahad A (2022) Analysis of e-commerce product review through machine learning—MS. STUTI M. MESHRAM, DR. NEERAJ SAHU. IJFMR 5(3)
8. Singh U, Saraswat A, Azad HK et al (2022) Towards improving e-commerce customer review analysis for sentiment detection. Sci Rep 12:21983
9. Yi S, Liu X (2020) Machine learning based customer sentiment analysis for recommending shoppers, shops based on customers' review. Complex Intell Syst 6:621–634

10. Xu F, Pan Z, Xia R (2020) E-commerce product review sentiment classification based on a naïve Bayes continuous learning framework. Inf Process Manag 57(5)
11. Das RK, Islam M, Hasan MM, Razia S, Hassan M, Khushbu SA (2023) Sentiment analysis in multilingual context: comparative analysis of machine learning and hybrid deep learning models. Heliyon 9(9)
12. Pulikonda VA, Vanukuru D, Navali MS, Motepalli B, Swathi K (2023) Exploring the applications, challenges, and issues of sentiment analysis. In: 2023 7th international conference on computing methodologies and communication (ICCMC), Erode, India
13. Lin X (2021) Sentiment analysis of e-commerce customer reviews based on natural language processing. In: Proceedings of the 2020 2nd international conference on big data and artificial intelligence (ISBDAI '20). Association for Computing Machinery
14. Loukili M, Messaoudi F, El Ghazi M (2023) Sentiment analysis of product reviews for e-commerce recommendation based on machine learning. Int J Adv Soft Comput Appl

AI-Driven Lung Nodule Detection System Using Deep Learning

Preeti Khanwalkar, Swetha Ravikanti, Swapnil Swaroop, Prateek Singh, Likhith Sairamk, and J. Sumanth

Abstract Lung cancer remains a major cause of death due to malignancy worldwide, and help in detecting early lung nodules has become an essential aspect of improving survival. This paper presents the deep learning technique for Lung Nodule Detection which improves the efficiency and the accuracy of detection from CT images. The diverse deep learning models, including Convolutional Neural Networks (CNNs), VGG16, ResNet, Xception, and DenseNet, are used to leverage and harness their strengths, such as feature extraction capabilities, depth management, and computational efficiency. The different models were evaluated based on the performance metrics like accuracy, precision, recall, F1-score, and the classification report which signifies that the Dense-Net Model outperforms in comparison to other models with 97% of precision and 87% accuracy.

Keywords Lung nodule · Deep learning · CT scan

1 Introduction

The lung cancer ranks among the most common and deadly types of cancers worldwide, contributing substantially to cancer-related fatalities. Identifying lung nodules—small tissue masses in the lungs—early is essential for prompt diagnosis and treatment. Nevertheless, conventional diagnostic approaches, like radiologists manually reviewing CT scans, can be slow and susceptible to human mistakes due to the intricate nature and large volume of data. Lung cancer is a major health issue worldwide, accounting for a major proportion of deaths compared to other types of cancers. Early diagnosis and treatment are important to the treatment process;

P. Khanwalkar (✉) · S. Ravikanti · S. Swaroop · P. Singh · L. Sairamk · J. Sumanth
Department of Electronics and Telecommunication Engineering, Dayananda Sagar College of Engineering, Bangalore, India
e-mail: preeti-tce@dayanandasagar.edu

S. Ravikanti
e-mail: swetha-tce@dayanandasagar.edu

H. Nagaraj et al. (eds.), *Advancements in Artificial Intelligence and Machine Learning*, Lecture Notes in Electrical Engineering 1572,
https://doi.org/10.1007/978-981-95-7449-0_27

image analysis of lung nodules in CT scans is very important. Traditional methods depend on visual analysis, and medical imaging data analysis has an exponentially high complexity that consumes much time and is prone to errors [1–4]. In the last years, AI deep learning has offered quite revolutionary solutions to medical imaging. A good number of the best successes that CNNs experienced were detected and segmented nodules from various studies that implemented architecture types VGG16 and MobileNetV2 among others in reference [3, 5–7]. Large-sized datasets help exploit complex patterns not achieved in such conventional methods, with high sensitivity as well as specificity [7–9].

2 Related Work

Recent progress in generative networks and its applications in the biomedical image segmentation in deep generative models has demonstrated considerable promise for image synthesis, detection, segmentation, and classification tasks. Segmenting medical images is regarded as a key challenge in the biomedical imaging domain. Generative Adversarial Networks (GANs): GANs have revolutionized medical imaging in addressing the challenges of segmentation, reconstruction, and data augmentation. The quality of the medical images synthesized through GANs has proved crucial in mitigating the problem of data scarcity [10–15, 20]. A thorough survey has been carried out on the use of GAN networks for medical image segmentation, concentrating mainly on different GAN-based models, performance evaluation metrics, loss functions, datasets, augmentation techniques, implementation of papers, and available source codes. Additionally, this paper offers an in-depth review of the application of GAN networks for segmenting various human diseases.

Investigation into the automatic detection of lung cancer through deep learning algorithms has shown promising outcomes. However, the intricate characteristics of tumor borders and potential shifts in tumor locations continue to pose significant challenges in diagnosing patients with lung tumors from computed tomography (CT) images. Transference or Knowledge Pre-trained model: Use of a pre-trained mobileNetV2 with VGG16 fine-tuned on our lung nodule detection also gave better performance with fewer computer requests needed [3, 8, 9]. Authors have created an enhanced hybrid neural network by combining the MobileNetV2 and UNET architectures for the purpose of semantic segmentation of malignant lung tumors in CT scans. The technique of transfer learning was applied, incorporating the pre-trained MobileNetV2 as the encoder within a traditional UNET framework to facilitate feature extraction. One-Stage Detection Models: Single-stage detection frameworks, such as ELCT-YOLO, can detect nodules at a much more efficient speed while scanning across scales and edge complexities [2, 12, 16–19]. Multi-Resolution Features: The incorporation of multi-resolution features in the detection pipeline improves the model's ability to identify nodules of different sizes and shapes [6, 18]. The model underwent training and fine-tuning using the training dataset obtained from the Medical Segmentation Decathlon.

Generative adversarial networks (GANs) and their variants have opened up numerous exciting avenues to address established and difficult problems in medical image analysis, including tasks like medical image de-noising, reconstruction, segmentation, data simulation, detection, and classification. Additionally, their capability to generate images with unprecedented realism raises the possibility of addressing the persistent issue of limited labeled data in the medical field through the use of these generative models. This review article presents a comprehensive overview of recent research on GANs in medical applications, thoroughly discusses the limitations and prospects of the proposed techniques, and outlines potential directions for future research.

3 Proposed Methodology

The Lung Cancer Detection System development process employs an agile methodology, prioritizing iterative sprints and consistent feedback mechanisms to adapt to user requirements and technological progress. It starts with gathering and analyzing requirements, which includes a thorough evaluation of user needs, market dynamics, and the latest advancements in medical imaging and deep learning. These requirements are ranked according to their feasibility and potential effect on user experience, laying the groundwork for a concentrated and effective scope.

Data augmentation techniques are applied such as rotation, scaling, and flipping of the images to improve dataset diversity and prevent overfitting [8, 13]. Once testing is successfully completed, the system is launched in a production environment and closely observed to ensure seamless operation in real-world scenarios. After deployment, user feedback from healthcare professionals and patients is actively collected to evaluate system usability and efficacy. This feedback, along with advancements in technology, drives continuous improvement and enhancement of the system. The iterative nature of the process ensures that the Lung Cancer Detection System stays effective, dependable, and attuned to the changing requirements of its users. The description of different phases is discussed as follows.

Data Acquisition and Preparation: Assemble and prepare a detailed dataset of lung cancer images, making sure it is accurately labeled and separated into three different sets, training, validation, and test sets to enable efficient model training and assessment.

Model Selection and Fine-Tuning: Employ the VGG16 neural network architecture, which has been pre-trained on the ImageNet dataset, and adjust it to specifically cater to the lung cancer image dataset, enhancing its performance for precise classification.

Data Augmentation and Enhancement: Utilize data augmentation methods on the training dataset to enhance its diversity, bolster the model's ability to generalize, and minimize the likelihood of overfitting.

In the design phase, the wireframes are created, and mock-ups and prototypes are used to represent both the user interface and system architecture visually. Close collaboration with designers and stakeholders guaranteed the design meets user expectations and work objectives. Prototypes are tested and improved based on feedback, ensuring that any problems are tackled before proceeding to development. This method guarantees that the system satisfies both functional and aesthetic demands. The objected lung nodule detection system integrates insights from recent studies and builds a scalable, high-performance AI solution. It adopts the VGG16 architecture, pre-trained on ImageNet, but modified to specifically accommodate lung nodule detection. It ensures efficient feature-extraction as well as robust classification [3, 5, 19]. The model incorporates interpretability features like attention maps that would help in understanding its reasoning behind decision-making. This also is backed with recent studies indicating a need for explainable AI in health care [12, 16, 20].

The work focuses on developing a system that leverages AI to identify lung nodules through deep learning techniques. The goal is to enhance the early detection of lung nodules, which is vital for diagnosing and treating lung cancer. This will be achieved by employing convolutional neural networks (CNNs); the system offers accurate and efficient analysis of medical imaging data, greatly enhancing the reliability of nodule detection in comparison to traditional techniques.

Figure 1 represents the Lung Nodule Detection System using deep learning. The process starts by data selection, wherein the relevant raw data are gathered and screened. Following that is data preprocessing that deals with cleaning, missing value handling, and data standardization. Feature engineering converts raw data into meaningful inputs; techniques such as feature selection and encoding are made in an effort to increase model performance. In the case of model training, the algorithms learn patterns on processed data optimized using a training and validation set. The last is deployment in which the trained model is utilized to make predictions in actual systems while under continuous monitoring and updating. This holistic approach leads to accurate, reliable, and scalable solutions by deep learning.

The process starts with CT scan data preprocessing, followed by model selection, where different architectures like CNN, VGG16, ResNet, Xception, and DenseNet are used for detection. The trained models generate predictions, which are then processed through a GUI dashboard for user image input and diagnosis. Finally, the system provides the final result based on the model predictions. Different steps starting from requirement analysis, system design, deployment, testing, debugging, maintenance, and updates are described as follows.

Requirement Analysis

For regular maintenance and update of the system, System keeps on checking performance parameters and keeps on regularly training the system on updated data. Set up logging and monitoring tools to observe the system's performance and detect possible problems like delayed response times or inaccuracies in predictions. Consistently refresh the model with new data to enhance its precision and reliability, making certain it adjusts to any shifts in medical imaging practices or population demographics.

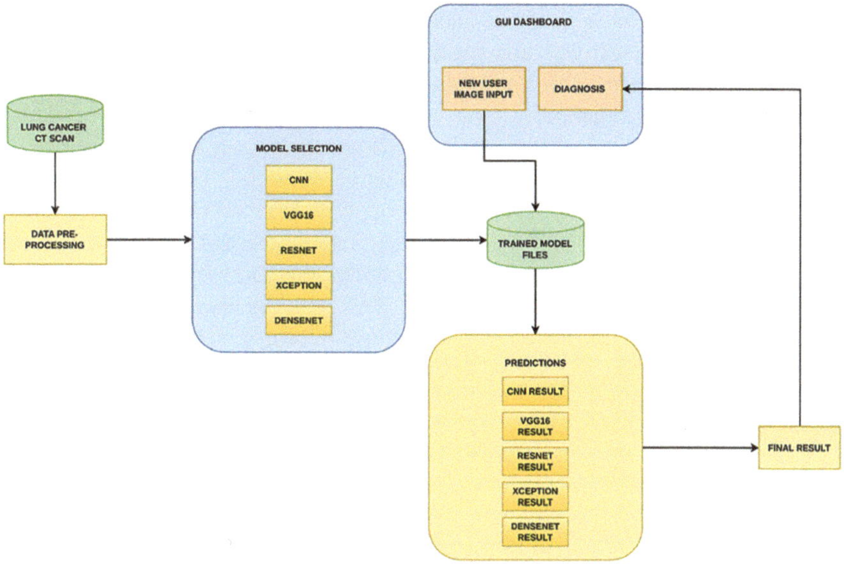

Fig. 1 Proposed methodology of lung nodule detection

System Design

The design and prototyping stage of this work centers on establishing a solid framework for the implementation and testing of the deep learning model. First, a conceptual blueprint delineates the model architecture, which integrates the VGG16 base model alongside tailored top layers for the purpose of lung cancer classification. The model architecture is based on VGG16, and more fully connected layers are applied for classification. The design is refined through testing on validation sets iteratively as suggested by prior research [6, 8, 13].

Development

The deployment and launch stage signifies the shift of the lung cancer detection model from a developmental prototype to a practical application ready for clinical or research utilization. Transfer learning is used to start the weights of VGG16, which reduces the training time and enhances convergence [3, 9, 11]. The model is coded in Python and TensorFlow and uses Flask as the framework for deployment for real-time prediction [5, 18]. This stage includes implementing the model on a scalable cloud platform or an on-site server, guaranteeing that it is reachable to end-users like healthcare practitioners.

Testing and Debugging

The testing and quality assurance stages are essential in the lifecycle of the lung cancer detection work, as they help guarantee the reliability and effectiveness of the implemented model. This stage requires thorough evaluation with both internal validation datasets and separate test sets to measure the model's performance is based on accuracy, sensitivity, and specificity. Accuracy, precision, recall, and F1-score: Performance metrics are measured [14, 19], concerning the reliability of these systems under varied situations.

Deployment

To deploy and launch lung cancer detection work utilizing the VGG16 model, several important steps need to be followed. First, make sure to save and serialize the trained VGG16 model in a format such as HDF5 or Saved Model. The system is deployed on a cloud platform, which enables the uploading of CT scans by healthcare professionals and receiving predictions. Continuous monitoring ensures stability and reliability [11, 18]. Then, establish a server environment, usually employing a web framework like Flask or Django for Python, to manage incoming requests and provide the model predictions.

Maintenance and Updates

Keeping lung cancer detection work up to date requires ongoing observation, performance enhancement, and frequent revisions. Set up logging and monitoring tools to observe the system's performance and detect possible problems like delayed response times or inaccuracies in predictions. Consistently refresh the model with new data to enhance its precision and reliability, making certain it adjusts to any shifts in medical imaging practices or population demographics.

The results of this work include a well-optimized and trained deep learning model that can pinpoint lung nodules with a high degree of accuracy, minimizing both false positives and negatives. This system not only assists radiologists by offering additional diagnostic insights but also speeds up the diagnostic workflow. Moreover, the execution of this work showcases the viability of incorporating AI into medical imaging processes, paving the way for wider applications in the healthcare sector. The primary benefit lies in the improvement of patient results via prompt interventions and better resource distribution in healthcare environments.

4 Results and Discussion

In order to measure models strengths and weaknesses, and to select the most effective model so as to validate the model's real-world applicability the performance analysis of models is carried out. The chest CT scan data is collected from Kaggle [15], and analysis is performed with different deep learning models. In the domain of lung cancer diagnosis, such evaluations ensure that deep learning models can deliver accurate, reliable, and timely predictions, ultimately aiding radiologists in critical decision-making. Figure 2 represents the samples of different classes of CT scan images available in the dataset.

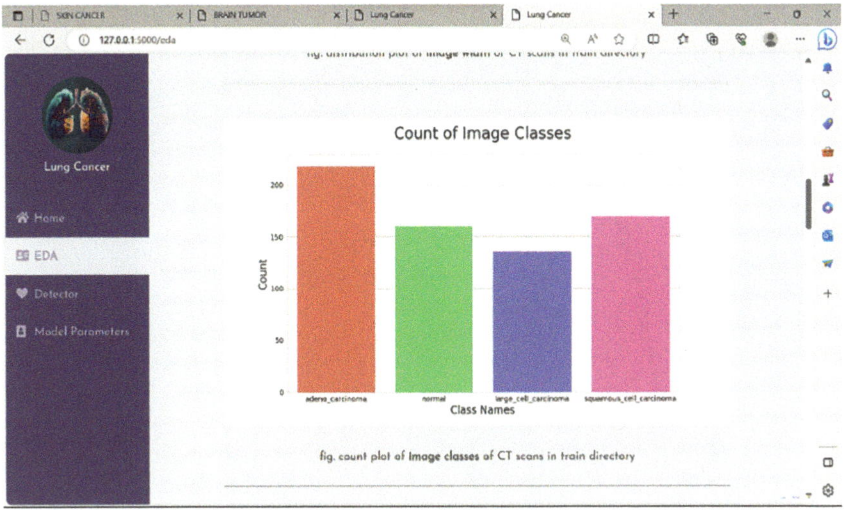

Fig. 2 Graphical representation of the different classes available in the dataset

Further, by applying different models over the given dataset, lung nodule is detected. The sample results of lung nodule detection of the CT scan image of lungs are shown in Fig. 3.

Furthermore, testing and validation accuracy is obtained for the given dataset as shown in Fig. 4. The classification reports are generated which offer valuable insights into a model's ability to perform across different classes, such as distinguishing subtle

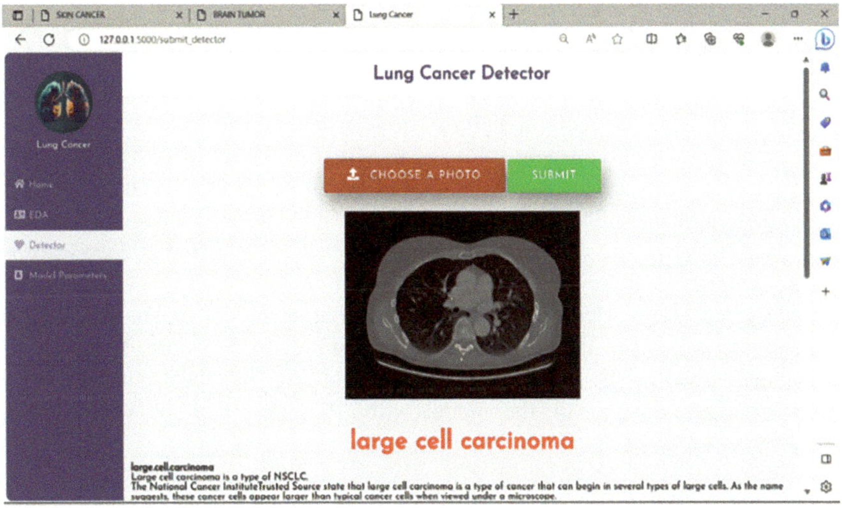

Fig. 3 Sample results of lung nodule detection of the CT scan image of lungs

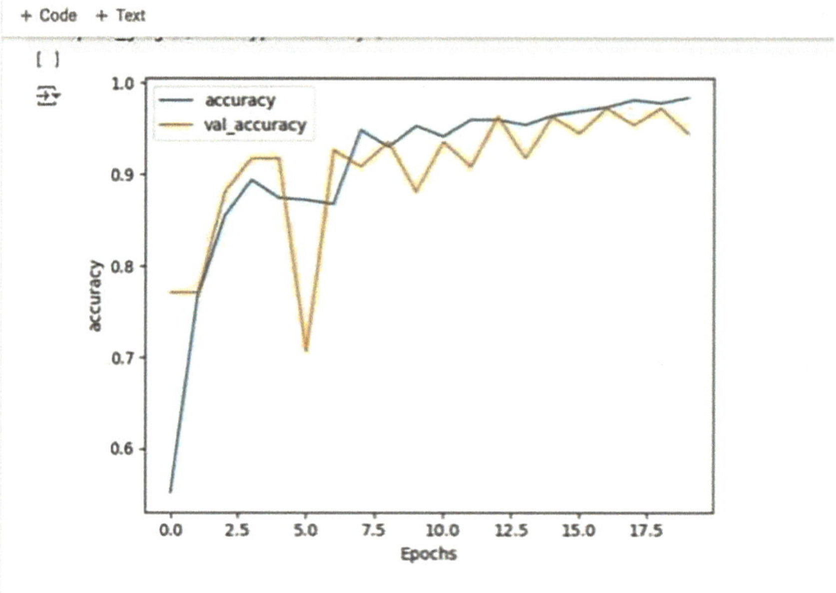

Fig. 4 Testing and validation accuracy

variations between benign and malignant nodules. This capability is essential for reducing diagnostic errors, improving early detection rates, and enhancing patient outcomes. By guiding decisions on model deployment and optimization, classification reports play a pivotal role in leveraging deep learning for advancements in healthcare diagnostics.

A classification report provides a comprehensive summary of key classification metrics, including precision, recall, F1-score, and support, for each class in the dataset. Precision measures the accuracy of positive predictions, recall quantifies the model's ability to correctly identify positive instances, and F1-score harmonizes precision and recall into a single value, especially useful when dealing with imbalanced datasets. The support metric indicates the number of true instances for each class. The sample classification report of VGG16 and DenseNet models is shown in Figs. 5 and 6. Furthermore, we compare the performance of different deep learning models as shown in Table 1.

Table 1 represents a comparison between different model parameters used in the proposed work, and it compares the precision, recall, F1-score, and accuracy of different models. The experimental results indicate that the DenseNet model outperforms in comparison to other models with 97% of precision and 87% accuracy.

Classification Report:

	precision	recall	f1-score	support
adenocarcinoma	0.93	0.47	0.62	120
large.cell.carcinoma	0.49	1.00	0.65	51
normal	0.96	0.98	0.97	54
squamous.cell.carcinoma	0.77	0.81	0.79	90
accuracy			0.74	315
macro avg	0.79	0.81	0.76	315
weighted avg	0.82	0.74	0.74	315

Fig. 5 Classification report of VGG16 model

Classification Report :

	precision	recall	f1-score	support
adenocarcinoma	0.97	0.75	0.85	120
large cell carcinoma	0.61	1.00	0.76	51
normal	1.00	0.98	0.99	54
squamous cell carcinoma	0.93	0.89	0.91	90
accuracy			0.87	315
macro avg	0.88	0.91	0.88	315
weighted avg	0.91	0.87	0.87	315

Fig. 6 Classification report of DenseNet model

Table 1 Comparative results of different models

Models	Precision	Recall	F1-score	Accuracy
CNN	1.0	0.56	0.45	0.29
VGG16	0.93	0.47	0.62	0.74
ResNet	0.92	0.46	0.61	0.70
Xception	0.38	0.27	0.31	0.26
DenseNet	0.97	0.75	0.85	0.87

5 Conclusion

This work diligently addressed the complex task of Lung Cancer Detection via CT scan images using state-of-the-art deep learning architectures. The presented work is grounded in the pursuit of technical excellence, striving to advance the capabilities of medical image analysis. Throughout this work, a meticulous data preprocessing pipeline is used which involves image resizing, normalization, and data augmentation. This preparation ensured that the dataset was suitably conditioned for training deep learning models. Simultaneously, exploratory data analysis (EDA) phase unveiled vital insights into the dataset's distribution, class balance, and key statistical attributes, informing critical decisions in model selection and evaluation. The diverse deep learning architectures, including Convolutional Neural Networks (CNNs), VGG16, ResNet, Xception, and DenseNet, are used to leverage to harness their strengths, such as feature extraction capabilities, depth management, and computational efficiency. The different models were evaluated based on the performance metrics like accuracy, precision, recall, F1-score, and the classification report which signifies that the Dense-Net Model outperforms in comparison to other models with 97% of precision and 87% accuracy. These metrics provided a comprehensive assessment of the models' diagnostic abilities, enabling us to discern their relative merits.

References

1. Ahmed I et al (2022) Generative adversarial networks in biomedical image segmentation. J Biomed Imaging Bioinform 12(4):123–134
2. Ji Z et al (2023) ELCT-YOLO: Efficient one-stage model for lung tumor detection. Med Imaging Anal J 15(2):56–68
3. Riaz Z et al (2023) Lung tumor image segmentation using MobileNetV2 and transfer learning. J Med Image Process 18(3):102–115
4. Kazeminia S et al (2020) GANs for medical image analysis: A comprehensive survey. Med Image Anal 54, Article 101762
5. Liao F et al (2019) Deep learning-based detection and segmentation of pulmonary nodules. IEEE Trans Med Imaging 38(6):1358–1369
6. Setio AAA et al (2016) Pulmonary nodule detection using multi-view CNNs. IEEE Trans Med Imaging 35(5):1160–1169
7. Shen W et al (2015) Multiscale CNNs for lung nodule classification. IEEE Trans Med Imaging 34(5):1096–1107
8. Dou Q et al (2016) Automated pulmonary nodule detection via 3D CNNs. IEEE Trans Med Imaging 35(5):1184–1197
9. Huang X et al (2017) 3D U-net-based lung nodule detection. IEEE Trans Med Imaging 36(4):996–1005
10. Li X et al (2020) Deep reinforcement learning for pulmonary nodule detection. IEEE Trans Med Imaging 39(2):409–420
11. Nishio M et al (2020) Radiomics-based deep learning for nodule diagnosis. IEEE Trans Med Imaging 37(6):1391–1401
12. Shen W et al (2015) Multi-scale convolutional neural networks for lung nodule classification. In: International conference on information processing in medical imaging (IPMI)

13. Choi H et al (2015) Multitask learning for pulmonary nodules. IEEE Trans Med Imaging 38(5):1135–1147
14. Huang X et al. 3D U-net-based lung nodule detection using multi-scale features. MICCAI Workshop
15. Kaggle. https://www.kaggle.com/datasets/mohamedhanyyy/chest-ctscan-images
16. Tang Y et al (2020) Semi-supervised learning for nodule detection with CNNs. IEEE Trans Med Imaging 39(8):2341–2352
17. Xie Y et al (2021) Explainable AI for pulmonary nodule diagnosis. IEEE Trans Med Imaging 40(9):2511–2523
18. Jin HK et al (2018) AI-based CAD for pulmonary nodules. IEEE Trans Med Imaging 37(5):1142–1153
19. Lee J et al (2021) Unsupervised pretraining for nodule detection. IEEE Trans Med Imaging 40(3):789–800
20. Liao F et al (2019) Deep learning-based detection and segmentation of pulmonary nodules in CT images. IEEE Trans Med Imaging 38(4):919–931
21. Ozdemir O et al (2019) Attention-based deep learning for pulmonary nodule classification. IEEE Trans Med Imaging 38(3):679–688
22. Nishio M et al (2018) Radiomics-based deep learning for pulmonary nodule diagnosis. Radiology
23. Choi H et al (2019) Multi-task learning-based detection of pulmonary nodules. Int J Comput Assist Radiol Surg
24. Ozdemir O et al (2019) Attention-based deep learning for pulmonary nodule classification. IEEE Trans Med Imaging
25. Dou Q et al (2020) Integration of multi-resolution features for accurate nodule detection. Nat Med 26(3):413–420
26. Lee S et al (2021) Unsupervised pretraining for pulmonary nodule detection. IEEE Trans Biomed Eng 68(9):2821–2829

Hybrid Deep Learning Approach for Accurate Suspicious Human Activity Recognition Using Autoencoders and 3D CNNs

S. R. Likhith and R. Anjali

Abstract Advancement of modern deep learning architecture for surveillances will ensure a high potential increase in public safety and security. This paper proposes using up-to-date techniques such as inflated 3D Convolutional Neural Networks (I3D CNN) and Autoencoders to develop a framework system for real-time detection of suspicious human activity. The approach offers increased performance of the system in detecting suspicious activities while addressing the variability imposed by certain contexts such as poor lighting. Using real-scenario datasets, the model ensures very reliable operation and reduced false positives, thus enhancing the productivity of security personnel. When hazardous activities are detected, real-time SMS alerts are triggered to devices for immediate intervention. Experimental results prove that the model outperforms the existing techniques with respect to accuracy and dependability. This is a major step toward the advancement of intelligent surveillance systems, which will assist in crime prevention and develop public safety through automatic detection and real-time actionable insights. Activity recognition and surveillance technology are set on a new threshold by the proposed framework, which is also promising for application in many scenarios.

Keywords Suspicious human activity recognition · Autoencoders · 3D Convolutional Neural Network (3D CNN) · Inflated 3D Convolutional Neural Network (I3D CNN)

S. R. Likhith (✉) · R. Anjali
Presidency School of Computer Science and Engineering, Presidency University, Bengaluru, India
e-mail: anandlikith111@gmail.com

H. Nagaraj et al. (eds.), *Advancements in Artificial Intelligence and Machine Learning*, Lecture Notes in Electrical Engineering 1572,
https://doi.org/10.1007/978-981-95-7449-0_28

1 Introduction

Computer vision and AI growth have made video surveillance the most essential need for identification of abnormality within the public domain. Suspicious human activity recognition (SHAR) is a unique facet of surveillance systems that attempts to identify possible threats through observation of human conduct [1]. The older systems are either monitored manually or by conventional methods like motion detection which is UIS and involve so much human interference [2]. It can, however, auto forward it to get more accurate results so that the number of false alarms reduces.

Indeed, during the past few decades, surveillance systems have evolved mainly to meet new requirements for more effective policing, protection of assets and property, and maintaining social order among others. Indian cities are becoming more complex and urbanized, hence the techniques of scanning which have underpinned traditional surveillance will not be adequate. Since conventional systems typically trail way behind hundreds of thousands of data produced by several thousand cameras installed in public places, this alone results in poor response and delayed threats. Such a call for an efficiency, scalability, and precision in a solution has never been louder in this day and age.

Substantial artificial intelligence is known as deep learning, which has enabled real-time video surveillance of human behavior. Such systems rely on a set of calculations to interpret visual data, which supposedly identifies acts that are dangerous to the society in one way or another, harmful [3]. Of course, this is now machine learning, and such surveillance conducted today can recognize patterns of activities and pick out anomalies of its normal workflow and shoot out alerts based on a setup criterion. This transformed the complete process of monitoring from a manual method to the use of an automated system, which was a significant step in surveillance technology, which could also have made timely response and prevention of criminal activities.

Hence by chance, powerful techniques in the form of hybrid approaches in deep learning appear at the right time. They mitigate drawbacks inherent in the conventional surveillance systems which cannot capture the real-time changing nature of human activity characteristics with a good degree of accuracy. Such hybrid models and methodologies, similar to these, can perfectly fit into the dynamism of the human activities. The second way of combining the two networks would be an integration of CNN to obtain spatial features and LSTM for temporal representation [4]. Both approaches enable the system to capture both the type and timing of human behavior, leading to a more comprehensive understanding.

This paper would also like to know more about how integration of hybrid deep learning methodologies could well improve the effectiveness of suspicious HAR systems. It involves some of the primary prone activities of concern which are classified as three categories, namely normal, violence, and weaponized activities; the videos are collected from various sources. A closer look at these activities makes it possible for this report to describe how hybrid deep learning systems enhance detection rates while simultaneously reducing false positives where necessary. This will equip them to well coordinate with police and other security forces in order to maintain safety of lives and property at all times.

1.1 Research Contribution

This research is vital in intelligent surveillance systems because it tries to break through the limitation of approaches presented in detecting suspicious human activities. This work will present a novel framework that can be used in real-time detection of suspicious activities based on advanced deep learning techniques like 3D Convolutional Neural Networks (3D CNNs) and Autoencoders. It overcomes the difficulties such as low illumination and unpredictability in the context by utilizing contrast-enhanced datasets which has been collected from various sources, and increased reliability and false positives. Speed computations and quick results are obtained. Higher performance of the model further increases the efficiency of the security personnel and the accuracy of the detection while generating real-time SMS notifications for immediate intervention in the case of dangerous activity. The results of the experiments show that the suggested hybrid systems outperform existing systems in terms of accuracy and dependability.

2 Related Work

The present work represents an alternative approach in the recognition of abnormal human activity in video frames, using deep learning strategies [5]. The work is structured to systematically detail the procedure involved—data collection, data preparation, annotation, data preprocessing, feature extraction, and model development. Hybrid architecture comprises the fusion of CNN and LSTM. Such architecture is used to consider both spatial features and temporal features of video frames. The implemented algorithm will be evaluated based on various performance measures in terms of accuracy, recall, precision, and F1-measure, which yield an accuracy of 90.14%. However, there are some drawbacks in adopting the hybrid approach to activity recognition with the CNN-LSTM model. The limitations include the size of the dataset being small, not considering computational complexity, and not comparing results with any of the existing techniques. The concluding remark of the paper stresses the effectiveness of algorithms that may be used for detecting suspicious activity.

This paper introduces a detailed literature survey of video action recognition techniques in the context of sports analytics, dwelling on datasets, methodologies, and applications at both the athlete and team levels within the sport domain [6]. Approaches have been classified into traditional hand-crafted models and modern deep learning models, which include 2D and 3D CNNs, two-stream models, skeleton-based models, and group activity recognition frameworks. Some sports datasets reviewed include SoccerNet, NCAA, and UCF Sports, among many others. This survey covers methods and challenges of video recognition in sports entirely but misses a quantitative meta-analysis of the performances of the methods and discussions on cross-sport transfer learning. In fact, the paper did not look into transformer-based methods very extensively. However, this paper is highly valuable as a starting point to help researchers enter the domain, providing crucial insight into current trends and research directions.

This survey paper takes into consideration, in extensive detail, techniques that machine learning is applied toward anomaly detection in surveillance videos—namely supervised, semi-supervised, and unsupervised methods. Here, existing anomaly detection methods from the previous literature are categorized based upon their advantages and disadvantages in terms of applications that cut across domains in surveillance [7]. Datasets such as UCF Sports and SoccerNet hold a lot of importance concerning training and benchmarking. Lacking adequate quantitative comparisons on benchmark datasets, except very briefly mentioning real-world deployment challenges, privacy issues, and ethical considerations, the paper has a good overview of the strengths and weaknesses of various approaches at good accuracy.

To prevent "over-exploiting," only one temporal transformation, temporal-aware contrastive network (TAC-Net), is proposed in the study for UAVS video anomaly detection [8]. In particular, it utilizes contrastive learning to learn features of normal patterns and uses a 3D encoder for calculating anomaly scores. The proposed approach achieves the best accuracy on benchmarked data and eliminates shortcomings of prior techniques. But it is slower, more sensitive to hyperparameters, and may be less effective in handling scales in videos. In summary, it has been shown that self-supervision can be effectively used in combination of Self-Supervised Contrastive Learning for capability of anomaly detection in the surveillance systems.

The work presents a novel deep learning method to human action recognition using drone imagery that can be translated into applying RGB video data. The methodology also includes other sophistications like bilateral filtering that is used in attenuating noises, quick shift segmentation for silhouette extraction, and extraction of the key body points and movement [9]. The approach involves classification using convolutional neural networks (CNNs) and the approach delivers best performance with benchmark datasets with good accuracy; in general the model is good but fails to work well with complex structures, and can hardly follow fast movements or changing angles.

A lightweight convolutional neural network (CNN) model for human activity recognition using Wi-Fi channel state information (CSI) data is developed in this work [10]. The methodology also uses a few data augmentation options to handle problems concerning small datasets and unbalanced classes. By preprocessing the

CSI data and creating a proper CNN architecture, the study shows enhanced performance compared to prior studies and as such is apt for real-time use on any such target devices with limited resources. In sum, the current study contributes a substantial enhancement into privacy-preserving human activity identification enabling smart home and healthcare applications.

The research brings a solution based on deep learning to improve security in real-time video surveillance when addressing face mask detection systems and protecting them against adversarial attacks [11]. Implementing the white-box Fast Gradient Sign Method (FGSM) attack, the study compares the effects of various epsilon values on model performance employing a custom dataset of 14,535 images. Therefore, the results demonstrate the presence of adversarial threats to detection systems while proving the efficacy of the proposed mitigation approach. Still, this decision limits the study area to one type of attack, and the custom dataset may not be typical for the real environment.

This work also employs deep leaning used for human activity recognition with features taken from video frames using pre-trained CNN (VGG16), and the temporal information is learn by Bi-GRU [12]. In feature selection using the random forest algorithm, the method also gains the lowest computational complexity while also attaining the best benchmark accuracy. However, because our approach is based on a pre-trained CNN, generalization to new domains may be limited, and performance overall is poor for at least one activity class. These limitations of the approach are again compounded by the fact that they are highly computational thus not advisable for IoT devices. In general, this research shows how future work can utilize multiple deep learning methods for advanced video comprehension tasks.

This work provides an overview and an evaluation of the current state of the art in MOT approaches for real-time action recognition in surveillance proposed by introducing the SurvTrack dataset based on VIRAT dataset [13]. In this paper, the performance of five MOT trackers including ByteTrack, OC-SORT, the Deep OC-SORT, the BoT-SORT, and the StrongSORT is analyzed based on the metrics. The results show the potential of the approach for detection of reliable samples and give information about the influence of detector confidence threshold and input resolution of images on the performance.

The paper introduces a technique for partitioning, numbering, and identifying routine physical movements from videos based on the 3D skeleton data obtained from the VIBE pose estimation model [14]. Having employed both PCA and a movement sequence algorithm, it reaches good levels of action. However, the approach relies on accurate pose estimation, as was mentioned earlier, which might fail in complex environments to provide all the required angles and coordinates, though more extensive testing in various setting is needed to improve external validity of the framework for fitness tracking and rehabilitation.

This work proposes a new method of using horizontal stacking of ridge regression, LightGBM, and gradient boosting machine (GBM) models [15]. The used methodology involves feature-domain extraction with fixed-window segmentation as the data preparation process. Nonetheless, the need for bigger quantities of fine training data, and non-interpretability of the stacked model can degrade the model's usefulness in constrained or real-time settings.

This research uses YOLOv3 for detecting precisely humans in video frames and drawing boundary boxes, but for the feature extraction Local Binary Pattern and Gray Level Co-occurrence Matrix (GLCM) are used [16]. Therefore, a redesigned LSTM network is used for the classification of the activities, further enhanced by a newly introduced HSM-CSO algorithm. The system identifies many missed skiing switch actions of lawbreakers and has high accurate rate of recognizing abnormal activities from the ATM surveillance videos, but these videos collected from YouTube may not illustrate real ATM scenarios and the complete process may have very high computational cost and thereby might not be extremely useful for real-time detection.

This makes a complete surveillance system to advance the human tracking and anomalous behavior detection based on deep learning [17] such that it mitigates false positivity and enhances the possibility of tracking by introducing new methods of object filtering. For object detection, the system uses YOLOv3, and the scene and behavior recognition system is based on a prompt-based recognition that leverages CLIP. However, it fails to work well in high density crowd situations and the control algorithm does not have the learning feature for the past detections and needs to be trained manually to perform well in future detections.

This work proposes a new HAR approach for using the mobile and the wearable sensors based on deep learning-based models including DCNNs and LSTMs [18]. For feature extraction, fuzzy logic-based genetic algorithm (FGA) is implemented; meanwhile for feature selection, particle swarm optimization (PSO) with the ant colony optimization (ACO) operators is employed. This type of model affords a total top recognition rate in other open datasets namely UCI-HAR, PAMAP2, and WISDM. It requires a vast amount of computing power and time for model training and testing, which might render the model unsuitable for real-time applications, in particular, noisy or incomplete data.

The framework put forward here discusses a CNN-based model used on the Edge AI devices for real-time activity recognition [19]. The system comprises three components: One involves capturing the data through surveillance cameras, and extracting the features with the aid of a mini-Convolutional Neural Networks model; the other is a classification of activity through the machine learning algorithms assessed using a created dataset of 10,000 frames belonging to six activities' classes. Due to low weight, the implemented approach improves computational speed; however, the usage of a custom dataset reduces the real-world application capability of the model, and the study does not discuss the computational complexity in detail.

3 Proposed System

The proposed methodologies are hybrid deep learning frameworks combining 3D Convolutional Neural Networks (3D CNNs), Autoencoders, and Inflated 3D CNNs (I3D CNN) for robust and efficient suspicious human activity recognition (SHAR) as shown is Fig. 1. The methodology initiates with the data preparation, in which video data is collected from various sources, namely HMDB51, UCF Crime dataset, and Smart-City CCTV Violence Detection Dataset (SCVD). The data is preprocessed. It divides the data into an 80–20% ratio for training and testing while normalizing, resizing, and encoding the video frames in order to ensure consistency in the features. The Hybrid model takes the feature-extracting capabilities of the Autoencoder to identify the anomalies, and the other 3D CNN focuses mainly on the spatial and temporal activity recognition. Real-time deployment is made possible through event detection thresholds that allow for an SMS alert system powered by the Twilio API, which ensures prompt response upon the identification of suspicious activity. This comprehensive approach ensures accuracy, robustness, and real-time capability, addressing critical gaps in existing surveillance methodologies. Due to the event-detection thresholds, it makes a real-time deployment of the system possible as it would actuate the SMS alert system powered by the Twilio API in real time that would respond right away as soon as the suspicious activities are detected. This methodology is therefore completely accurate as well as robust, giving it a capability for true real-time functionality to correct the critical gaps within surveillance methodologies.

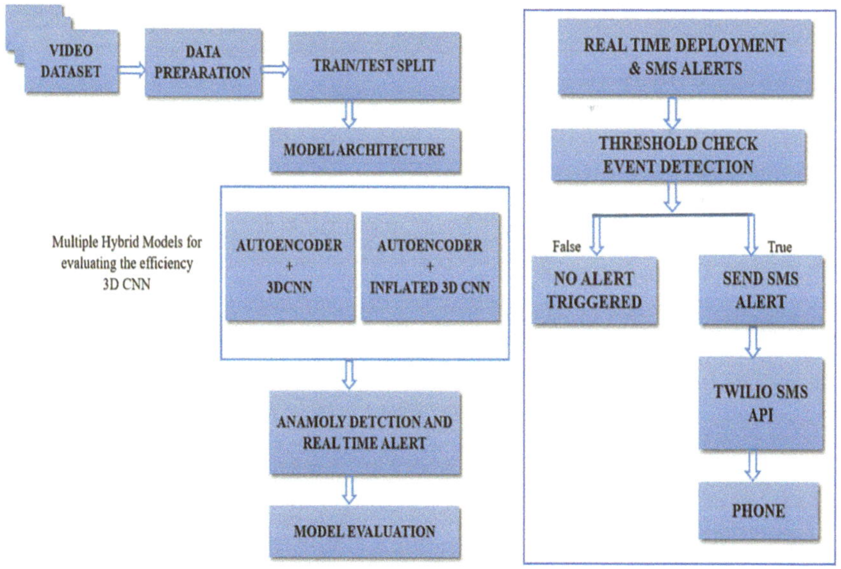

Fig. 1 General work flow of the model

3.1 Dataset

The dataset selected for this research project was taken from various sources, namely HMDB51, the Smart-City CCTV Violence Detection (SCVD) dataset, and the UCF Crime dataset. These provide a rich and diverse foundation for developing a robust suspicious human activity recognition system. The HMDB51 dataset consists of 7000 video clips across 51 action categories. The UCF Crime Dataset consists of videos of 14 action classes of suspicious criminal activities, and the SCVD Dataset consists of approximately 500 video clips. A total of 650 videos were selected from these datasets ensuring balanced representation of the real-world scenarios, including low light conditions. The collected dataset was categorized into three categories: Normal, Violence, and Weaponized. The "normal" class consists of 174 training and 43 testing videos, the "violence" class consists of 173 training and 43 testing videos, and the "weaponized" class consists of 174 training and 43 testing videos. The dataset was split into 80–20% Train-Test ratio, with 521 videos for training and 129 for testing. Preprocessing steps such as resizing the videos to standardize dimensions 64×64 or 96×96 pixel based on certain analysis using OpenCV's `resize` function and frame extraction of 16 frames per video were applied to optimize the videos for the hybrid deep learning model which is depicted in Figs. 2, 3, and 4.

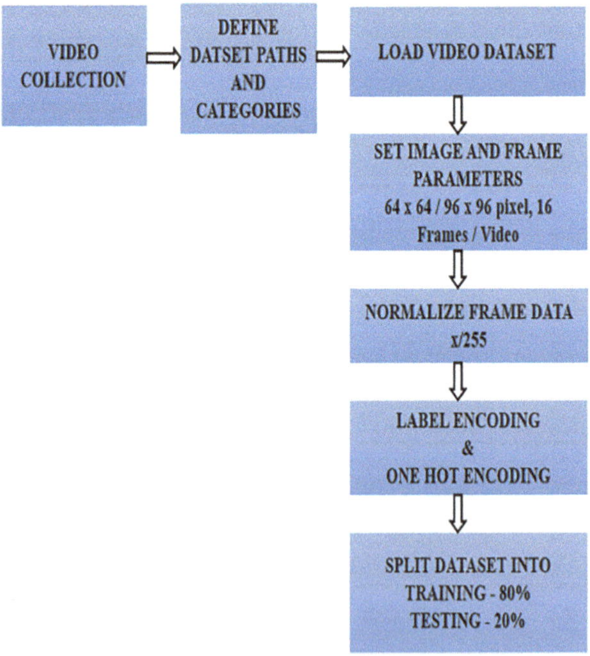

Fig. 2 Data preprocessing

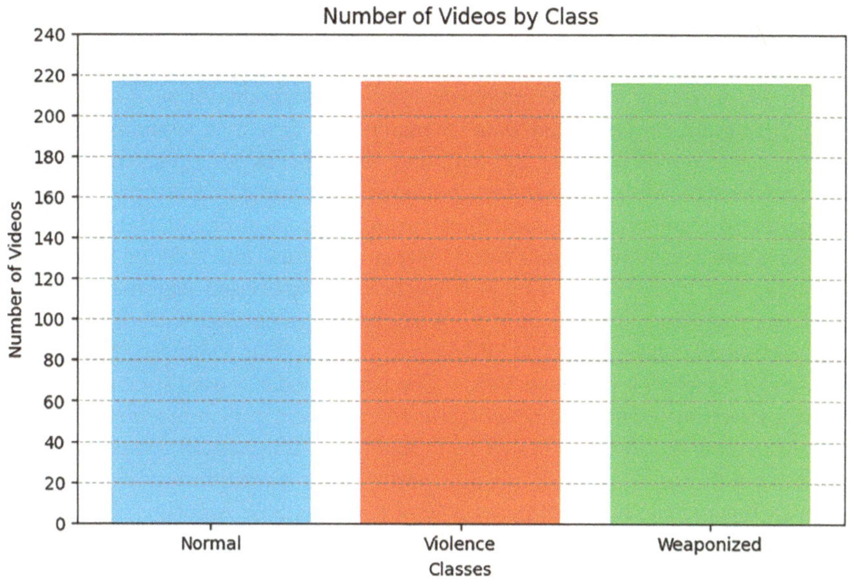

Fig. 3 Dataset distribution

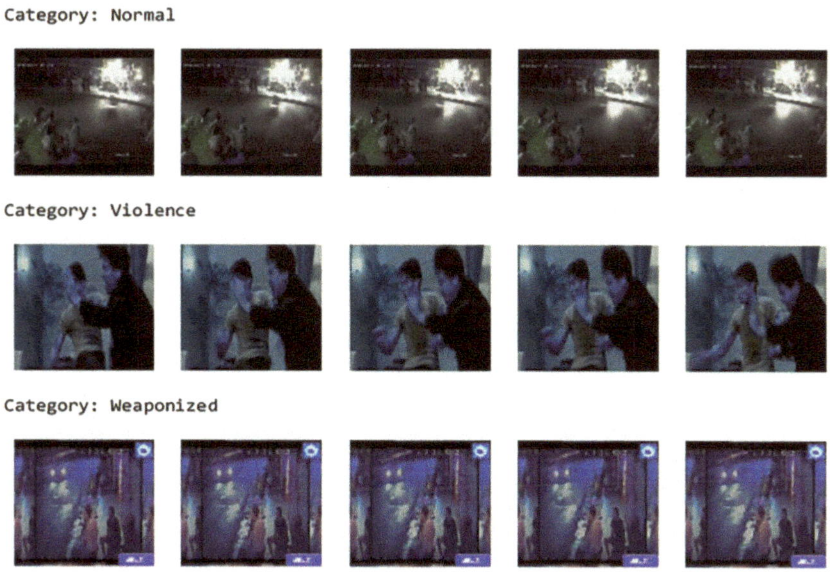

Fig. 4 Input videos converted into frames

4 Methodology

This work involves the development and deployment of a hybrid deep learning framework for video data anomaly detection and supports real-time alerting capabilities. It starts by initiating the process with the data preprocessing, where the video dataset gets divided into individual frames; preprocessing techniques like resizing and normalization are done to ensure that models train uniformly and are more robust. Video dataset is divided into training and testing subsets so that it can be utilized with the help of supervised learning. It has an architecture of two hybrid models: (i) Autoencoder integrated with 3D Convolutional Neural Network (3D CNN) [20] and (ii) Autoencoder combined with Inflated 3D Convolutional Neural Network [21]. The Autoencoder is used for unsupervised feature extraction learning compact latent representations of input frames, which is further forwarded to the 3D CNN and I3D models for extracting spatiotemporal features. Following is the table which consists of the algorithm used which determines each step as to how the model is working. The table Algorithm 1 structures the systematic framework implemented in this research by describing the specific steps undertaken in this research process. The approach initiates with data selection through various sources based on certain analysis, then the data preprocessing is performed to structure the data for further analysis and is divided into train-test split. Then the models are trained; there are two hybrid models introduced. Then the model is evaluated using accuracy, precision, recall, F1-score. Further, the trained model is saved and a new unseen video is provided to the model to detect if any suspicious activity and a threshold of 85% is considered; if the probability of the activity detected is more than the threshold, an SMS alert is sent to the phone number provided.

Algorithm 1: Proposed algorithm

Data Selection: Video data collected from various sources
Data Preprocessing: Resize: 64 × 64 or 96 × 96 pixels, 16 frames per video, Normalization: Pixel values of frames are normalized to [0, 1]. Formula: $X_{norm} = X/255$. Label Encoding: Labels are one-hot encoded
Train/Test split: Training Set = 80% of the dataset, Testing Set = 20% of Dataset
Models Implementation
1. Autoencoder:
Encoder
Compress video frames into latent features using 3D convolutions and max pooling
Convolution operation: $Zi + 1 = \sigma\ (W * X_1 + b_i)$, $Wi \in R^{k1\times k2\times k3}$, $b_i \in R$
Max pooling: P = max (Z, pool_size)
Decoder
Reconstruct the video from latent features using upsampling and 3D deconvolutions: $\widehat{X}$ = UpSampling3D(P)
Loss function for reconstruction: $L_{reconstruction} = (\frac{1}{N})\sum_{i=1}^{N}(X_i - \widehat{X}_i)^2$
2. Inflated 3D Convolutional Layers
Inflated Convolutions
Extend 2D kernels to 3D by inflating the temporal dimension
W3D = tile (W2D, [temporal_depth, 1, 1, 1])
Global Average Pooling
Extract global spatiotemporal features: $GAP = (\frac{1}{N})\sum_{i=1}^{N} Z_i$
Classification Layer
Map features to class probabilities using SoftMax activation:
$\widehat{y_k} = \frac{e^{zk}}{\sum_{j=1}^{K} e^{zj}}, k \in [1, K]$
Loss function:
$\sum_{i=1}^{N}\sum_{k=1}^{K} y_{i,k}\log(\widehat{y}_{i,k})$
3. 3D CNNs
- Extracts spatiotemporal features from the compressed output of the encoder
- Layers consist of 3D convolution, max pooling, and global average pooling

Formula: $y = f_{3DConv(z)}$ = SoftMax (Dense (GlobalAvgPool3D(z)))
- Loss for classification:

Formula: $L_{classification} = -\frac{1}{N}\sum_{i=1}^{N}\sum_{j=1}^{C} y_{i,j}\log(\widehat{y}_{i,j})$
4. Hybrid Model
Autoencoder + Inflated 3D CNNs
Autoencoder + 3D CNNs
Use the encoder's latent features as input to the I3D model: $Z_{encoded} = f_{encoder}(X)$
Pass latent features to inflated I3D for classification: $\widehat{y} = f_{I3D}(Z_{encoded})$
5. Compile the Hybrid Model
Optimized using Stochastic Gradient Descent (SGD) with a learning rate of 0.01
6. Evaluation
$\text{Accuracy} = \frac{\text{True Positive} + \text{True Negative}}{\text{Total Samples}}$
$\text{Precision} = \frac{\text{True Positive}}{\text{True Positive} + \text{False Positive}}$
$\text{Recall} = \frac{\text{True Positive}}{\text{True Positive} + \text{False Negative}}$
$\text{F1} - \text{Score} = 2 \times \frac{\text{Precision} \times \text{Recall}}{\text{Precision} + \text{Recall}}$
SMS Alert using Twilio API: If the probability of the suspicious activity recognized by the model is more than 85% then SMS alert is sent to the provided phone number

A. Hybrid Model Combining Autoencoder and 3D CNNs

The proposed hybrid model combines Autoencoder and 3D CNNs to solve suspicious human activity recognition in videos. The autoencoder has an encoder and a decoder block. The encoder extracts features by applying the 3D convolutional layers followed by max pooling, reducing the spatial and temporal dimensions of the input videos. The decoder reconstructs the original input video, using upsampling and transposed convolutions. The last layer of the decoder uses a sigmoid activation function to ensure that the reconstructed output is identical to the original input. The autoencoder is trained using the Adam optimizer and MSE loss for the unsupervised learning. The 3D CNN component processes spatiotemporal data by using 3D Convolutions, capturing both spatial and temporal features across video frames. Max pooling layers are used to decrease the spatial dimensions and employ global average pooling to collapse the learned features into a vector, followed by softmax layer for classification into predefined activity classes such as normal, violence, and weaponized. The hybrid model makes the best use of unsupervised feature extraction from Autoencoder and the ability to perform temporal feature extraction in a 3D CNN model. This gives rise to a robust framework to classify activities in the videos. For classification, the model makes use of Stochastic Gradient optimizer consisting of learning rate of 0.01 with categorical cross-entropy as the loss function optimized on accuracy.

B. **Hybrid Model Combining Autoencoder and Inflated 3D CNN**

This method integrates an autoencoder and an inflated 3D CNNs to overcome the challenges of suspicious human activity recognition from video data. This hybrid model uses the strength of unsupervised learning for feature extraction and spatiotemporal processing to classify various activities. The model utilizes an Autoencoder to perform unsupervised learning, extracting meaningful features from input video data. The encoder part of the autoencoder compresses the video data into a lower dimensional representation, which captures the essential spatial and temporal patterns. To enhance the model's generalization, dropout layers are incorporated to prevent overfitting. The decoder reconstructs the input from the encoded features. The extracted features are passed to the Inflated 3D CNNs for further spatiotemporal learning. The I3D CNN model employs a series of 3D convolutional layers that are inflated from traditional 2D convolutional layers which allows the model to capture temporal dynamics across frames. These layers are vital for understanding the motion and patterns in the temporal dimension which makes them well suited for activity recognition in video data. The I3D CNN model then applies pooling and dropout techniques to further refine the learned features before a final classification layer which predicts the activity class. The hybrid model is trained using a categorical cross-entropy loss function with Stochastic Gradient Descent optimizer which provides a robust optimization strategy to handle the multi-class classification problems. The learning rate is set to 0.001 with momentum of 0.7 to accelerate convergence while maintaining stability. This method is highly effective for activity recognition tasks as it benefits from both unsupervised feature learning and powerful spatiotemporal analysis (Table 1).

Table 1 Model parameters

Model	Layers	Epoch	Activation function	Loss Function	Optimizer
Autoencoder and 3D CNN	Conv3D, MaxPooling3D, UpSampling, GlobalAveragePooling, Dense	240	Softmax	Categorical cross entropy	Stochastic gradient descent
Autoencoder and I3D CNN	Layer_2d, Conv3D, MaxPooling3D, UpSampling, GlobalAveragePooling, Dense	250	Softmax	Categorical cross entropy	Stochastic gradient descent

5 Results and Discussion

In this stage, we look into the experimental results from the application of hybrid models aimed at recognizing suspicious human activity in video data. The models are effectively evaluated to understand its overall performance. A complete evaluation of each model is conducted using various evaluation metrics consisting of F1-score, recall, precision, accuracy, and confusion matrix. Table 2 displays the accuracy obtained by both the models.

A. **Hybrid Model Combining Autoencoder and 3D CNN**

The results acquired from the hybrid model that combines Autoencoder and 3D CNN architectures highlight the model's significant capability for Suspicious Human Activity Recognition as shown in Fig. 5a, b, c. The model obtained a training accuracy of 99.92% and testing accuracy of 96.90%. The training and validation accuracy graphs show a clear upward trend in training accuracy, which progressively increases to nearly 99.92% over 240 epochs. This concludes that the proposed hybrid model has effectively learned from the dataset provided. However, the validation accuracy demonstrates notable fluctuations, representing variability in model performance on unseen data across iterations. These undulations describe the differences in the validation set's complexity or instances of minor overfitting during certain training phases. Despite these variations the overall performance between the training and validation curves confirms that the model summarizes well. This showcases its robustness in handling spatiotemporal activity data. Further into the classification report, performance metrics for this model, across three categories for activities, are explained as "Normal, Violence, and Weaponized." In the class "Normal," precision is calculated

Table 2 Comparative results of the hybrid models

Model	Accuracy (%)
Autoencoder and 3D CNN	96.90
Autoencoder and I3D CNN	97.67

to be 0.93, recall being 0.98, and the F1-score comes out to be 0.95; this gives an impression of effective activity detection with minimal false positives in the model. As per results, the "Violence" class achieves a precision of 0.98 along with a recall of 0.95 and finally an F1-score of 0.96 to indicate that the model effectively picks up aggressive actions despite only a few instances appearing as "Normal." Critical for security applications, it is seen that the class "Weaponized" goes perfectly precise at 1.00, with its recall at 0.98 and an F1-score of 0.99. This excellent performance indicates that the model is highly sensitive and reliable in detecting armed or weapon-related suspicious activities, thus minimizing false negatives for this crucial class. Overall, the hybrid model of Autoencoder and 3D CNN proves to be a robust architecture for this task. The minor fluctuations in validation accuracy could be overcome with techniques like early stopping, dropout, or hyperparameter optimization to improve the model's stability. These results represent the model's high relevance and applicability in real-time surveillance systems which offer both precision and reliability for enhancing public safety and in identifying potential threats effectively.

B. **Hybrid Model Combining Autoencoder and Inflated 3D CNN**

The results acquired from the hybrid model that combines Autoencoder and inflated 3D CNN architectures highlight the model's significant capability for Suspicious Human Activity Recognition as shown in Fig. 6a, b, c. The model obtained a training

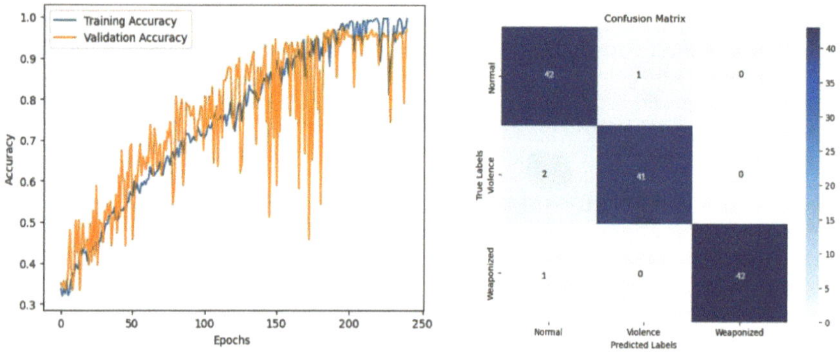

(a) (b)

```
5/5 ━━━━━━━━━━━━━━━━━━━━ 1s 115ms/step
Classification Report:
              precision    recall  f1-score   support

      Normal       0.93      0.98      0.95        43
    Violence       0.98      0.95      0.96        43
  Weaponized       1.00      0.98      0.99        43

    accuracy                           0.97       129
   macro avg       0.97      0.97      0.97       129
weighted avg       0.97      0.97      0.97       129

Confusion Matrix:
[[42  1  0]
 [ 2 41  0]
 [ 1  0 42]]
```

(c)

Fig. 5 **a** Model accuracy **b** Confusion matrix for the model **c** Classification report

accuracy of 100% and testing accuracy of 97.67%. The training and validation accuracy curves indicate a steady improvement, with convergence around epoch 150, reaching close to 1.0 accuracy by 250 epochs for both metrics. This represents a well-regularized model with less overfitting. The classification report represents high precision, recall, and F1-scores for all three classes. The classification report is represented in Fig. 6. Further into the classification report, performance metrics for this model, across three categories for activities, are explained as "Normal, Violence, and Weaponized." In the class "Normal," precision is calculated to be 0.95, recall being 0.98, and the F1-score comes out to be 0.97; this gives an impression of effective activity detection with minimal false positives in the model. As per results, the "Violence" class achieves a precision of 0.98 along with a recall of 0.95 and finally an F1-score of 0.96 to indicate that the model effectively picks up aggressive actions despite only a few instances appearing as "Normal." Critical for security applications, it is seen that the class "Weaponized" goes perfectly precise at 1.00, with its recall at 1.0 and an F1-score of 1.0. To conclude, the hybrid model excels in achieving high accuracy and reliability across the performance metrics. This represents its capability in real-time implementation of this model in suspicious human activity detection.

5.1 Comparative Analysis of Both the Hybrid Models

The Autoencoder + 3D CNN (A) and Autoencoder + Inflated 3D CNN (B), both the hybrid models demonstrate exceptional potential in suspicious human activity recognition but they also exhibit notable differences in their performances and features which are discussed here. Model A represents fluctuations in validation accuracy representing overfitting and variability in performance on unseen test dataset. There is slight misclassification in the normal and violence categories. Its susceptibility to fluctuations needs good regularization and tuning.

In contrast Model B consistently converges which indicates a regularized model. The model enhances the learning process by inflating 2D kernels into 3D which allows it for more extensive spatial and temporal feature extraction. This results in faster convergence, improved generalization, and robustness. Although the hybrid model of autoencoder and 3D CNN model performs exceptionally well with balance across metrics, the autoencoder and inflated 3D CNN model exhibits slight meaningful improvements in accuracy, robustness, and convergence making it more suitable for critical applications in human activity recognition in videos in different light intensity conditions and real-time scenarios.

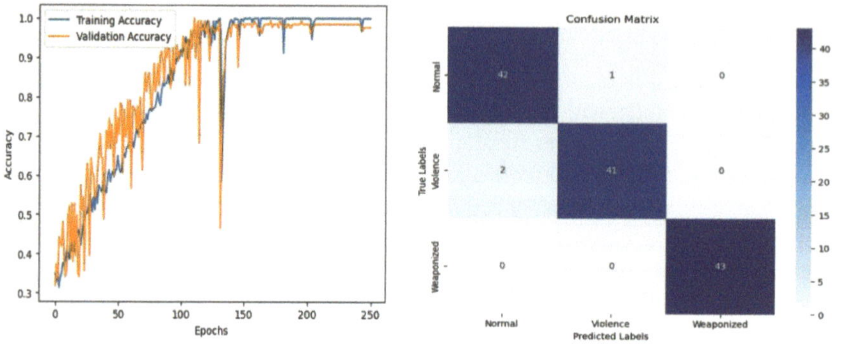

(a) (b)

```
5/5 ━━━━━━━━━━━━━━━━━━━━ 1s 115ms/step
Classification Report:
              precision    recall  f1-score   support

      Normal       0.95      0.98      0.97        43
    Violence       0.98      0.95      0.96        43
  Weaponized       1.00      1.00      1.00        43

    accuracy                           0.98       129
   macro avg       0.98      0.98      0.98       129
weighted avg       0.98      0.98      0.98       129

Confusion Matrix:
[[42  1  0]
 [ 2 41  0]
 [ 0  0 43]]
```

(c)

Fig. 6 Results of hybrid model autoencoder and inflated 3D CNN. **a** Model accuracy **b** Confusion matrix for the model **c** Classification report

5.2 Model Prediction

When an unseen video from YouTube is provided to the trained model both the models were successful in predicting the suspicious activity and send SMS Alerts to the provided phone number. The model was able to predict the violence activity with the probability of 99.61%. This result is obtained by examining the frames of the YouTube video and categorizing of the data by understanding the patterns from the training data. The automated analysis and classification of the data can be attained using these methods of deep learning, making it highly efficient in applications like action recognition and video surveillance (Fig. 7).

Table 3 represents a comparison of the proposed hybrid models with some of the methods used in the literature for recognition of suspicious human activity. It can be noticed from the compared methods that CNN-LSTM model has attained 76.50% accuracy. The Conv3D architecture improves drastically to an accuracy of 88.23%, whereas Time-Distributed CNN continues to improve and delivers an accuracy of

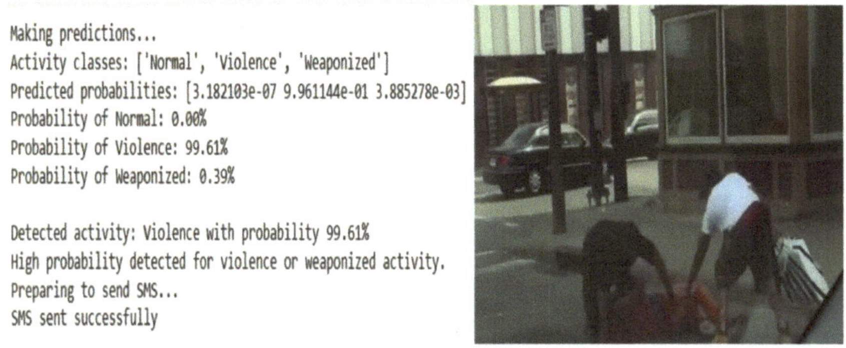

Fig. 7 Unseen YouTube video prediction

Table 3 Comparative results of the proposed model and existing models

Existing [1]	Model	Accuracy (%)
	CNN-LSTM	76.50
	CONV3D	88.23
	Time-distributed CNN	90.14
Proposed approach	Autoencoder + 3D CNN	96.90
	Autoencoder + I 3D CNN	97.67

90.14% because of extraction of local features over the temporal space, though both approaches improve and outperform the best state-of-the-art model existing. An accuracy as high as 96.90% is achieved in the proposed approach when dimensionality reduction is integrated along with feature extraction in an Autoencoder comprising 3D CNN. Notably, by further increasing the performance of Autoencoder up to 97.67% using an inflated 3D CNN, I3D CNN outperforms all the comparisons. The result shows that this hybrid architecture effectively captures intricate spatiotemporal dynamics and can be used as a robust and reliable solution for complex classification tasks.

6 Conclusion

This research project has successfully implemented the hybrid models to enhance surveillance systems for the detection of suspicious human activity. The hybrid model delivers high-speed computation, greater accuracy, and SMS warning issued when suspicious activities are detected, allowing fast responses to public threats but has certain drawbacks in performance in congested public environments; though the model is able to predict in low light conditions, it needs to be trained with a larger set of data to improve the performance. The future developments of this approach are training the model with larger datasets, real-time video feed incorporation, accurate identification in crowded areas, and combination with smart surveillance cameras. The idea has a great promise for enhancing public safety and security in various dimensions.

References

1. Perez M, Kot AC, Rocha A (2019) Detection of real-world fights in surveillance videos. In: ICASSP 2019–2019 IEEE international conference on acoustics, speech and signal processing, pp 2662–2666. IEEE
2. Bouchabou D, Nguyen SM, Lohr C, LeDuc B, Kanellos I (2021) A survey of human activity recognition in smart homes based on IoT sensors algorithms: Taxonomies, challenges, and opportunities with deep learning. Sensors 21(18):6037
3. Sakka S, Liagkou V, Stylios C (2023) Exploiting security issues in Human Activity Recognition Systems (HARSs). Information 14(6):315
4. Liu, Luo W, Lian D, Gao S (2018) Future frame prediction for anomaly detection—a new baseline. In: 2018 IEEE/CVF conference on computer vision and pattern recognition, Salt Lake City, UT, USA, pp 6536–6545. https://doi.org/10.1109/CVPR.2018.00684
5. Zaidi MM, Sampedro GA, Almadhor A, Alsubai S, Al Hejaili A, Gregus M, Abbas S (2024) Suspicious human activity recognition from surveillance videos using deep learning. IEEE Access
6. Wu F, Wang Q, Bian J, Ding N, Lu F, Cheng J, Xiong H (2022) A survey on video action recognition in sports: Datasets, methods and applications. IEEE Trans Multimedia 25:7943–7966
7. Choudhry N, Abawajy J, Huda S, Rao I (2023) A comprehensive survey of machine learning methods for surveillance videos anomaly detection. IEEE Access
8. Huang C, Wu Z, Wen J, Xu Y, Jiang Q, Wang Y (2021) Abnormal event detection using deep contrastive learning for intelligent video surveillance system. IEEE Trans Industr Inf 18(8):5171–5179
9. Azmat U, Alotaibi SS, Abdelhaq M, Alsufyani N, Shorfuzzaman M, Jalal A, Park J (2023) Aerial insights: Deep learning-based human action recognition in Drone imagery. IEEE Access
10. El Zein H, Mourad-Chehade F, Amoud H (2024) CSI-based human activity recognition via lightweight CNN model and data augmentation. IEEE Sensors J
11. Panigrahi GR, Sethy PK, Behera SK, Gupta M, Alenizi FA, Nanthaamornphong A (2024) Enhancing security in real-time video surveillance: A deep learning-based remedial approach for adversarial attack mitigation. IEEE Access
12. Ahmad T, Wu J, Alwageed HS, Khan F, Khan J, Lee Y (2023) Human activity recognition based on deep-temporal learning using convolution neural networks features and bidirectional gated recurrent unit with features selection. IEEE Access 11:33148–33159
13. Alikhanov J, Kim H (2023) Online action detection in surveillance scenarios: A comprehensive review and comparative study of state-of-the-art multi-object tracking methods. IEEE Access
14. Cheng SH, Sarwar MA, Daraghmi YA, İk TU, Li YL (2023) Periodic physical activity information segmentation, counting and recognition from video. IEEE Access 11:23019–23031
15. Dahal A, Moulik S (2024) The multi-model stacking and ensemble framework for human activity recognition. IEEE Sensors Lett
16. Kshirsagar AP, Azath H (2023) YOLOv3-based human detection and heuristically modified-LSTM for abnormal human activities detection in ATM machine. J Vis Commun Image Represent 95:103901
17. Jeon H, Kim H, Kim D, Kim J (2024). PASS-CCTV: Proactive Anomaly surveillance system for CCTV footage analysis in adverse environmental conditions. Expert Syst Appl:124391

18. Jameer S, Syed H (2023) A DCNN-LSTM based human activity recognition by mobile and wearable sensor networks. Alex Eng J 80:542–552
19. Aishwarya D, Minu RI (2021) Edge computing based surveillance framework for real time activity recognition. ICT Express 7(2):182–186
20. Kuppusamy P, Bharathi VC (2022) Human abnormal behavior detection using CNNs in crowded and uncrowded surveillance–a survey. Measure: Sensors 24:100510
21. Shin J, Miah ASM, Kaneko Y, Hassan N, Lee HS, Jang SW (2024) Multimodal attention-enhanced feature fusion-based weakly supervised anomaly violence detection. IEEE Open J Comput Soc

Facial Expression Analysis with Convolutional Networks

G. Nitheesh Kumar, P. Deepika, and K. S. Shreedhara

Abstract The facial expression analysis system presented in this work provides a flexible facial detection model based on mapping behavioral characteristics with physical features. The physical features of the human face related to various expressions such as happiness, sadness, fear, anger, surprise, and disgust are associated with geometric structures that have been restored as the basis matching template for the identity system. Facial expression analysis is in pattern these days in view of its wide application zone. Expression recognition is popular due to its broad range of applications. In business, emotion detection has becoming more crucial due to advancements in the science of learning. Perceiving an individual's sentiments is done through feeling acknowledgment. Various techniques have been developed to identify emotional states in images.

Keywords Facial expression · Convolutional Neural Network (CNN) · Support Vector Machine (SVM) · Local Binary Patterns (LBP) · K-Nearest Neighbor (KNN) · Emotion detection

1 Introduction

The passionate state is expressed through outward appearances. The objective is to project a distinct picture of a person's excitement into the outside world. This viewpoint is intimately related to a person's internal condition since expressions offer a distinct viewpoint. In order to support or react to an individual's external look, observers might evaluate the person's spiritual condition [2]. In psychological healthcare offices, medical staff members search for this to help doctors create therapy applications which are readily available to patients. Therefore, it is essential to comprehending feelings and the causes behind them, as well as the range of emotions that are present in decision-making, healthcare settings, sentiment, and

G. N. Kumar (✉) · P. Deepika · K. S. Shreedhara
Department of CSE, UBDT College of Engineering, VTU, Davanagere, India
e-mail: nitheesh.kumar9408@gmail.com

H. Nagaraj et al. (eds.), *Advancements in Artificial Intelligence and Machine Learning*, Lecture Notes in Electrical Engineering 1572,
https://doi.org/10.1007/978-981-95-7449-0_29

educational institutions. One of the trickiest parts of computer vision difficulties is separating exterior appearances, which is why facial pattern recognition algorithms are used. Since it is hard to design a precise model, the representation should identify the characteristics such as happiness, anger, sadness, calmness, surprise, and so forth that set apart a particular feeling from resentment. Image processing is a type of signal processing in that the input and output signals are images. The most essential uses of image processing is facial expression recognition. Our emotions are revealed through our facial expressions [1]. Facial expressions are vital in interpersonal communication. Facial expression is a nonverbal scientific gesture that we express with our faces based on our emotions. Automatic face expression detection is vital in artificial intelligence and robotics, making it a generational need [3]. Personal identity and access control, videophone and teleconferencing, forensics, human-computer interaction, automated surveillance cosmetology, and other applications fall under this category.

2 Related Work

There is no set framework for recognizing the emotional states of people. The estimate is predicated on the different boundaries of an action part of the human tendency to judge emotions in others. A few methods for automated recognition of external appearances are shown in this article. Angle essential dynamic presence model, is used by remainder of system to identify the face focuses on the basic edge that certain modern frameworks physically name. Precise feature acquisition might not be feasible in many functional scenarios. The articulation check is carried out by articulate support vector machine (SVM) classifiers that support [7] local binary pattern (LBP) highlights [4]. It is absurd in most functional settings that the writers made it universally beneficial by physically noting the eye locations. The computer expression recognition toolbox (CERT) proposed by Larry, to detect face highlights, CERT forbids the insertion of unique letters into facial images. Instead, it employs SVM and multidimensional strategic relapse to identify external appearances. These methods specifically call for programming and computational work, for eliminating highlights from photos; the deep convolution neural network (DCNN) framework is frequently utilized [15]. DCNN uses multiple layers to achieve accurate element learning. The highlights shown here serve as channels, and these channels, as stated by bright and dim across the info picture, provide the highlights used by different layers of the organization. Techniques like Kazoo's models, which are based on convolution neural networks, have been offered as identifying external appearances. Still, they mostly use other facial datasets for model training. For instance, Google extracted facial highlights using Deep Conventional Activation highlight for generalized visual identification that is not necessary extensive preparation, even though this model is typically used to prepare extremely tiny picture datasets.

3 Methodology

Applications of artificial intelligence (AI), like machine learning (ML), use algorithms to access or evaluate information, make judgments from it, and then use those conclusions to guide actions. The term machine learning describes a machine's capacity to forecast future events without explicit programming by using past behavior or data. Deep learning (DL) uses structural layers to build artificial neural networks (ANN) that are able to learn on their own and form logical conclusions. An ANN has many neurons in each layer. Use deep learning to analyze large amounts of data to produce accurate findings with high output performance. Deep learning is used in picture description applications. The method of describing the information contained in a picture is known as image description.

3.1 Convolutional Neural Networks

Convolutional neural networks (CNN) categorize images by analyzing them for purposes such as object detection, machine vision, identification, and categorization [10]. CNN uses an algorithm to classify images then organizes them into top-to-bottom subcategories to extract important information.

Convolutional neural networks (CNN) are capable of processing input forms that roughly approximate two-dimensional (2D) matrices. CNN provides resources for handling images, and a 2D matrix is a straightforward way to represent images [11]. CNN is mainly employed for image classification, i.e., identifying if a picture shows superman, a jet, or a bird [5]. Figure 1 shows the working of a convolution neural network.

Convolution Layers: Convolution layers are the initial layers that are utilized to derive characteristics of images. Convolution layers pick up feature from a small input

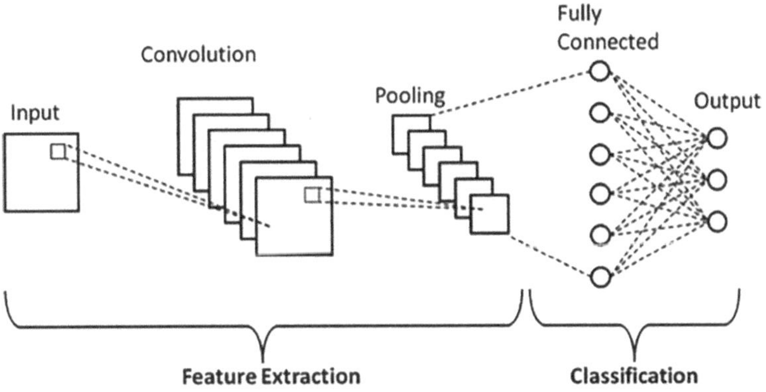

Fig. 1 Working of convolution neural networks

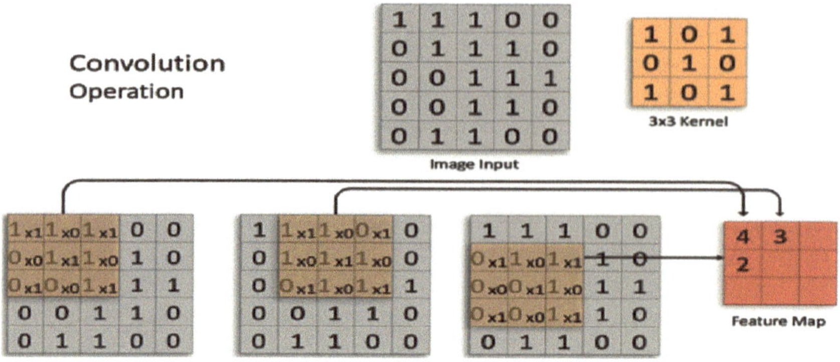

Fig. 2 Convolution operation

data series and maintain the link between pixels [12]. In mathematics, a convolution layer is made up of three parts: two inputs, a photo matrix, and a kernel or filter [8]. The steps listed below are the steps in the matrix multiplication process. The output matrix of the [13] Convolution Layers for a 5×5 image multiplied by a 3×3 filter is displayed below the feature that results referred to as convolved feature or feature map. Figure 2 explains about the Convolution operation.

Strides: The transfer of information from a single neuron to another is called a stride. During the array creation process, images shift from the matrix that is output to the input matrix, and the filter itself shifts one image at a time when the total amount of strides is one. When the total amount of strides hits 2, the filters are also changed to 2 pixels, and so forth. Convolved feature is explained in Fig. 3.

Padding: Convolutional neural network construction requires padding. After the procedure of convolution, the image is shrunk in size. Following each convolution layer stage in an image categorization challenge, the initial image is reduced. Overlapping occurs when the kernel travels through the middle layer more times than the outer layers as it proceeds over the original image. The concept of padding was developed as a solution to the overlapping issue. An additional layer known as padding can expand an image's boundaries without reducing or preserving the dimensions of the original image. Padding is represented in Fig. 4.

Maximum Pooling: Maximum Pooling, commonly referred to as max pooling, is a rule which prioritizes image elements by using a certain location [9]. Max pooling is a sample-based technique that converts continuous functions to discrete ones [6]. Its main goal is take an input and make assumptions about rejected sub-region data. Max pooling is explained in Fig. 5.

Average Pooling: Max pooling and average pooling are not the same. By dividing the input matrix into rectangular sections and calculating the average figures for each, average pooling reduces the matrix. This preserves data regarding less significant elements. Average pooling is represented in Fig. 6.

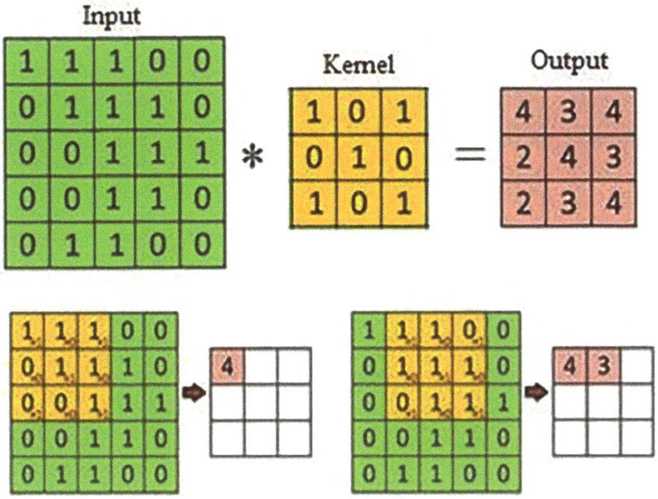

Fig. 3 Convolved feature

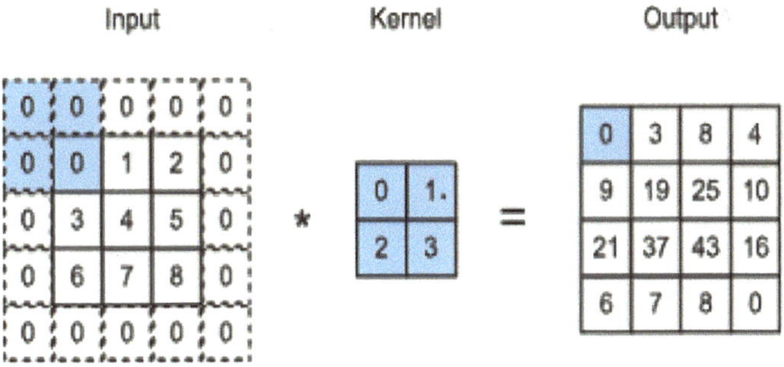

Fig. 4 Padding

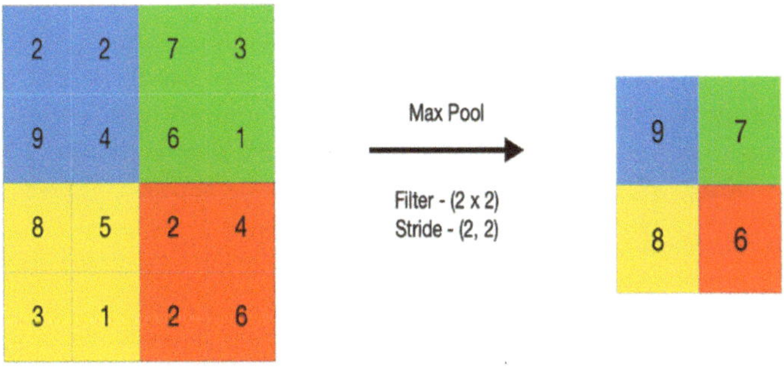

Fig. 5 Max pooling

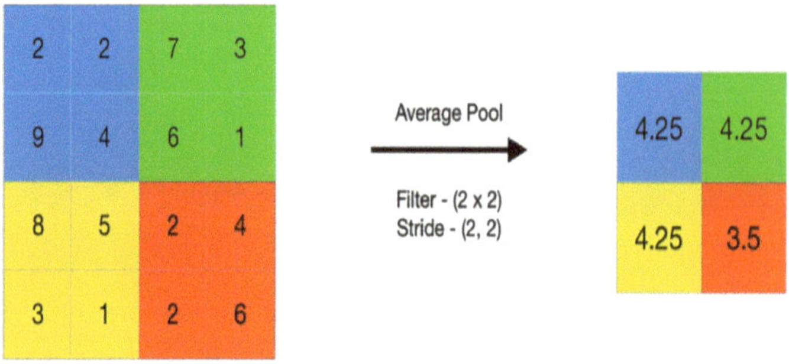

Fig. 6 Average pooling

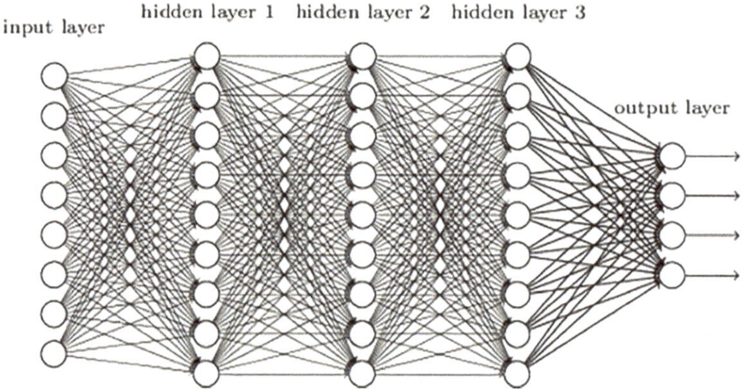

Fig. 7 Structure of neural network

Neural Network: The Neural Network's topology is depicted in Fig. 7. It consists of interconnected neurons and each neuron has three characteristics: (i) Weight, (ii) Bias, and (iii) Activation Function.

After being fed into the input layer, the input is transformed linearly by the neurons using the weights and biases in equation: **X = (weight x input) + bias**.

Applying an Activation Function to the aforementioned result, which is represented in equation, precedes the linear transformation: **Y = Activation (∑ (weight x input) + bias)**.

Activation Function

Rectified Linear Unit (ReLU): One common irregular activation function used in deep learning is ReLU. Since ReLU is unable to activate all neurons simultaneously, neurons will only become silent if the outcome of the linear transformation is less than zero. ReLU has this one major benefit above other activation functions (Fig. 8).

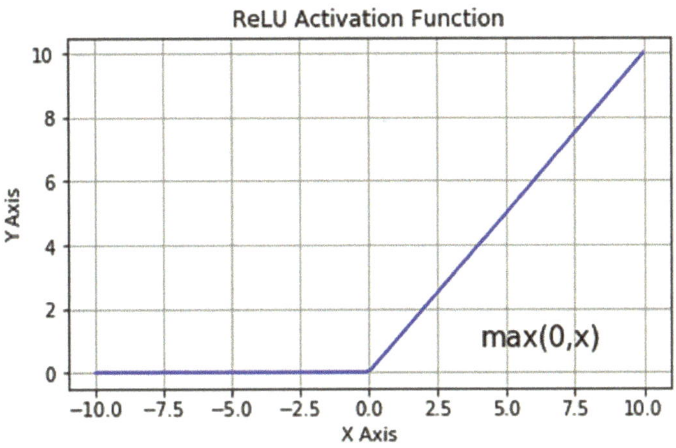

Fig. 8 Graphical representation of ReLU function

Fig. 9 FER-2013 dataset from Kaggle

4 Dataset

The project requires knowledge of Deep Learning, Python, TensorFlow, Keras, NumPy, OpenCV, and Install packages TensorFlow, Keras, Pillow, NumPy, and Tqdm to ensure appropriate operation. This experiment used the FER-2013 dataset, which was obtained via Kaggle [14]. Categorize each expression of a face into the following seven categories: angry, sadness, happiness, surprised, disgusted, fearful or neutral training set has 28,709 examples, whereas the public test set includes 3,589 examples (Fig. 9).

5 Comparative Study

See Table 1.

Table 1 Comparative analysis between the papers

Sl. no	Year	Classifier	Expressions
1	[1] 2023	CNN	Happy, angry, scared
2	[2] 2021	CNN	Happy, sad, surprise
3	[3] 2019	LBP	Happy, angry, sad
4	[4] 2022	SVM, KNN	Happy, scared, neutral
5	[5] 2022	SVM, KNN	Happy, sad, surprise
6	[6] 2021	CNN, KNN	Happy, neutral, sad
7	[7] 2021	LBP, CNN	Happy, sad, angry
8	[8] 2020	CNN	Happy, sad, angry
9	2024 This paper	CNN	Happy, sad, neutral, angry, surprise, fear

6 Results

The FER-2013 database, which contains seven expressions happy, anger, sad, disgust, neutral, fear, and surprise, was used to train our Convolutional Neural Network model. After being reduced to 48x48 pixels and transformed into grayscale, the identified face photos were fed into the CNN model. White lettering indicates the expected expression label, and the expression is shown by the red bar. At the 30 epochs, we obtained a 62% accuracy rate. Our approach predicts happy, angry, neutral, fear, and surprise faces quite well. However, it mixes with sad and disgust expressions; it predicts faces rather poorly.

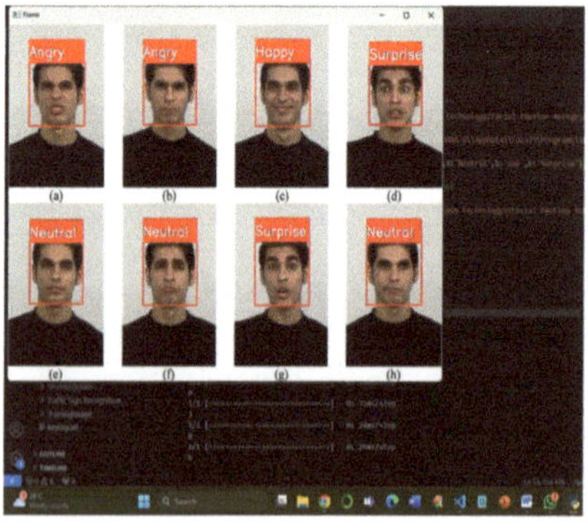

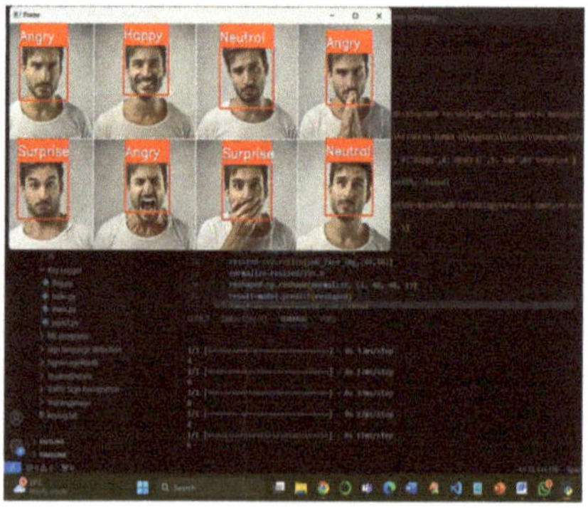

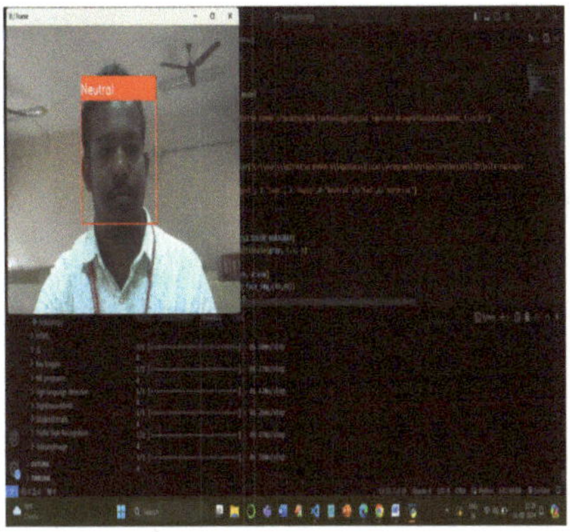

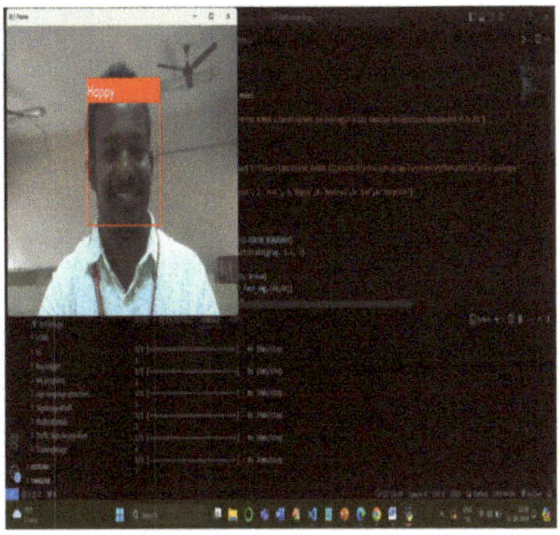

7 Conclusion

This study's facial expression analysis with the convolutional neural network method makes use of a versatile facial recognition model that associates physical characteristics with behavioral aspects. The human face corresponds to many expressions, including anger, sadness, happiness, surprise, disgust, fear, and neutrality. The FER-2013 dataset and Python source code were utilized in this investigation to identify facial expressions. The model is first trained using photos, after which it can identify the expressions in the photographs and does so in real time. This article discusses the facial expression analysis with convolutional neural network system under investigation.

References

1. Suresh V (2023) Facial emotion recognition using CNN. IJCRT 11. ISSN. 2320–2882
2. Khopkar A (2021) Facial expression recognition using CNN with Keras. Research gate 14:47–50
3. Singh S (2019) Face recognition system. IJERT 8. ISSN. 2278–0181
4. Lasri I (2022) Facial emotion recognition of students using convolutional neural network. Research gate
5. Kumar R (2022) Facial emotion detection. IJCRT 10:2320–2882
6. Badrulhisham NAS (2021) Emotion recognition using convolutional neural network. Iop
7. Umarani V (2021) Face emotion recognition. IJCRT 9. ISSN. 2320–2882
8. Mehendale N (2020) Facial emotion recognition using convolutional neural networks. Springer Nature
9. Dwijayanti S (2020) Facial expression recognition and face recognition using a convolutional neural network. ISRITI, 978–1–7281–8406–7
10. Haghpanah MA (2021) Real-time facial expression recognition using facial landmarks and neural networks

11. Begaj S (2021) Emotion recognition based on facial expressions using Convolutional Neural Network (CNN), IEEE Xplore
12. Huang ZY (2023) A study on computer vision for facial emotion recognition. Scientific Rep 41598–023–35446–4
13. Hu P (2024) LCANet: amodelforanalysis of students real-time sentiment by integrating attention mechanism and joint loss function, Springer
14. Zhou N (2021) A lightweight convolutional neural network for real-time facial expression detection, IEEE Access
15. Akhand MAH (2021) Facial emotion recognition using transfer learning in the deep CNN. Electronics 10:1036

Ethical Aspects of Generative Artificial Intelligence—A Comprehensive Scoping Review

Ekta Maini, Baljeet Maini, Dheeraj Marwaha, and Silpa Ajith Kumar

Abstract Generative artificial intelligence (GenAI) is swiftly reshaping various sectors by facilitating the production of diverse forms of content, including textual material, visual art, musical compositions, and programming code. Nevertheless, its extensive integration prompts considerable ethical dilemmas that traverse multiple domains, such as privacy concerns, intellectual property rights, the propagation of misinformation, inherent biases, and broader societal implications. This chapter investigates the ethical issues pertinent to GenAI. The PRISMA extension framework was employed to conduct this systematic review. A cumulative total of 209 scholarly articles from the Scopus, IEEE Xplore, and PubMed databases were selected during the screening phase. Following the application of screening and specific eligibility criteria, 53 studies were identified for in-depth examination. This chapter aspires to systematically delineate and analyze the prevailing literature regarding the ethical ramifications of GenAI, while also identifying principal themes and emergent discussions. Through the methodical review of academic publications, policy documents, and industry reports, this investigation underscores significant concerns such as risks associated with data privacy, the exacerbation of biases, the diminishing of human authorship, and the potential for the dissemination of disinformation. Furthermore, it evaluates regulatory responses, ethical frameworks, and industry standards designed to alleviate these challenges. The review furnishes a cohesive understanding of the prevailing ethical landscape, accentuating the necessity for collaborative and interdisciplinary methodologies to promote responsible innovation. This research acts as an essential resource for policymakers, scholars, and developers seeking to navigate

E. Maini (✉)
Amity University Hyderabad, Hyderabad, India
e-mail: ekta.marwaha@gmail.com

B. Maini
Adesh Medical College and Hospital, Kurukshetra, India

D. Marwaha
Amazon Develeopment Centre India Pvt. Ltd., Hyderabad, India

S. A. Kumar
Dayananda Sagar College of Engineering, Bengaluru, Karnataka, India

H. Nagaraj et al. (eds.), *Advancements in Artificial Intelligence and Machine Learning*,
Lecture Notes in Electrical Engineering 1572,
https://doi.org/10.1007/978-981-95-7449-0_30

the intricate ethical landscape of GenAI while ensuring its coherence with societal values.

Keywords Generative artificial intelligence · Ethics · Large language models · Society

1 Introduction

As its applications expand across industries such as healthcare, education, entertainment, and finance, GenAI promises substantial benefits, including enhanced productivity, personalized user experiences, and creative innovation [1]. However, alongside these transformative opportunities, significant ethical concerns have emerged.

By employing advanced machine learning paradigms, particularly those classified under deep learning, generative artificial intelligence systems possess the capability to produce novel content—including textual, visual, auditory, and audio-visual formats—that closely emulates the nuanced creativity associated with human cognition. Noteworthy instances of such systems encompass OpenAI's GPT models designated for text generation, DALL·E utilized for image generation, and various deepfake technologies adept at crafting hyper-realistic audio-visual outputs. While these technological advancements present substantial prospects across diverse sectors—augmenting creative processes, optimizing operational efficiencies, and facilitating innovative modalities of human–computer interaction—they concurrently engender a plethora of ethical dilemmas that necessitate immediate academic and regulatory review [2]. The extensive integration of generative AI engenders both unparalleled opportunities and significant hazards. On one side, it harbors the potential to transform industries through the automation of creative endeavors, the production of personalized content, and the democratization of access to creative instruments. Conversely, generative AI introduces considerable ethical quandaries that may erode public trust, intensify social disparities, and infringe upon individual rights. Ethical considerations traverse multiple dimensions, encompassing issues such as misinformation, bias, privacy infringements, manipulation, and accountability. In light of these challenges, regulatory entities across the globe are initiating the development of policies intended to oversee the maturation and application of AI technologies. The European Union AI Act represents one of the most thoroughgoing regulatory frameworks, with the objective of categorizing AI systems according to risk levels and enforcing ethical norms to protect human rights. Notwithstanding such initiatives, the regulatory framework remains disjointed, and existing legal structures frequently fall short of addressing the intricate ethical dilemmas introduced by generative AI. For instance, although data protection regulations such as the General Data Protection Regulation (GDPR) tackle privacy issues, they do not adequately encompass the ethical ramifications of AI-generated misinformation or the potential manipulation of human behavior through content produced by AI [3]. This highlights the necessity for contextually relevant ethical directives that can adapt to the specific challenges posed by various applications of generative AI. The ethical issues surrounding generative AI are intricate and multifaceted, necessitating a unified approach from scholars, policymakers, and industry representatives to mitigate emerging threats. By investigating the ethical ramifications and addressing prevailing research deficiencies, we can cultivate a more thorough comprehension of how to reconcile the transformative capabilities of generative AI with the essential need for ethical governance and societal welfare. This chapter aspires to contribute to this ongoing dialog by scrutinizing the ethical obstacles associated with generative AI, pinpointing significant research voids, and suggesting avenues for responsible innovation and regulation. As a scoping review, this study aspires to fill this void and to furnish a comprehensive overview for academics, AI practitioners, policymakers, journalists, and other pertinent stakeholders. While various reviews of AI ethics exist, a systematic review of the ethical implications of generative AI, with a focus on empirical studies and intervention strategies, has not been adequately explored. The purpose of this review is to systematically analyze existing literature on the ethical challenges posed by generative AI and to identify research gaps in addressing these concerns.

2 Methodology

This review was conducted with an aim to explore ethical concerns associated with generative AI. This review was carried out in accordance to Preferred Reporting Items for Systematic Reviews and Meta-Analyses (PRISMA) protocol. Only the peer-reviewed studies, conference papers, and reports on generative AI's ethical concerns were considered eligible for this study. Peer-reviewed studies, conference papers, and reports on generative AI's ethical concerns were excluded. Reputed databases like IEEE Xplore, Scopus, and PubMed were used for this study. Given that the search engines of various databases exhibit distinctive methodologies for configuring options, the search parameters were adjusted to conform to the specifications of each individual database. We searched Phil Papers, a repository for scholarly works in philosophy and related fields such as ethics. In addition, Elicit, the AI-driven paper search engine, was also used. The following search terms were used: "ethical concerns of generative AI," "bias in generative AI," "deepfakes ethics," "AI misinformation," and "accountability in AI." Our investigation was confined exclusively to scholarly articles disseminated from the year 2020 onward, given that generative artificial intelligence had not been extensively utilized in societal contexts prior to that timeframe. Two independent reviewers, EM and SR, screened titles and abstracts for relevance. Full-text articles were assessed for eligibility based on the predefined inclusion and exclusion criteria. Disagreements were resolved through consensus.

2.1 Identification of the Papers

Systematic screening and criteria for inclusion were instituted to assemble a collection of scholarly articles pertinent to this investigation. Initially, a cumulative total of 32 duplicate publications were systematically excluded from further consideration. Subsequently, the authors conducted a thorough examination of the abstracts and titles of the remaining manuscripts to discern peer-reviewed articles specifically pertinent to ethical dilemmas in generative artificial intelligence, thereby excluding editorial pieces, systematic literature reviews, meta-analyses, and commentaries. Papers not written in English ($n = 11$) and published before 2020 ($n = 18$) were excluded. The remaining 148 articles were subjected to a rigorous evaluation, culminating in the exclusion of those that were misaligned with the ethical dimensions of generative AI or deemed irrelevant for the objectives of this analysis. Contexts that extended beyond the purview of ethical considerations ($n = 36$), in addition to conceptual, theoretical, and review articles ($n = 59$), were systematically eliminated from further deliberation. Ultimately, a total of 53 scholarly publications were selected for incorporation into this discourse. Figure 1 delineates the PRISMA flow diagram that corresponds to this research initiative. It was observed that a considerable proportion of the articles were disseminated within the preceding two years, specifically between 2022 and 2024.

2.2 Analysis of the Papers

In order to comprehensively address the research inquiries, a total of 53 scholarly works were incorporated following a meticulous screening process, which included 18 empirical investigations, 28 review articles, and 7 policy documents. The methodologies employed across these studies exhibited considerable diversity, with 10 studies concentrating on issues of bias and fairness, 8 addressing the principles of transparency and accountability, 7 discussing privacy-related concerns, 3 focusing on the phenomenon of hallucinations, 5 exploring the regulatory environment, and 4 examining various miscellaneous topics. The review elucidated several critical ethical dilemmas associated with generative AI the perpetuation of biases, the absence of transparency, privacy concerns, and the imperative for robust regulatory frameworks. The extant literature advocates for a multidisciplinary strategy to tackle these challenges, integrating technical solutions with regulatory and ethical considerations.

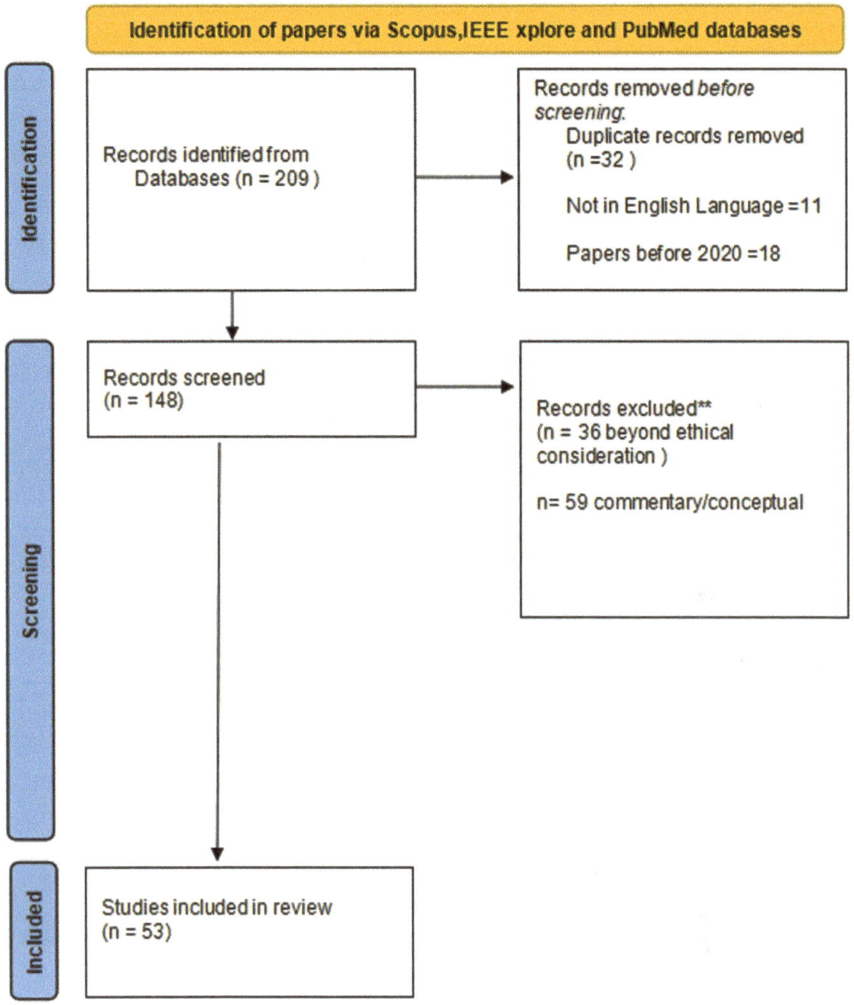

Fig. 1 PRISMA methodology for the review

3 Results and Discussions

Numerous investigations ($n = 10$) concentrated on the biases inherent within generative AI systems, particularly in the context of language models such as GPT. Several studies underscored the potential risks of artificial intelligence perpetuating stereotypes or marginalizing underrepresented voices. Proposed remedies encompassed enhanced diversity within datasets, rigorous algorithmic auditing, and the implementation of real-time bias detection mechanisms. Transparency in artificial intelligence systems emerged as another prominent theme ($n = 8$ studies). These investigations indicate that users ought to be sufficiently informed when engaging with content generated by AI systems. However, the literature revealed a deficiency in clear guidelines pertaining to accountability within AI decision-making frameworks. The review noted that regulatory measures, such as the EU AI Act, are emerging to address these issues, yet significant gaps persist in the enforcement of transparency standards. Privacy infringements, particularly in relation to data utilization within generative AI systems, were examined in 7 studies. Academics stressed the necessity for data anonymization and consent-driven frameworks to alleviate privacy risks, particularly when personal data is utilized in the training of generative models. Notions such as trustworthiness and accountability have diminished in significance, as a reduced number of articles engaged with these themes, while other topics, particularly fairness and safety, have gained prominence. Conversely, certain subjects have exhibited remarkable continuity, exemplified by ongoing discussions related to sustainability and transparency. In summary, our review illuminated ten categories of ethical topics, all of which will be elaborated upon in the subsequent sections, arranged in descending order of significance (refer to Fig. 2).

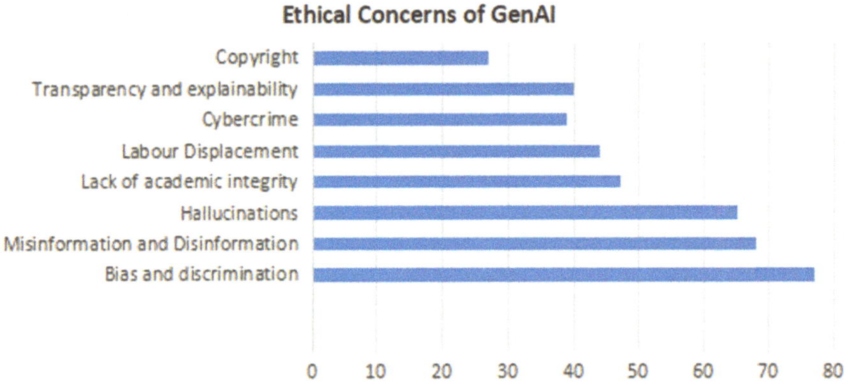

Fig. 2 Major ethical concerns associated with GenAI

3.1 Misinformation and Disinformation Creating Harmful Toxicity

Generative artificial intelligence possesses the capability to produce highly persuasive counterfeit content, including deepfakes, fabricated news articles, or manipulated images, which may propagate false information and distort public perception. This phenomenon can undermine trust in media institutions, destabilize democratic processes, and exacerbate political or social discord [4]. The generation of pornographic, unethical, deceitful, toxic, ferocious, or otherwise detrimental content constitutes a significant area of concern, particularly with respect to large language models (LLMs) and text-to-image generation frameworks. A multitude of empirical studies elucidates the inherent risks linked to the deliberate fabrication of disinformation, emphasizing the substantial threat these activities pose to the integrity of public discourse and the reliability of reputable media sources. Moreover, scholarly articles examine the potential for generative models to facilitate illicit activities or impersonation. Additionally, the academic literature scrutinizes the hazards presented by LLMs when providing guidance in high-stakes contexts, including healthcare, safety-related issues, and legal or financial matters.

3.2 Bias and Discrimination

Bias and discrimination in generative AI arise when models reflect or amplify biases embedded in training data, algorithmic structures, or deployment contexts. These biases can stem from non-representative datasets, algorithmic choices that favor certain groups, or cultural insensitivity in model design. For example, AI systems may generate outputs that reinforce gender, racial, or cultural stereotypes, such as associating specific professions with certain genders or presenting a narrow view of cultural norms. This leads to significant consequences, including the perpetuation of societal inequalities, discriminatory decision-making, erosion of public trust, and marginalization of minority groups [5]. Addressing these issues requires creating diverse and inclusive datasets, conducting regular bias audits, implementing human oversight, and building transparent models. Methods such as fairness assessments, differential privacy, and feedback loops can help ensure more equitable outputs. Furthermore, policies and regulations are crucial to enforce ethical standards and hold AI systems accountable.

The academic literature highlights fairness as a central concern in generative AI, particularly within large language models (LLMs) and text-to-image generation systems. It underscores how biases in training data lead to harmful effects, such as racism, sexism, and the reinforcement of societal stereotypes. Additionally, there are concerns about the further entrenchment of biases when newer generative models are trained on synthetic data derived from previous models [6]. The literature also critiques the centralization of AI research within a few well-funded organizations, which can limit access and innovation. There are significant challenges in ensuring diversity in AI development teams and providing equitable access to AI technologies, especially in developing nations or among economically disadvantaged groups. Lastly, concerns are raised about AI systems imposing values from one culture onto others, particularly when these systems are used globally.

3.3 Hallucinations

Hallucination in generative AI refers to the generation of inaccurate, fabricated, or misleading content by AI systems, which may sound plausible but is disconnected from reality. This issue is common in large language models (LLMs) and other generative systems, where AI produces false facts, invented statistics, or fictional quotes. The primary causes of hallucinations include limitations in training data, where models may replicate misinformation or incomplete information found online; the complexity of the models, which learn patterns without true understanding; and the lack of access to real-time data, leading to outdated or incorrect outputs [7]. Additionally, AI models often present hallucinated content with high confidence, further misleading users into believing it is factual. The implications of hallucinations are significant. They can contribute to the spread of misinformation, especially in sensitive fields like healthcare, law, or news, which can have harmful real-world consequences. Hallucinations also erode trust in AI systems, as users may question their reliability, particularly in applications requiring high accuracy. This loss of confidence can damage the reputation of organizations that use generative AI for content creation or customer service [8]. Moreover, hallucinations raise ethical concerns, as AI-generated falsehoods can perpetuate harm, spread bias, or mislead vulnerable populations. Addressing these issues requires improving model training, increasing transparency, and incorporating human oversight to ensure AI systems produce more reliable and accurate outputs.

3.4 Lack of Academic Integrity in Education and Research Writing

Education learning in contrast to conventional machine learning paradigms, the influence of generative artificial intelligence within the educational domain has garnered substantial scrutiny in the scholarly discourse [9]. In addition to challenges arising from the inability to differentiate between content produced by students and that generated by AI, which results in various avenues for academic dishonesty during online or written assessments, literature highlights the prospective advantages of generative AI in augmenting pedagogical and educational methodologies, particularly concerning individualized learning strategies Nevertheless, certain studies propose that generative AI may engender diminished effort or a propensity for complacency among students. Furthermore, a significant emphasis within the literature pertains to the enhancement of literacy and education regarding generative AI systems themselves, which includes the instruction of prompt engineering methodologies [10].

3.5 Cybercrime

Cybercrime is intricately linked to discourses regarding security and detrimental content; the domain of cybersecurity rigorously examines the exploitation of generative AI in fraudulent online endeavors. A significant emphasis is placed on social engineering attacks exemplified by the employment of generative AI for the impersonation of individuals, the fabrication of fictitious identities, the replication of vocal patterns, or the formulation of phishing communications [11]. A further prominent issue is the utilization of large language models (LLMs) for the generation of malicious code or unauthorized access. The emergence of generative AI has engendered fresh avenues for cybercrime, as its capacity to produce authentic content, including deepfakes and phishing emails, is exploited by nefarious actors. The catalysts for AI-driven cybercrime encompass the manipulation of AI capabilities to generate persuasive malicious content, the absence of stringent security protocols within AI systems, and the automation of attacks, which facilitates the execution of extensive operations by cybercriminals [12]. The ramifications are extensive, with generative AI being harnessed to disseminate misinformation, devise intricate phishing schemes, tarnish reputations, and undermine cybersecurity. It also presents obstacles for legal frameworks and enforcement mechanisms, thereby complicating efforts to trace and thwart such criminal activities. To alleviate these risks, it is imperative to establish more robust security protocols, advocate for ethical AI development, and implement regulatory measures aimed at safeguarding individuals and organizations from the escalating menace of AI-driven cybercrime.

3.6 Labor Displacement

The academic literature consistently underscores apprehensions that generative artificial intelligence systems may negatively influence the economy, potentially culminating in widespread unemployment. Labor displacement attributed to generative AI encompasses the prospective elimination of employment opportunities and significant transformations in employment paradigms as AI technologies undertake tasks traditionally executed by humans [13]. As generative AI systems attain heightened capabilities, they are increasingly able to execute a diverse array of tasks, including content generation, customer service, design, and even software development, with a scale and efficiency that may surpass human workers. This phenomenon could precipitate job displacement across industries such as media, marketing, programming, and customer support, where AI tools can automate repetitive or cognitively intensive tasks such as composing articles, generating graphics, and responding to inquiries. The anxiety surrounding extensive job losses is intensified by the escalating sophistication of AI, which is progressively adept at performing tasks previously believed to necessitate human creativity or discernment. Nevertheless, the effects of displacement are not homogeneous; specific occupations that demand creativity, empathy, and intricate problem-solving capabilities may exhibit resilience against automation, while low-skilled or routine positions are more susceptible to such threats. The ramifications of labor displacement are substantial, encompassing economic upheavals, income inequality, and augmented strain on social safety nets. Workers in sectors susceptible to automation may confront unemployment or the necessity for reskilling to adapt to emerging roles. Although generative AI has the potential to facilitate job creation in areas such as AI development, data analysis, and oversight of AI systems, the transition to an AI-centric workforce necessitates meticulous planning. It is imperative that governments and organizations allocate resources toward retraining initiatives, universal basic income, and strategies that bolster workers impacted by technological disruption, thereby ensuring that the advantages of AI advancements are equitably distributed throughout society.

3.7 Transparency and Explainability

A significant challenge arises from the "black box" characteristic inherent to numerous generative models, where the internal mechanisms of the AI remain opaque, even to the developers tasked with their creation. This deficiency in visibility can precipitate a variety of complications, including the exacerbation of biases in AI-generated outputs, given that users are unable to trace the decision-making processes or identify problematic patterns. In critical contexts, such as medical diagnoses or legal determinations, the incapacity to elucidate AI-generated decisions raises substantial concerns, as it undermines accountability and engenders ethical and legal dilemmas. In response to these challenges, scholars are actively engaged in the development of more interpretable AI systems and the formulation of methodologies that can articulate AI decisions in human-comprehensible terms, such as feature importance or decision trees. Furthermore, proposed regulatory frameworks advocate for the enforcement of transparency in AI systems, guaranteeing that users have access to explanations for AI outputs and the ability to verify their fairness. Enhancing transparency and explainability in generative AI is imperative for fostering public trust, ensuring ethical deployment, and mitigating risks associated with bias, discrimination, and erroneous decisions in vital applications. From an organizational perspective, transparency pertains to practices that include informing users about the capabilities and limitations of models as well as compliance with documentation and reporting mandates concerning data collection processes or risk assessments [14].

The issues of transparency and explainability in generative AI are paramount, particularly as these advanced technologies increasingly permeate decision-making frameworks across a multitude of sectors, including healthcare, finance, and law. Transparency is defined as the capacity to comprehend the operational dynamics of AI models, which encompasses the datasets utilized for training, the algorithms implemented, and the underlying decision-making processes. Conversely, explainability involves furnishing lucid and interpretable rationales for how a model derives specific outputs or predictions. The absence of transparency and explainability in generative AI systems, particularly in the context of intricate architectures such as deep learning networks, results in substantial challenges for users in terms of establishing trust and verifying the decisions generated by these technologies.

3.8 Copyright

Ethical considerations associated with copyright in the context of generative artificial intelligence are intricate and multifaceted, chiefly centered on the interactions between AI models and prevailing intellectual property legislation. Generative AI systems undergo training utilizing vast datasets, frequently encompassing copyrighted materials such as literature, musical compositions, visual art, and software code. In numerous instances, this content is harvested from the internet without securing explicit authorization from the original creators, thereby provoking inquiries regarding the classification of such usage as copyright infringement [15]. The absence of transparency in data procurement further intensifies these issues, complicating the assessment of whether outputs generated by AI are derivative works or authentic creations. A subsequent complication emerges during the output generation phase. Content produced by AI may occasionally closely replicate the stylistic elements, structural components, or even particular attributes of copyrighted works, engendering potential legal conflicts concerning authorship and originality. For example, should an AI-generated musical composition exhibit a remarkable similarity to a pre-existing piece, the question arises as to who possesses the rights to this new creation—the AI developer, the end-user, or the original artist whose work served as an influence on the model. The advent of generative artificial intelligence introduces challenges pertaining to disruptions of established copyright paradigms [16]. The scholarly discourse frequently addresses infringements of copyright and intellectual property rights resulting from the unauthorized acquisition of textual or visual training datasets Furthermore, an additional concern pertains to the propensity of generative models to memorize or plagiarize protected content.

4 Conclusion

This PRISMA scoping review critically investigated the ethical dilemmas associated with generative AI (GenAI), pinpointing pivotal concerns such as bias and equity, transparency, breaches of privacy, dissemination of misinformation, and manipulation. The review elucidates the increasing acknowledgment of these ethical predicaments while emphasizing that viable solutions are yet to be established. Although advancements have been made in tackling bias and transparency, numerous methodologies for mitigating bias are still in nascent phases, and the accountability frameworks for content generated by AI remain inadequately developed. Privacy apprehensions, especially those pertaining to data utilization and user consent, necessitate the implementation of more robust protective measures. In summation, although the body of literature concerning GenAI ethics has broadened, significant gaps persist in comprehending the real-world repercussions and in formulating effective regulatory frameworks. Additional empirical research is imperative to explore these matters and to establish pragmatic ethical guidelines that prioritize human rights and societal welfare. This review serves as a foundational basis for prospective inquiries aimed at unraveling the complexities of generative AI within the societal context.

References

1. Anderljung M, Barnhart J, Korinek A, Leung J, O'Keefe C, Whittlestone J et al (2023) Frontier AI regulation: managing emerging risks to public safety, pp 1–51. arXiv:2307.03718
2. Gupta M, Akiri C, Aryal K, Parker E, Praharaj L (2023) From ChatGPT to ThreatGPT: impact of generative AI in cybersecurity and privacy, pp 1–27. arXiv:2307.00691
3. Yan L, Sha L, Zhao L, Li Y, Martinez-Maldonado R, Chen G et al (2023) Practical and ethical challenges of large language models in education: a systematic scoping review. Br J Educ Techol 1–23
4. Derner E, Batistič K (2023) Beyond the safeguards: exploring the security risks of ChatGPT, 1–8
5. Firt E (2023) Calibrating machine behavior: a challenge for AI alignment. Ethics Inf Technol 25(3):1–8
6. Fraser KC, Kiritchenko S, Nejadgholi I (2023) Diversity is not a one-way street: pilot study on ethical interventions for racial bias in text-to-image systems, 1–5
7. Gill SS, Kaur R (2023) ChatGPT: vision and challenges. Internet Things Cyber-Phys Syst 3:262–271
8. Gupta M, Akiri C, Aryal K, Parker E, Praharaj L (2023) From ChatGPT to ThreatGPT: impact of generative AI in cybersecurity and privacy, 1–27
9. Hagendorf T, Bossert LN, Tse YF, Singer P (2023) Speciesist bias in AI: how AI applications perpetuate discrimination and unfair outcomes against animals. AI Ethics 3(3):717–734
10. Kasirzadeh A (2024) Two types of AI existential risk: decisive and accumulative, 1–31
11. Korinek A, Balwit A (2022) Aligned with whom? Direct and social goals for AI systems. SSRN Electron J 1–24
12. Mökander J, Axente M, Casolari F, Floridi L (2021) Conformity assessments and post-market monitoring: a guide to the role of auditing in the proposed European AI regulation. Minds Mach 1–28

13. Noy S, Zhang W (2023) Experimental evidence on the productivity effects of generative artificial intelligence. Science 381(6654):187–192
14. Panagopoulou F, Parpoula C, Karpouzis K (2023) Legal and ethical considerations regarding the use of ChatGPT in education, 1–11
15. Ray PP (2023) ChatGPT: A comprehensive review on background, applications, key challenges, bias, ethics, limitations and future scope. Internet Things Cyber-Phys Syst 121–154
16. Sætra HS, Danaher J (2022) To each technology its own ethics: the problem of ethical proliferation. Philos Technol 35(4):1–26

PulseGrid: AI-Powered Health Monitoring System for Workplace Optimization

B. G. Mamatha Bai, R. M. Darshan, L. Anushree, K. Amogha, and Acharya Sthuthi Jagdish

Abstract This chapter presents the development and implementation of the real-time employee health monitoring system which utilizes Internet of things (IoT) and machine learning (ML) to improve workplace safety and efficiency. The system continuously collects key health parameters which include oxygen levels, body temperature, and heart rate by the sensors, which are uploaded to the ThingSpeak cloud for processing then into analysis. Multiple ML models, including Random Forest, XGBoost, and Logistic Regression, were tested to classify health conditions and detect issues. The Extra Trees Classifier performed the best, achieving accuracy and precision of 98.5%. This system ensures continuous health tracking and helps identify health risks early. This can further be extended to a PulseGrid Web application and alert system that acts as an easy interface with user and include continues learning of model for accurate real-time prediction.

Keywords IoT · Machine learning · Health monitoring · Workplace safety · Real-time analysis

1 Introduction

In modern workplaces, ensuring employee health and safety is essential for maintaining productivity and fostering a sustainable working environment. However, traditional health monitoring methods often lack real-time capabilities, resulting in delayed responses to potential health risks. The integration of IoT technologies offers a promising solution to these limitations by facilitating continuous, real-time monitoring of employee health parameters. A clever system for ambulance navigation and health monitoring uses IoT sensors and mobile applications to send patient information to hospitals instantly. This setup allows for quick emergency responses

B. G. Mamatha Bai (✉) · R. M. Darshan · L. Anushree · K. Amogha · A. S. Jagdish
Computer Science and Engineering, Nitte Meenakshi Institute of Technology (NMIT), Nitte (Deemed to be University), Bengaluru, Karnataka, India
e-mail: mamathamane@gmail.com

H. Nagaraj et al. (eds.), *Advancements in Artificial Intelligence and Machine Learning*, Lecture Notes in Electrical Engineering 1572,
https://doi.org/10.1007/978-981-95-7449-0_31

and makes it easy to track and navigate ambulances, leading to better healthcare services [1]. This chapter looks at how wearable health technology has changed over time, improved, and made a big difference in active healthcare, managing long-term illnesses, and keeping an eye on patients in real time by putting different systems together and sharing real-life examples [2]. This chapter presents an IoT-driven employee health monitoring system that incorporates advanced sensors, including LM35 for temperature, AD8232 ECG for heart rate, and MAX30100 for SpO2, to collect real-time health metrics. Data is transmitted via ESP32 to the cloud for preprocessing, involving cleaning, transformation, and splitting for machine learning analysis. Various algorithms, including Support Vector Classifier (SVC), Logistic Regression (LR), Random Forest (RF), and XGBoost (XGB) are examined and assessed, with Extra Trees Classifier (ETC) and XGB demonstrating superior accuracy for anomaly detection and classification. The system enables continuous monitoring, early intervention, and actionable insights to improve workplace safety and productivity. It outlines the complete workflow from sensor integration to cloud-based processing and model training, emphasizing scalability and reliability. The proposed framework addresses the limitations of single-model dependency by integrating multiple algorithms for robust decision-making. Additionally, it offers real-time health categorization, ensuring timely responses to potential emergencies. Future enhancements include developing a web application, an alert system, and continuous ML model learning to adapt to evolving workplace health needs.

2 Background

2.1 Background Study

The demand for a smart health monitoring system stems from the growing need for real-time health tracking, particularly for prior detection of health issues and preventive care. The researchers implemented the K-star classification method, which outperformed SVC, Multi-Layer Perceptron (MLP), RF, and J48 classifiers, achieving an accuracy of 95%, precision of 94.5%, recall of 93.5%, and an F-score of 93.99% [3].

The [4] research looks at the way IoT technology changed healthcare during the COVID-19 pandemic. It highlights how patient monitoring became better, treatments became more accurate, and healthcare systems became more effective, thanks to new applications and solutions that rely on sensors. Khan et al. [5] focuses on IoT-based system which uses SVM algorithm with accuracy rate of 96%. The authors in [6] developed KNN algorithms, while in [7] they applied various deep learning algorithms for better analysis.

Paraschiv et al. [8] introduced the RO-Smart Aging system; an IoT-based solution designed for the remote health monitoring of elderly individuals. The [9] model uses a method called CSO-CLSTM and reached diagnostic accuracies of 96.16% for heart disease and 97.26% for diabetes. This development helps smart healthcare systems by improving how data is collected, processed, and classified. The performance of their proposed estimation model was evaluated using R^2 and mean square error (MSE), with the best values achieved using the XG Boost regressor at 0.967 and 0.0033, respectively [10].

Elkahlout et al. [11] highlights the security challenges posed by IoT systems that necessitate remote access to personal data and storage. Shalini [12] proposed a system that alerts a doctor via SMS in case of unforeseen changes in a patient's heartbeat and blood pressure. The system utilizes a GSM module to send messages and provides live updates on the patient's heartbeat and blood pressure.

In [13], authors emphasized remote monitoring of SpO2 (MAX30100), heart rate, and body temperature (DS18B20). Additionally, their system allows on-demand location tracking of the patient using the SIM7600E GSM and GNSS HAT (Hardware Attached on Top) module, with a Raspberry Pi 4B functioning as the microcontroller to collect data from various health sensors.

Finally, Fazal M presented an NB-IoT system based on NTN SATCOM, integrated with a back-end intelligent system, offering enhanced capabilities for smart health monitoring [14].

2.2 *Motivation*

Continuous health monitoring in workplaces is essential for ensuring employee safety and maintaining productivity. Traditional systems often adopt a reactive approach, addressing health concerns only after they escalate, leading to undetected risks and workplace hazards. The lack of real-time data collection and analysis amplifies these challenges, highlighting the need for proactive solutions. By integrating IoT and machine learning, this project seeks to provide continuous health assessment and anomaly detection in real time, ensuring timely medical interventions and reducing health-related incidents. Insights from F Ahmod's study on health monitoring using NB-IoT and LTE-M emphasize the transformative potential of IoT technologies in enabling robust, real-time tracking of critical health parameters, even in high-stress environments [15]. Furthermore, the struggles faced by delivery agents in India, as highlighted by Al Jazeera, underscore the importance of addressing health risks in industries characterized by high operational demands and physical stress [16]. This project, inspired by such challenges, seeks to enhance workplace safety and operational efficiency through advanced, technology-driven health monitoring systems.

3 System Model

The proposed model depicted in Fig. 1 has two stages, namely the data collection phase and the ML model development phase for tracking and estimation of health.

3.1 Data Acquisition

The data is collected using various sensors and preprocessed for further analysis by ML model. This collected data is used to monitor health.

Sensors

Sensors have the capacity to measure and identify changes in the physical, chemical, or biological world and translate those changes into electrical impulses for study. They are used to measure parameters like body temperature, heart rate, and other health indicators to gain insights into human health. The following sensors are used in the proposed model.

Body Temperature Sensor (LM35): Fig. 2 is a precision integrated-circuit temperature sensor. It outputs a voltage that can be converted to temperature in degrees Celsius, making it straightforward to interface with microcontrollers and other circuits.

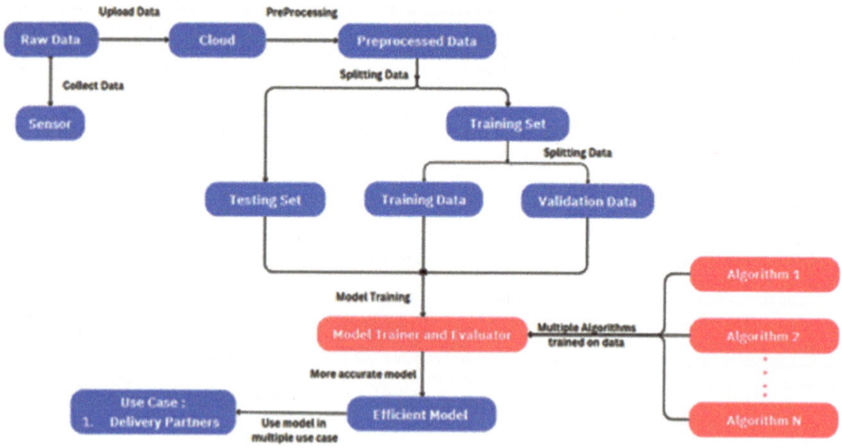

Fig. 1 Proposed model flow diagram

Fig. 2 LM 35 (Body temperature sensor)

Heart Rate and ECG (AD8232 ECG Module): The AD8232 ECG module in Fig. 3 records the heart's electrical activity, providing amplified and filtered signals for accurate ECG monitoring. It is used in health systems to assess heart health, detect abnormalities, monitor stress levels, and support fitness tracking.

Oxygen Saturation (SpO2) Sensor: Fig. 4 measures SpO2 (blood oxygen saturation) and heart rate using optical methods. It is widely used in wearable health devices like pulse oximeters and fitness trackers for real-time monitoring.

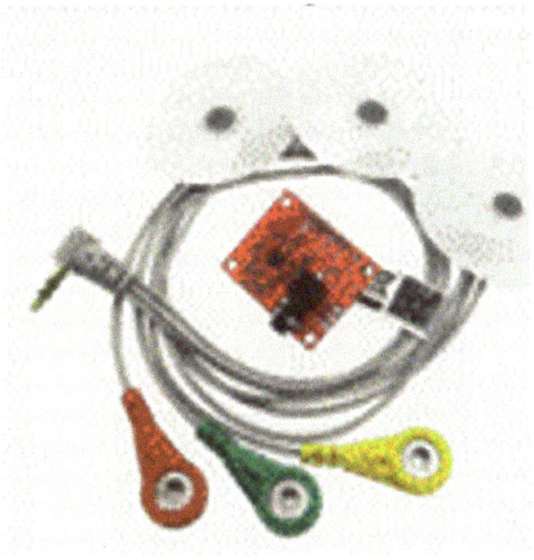

Fig. 3 AD8232 ECG module (Heart rate and ECG)

Fig. 4 Oxygen saturation (SpO2) sensor

Fig. 5 Galvanic skin response (GSR) sensor

Galvanic Skin Response (GSR) Sensor: GSR sensor in Fig. 5 is used to measure the skin's ionic conductance by the activity of the sweat gland. It is commonly used to monitor stress levels, emotional states, and physiological responses in health and fitness applications.

*Bluetooth or Wi-Fi Module (*ESP32): The ESP32 in Fig. 6 is a versatile microcontroller with built-in Wi-Fi and Bluetooth, ideal for sending sensor data to the cloud or a local server. It enables real-time analysis, predictions, and IoT integration in health monitoring and other applications.

Arduino Uno: Fig. 7 is a microcontroller board, widely used for prototyping and DIY projects. It interfaces with sensors, processes data, and controls devices, making it ideal for small-scale IoT and automation applications.

Cloud

The data is collected from the sensors and stored in ThingSpeak, a cloud platform designed for secure and scalable data storage. Once the data is securely stored, it can be accessed for further processing, analysis, and integration with other relevant parameters. This data is further used as input for a ML model, which generates meaningful insights, such as health predictions and alerts. The integration of sensor data with the ML model helps in providing real-time health monitoring, personalized recommendations and early detection.

Fig. 6 ESP32

Fig. 7 Arduino Uno

Preprocessing Data

The raw data collected from sensors has outliers, noise, and various other issues that must be processed before inputting to the ML model for its better performance. Fig. 8 shows the steps involved in preprocessing.

Data Cleaning: Data cleaning is a crucial step to maintain the quality of the dataset used in the health monitoring system. This process involved addressing missing values by using techniques such as interpolation for continuous variables like temperature and heart rate. Outliers, which could arise due to sensor malfunctions or anomalous readings, were detected using statistical methods like the interquartile range (IQR) and handled appropriately. Smoothing techniques were utilized to reduce the noise in the data, which was generated by variations in sensor readings. Duplicate records, which could skew the analysis, were identified and removed to maintain dataset integrity.

Data Integration: It was performed to combine multiple data sources into a cohesive dataset. While sensor data provided critical health metrics such as heart rate,

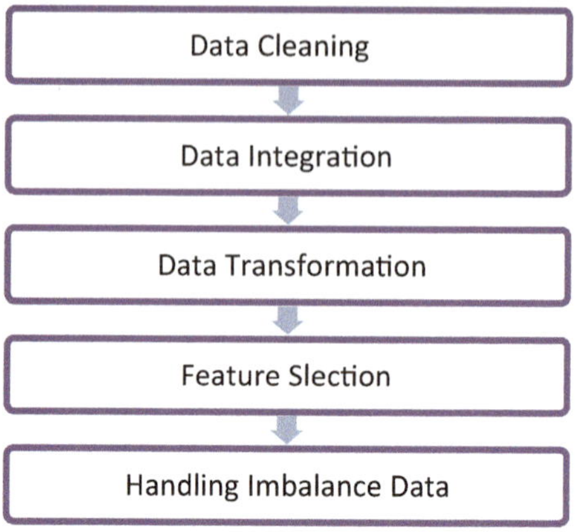

Fig. 8 Data preprocessing

oxygen saturation, and galvanic skin response (GSR), additional demographic information like age, height, and weight was obtained from the user input. This ensured a comprehensive picture of the user's health by merging data from these different sources.

Data Transformation: Data transformation involved encoding, standardization, and feature engineering to prepare the data for analysis. Label encoding was applied to categorical variables like gender to convert them into numerical formats usable by machine learning algorithms. Standardization by standard scalar was performed on continuous variables such as heart rate and temperature to bring them to a uniform scale, improving model performance. Feature engineering played a key role in deriving new insights, BMI was calculated using height and weight, and a weighted health score was created to provide a single indicator of overall health which is calculated using all parameters and their weights. These transformations enhanced the quality and usability of the dataset.

Feature Selection: Feature selection was employed to streamline the dataset by removing less relevant columns, such as height and weight, which were redundant after calculating BMI. This step was essential to reduce dimensionality, improve computational efficiency, and minimize noise in the data. By focusing only on the most critical features, the system was optimized for better predictive accuracy.

Handling Imbalance Data (SMOTE): Handling imbalanced data was necessary to address the unequal distribution of health status, there were more of healthy data compared to unhealthy, which could lead to biased model predictions. Synthetic minority oversampling technique (SMOTE) was used to balance the dataset. This technique makes sure that the model could effectively identify health risks without being overwhelmed by the majority class, thereby improving its sensitivity and overall performance.

Table 1 Attributes

Sl. no	Attributes
1	Age
2	Gender
3	Weight (in KG)
4	Height (in CM)
5	Temperature (in °C)
6	Heart rate (in BPM)
7	Heart rate variability (ms)
8	GSR (microsiemens)
9	Oxygen saturation
10	BMI
11	Health score weighted

Dataset

The dataset has 11 parameters as given in Table 1, which are used for health monitoring, out of which age, weight, height, and gender were collected from users. Parameters like temperature, heart rate, heart rate variability, GSR (microsiemens), and oxygen saturation were collected from sensors, and other parameters were feature engineered. The dataset has 5496 number of data points. Each target is having 1832 number of data points as they were balanced. The dataset is collected using sensors through hardware implementation as shown in Fig. 9 and further preprocessed for model.

Data Splitting

Data splitting in Fig. 10 is essential for evaluating model performance and ensuring that it generalizes well to new data.

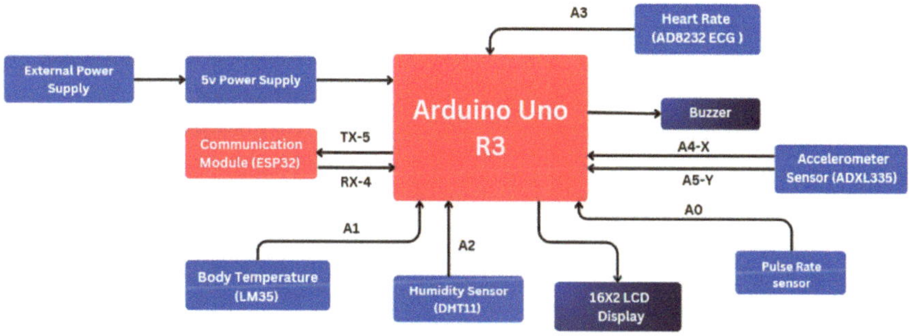

Fig. 9 Hardware block diagram

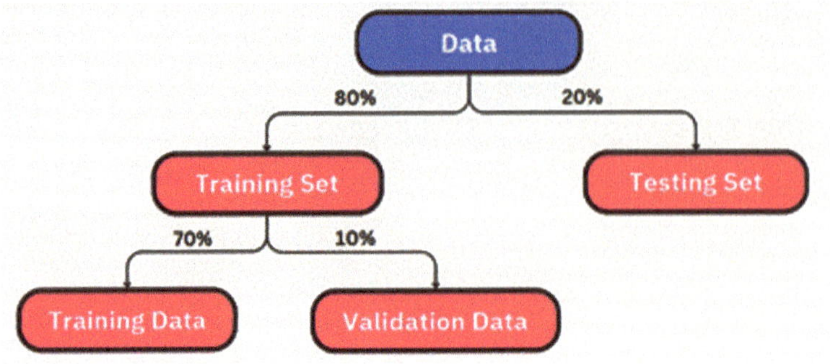

Fig. 10 Data splitting

(A) *Train*: The ML model is trained using this portion of the data. In order to reduce the error, the model modifies its parameters based on the features and associated labels in this set.

(B) *Test*: Following training and validation, the model's performance is evaluated using the test data. Once the model has been trained and adjusted, this set is utilized just once.

(C) *Validation*: During training, this subset is utilized to assess the model's performance and adjust hyperparameters (e.g., out of bag error, early stopping). By giving input on the model's generalizability to new data, it helps avoid overfitting.

3.2 *Machine Learning Model Development Phase*

This phase includes multiple ML models where trained and tested to select the most efficient model for health analysis. By inputting the preprocessed data, six ML models were used.

Multiple Model

Support Vector Classifier (SVC): SVC is a supervised learning algorithm. In our project, SVC plays a critical role in classifying health data by finding the optimal hyperplane that segregates different classes of health conditions. Its ability to handle high-dimensional data ensures accurate predictions even with a small dataset. Furthermore, the kernel trick allows SVC to work efficiently with nonlinear data, making it particularly useful for analyzing complex relationships between health parameters. This precision contributes to timely health risk detection, ensuring delivery personnel receive alerts before their condition worsens.

Logistic Regression (LR): LR is a straight forward, effective algorithm for binary classification problems. In our project, it served as a baseline model to classify whether a delivery partner's health condition is safe or at risk. Its ability to provide interpretable coefficients made it a valuable tool for understanding the relationship between health parameters and their influence on the outcome. But this algorithm is less complex than other algorithms.

XGBoost Classifier (XGB): XGB (Extreme Gradient Boosting) is a powerful gradient-boosting algorithm known for its scalability and high performance. It was employed in our project to refine predictions and handle large, imbalanced datasets efficiently. XGBoost improves model accuracy by sequentially correcting the errors of prior models and applying regularization techniques to prevent overfitting. This made it particularly effective for distinguishing subtle variations in health data, such as abnormal oxygen levels or erratic heart rates. By leveraging XGBoost, our system achieved enhanced accuracy, ensuring reliable health status predictions for delivery personnel.

Random Forest (RF): RF is a versatile ensemble learning algorithm that aggregates the results of multiple decision trees to improve accuracy and reduce variance. In our project, Random Forest Classifier was instrumental in analyzing diverse health parameters by splitting the data into multiple subsets and making predictions based on majority voting. Its inherent robustness and ability to handle missing or noisy data ensured consistent performance even with real-world sensor readings. This algorithm provided a reliable framework for identifying health anomalies, contributing to the system's overall effectiveness.

AdaBoost Classifier: AdaBoost (Adaptive Boosting) is a boosting algorithm that combines multiple weak classifiers to form a strong classifier by focusing on previously misclassified data points. In our health monitoring system, AdaBoost helped enhance the accuracy of predictions by iteratively training models and assigning higher weights to challenging samples. This adaptability made it particularly effective for detecting less obvious health anomalies, such as mild fluctuations in heart rate or temperature that could indicate early signs of distress. By incorporating AdaBoost, we ensured the system remained sensitive to subtle health risks.

Extra Tree Classifier (ETC): ETC is an ensemble learning technique that constructs multiple decision trees to classify data. Unlike traditional decision trees, it introduces randomness by selecting cut points randomly, which helps to improve model robustness and reduce overfitting. In this project, the Extra Tree Classifier was used to analyze the health parameter dataset and identify key patterns that indicate potential health risks. Its ability to handle noisy data and compute results quickly ensures efficient real-time health monitoring. This makes it a suitable choice for integrating with IoT-based systems, where speed and accuracy are critical.

Evaluation

Evaluation is the process of determining how well the model performs. Using predetermined criteria like accuracy, precision, recall, F1-score, and others, it compares the model's predictions to the actual values in the dataset.

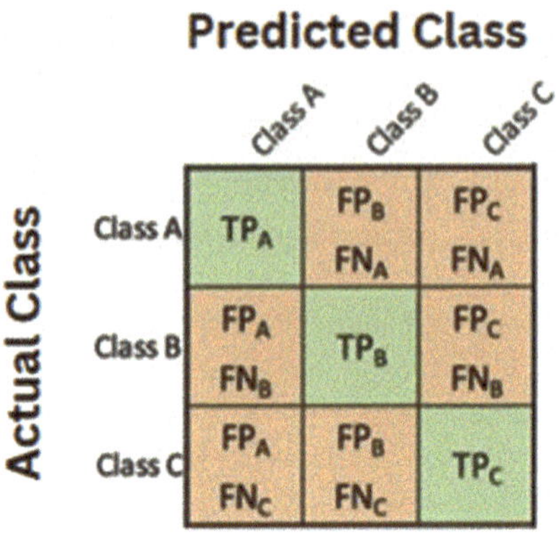

Fig. 11 Confusion matrix

Confusion Matrix: Fig. 11 is a tabular representation. It provides a clear visual representation of the model's performance in the confusion matrix by displaying the counts of values. For our health monitoring system, the confusion matrix helps us analyze specific errors, such as instances where the model incorrectly classifies a healthy delivery partner as at risk or vice versa. This insight aids in refining the model for better performance.

- True Positive (TP) - The predicted and actual class are both positive.
- True Negative (TN) - The predicted and actual class are both negative.
- False Positive (FP) - The predicted class is positive, but actual class is negative.
- False Negative (FN) - The predicted class is negative, but actual class is positive.

Accuracy: Accuracy in Eq.1 evaluates the model's overall accuracy by dividing the number of properly predicted observations by the total number of observations. It shows how well the health monitoring system categorizes health states as safe or at risk in our project. Although accuracy is a valuable metric, it may overemphasize the majority class in imbalanced datasets, making it misleading. For a more thorough assessment, we therefore supplemented it with additional parameters like precision and recall.

$$\text{Accuracy} = \frac{TP + FP}{(FP + TP + FN + TN)} \tag{1}$$

Precision: The ratio of accurately predicted positive observations to all anticipated positives is measured by Eq. 2. For our system, precision highlights how accurately the model identifies delivery personnel who are truly at risk among all those flagged. A high precision score ensures that the system minimizes false positives, avoiding unnecessary health alerts, which is critical in maintaining its reliability for real-world deployment.

$$\text{Precision} = \frac{TP}{(FP + TP)} \tag{2}$$

Recall: In Eq. 3, recall, often referred to as sensitivity, determines the proportion of accurately anticipated positive observations to all actual positive observations. In the context of our project, it shows the system's ability to detect delivery personnel who are genuinely at health risk. A high recall score is essential for ensuring no critical cases are missed, as overlooking health anomalies could have serious consequences.

$$\text{Recall} = \frac{TP}{(FN + TP)} \tag{3}$$

F1-Score: The F1-score in Eq. 4 is a balanced metric that takes into account both false positives and false negatives. It is calculated as the harmonic mean of precision and recall. It is particularly useful in our project where the cost of misclassification (false negatives or positives) is significant. A high F1-score ensures that the model achieves a balance between detecting true health risks and avoiding false alarms, making it ideal for this application.

$$F1\text{Score} = \frac{(2 * \text{Precision} * \text{Recall})}{(\text{Precision} + \text{Recall})} \tag{4}$$

4 Result and Discussion

Six machine learning algorithms, namely SVC, LR, ETC, AdaBoost Classifier, RF, and XGB Classifier, were evaluated to predict health status using sensor dataset. Accuracy, precision, recall, and F1-score were used to evaluate these algorithms' performance, as indicated in Table 2 and Fig. 12.

Table 2 Evaluation result

Algorithm	Accuracy	Precision	Recall	F1-Score
ETC	0.985401	0.985559	0.986671	0.986711
XGB	0.963156	0.960461	0.963164	0.962161
RF	0.950671	0.951772	0.953412	0.952721
LR	0.737226	0.731868	0.787273	0.786848
AdaBoost	0.660584	0.646748	0.674545	0.612197
SVC	0.359489	0.352411	0.356161	0.351168

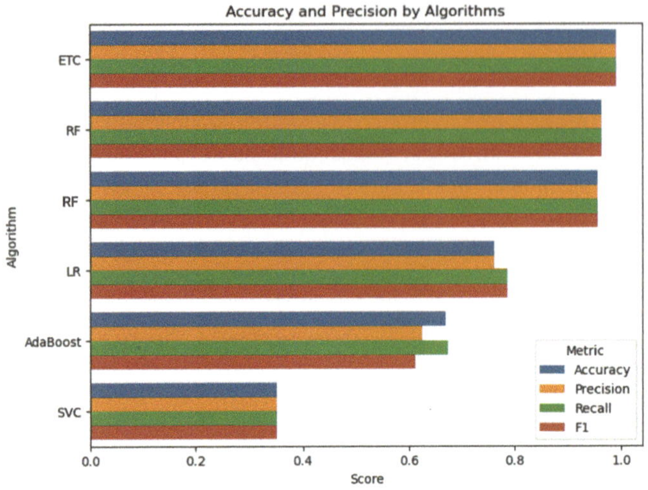

Fig. 12 Model comparison graph

The ETC achieved near-perfect accuracy, precision, recall, and F1-score of 0.98, outperforming all other models. Its success can be due to its capacity to manage complex, hierarchical feature interactions, which makes it especially useful for datasets that contain a combination of continuous (like oxygen saturation) and numerical (like BMI) data types. The ensemble approach of Extra Trees, which combines multiple decision trees with randomized splits, further enhances its robustness and reduces overfitting, making it an ideal choice for health datasets with structured information.

The RF and XGB Classifiers also performed exceptionally well, with accuracies and F1-scores above 0.97. By averaging predictions over several trees, RF's ensemble process lowers variance and enhances generalization. This makes it particularly effective in capturing subtle relationships between health parameters. Similarly, XGB's gradient-boosting approach is well-suited for datasets with nonlinear feature interactions and imbalanced classes, allowing it to model complex relationships like those between oxygen saturation and health status effectively.

The LR algorithm achieved moderate performance, with an accuracy and F1-score of 0.73. While it is computationally efficient and interpretable, its assumption of linear relationships between features and target variables likely limited its performance. Health datasets often exhibit nonlinear interactions, such as the relationship between BMI, age, and health status, which LR cannot capture effectively without extensive feature transformations.

The AdaBoost Classifier performed poorly, with an accuracy of 0.66 and an F1-score of 0.61. Although AdaBoost excels at improving weak learners, it is highly sensitive to noisy or imprecise data, which is common in health datasets due to sensor variability and natural fluctuations in parameters like heart rate or body temperature. This sensitivity may have caused its lower predictive performance.

The SVC performed the worst, with an accuracy and F1-score of 0.35. SVC typically works best for smaller datasets with well-separated classes. However, the health dataset likely contained overlapping health categories and nonlinear patterns, making it difficult for SVC to find an optimal decision boundary without advanced kernel techniques or significant feature engineering. Additionally, the high-dimensionality of the dataset may have contributed to its suboptimal performance.

Overall, the ETC emerged as the most suitable model for this application, followed closely by RF and XGB. These models demonstrated the ability to handle the complex and nonlinear interactions between health parameters effectively. In contrast, algorithms like LR, AdaBoost, and SVC were less suited for this dataset due to their limitations in handling noise, overlapping classes, and nonlinear relationships.

5 Conclusion

In comparison with existing research that often relies on a single or very few machine learning model or fuzzy logic for health monitoring, our approach stands out by utilizing a multi-model analysis, particularly the Extra Trees Classifier, alongside other models like LR and SVC. Many studies have used publicly available dataset and shown that fuzzy logic or individual models may fall short when it comes to capturing the complex, nonlinear relationships present in health data. For instance, fuzzy logic approaches, while interpretable and flexible, may struggle with accuracy when the data exhibits intricate patterns that are better modeled by advanced machine learning algorithms.

On the other hand, ensemble models like ETC leverage the collective decision-making of multiple decision trees, which allows them to better handle the high-dimensionality and noise often present in health sensor data. Unlike single models or fuzzy logic, this multi-model approach is more robust, leading to superior performance in both predictive accuracy and generalizability.

Additionally, by applying hyperparameter optimization to these models, we enhance their predictive capabilities even further, ensuring that the system can make more accurate, real-time health assessments. This advancement is especially critical in the context of personalized health monitoring, where the system must adapt to individual health profiles and respond to subtle changes in sensor data. In contrast, simpler models or fuzzy logic-based systems are often less adaptable and may not perform as well across diverse individuals or a wide range of health conditions.

Therefore, our multi-model approach with the use of real-time collected dataset offers a more advanced and scalable solution, with the potential to transform how wearable devices or mobile health apps monitor and predict health conditions, making health management more personalized, proactive, and accessible. It gives an insight on which model would perform better on health dataset.

References

1. Poongodi M, Sharma A, Hamdi M et al (2021) Smart healthcare in smart cities: wireless patient monitoring system using IoT. J Supercomput 77:12230–12255. https://doi.org/10.1007/s11227-021-03765-w
2. Tariq MU (2024) Advanced wearable medical devices and their role in transformative remote health monitoring. Transformative approaches to patient literacy and healthcare innovation. IGI Global, 308–326
3. Akhbarifar S, Javadi HHS, Rahmani AM et al (2023) A secure remote health monitoring model for early disease diagnosis in cloud-based IoT environment. Pers Ubiquit Comput 27:697–713. https://doi.org/10.1007/s00779-020-01475-3
4. Javaid M, Khan IH (2021) Internet of Things (IoT) enabled healthcare helps to take the challenges of COVID-19 Pandemic. J Oral Biol Craniofacial Res 11(2):209–214
5. Khan MN et al (2024) Real-time health monitoring with IoT. Int J Fundamental Med Res (IJFMR) 6(1):227–251
6. Vimal SP, Vadivel M, Baskar VV, Sivakumar VG, Srinivasan C (2023) Integrating IoT and machine learning for real-time patient health monitoring with sensor networks. 2023 4th International conference on smart electronics and communication (ICOSEC), Trichy, India, pp 574–578. https://doi.org/10.1109/ICOSEC58147.2023.10275890
7. Wu X et al (2023) Internet of things-enabled real-time health monitoring system using deep learning. Neural computing and applications, 1–12
8. Paraschiv E-A, Tudora E, Tîrziu E, Alexandru A (2021) IoT & cloud computing-based remote healthcare monitoring system for an elderly-centered care. 2021 International conference on e-health and bioengineering (EHB), Iasi, Romania, pp 1–4. https://doi.org/10.1109/EHB52898.2021.9657585
9. Mansour RF, Amraoui AE, Nouaouri I, Díaz VG, Gupta D, Kumar S (2021) Artificial intelligence and internet of things enabled disease diagnosis model for smart healthcare systems. IEEE Access 9:45137–45146. https://doi.org/10.1109/ACCESS.2021.3066365
10. Mishra S, Singh A (2024). IoT ML driven holistic health monitoring and fitness assessment empowering proactive wellbeing management. 2024 IEEE students conference on engineering and systems (SCES), Prayagraj, India, pp 1–6. https://doi.org/10.1109/SCES61914.2024.10652387
11. Elkahlout M, Abu-Saqer MM, Aldaour AF, Issa A, Debeljak M (2020) IoT-based healthcare and monitoring systems for the elderly: a literature survey study. 2020 International conference on assistive and rehabilitation technologies (iCareTech), Gaza, Palestine, pp 92–96. https://doi.org/10.1109/iCareTech49914.2020.00025
12. Shalini VB (2021) Smart health care monitoring system based on internet of things (IOT). 2021 international conference on artificial intelligence and smart systems (ICAIS), Coimbatore, India, pp 1449–1453. https://doi.org/10.1109/ICAIS50930.2021.9396019
13. Mohammed B, Hasan D (2023) Smart healthcare monitoring system using IoT. Int J Interactive Mobile Technol (iJIM) 17:141–152. https://doi.org/10.3991/ijim.v17i01.34675
14. Mahmood F (2023) Smart autonomous location tracking & health monitoring of war fighters using NB-IoT/LTE-M with SATCOM. 2023 IEEE future networks world forum (FNWF), 1–6

15. https://www.aljazeera.com/economy/2022/2/23/as-indians-expect-groceries-in-10-mins-delivery-agents-struggle.
16. https://www.google.com/amp/s/m.economictimes.com/magazines/panache/from-hospital-bed-a-25-year-old-it-professional-warns-about-real-price-of-sacrificing-sleep-and-no-diet-plan/amp_articleshow/113013044.cms

Implementation and Comparison of Deep Learning Algorithms for Kidney Cancer Detection

Bushra Siddiqui, Chandrama Saha, Harsh Bhoi, and P. Sudhanya

Abstract This work focuses on leveraging advanced machine learning techniques, specifically convolutional neural networks (CNNs), to train histopathological images and detect kidney cancer. Early detection of cancer cells is essential for successful treatment and better patient outcomes. However, traditional diagnostic methods are often time-consuming and require significant expertise, leading to potential delays in diagnosis. This research aims to address these challenges by implementing machine learning algorithms for accurate and efficient detection of cancer cells. CNNs offer powerful image analysis capabilities, enabling precise identification of malignant cells from histopathological images with high accuracy. CNNs can automatically extract relevant features from images, reducing the need for manual feature selection and improving overall diagnostic performance. These models are highly scalable and can be trained on large datasets, enhancing their ability to generalize across different cases. The accuracy obtained in this work is 97.53%, demonstrating the effectiveness of CNNs in detecting kidney cancer with high precision. By leveraging CNNs, this work aims to improve diagnostic accuracy, speed, and reliability in cancer detection.

Keywords Kidney cancer · Cancer detection · Machine learning · Convolutional neural networks (CNNs) · Medical imaging · Neural networks

B. Siddiqui · C. Saha · H. Bhoi · P. Sudhanya (✉)
Department of Electronics and Communication Engineering, SRM Institute of Science and Technology, Kattankulathur, India
e-mail: sudhanyp@srmist.edu.in

B. Siddiqui
e-mail: bi5286@srmist.edu.in

C. Saha
e-mail: cp2223@srmist.edu.in

H. Bhoi
e-mail: hd0719@srmist.edu.in

H. Nagaraj et al. (eds.), *Advancements in Artificial Intelligence and Machine Learning*, Lecture Notes in Electrical Engineering 1572,
https://doi.org/10.1007/978-981-95-7449-0_32

1 Introduction

Kidney cancer is a condition in which an abnormal cell growth occurs in the lining of the proximal convoluted tubule (extremely small tubes in the kidney that carry the primary urine). Renal cell carcinoma (RCC) is the most common type of kidney cancer in adults, accounting for about 90% of cases [1]. The French National Cancer Institute states that kidney cancer is usually discovered by chance during an ultrasound or abdominal scan because kidney cancer typically shows no symptoms in its early stages. However, as the tumor grows, symptoms like urinal blood, lower back pain, abdominal lumps, constant fatigue, and rapid, unexplained weight loss may occur. Kidney cancer affects around 400,000 people globally each year, mainly men and people above 75 years of age, and has a mortality rate of approximately 175,000 annually, making it the 12th most common cancer worldwide with a combined survival rate of 78% [2].

Identifying kidney cancer is often challenging due to the complexity of imaging data and the subtle differences between cancerous and non-cancerous cells. In the case of cancer diagnosis, the earlier the detection, the more effective the treatment will be. But, traditional diagnostic methods, like biopsy, cystoscopy, or radiology scans like MRI or CT scans [3], require specialized skills, equipment, and prompt availability of doctors and medical professionals. They are also more time-consuming, sometimes taking multiple days for the results to arrive. Moreover, early-stage cancer cells can be difficult to detect due to their small size or similarity to healthy tissue, leading to delays in diagnosis or inaccurate results.

Machine learning, particularly through techniques like convolutional neural networks, can considerably enhance the accuracy of cancer detection by automating and enhancing image analysis. A convolutional neural network (CNN) is configured to utilize various layers like convolutional layers, pooling layers, and fully connected layers to independently and learn spatial feature hierarchies adaptively via backpropagation [4]. CNNs mimic the human visual system and are highly effective in recognizing spatial patterns in data, such as edges, textures, and shapes in images. They can process enormous amounts of data quickly by identifying patterns and features that the human eye may miss during manual detection. By training models on these large medical image datasets, machine learning algorithms can learn to differentiate between healthy and cancerous kidney cells with high precision, leading to early and accurate diagnosis.

Machine learning models such as convolutional neural networks (CNNs) are particularly suited for tasks that require high computational power, such as processing medical imaging data for cancer detection. These models can detect patterns in images, enabling faster and more accurate diagnostics compared to traditional methods. By training models on vast datasets of medical images, machine learning algorithms can learn to differentiate between healthy and cancerous kidney cells with high precision, leading to early and accurate diagnosis.

2 Literature Review

Various scholarly works have been examined in the domain of interest. For instance, [2] takes a deep dive into the statistics and impact of kidney cancer, while mentioning the effective detection and treatment methods. Survey article [5] discusses the use of various ML and deep learning techniques like convolutional neural networks (CNNs), artificial neural networks (ANNs), and support vector machines (SVMs) for kidney cancer classification and prediction, and provides an emphasis on the importance of smart healthcare technologies, like the Internet of Medical Things (IoMT), for early detection and diagnosis, thereby aiding in timely and effective intervention.

In [6], the authors provide a comprehensive review of imaging-based deep learning applications in kidney diseases, including both neoplastic and non-neoplastic conditions. The paper highlights several key areas where deep learning has shown promise, such as renal tumor segmentation, differential diagnosis, tumor staging/grading, surgical planning, and prognosis prediction. For kidney cancer specifically, the authors discuss studies using convolutional neural networks and other deep learning approaches to accurately segment kidney tumors, classify tumor types, and predict outcomes.

In 2024, Rahat et al. conducted a comparative analysis of machine learning methods for the early detection of chronic kidney diseases [6]. The researchers implemented and compared several ML algorithms like random forest, XGBoost, logistic regression, AdaBoost, and proposed a new hybrid model. Their hybrid model, which combines multiple algorithms using stacking classifiers, achieved the highest accuracy of 95% in predicting chronic kidney disease. Similarly, T. Mahmood et al. presented a novel approach to classify renal cell carcinoma using a hybrid model that integrates vision transformers and CNNs [7]. Their model consists of two main components: an adaptive convolution (AC) block for feature extraction and spatial representation, and a dynamic attention (DA) block with a transformer encoder to refine the learnings. Their model showed a 90.62% accuracy, demonstrating the superiority of the model compared to existing methods. These studies demonstrate the potential of hybrid models in improving the accuracy of early kidney disease detection.

In [8], an effective method for diagnosing kidney tumors using deep learning models was presented. The authors proposed several 2D-CNN models, including CNN-6, ResNet50, and VGG16, for kidney tumor detection, along with a CNN-4 model for classification. The study utilized a novel dataset consisting of 8,400 CT scan images from 120 patients, and their proposed models achieved high accuracy, with CNN-6 and ResNet50 reaching up to 97% and 96%, respectively. Likewise, some researchers deployed the European Deep Health toolkit's convolutional neural network with the European Computer Vision Library (ECVL) and the European Distributed Deep Learning Library (EDDL), and utilized U-net architecture for image segmentation and ResNet101 for classification for RCC diagnosis, and this model achieved a detection accuracy of 84%, classification accuracy of 88.5%, and a mean dice score of 0.675 for tumor segmentation [9], demonstrating its potential to enhance kidney cancer diagnosis.

An automated system to classify kidney cancer grades accurately using deep learning neural networks (DLNNs) was designed and implemented on FPGAs where histopathological images of renal cell carcinoma (RCC) from Kasturba Medical College (KMC) were collected and the performance of deep learning architectures on an FPGA board was evaluated using tools like Google Colab, Xilinx Vivado, and Vitis AI [10]. The results indicated that Inception-V3 achieved a superior accuracy for the detection of kidney cancer, while DenseNet-169 performed a better grading of kidney cancer, demonstrating the potential of FPGA-based implementations in enhancing medical diagnostic capabilities.

3 Methodology

The machine learning algorithm employed in this work is convolutional neural network (CNN) using Python, for kidney cancer detection. This network was initially designed to classify handwritten digits and without the need for preprocessing, it was able to directly recognize visual patterns from input images [11]. Due to the unavailability of training data and computing power, this model was unsuccessful in performing well in complex problems. This model gained popularity when Krizhevsky et al. came up with an advanced version of CNN [12] that succeeded in bringing down the error rate in the LSVRC competition.

Convolutional neural network (CNN) is often referred to as ConvNet, which comprises a deep architecture [13]. The first and the basic network architecture was LeNet, proposed in 1998 [14]. It employs an architecture in which the five convolutional layers are directly connected to the pooling layer. The CNN architecture can be depicted as a hierarchical feature extraction, in which the trained filters belonging to the first layer capture basic edges of the second layer, the filters in the subsequent layers are capable of detecting parts of objects, and the filters in the final layers identify recognize these objects [11]. The major elements of CNN are the mainly the various layers such as input, convolutional, pooling, fully connected, and output [14]. The layers present in a CNN is demonstrated in Fig. 1.

When the individual neurons in the subsequent layer are linked to certain neurons in the preceding layer, this structure or correlation is called the receptive field [13]. The convolutional layer consists of the filter size, the size of the stride, and the filter numbers. The size of the filter can vary. The stride (for filter movement) [11] size of the feature map can be downsized, which helps in reducing the computation load. Filter number indicates the distinctive feature detectors, operating on the input. Convolution aims at shifting or sliding the filter throughout the input vector to produce a feature map [15]. Feature map reflects the degree of match. The output of the next layer is computed with the help of a mathematical formula, which considers the weights of the two layers mainly the input and the hidden and the bias value. The output volume is produced by stacking together the separate 2D activation maps, which are

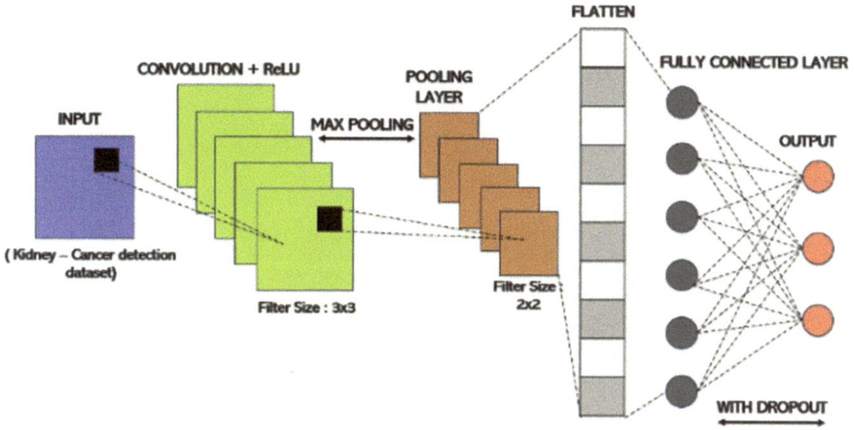

Fig. 1 Architecture of a convolutional neural network (CNN)

produced when the filters are convolved with the feature maps. The backpropagation method is employed, which allows us to learn the parameters of the model in a very efficient way.

Pooling layers are employed to reduce the spatial dimension of the data, which reduces the computational complexity and combats overfitting. The reduction takes place without any loss of information. There are many pooling operations, some of them are max pooling, stochastic pooling, multi-scale orderless pooling, average pooling, and spatial pyramid pooling. The pooling method utilized in the Python code for implementation is max pooling. Max pooling considers a particular filter size and considers the maximum value and down samples of the feature map. This method has proven to outperform many existing complex architectures. The fully connected (FC) layer is the last part of CNN, in which the neurons of two layers are fully connected. There cannot be the existence of a convolutional layer after the FC layer because these neurons are not spatially arranged. Activation functions are mathematical functions, which are utilized to introduce nonlinearity into the CNN. With the help of activation functions, the network learns complex patterns and relationships in data. There are various activation functions such as sigmoid, rectified linear unit (ReLU), and Tanh. The activation function adopted in the Python code for implementing CNN is ReLU. This function outputs zero for negative inputs, and positive inputs are passed unchanged. It is easy to compute and involves a single threshold computation.

The dataset given in Fig. 2 used in this study pertains to the detection of kidney cancer. It was sourced from the Kaggle website and contained images of healthy kidney cells as well as kidney cells affected by tumors. This data was gathered from the PACS (Picture Archiving and Communication System) from patients diagnosed with kidney tumors in various hospitals in Dhaka, Bangladesh. These were initially represented as DICOM (Digital Imaging and Communications in Medicine) images, which were then converted to lossless JPEG image format. After the conversion,

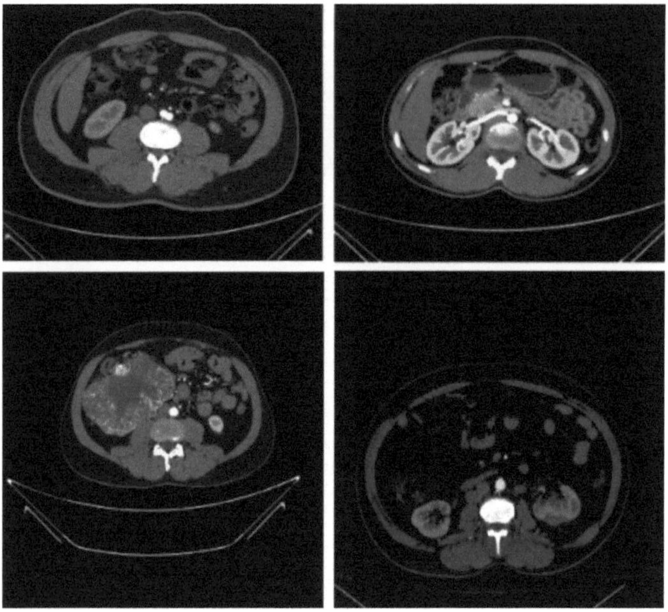

Fig. 2 Sample images from the dataset

the dataset was subjected to verification by a radiologist and a medical technologist, to validate the correctness and accuracy of the data [16]. This image classified as normal includes CT images of healthy kidneys without any visible abnormalities. These images are without any signs of cysts, tumors, or stones. These images are included in the dataset because they are a reference for what healthy kidneys look like. The images classified as tumor in the dataset show the presence of tumors in the kidneys.

The design flow of this work is as follows: it began with the extraction of the dataset. A total of 7360 images from the dataset were subjected to preprocessing, where 5887 images were in the training set and 1473 images in the validation set. This preprocessing step involved for the training set is rescaling. The dataset was resized to 16×16 pixels, and was also segmented into batch size of 32, which indicates the number of images that can be processed simultaneously (Fig. 3).

The model used here is a basic CNN with a structure like the VGG family of networks. The method used for defining our model is sequential, which will connect all the layers sequentially. The different layers were declared in our Python code with the aid of the layers class from Keras in TensorFlow. TensorFlow and Keras are open-source libraries for machine learning and artificial intelligence. These libraries offer a Python interface for building artificial neural networks. The Conv2D class was used for the convolutional layers, dropout class for the dropout layers, and the dense class was used for the fully connected layers, which includes the output layer. A single convolutional layer was utilized in our model with four filters and a constant

Fig. 3 Block diagram of the design flow

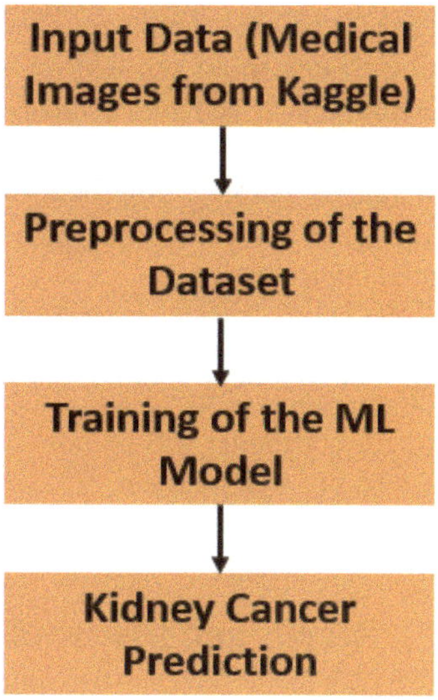

kernel size of 3 × 3. The input shape must be specified only in the first convolutional layer to establish a connection between the layer and the input layer.

The rectified linear unit (RELU) function was applied in the convolutional layer. This filter is used for breaking down the linearity, since the images and the datasets exhibits nonlinear behavior. The dropout layer is also included with a dropout rate of 0.25, which means that 75% of the neurons in the layer are active (working), while 25% are randomly ignored for each pass. This layer helps to prevent overfitting and improves the model reliability. Flattening of the multi-dimensional tensor into a one-dimensional vector by stacking the elements is done after the convolutional layers. This stacked output is given as the input to the artificial neural network (ANN) which is demonstrated with the aid of dense class from Keras of TensorFlow. This dense represents the output layer with 2 units and SoftMax activation function.

Once the building of the CNN model is accomplished, the compilation of the model occurs. The training and testing of the model were done using the Google Colab (Collaboratory) platform. Google Colab is cloud-based platform that offers a free environment for developing and executing machine learning (ML) models, including those used in cancer prediction and detection. It provides complimentary access to graphics processing units (GPUs) and tensor processing units (TPUs) [17], which are essential for training deep learning models such as CNN. These resources accelerate training, especially when working with large datasets like medical images by leveraging Google's computational infrastructure. Google Colab supports various deep learning frameworks like TensorFlow, Keras, PyTorch, and Scikit-learn, and allows users to write and execute Python code within a browser, making it highly accessible for data scientists, researchers, and students.

The compilation of the model is achieved with the help of an Adam optimizer. Adam optimizer is utilized to adjust the weights of a neural network to make it learn better. This optimizer provides several benefits as it combines the advantages of the momentum and RMSprop optimizer in terms of speed and accuracy. While compiling, the categorical cross-entropy (CCE) loss function is used. By minimizing this loss, we are training the model to make its predictions as close to the correct answer as possible, improving the accuracy. Upon completion of the training, the testing of the model was achieved. A total of 100 images were given, and the prediction was done.

So far, we have deployed our ConvNet model and obtained a desirable accuracy. Future steps could be to convert our Python-based model into a high-level language (like C or C++) using the 'hls4ml' package, and then run it using Xilinx Vitis-AI, where our model will be converted into hardware description language (HDL) for FPGA board simulation.

4 Results and Discussions

The mean accuracy of the training set and the validation set was achieved with the help of a python library NumPy, which works excellently with large matrices and dimensional arrays. It comprises of high-level mathematical functions. The model was trained for different number of epochs. Higher value of accuracies was obtained for training and validation for higher epochs. The values of these mean accuracy were obtained and displayed in Fig. 4. There should not be much difference between the mean accuracy values, because larger differences indicate overfitting of the model.

The mean training and mean validation loss were also obtained. Higher value of accuracies results in lower losses. In both the cases, the validation loss is slightly less than the training loss, which suggests that the model is generalizing well and is not overfitting to the training data. These loss values are displayed in Fig. 5.

The plot of the mean accuracy values and the mean loss values were obtained by using the Matplotlib, which is a plotting library for Python. The plots are displayed in Fig. 6.

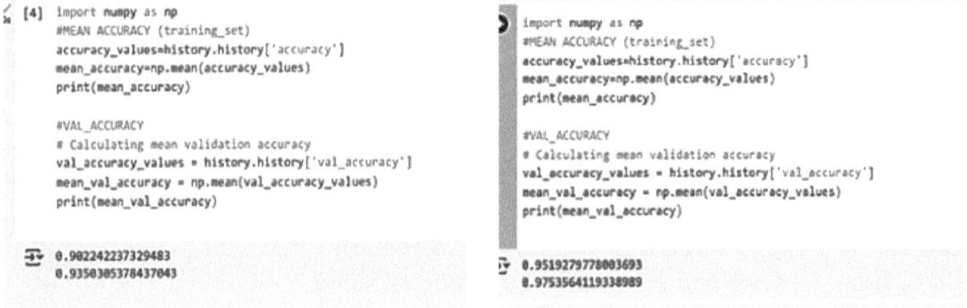

Fig. 4 Mean accuracies

```
[13] import numpy as np
     #MEAN_LOSS (training_set)
     loss_values=history.history['loss']
     mean_loss=np.mean(loss_values)
     print(mean_loss)

     #VAL_LOSS
     # Calculating mean validation accuracy
     val_loss_values = history.history['val_loss']
     mean_val_loss = np.mean(val_loss_values)
     print(mean_val_loss)

     0.1860578663647175
     0.13082293942570686
```

```
import numpy as np
#MEAN_LOSS (training_set)
loss_values=history.history['loss']
mean_loss=np.mean(loss_values)
print(mean_loss)

#VAL_LOSS
# Calculating mean validation accuracy
val_loss_values = history.history['val_loss']
mean_val_loss = np.mean(val_loss_values)
print(mean_val_loss)

0.12804880999028684
0.08572959005832673
```

Fig. 5 Mean losses

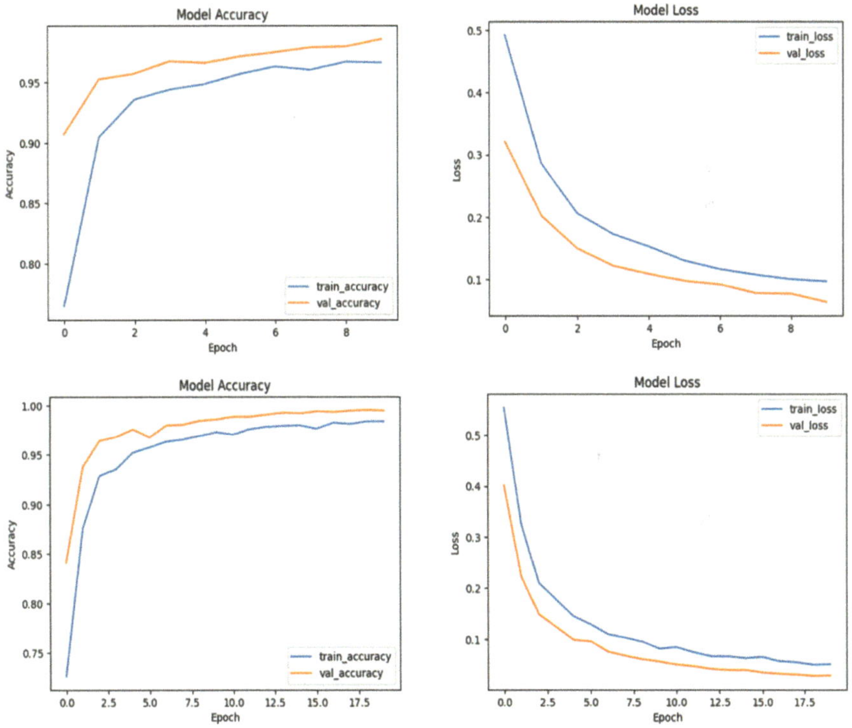

Fig. 6 Model accuracy and loss for 10 and 20 epochs

To evaluate the model's performance, a confusion matrix was generated, displayed in Fig. 7, along with the mean accuracy. Scikit-learn is another open-source ML library for Python, which was utilized to obtain the confusion matrix.

The predicted results are displayed in Fig. 8. The results reveal the similarity in majority of the actual and the predicted values.

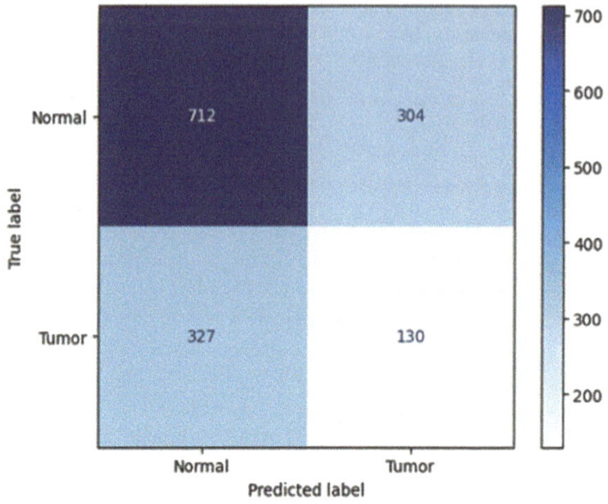

Fig. 7 Confusion matrix of the trained model

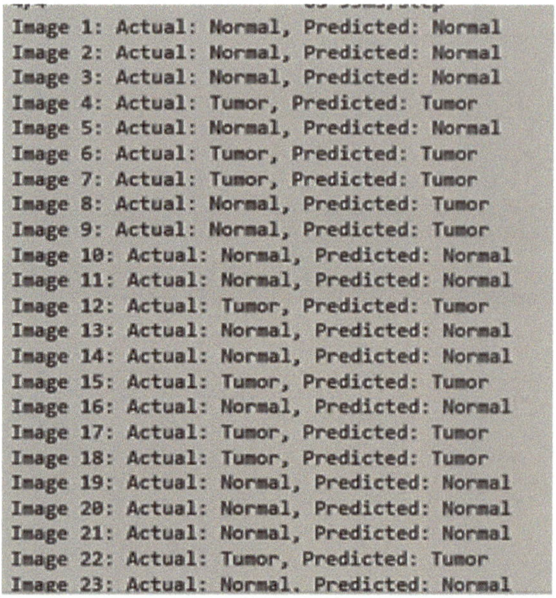

Fig. 8 Preview of the prediction

The difference between predicted values and actual values is very less due to an accuracy of over 95%. When compared to existing deep learning models used for kidney cancer detection, our model exhibits a significant improvement over several others as given in Table 1. The study by Alzu'bi et al. presented VGG16, ResNet50, and 2D-CNN models for kidney tumor detection [8]. Our model's performance is competitive with these state-of-the-art methods, and it performs significantly better than other primitive deep learning models such as DarkNet-53 (78.2%) and AlexNet (50.78%) [18]. Another study about the European Deep-Health toolkit,

Table 1 Comparison of accuracies

Deep Learning Model	Accuracy (%)
AlexNet	50.78
DarkNet-53	78.2
Hybrid Model (U-net + ResNet101)	88.5
RCCGNet	90.14
CNN (Proposed Method)	**97.53**

which utilized U-net for image segmentation and ResNet101 for classification, showed a kidney segmentation accuracy of 84% and an RCC classification accuracy of 88.5% [9]. Furthermore, a hybrid approach integrating convolution transformers achieved 90.62% accuracy in grading renal cell carcinoma [7], while a fully automated deep learning framework, RCCGNet, deployed in [5] achieved 90.14% accuracy. Once again, our model surpassed these approaches. A learning rate of 0.001 was followed for these comparisons.

These comparisons highlight the superior accuracy and potential of our model in improving kidney cancer detection. The notable performance of our model not only surpasses previous benchmarks but also illustrates the effectiveness of deep learning in medical image analysis. This enhancement can be attributed to the use of a deeper CNN architecture, which allowed for more complex feature extraction and a better understanding of the patterns present in histopathological images.

The novelty of this project lies in the implementation of a simpler model which provides higher accuracy when compared to existing models, marking a significant advancement in the field. This approach not only showcases the feasibility of implementing complex machine learning algorithms but also paves the way for real-time processing capabilities in clinical settings.

5 Conclusion

This research demonstrates the effectiveness of using CNN for kidney cancer detection, with an accuracy of 97.53% making it competitive with existing deep learning approaches. Our model's deeper architecture, combined with the computational efficiency of CNN, positions it as a viable tool for real-time medical diagnostics.

The results suggest that our CNN model can provide a scalable, energy-efficient alternative to traditional GPU-based deep learning models in the medical field, particularly for real-time applications like cancer detection. While challenges such as dataset availability and computational demands remain, the findings point to significant opportunities for further research and application, like extending the model's application to different types of cancer and exploring ways to optimize the model further, particularly in terms of reducing hardware constraints and increasing dataset diversity.

For our future progress, may also implement our model using field-programmable gate arrays (FPGAs). This is because FPGAs are integrated circuits (ICs) with a reprogrammable hardware architecture and can execute algorithms at high speeds, with low latency and power consumption. Unlike traditional processing units like CPUs, GPUs, or ASICs, the internal circuitry of an FPGA is not permanently fixed, allowing it to be reconfigured as needed [19]. This capability makes them particularly suited for tasks that require high computational power, such as running machine learning algorithms on medical imaging data. Moreover, FPGAs are reconfigurable hardware platforms, i.e., they can be reprogrammed to suit our needs, whenever required. Hence, by combining machine learning with FPGA technology, we can aim to revolutionize kidney cancer detection, offering a faster, more efficient, and scalable diagnostic too.

References

1. Tyagi AK (2021) Data science and data analytics: opportunities and challenges
2. Cirillo L, Innocenti S, Becherucci F (2024) Global epidemiology of kidney cancer. Nephrology dialysis transplantation, pp 920–928
3. Hussain S, Mubeen I, Ullah N, Shah SS, Khan BA, Zahoor M, Ullah R, Khan FA, Sultan MA (2022) Modern diagnostic imaging technique applications and risk factors in the medical field: a review. Biomed Res Int
4. Yamashita R, Nishio M, Do RK, Togashi K (2018) Convolutional neural networks: an overview and application in radiology. Insights Imaging, pp 611–629
5. Gujarathi SM, RS, Angadi SK (2024) A survey of kidney cancer analysis using machine learning and deep learning algorithms. J Electr Syst, pp 2491–2501
6. Zhang M, Ye Z, Yuan Z (2024) Imaging-based deep learning in kidney diseases: recent progress and future prospects. Insights in Imaging,
7. Mahmood T, Wahid A, Hong JS, Kim SG, Park KR (2024) A novel convolution transformer-based network for histopathology image classification using adaptive convolution and dynamic attention. Eng Appl Artif Intell
8. Alzu'bi D, Abdullah M, Hmeidi I, AlAzab R, Gharaibeh M, El-Heis M, Almotairi KH, Forestiero A, Hussein AM, Abualigah L (2022) Kidney tumor detection and classification based on deep learning approaches: a new dataset in CT scans. J Healthcare Eng
9. Bolocan VO, Secareanu M, Sava E, Medar C, Manolescu LS, Cătălin Rașcu AȘ, Costache MG, Radavoi GD, Dobran RA, Jinga V (2023) Convolutional neural network model for segmentation and classification of clear cell renal cell carcinoma based on multiphase CT images. J Imaging
10. Lal S, Chanchal AK, Kini J, Upadhyay GK (2024) FPGA implementation of deep learning architecture for kidney cancer detection from histopathological images. Multimedia Tools Appl
11. Geetha M, Neena A (2017) A review on deep convolutional neural networks. In: International conference on communication and signal processing, IEEE Advancgin Technology for Humanity
12. Krizhevsky A, Sutskever I, Hinton GE (2012) ImageNet classification with deep convolutional neural networks. Adv Neural Inf Process Syst
13. Nebauer C (1998) Evaluation of convolutional neural networks for visual recognition. In: IEEE transactions on neural networks
14. Shuang Cong YZ (2022) A review of convolutional neural network architectures and their optimizations. Springer, Artificial Intelligence Review
15. Lee KB, Cheon S, Kim CO (2017) A convolutional neural network for fault classification and diagnosis in semiconductor manufacturing processes. IEEE transactions on semiconductor manufacturing, pp 135–142

16. Sasikaladevi N, Revathi A (2024) Digital twin of renal system with CT-radiography for the early diagnosis of chronic kidney diseases. Biomedical signal processing and control
17. Canesche M, Bragança L, Neto OP, Nacif JA, Ferreira R (2021) Google Colab CAD4U: hands-on cloud laboratories for digital design. In: IEEE international symposium on circuits and systems (ISCAS), Daegu, South Korea
18. Taha ET, Mustafa TE (2024) Enhanced deep learning based decision support system for kidney tumour detection. BenchCouncil transactions on benchmarks, standards and evaluations
19. Sasikaladevi N, Revathi A (2024) Digital twin of renal system with CT-radiography for the early diagnosis of chronic kidney diseases. Biomedical signal processing and control

Deep-Learning-Based Physical Rehabilitation Activity Classification Using Kinect Sensor Data

Aparna Haldar, Uzzal Biswas, Mridul Paul, Abdullah-Al Nahid, and Abu Shamim Mohammad Arif

Abstract Though people want to live a healthy life, they experience some form of disabilities at some stages due to various diseases or injuries. Physical rehabilitation is an effective approach to recover from injuries and get back to normal life. Conventional rehabilitation approach is usually expensive and has limited training sessions that lead to slow progress in condition. A recent technological development brings a revolution by introducing sensor-based automatic rehabilitation. Kinect sensor is one of the well-known skeletal tracking devices that can monitor activity and provide feedback regarding the progress and movement adjustments. To track progress, it is necessary to perform correct classification of patient gestures. In this study, a deep learning technique, convolution neural network (CNN), has been applied on Kinect sensor captured IntelliRehab DS (IRDS) dataset to perform accurate classification of patient gestures. Different optimization algorithms were also applied to optimize CNN, where Adam-based CNN shows the highest accuracy of 89.34%.

Keywords Physical rehabilitation · Kinect sensor · Activity monitoring · Activity classification · Deep learning

A. Haldar (✉) · A. S. M. Arif (✉)
Computer Science and Engineering Discipline, Khulna University, Khulna, Bangladesh
e-mail: aparnahaldarku@gmail.com

A. S. M. Arif
e-mail: shamim@cse.ku.ac.bd

U. Biswas · M. Paul · A.-A. Nahid
Electronics and Communication Engineering Discipline, Khulna University, Khulna, Bangladesh
e-mail: uzzal.biswas@ece.ku.ac.bd

A.-A. Nahid
e-mail: nahid@ece.ku.ac.bd

H. Nagaraj et al. (eds.), *Advancements in Artificial Intelligence and Machine Learning*, Lecture Notes in Electrical Engineering 1572,
https://doi.org/10.1007/978-981-95-7449-0_33

1 Introduction

Many people experience some sort of disability at different stages of their whole life. According to World Health Organization (WHO), an estimated 1.3 billion people, which is the 16% of total world's population, are now living with significant disability [1]. Disabilities may happen due to accident or injuries like sports injury, road accident, as well as several disease conditions like ischemic heart disease, stroke, which are the major causes of death nowadays [2, 3]. After these adverse events (i.e., accidents, injuries, and diseases), the patients usually go through surgery and have moderate-level impairments. As a result, they need special care and rehabilitation to get back to their normal activities. Therefore, for postoperative recovery and treatment, more specifically to restore patient's physical, sensory, and physiological conditions, participation in physical rehabilitation is usually compulsory after surgery. However, providing the patient's access to a clinician in every single session is unrealistic, time-consuming, and highly expensive [4].

In current healthcare system, the primary portion of the rehabilitation program is performed by the direct supervision of a clinician in hospital inpatient setting while the secondary portion is conducted on outpatient setting, where the patient complete predefined exercise at home. A report stated that 90% of all rehabilitation sessions are conducted in outpatient setting [5]. In this setting, patients are requested to perform prescheduled exercise, record their progress, and follow-up on the clinic frequently in a certain interval to assess their functional improvement. Several studies reported that, in this home-based rehabilitation, the patient nonadherence to recommended exercise is a significantly high, which is ranging from 34 to 74% [6]. This low adherence rate usually increased the duration of the treatment time and related healthcare costs. A study reported that the lack of continuous feedback and oversight of patient exercise progress by a healthcare expert are the two main factors of this high nonadherence rate in home-based rehabilitation [7]. Because it is recommended that the rehabilitation session should always be supervised by an expert clinician, who can observe the patient physically and perform correct evaluation of the progress.

The development in various wearable sensors and virtual devices over the last few years played a significant role to overcome the limitations in home-based rehabilitations where the patients may perform their recommended exercise and body parts movements at home and the expert clinician can supervise and monitor the session from hospital setting [8]. In addition, the use of those sensors may reduce the transportation issues from patient's home to hospital setting for training and follow-up and save time and travel cost. Virtual reality is one of the most well-known rehabilitation tools that can compare rehabilitation in between real and virtual worlds. Nowadays, many clinical research experts and therapists have utilized the use of virtual reality to recover and assess patients' disabled skills and functionalities [9, 10]. Over the last decade, the development in machine learning and wearable sensors has created the feasibility of a new technology-assisted physical rehabilitation system based on Kinect sensor. Kinect sensor was launched by Microsoft in 2010, which is a line of motion sensing devices that can capture patients' movements, postures and speech using a 3D depth camera [11]. The camera of Kinect sensor sends near-infrared rays to identify patient movements. These non-wearable sensors can also effectively capture the joint positions and angle trajectories of the movements and provide the longitudinal time-series data efficiently. Various kinematic metrics such as range of motion and mean error can be determined from these joints rotational data.

As the number of elderly and disabled people are increasing day by day, nowadays, robots are also being used in rehabilitation program to help the patients to move their different body parts such as entire body, upper and lower limb [12]. Robots are also assisting stroke patients to perform recommended exercise by applying force in certain body parts [13]. In these rehabilitation programs, as the activity monitoring and classification is a complex problem, therefore, to ensure proper classification, observation, and treatment plan, machine learning has been playing a crucial role for the last few years [14]. Machine learning models usually employ complex algorithms to develop more accurate and robust models to assess the rehabilitation activity. It not only classifies patient activity but also predicts patient progress. However, most of the machine learning-based rehabilitation research has been conducted using upper body movements and vision-based gesture detection, which may not perform well for the accurate recognition of human activity. For complex human activity recognition, deep learning approach shows superior performance than conventional machine learning approaches [15]. Therefore, in this study, we focused on Kinect sensor and deep learning-based automatic rehabilitation monitoring and activity classification model.

The remainder of the chapter is organized as follows. A brief overview of the related literature is outlined in Sect. 2; the dataset used and the proposed method are described in Sect. 3; results and discussion of the study findings are included in Sect. 4; and Sect. 5 represents conclusion and future direction of the proposed work.

2 Literature Review

Though rehabilitation is serving injured and disabled people from ancient times, it has been widely used for the recent years. As rehabilitation is a vast research field, a lot of work has been conducted on machine-learning-based activity monitoring by prolific scholars, where some works are related to our objective and motivation. However, there is limited work on automatic rehabilitation monitoring and deep learning-based gesture classification. Some of the recent research on rehabilitation activity monitoring, and classification are discussed below.

In 2017, Zhi et al. [16] performed automatic detection of compensation during robotic stroke rehabilitation therapy. Data were recorded from 9 stroke patients and 10 healthy individuals, who performed their exercise with the help of a table-top rehabilitation robot. They used support vector machine (SVM) and recurrent neural network (RNN) to detect the lean-forward (LF), truck-rotation (TR) and shoulder elevation (SE) compensations. Though both classifiers showed good AUC and F1 score for detecting the LF compensation, they had a moderate and very poor detection performances for the TR and SE compensations. To improve the detection performance, they recommended the predefinition of motion range, adjustment of camera placement, and full therapy sessions recording.

Miron et al. [17] conducted a study on action correctness classification in physical rehabilitation exercises in 2018. The study used University of Idaho-Physical Rehabilitation Movement Data (UI-PRMD) dataset and applied temporal convolution neural network (Res-TCN) to perform the classification of activity correctness. However, the study had the limitation of high variability in classifying the correctness for various movements. Therefore, they recommended the utilization of data augmentation technique and application of more advanced approaches like convolution neural networks (CNNs) in future study.

In the next year, a study on gesture correctness estimation with deep neural networks and rough path estimators was performed by Sadawi et al. [15]. The study collected the rehabilitation data from 17 patients and 14 healthy individuals using Kinect sensor as a as part of the project from Perkeso Tun Razak Rehabilitation Centre, Malacca, Malaysia. They applied convolution neural networks (CNNs), long short-term memory (LSTM) neural networks, and rough path-based classifiers to perform activity correctness estimation. The study achieved an average accuracy of 85.24%, 85.63%, and 94.67%, respectively, for three different techniques across all features. However, they had the limitation of insufficient training data for some samples, and they used one data augmentation method.

In 2020, Liao et al. [18] conducted a study on automatic assessment of physical rehabilitation exercise using a neural network (NN) and LSTM-based deep learning framework. The study used Vicon optical tracking system to record 10 physical rehabilitation exercises from 10 healthy patients and developed the University of Idaho-Physical Rehabilitation Movement Data (UI-PRMD) dataset. Though the exercise quality scores of the proposed framework were close to the ground truth quality scores of the movements, the study had several limitations. They performed the validation on the rehabilitation exercises captured by expensive optical tracking system from 10 healthy individuals without any real patients. Moreover, the largest segment of the validation was performed without any ground truth assessment of quality score labeled by the clinicians. The partial validation of deep squat exercise was conducted against KIMORE dataset, where data were captured by low-cost sensors.

A study on patient identification based on physical rehabilitation movements using skeleton data was investigated by Kim et al. [19] in 2021. The study utilized IRDS dataset, which was captured by using Kinect sensor from 15 patients and 14 normal subjects for 9 different gestures. A pre-trained ResNet backbone was used to identify the patient based on different gestures. The ResNet model achieved an accuracy of 98.59% for shoulder flexion left gestures to identify the patient though the performance was bit lower for other gestures. Furthermore, the study had the limitation of insufficient data, as the IRDS dataset has 300 videos for each gesture, which are insufficient to train ResNet.

In 2024, Zaher et al. [20] investigated the potential of RNN and CNN models for the accurate classifications of rehabilitation exercise on multi-datasets. Two different datasets KIMORE and UI-PRMD were utilized for the investigation of exercise classification. They applied different deep learning models such as Bi-LSTM, LSTM, CNN, and CNN-LSTM and performed hyperparameter tuning using random search algorithm. The outcome represented that CNN model shows best performance among all models for exercise classifications with 93.08% accuracy on KIMORE dataset and 99.7% accuracy on UI-PRMD dataset. However, the best performance of CNN model comes with reliance on preprocessed dataset, which is limiting the adaptability and generalizability of the model on real-world situations where the precise cropping may not be readily available.

Recently, Tahsin et al. [21] conducted a study on machine learning-based stroke patient rehabilitation stage classification using Kinect data. The study investigated Toronto Rehab Stroke Pose Dataset (TRSP), which was captured by using Kinect sensor from rehabilitation patients. Several conventional classifiers such as extreme gradient boosting (XGB), random forest (RF), decision tree (DT), K-nearest neighbor (K-NN), and Gaussian Naïve Bayes (GNB) were applied for activity classification, where XGB shows best performance with 92% accuracy. The study used hand-crafted features for activity classifications. Furthermore, they recommended to use deep learning model to improve the classification accuracy.

Among all the previous studies on physical rehabilitation activity, classifications usually applied conventional machine learning algorithms except Zaher et al. [20] who used deep learning model on two other datasets. All the other studies who used deep learning models mostly focused on gesture correctness other than activity classification. Furthermore, there is very limited work on IRDS dataset, while no study was conducted for activity classification. After the investigation on previous works, we have motivated Kinect-based automatic physical rehabilitation along with feedback strategy. In this study, we are going to classify physical rehabilitation activity using a powerful deep learning model with real-time monitoring for better performance.

3 Materials and Methods

The main aim of our study is the Kinect sensor-based automatic physical rehabilitation using a CNN deep learning approach. To do this, initially, we performed a few preprocessing steps to fit the raw data image into the CNN model. Then, we divided the dataset into different considerable ratios to train and test the model. Finally, the model was optimized using different optimizers for better performance. The flow diagram of the proposed methodology is shown in Fig. 1.

3.1 Dataset

IRDS dataset has been used in this study, which captures 3D data from 25 body skeleton joints using Kinect sensor at a frame rate of 30 frame per second. This dataset was collected from [22]. The dataset contains 9 gestures of 29 subjects where

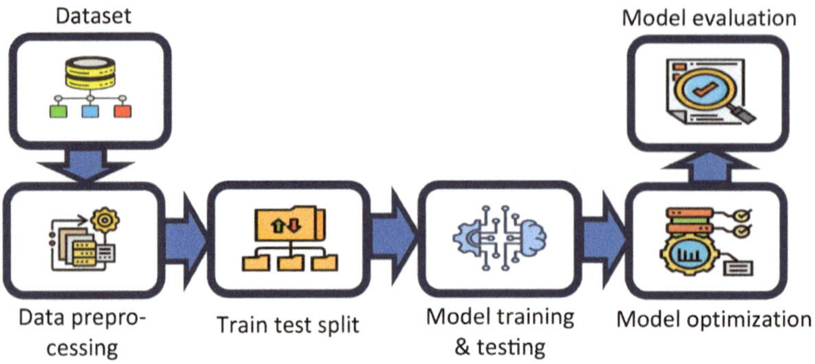

Fig. 1 Overall process of the proposed methodology

15 (IDs 201 to 216) were real patients and 14 (IDs 101 to 107 & IDs 301 to 307) were healthy individuals. Two independent annotators annotated the performed exercise in the dataset according to movement type and labeled the data based on the activity which was performed correctly or not. The nine different gestures consist of nine different classes in the dataset such as elbow flexion left, elbow flexion right, shoulder flexion left, shoulder flexion right, shoulder abduction left, shoulder abduction right, shoulder forward elevation, side tap left, and side tap right, which are given in Table 1. The distribution of all these nine different gestures class in the dataset is relatively balanced as shown in Fig. 2. The number of gesture repetitions were varied among the subjects, which results in a total of 2577 gesture sequences recording, where 1215 sequences were performed during standing, 957 sitting on chair, 359 sitting on wheelchair, and 51 with the help of a support stand frame. Finally, the date was stored in CSV files, where each file contains the exercise of a single person.

Table 1 Nine different gestures class and name in the dataset

Index of gesture	Gesture name
0	Elbow flexion left
1	Elbow flexion right
2	Shoulder flexion left
3	Shoulder flexion right
4	Shoulder abduction left
5	Shoulder abduction right
6	Shoulder forward elevation
7	Side tap left
8	Side tap right

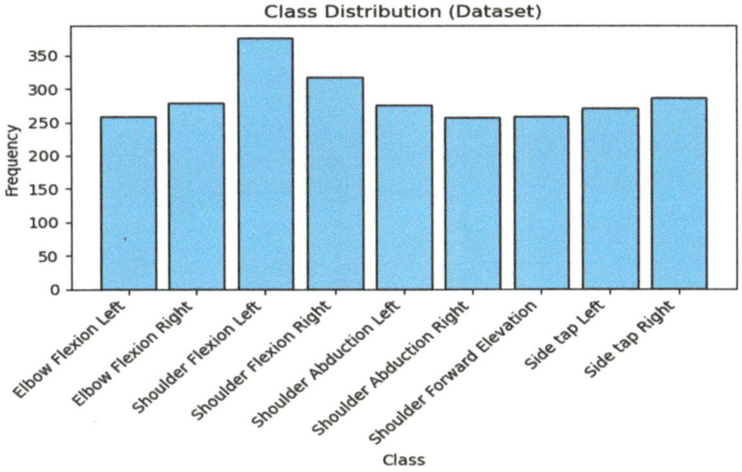

Fig. 2 Distribution of nine gestures class in the dataset

3.2 Data Preprocessing

Though the data from Kinect sensors represents the 25 body joints skeleton information in 3D axis, Duan [19] reported that 2D skeleton showed better performance than the 3D skeleton. Therefore, we reproduced all the images for each gesture from three-dimensional axis to two-dimensional axis. For each gesture, as the original dataset contains the movement information of different locations in different frames, we recombined all the information for each gesture from different frames into a single frame, as shown in Fig. 3 for shoulder flexion left gesture. Moreover, as the longest dimension of the image may not fit into the target dimension (128 × 128), therefore, according to the requirement of CNN model, all the images were resized to match the target size, as shown in Fig. 4 for side tap right gesture.

3.3 Train–Test Split

Data splitting is a strategy to divide the collected data into two or more subsets for training, testing, and validation purposes. It is important for investigating the model performance and generalization capabilities, facilitating to avoid overfitting and underfitting. In this study, the datasets are divided into three different subset combinations such as 70:30, 75:25, and 80:20, respectively, for training and testing the model.

Fig. 3 Regeneration of images from 3 to 2D for shoulder flexion left gesture

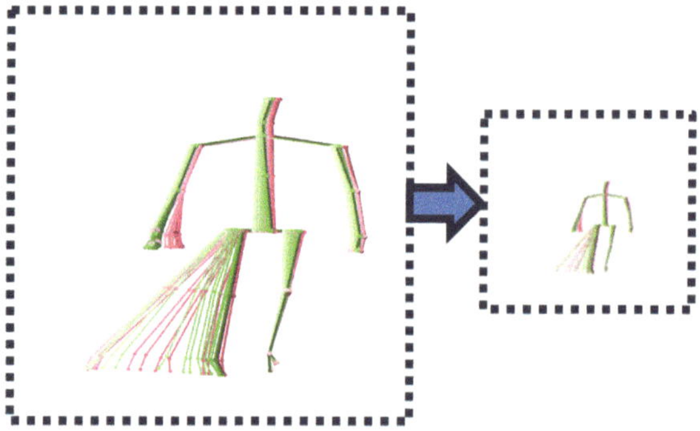

Fig. 4 Resize image to match the longest dimension into target size for side tap left gesture

3.4 Deep Learning Model: CNN

CNN is a deep learning network architecture that can learn directly from data, which is especially useful for identifying the pattern in images to determine objects, categories, and classes [15]. CNN architecture may have tens of hundreds of layers that can learn different features of an image. Here, various filters are applied to the training images at different resolutions, and the output of each convolved image is considered as the input of the next layer. The filter usually starts with simple features and then increases the complexity to identify the features that uniquely define the image. There are three main layers in the model, where the input layer takes the gesture images as input and output layers classify the image into one of nine gesture classes. In between these two layers, there are three main hidden layers that perform operations to learn the specific features of the images. Among these hidden layers, convolution layer passes the input images through a set of convolution filters to identify certain features. The rectified linear unit allows faster and efficient training by mapping the

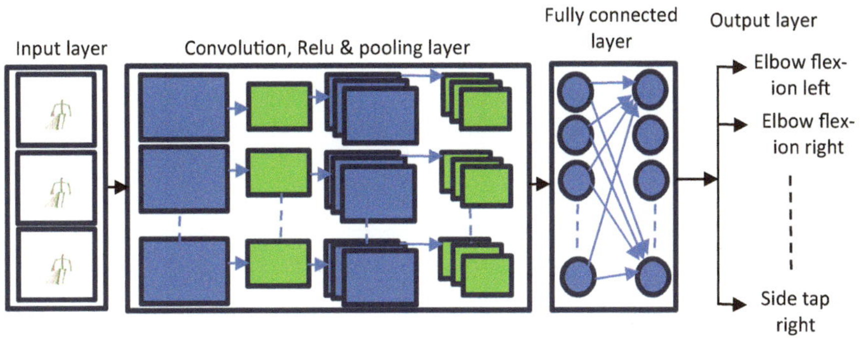

Fig. 5 CNN architecture for rehabilitation activity classification

negative values to zero and keeping only positive values and forwarding them to the next layers. Finally, the output is simplified by pooling layer that performs nonlinear down sampling to reduce the network learning parameters. Before output layer, there is a fully connected layer that generates output vectors of class dimensions to represent the probabilities for each gesture class of an image that is being classified. As our image is RGB, three input channels were utilized. A total of 12 filters were applied in the first convolution layer; second layers take 12 input channels and generate 24 outputs. Finally, max pooling was applied with kernel size 2. A CNN architecture is shown in Fig. 5.

3.5 Hyperparameter Optimization

There are few parameters in every model that control the performance of the model. These parameters can be optimized to achieve the better result of the model. In this study, we have used four different optimizing algorithms such as Adam, Nadam, Adagrad, and Adadelta to tune the hyperparameters and optimize the performance of CNN classification model [23]. These are adaptive learning algorithms that can adapt into different dimensions and layers to optimize the model performance. Adam optimization is a stochastic gradient descent method, whereas Nadam is a new variant of Adam that incorporates the Nesterov momentum into Adam. Adagrad is an optimizer with parameter-specific learning rate that is adapted based on the frequency of the parameter being updated during training. Adadelta is also stochastic gradient descent method.

3.6 Model Evaluation

Various performance metrics have been used in this study to quantitatively evaluate the performance of the CNN classifier such as accuracy, precision, and F1 score. These metrics can be defined as follows:

$$\text{Accuracy} = \frac{TP + TN}{TP + TN + FP + FN} \quad (1)$$

$$\text{Precision} = \frac{TP}{TP + FP} \quad (2)$$

$$F1\text{score} = \frac{TP}{\left(TP + \frac{1}{2}(FP + FN)\right)} \quad (3)$$

In the above equations, TP is true positive, TN is true negative, FN is false negative, and FP is false positive.

ROC Curve: A receiver operating characteristic curve which is a plot of true positive rate against false positive rate has been used in this study to qualitatively evaluate the performance of CNN classifier. As there are multiple activity classes, a multiclass ROC has been plotted here by combining the ROC for each class.

4 Results and Discussions

We have applied a CNN-based deep learning model on IRDS dataset to perform the classification of nine different gestures. The CNN model is then optimized using Adam, Nadam, Adagrad, and Adadelta optimizers, where Adam shows the better performance for classification. The comparison of performance and analysis of the model is performed based on different performance metrics, confusion matrix, and ROC curve. The performance is investigated initially for basic CNN model and then for optimized CNN model.

4.1 Performance Analysis with CNN Model

To determine the generalized performance on IRDS dataset and to perform the accurate classification of nine exercise gestures, we have investigated the performance of basic CNN model with default Adam optimizer on three different (70:30, 75:25, and 80:20) train–test split dataset combinations. Table 2 presents the performance of basic CNN classifier on three train–test subset combinations, where we have made a comparison between different performance metrics such as accuracy, precision, and

F1 score. From the table, we have achieved the maximum classification accuracy of 89.34% for 80:20 dataset combinations and 86.36% for 75:25 combinations, where other subset has a bit lower accuracy. Meanwhile, the precision and F1 score were almost similar for both 80:20 and 75:25 dataset combinations. As the basic CNN with Adam optimizer provides the generalized and best classification performance for 80:20 train-test dataset combination, we have selected this combination for further analysis of CNN model.

The multiclass ROC curve for the basic CNN model on 80:20 dataset combination is shown in Fig. 6, where nine curves represent the nine different classes.

Each color in the ROC curve represents the classification accuracy for the corresponding gesture. The classes are labeled as elbow flexion left (0), elbow flexion right (1), shoulder flexion left (2), shoulder flexion right (3), shoulder abduction left (4), shoulder abduction right (5), shoulder forward elevation (6), side tap left (7), and side tap right (8). The area under the curve (AUC) represents classification performance,

Table 2 Performance metrics of basic CNN model for nine gestures

Dataset combinations train: test	Accuracy (%)	Precision (%)	F1 Score (%)
70:30	83.33	84	84
75:25	86.36	89	88
80:20	89.34	88	88

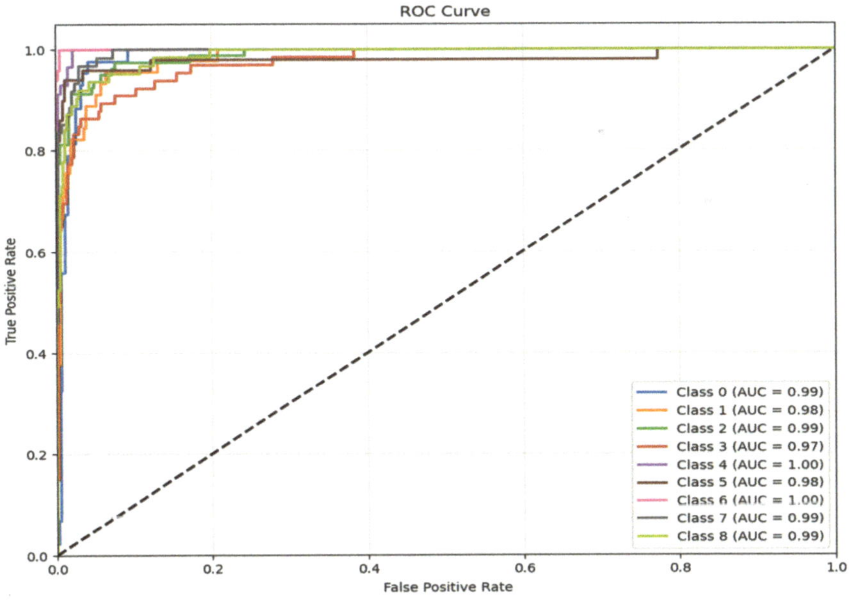

Fig. 6 ROC curve of basic CNN model with Adam optimizer for nine (0–8) gestures

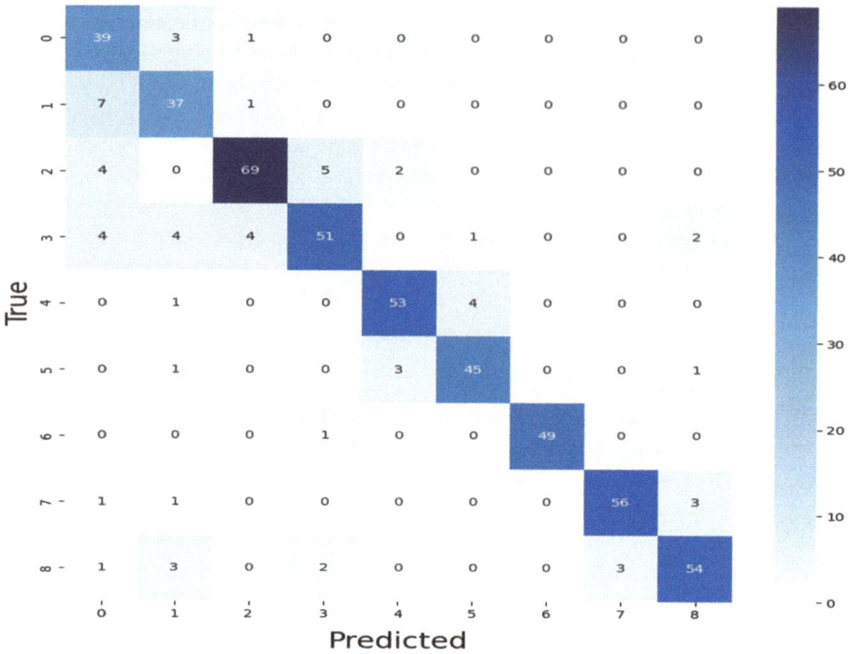

Fig. 7 Confusion matrix of basic CNN model with Adam optimizer for nine (0–8) gestures

whose value higher is better. The CNN model shows the best performance for class 4 and 6, though the difference with other classes is not much.

A confusion matrix represents the summary of a number of current and incorrect classifications made by the classifier. In multiclass problems, the dimension of the confusion matrix represents the number of classes, and the entries determine the correct and incorrect prediction. Figure 7 represents the confusion matrix for CNN classifier, where we can see for class 6, the classifier correctly classified all 49 gesture sequences, while the performance of other classes was also very close.

4.2 *Performance Analysis with Optimized CNN Model*

To tune the hyperparameters and optimize the performance of CNN model, we applied four different optimization techniques (Adam, Nadam, Adagrad, and Adadelta) with the learning rate 0.001 to the previously selected 80:20 dataset combinations. The performance of all different optimizers along with their learning rate is given in Table 3. From the table, we can see that the Adam optimizer still represents the best performance with 89.34% accuracy, 88% precision, and 88% F1 score than the other optimizers' 80:20 dataset combination. Therefore, we have selected the

Table 3 Learning rate and performance metrics of Adam, Nadam, Adagrad, and Adadelta optimizer

Optimizer name	Learning rate	Accuracy (%)	Precision (%)	F1 Score (%)
Adam	0.001	89.34	88	88
Nadam	0.001	86	88	87
Adagrad	0.001	66.47	67	66
Adadelta	0.001	13.57	2	3

Table 4 Comparison of our work with other related works

Work	Dataset	Used model	Accuracy (%)	Precision (%)	F1 Score (%)
[17]	TRSP	SVM	–	86	87
[21]	KIMORE	CNN	93.08	93.07	91.79
[21]	UI-PRMD	CNN	99.70	99.70	99.70
[22]	TRSP	XGB	92	–	92
Our study	IRDS	CNN	89.34	88	88

Adam-based optimized CNN model as the best classifier for physical rehabilitation activity classification.

As there is no study on activity classification on utilized dataset, we cannot directly compare our findings with others on the same dataset. However, we have investigated our performances with respect to other studies on related datasets. Table 4 presents the comparison of our work with other related works.

For activity classification, a few studies used conventional machine learning approaches, and one achieved the maximum accuracy of 92% for XGB on a different TRSP dataset. Though one study utilized a deep learning model and achieved good accuracy, precision, and F1 score, they used two other datasets. In our study, we applied a deep learning model on the new IRDS dataset for the first time and achieved 89.34% accuracy with higher precision and F1 score. Moreover, in this study, we investigated optimizing performance of CNN model with four different optimizers and acquired best performance for Adam optimizer. However, our proposed model relies on a little bit of preprocessing of the input image, which might be an issue in some cases where preprocessing is unavailable for input image.

5 Conclusion

Powerful deep learning-based and Kinect-sensor-assisted physical rehabilitation activity classification and feedback approach could be great addition to automatic rehabilitation system. Accurate classification of patient exercise is very crucial in automatic rehabilitation. In this study, a deep learning model has been applied on a new IRDS dataset, which achieved an accuracy of 89.34%. The hyperparameters of the model are optimized with four optimizing techniques. The proposed model may be a great help in real-time gesture monitoring system to efficiently classify the exercise and assist the patients to get back to their normal life. In future, the model can be validated on other available datasets, and its accuracy can be increased with more advanced deep learning model and optimization technique.

References

1. World Health Organization. "Disability," https://www.who.int/news-room/fact-sheets/detail/disability-and-health.
2. Boehme AK, Esenwa C, Elkind MS (2017) Stroke risk factors, genetics, and prevention. Circ Res 120(3):472–495
3. Khazaie H, Zakiei A, Rezaei M, Brand S, Komasi S (2020) The role of traffic and road accidents in causing disabilities in Iran. Iran J Public Health 49(9):1804–1805
4. Machlin SR, Chevan J, Yu WW, Zodet MW (2011) Determinants of utilization and expenditures for episodes of ambulatory physical therapy among adults. Phys Ther 91(7):1018–1029
5. Komatireddy R, Chokshi A, Basnett J, Casale M, Goble D, Shubert T (2014) Quality and quantity of rehabilitation exercises delivered by a 3-d motion controlled camera: a pilot study. Int J Phys Med Rehabil 2(4)
6. Niyonsenga J, Uwingeneye L, Musabyemariya I, Sagahutu JB, Cavallini F, Caricati L, Eugene R, Mutabaruka J, Jansen S, Monacelli N (2024) The psychosocial determinants of adherence to home-based rehabilitation strategies in parents of children with cerebral palsy: a systematic review. PLoS ONE 19(6):e0305432
7. Miller KK, Porter RE, DeBaun-Sprague E, Van Puymbroeck M, Schmid AA (2017) Exercise after stroke: patient adherence and beliefs after discharge from rehabilitation. Top Stroke Rehabil 24(2):142–148
8. De Fazio R, Mastronardi VM, De Vittorio M, Visconti P (2023) Wearable sensors and smart devices to monitor rehabilitation parameters and sports performance: an overview. Sensors (Basel) 23(4)
9. Qiu Q, Cronce A, Patel J, Fluet GG, Mont AJ, Merians AS, Adamovich SV (2020) Development of the home based virtual rehabilitation system (HoVRS) to remotely deliver an intense and customized upper extremity training. J Neuroeng Rehabil 17(1):155
10. Arntz A, Weber F, Handgraaf M, Lällä K, Korniloff K, Murtonen KP, Chichaeva J, Kidritsch A, Heller M, Sakellari E, Athanasopoulou C, Lagiou A, Tzonichaki I, Salinas-Bueno I, Martínez-Bueso P, Velasco-Roldán O, Schulz RJ, Grüneberg C (2023) Technologies in home-based digital rehabilitation: scoping review. JMIR Rehabil Assist Technol 10:e43615
11. Vaughan N, Gabrys B, Dubey VN (2016) An overview of self-adaptive technologies within virtual reality training. Comput Sci Rev 22:65–87
12. Duncan CC, Edgley SR (2019) 13—upper limb Orthoses for the stroke- and brain-injured patient. Atlas of orthoses and assistive devices (Fifth Edition), Webster JB, Murphy DP (eds), pp 146–156.e2, Philadelphia: Elsevier
13. Chang WH, Kim YH (2013) Robot-assisted therapy in stroke rehabilitation. J Stroke 15(3):174–181
14. Chaturvedi S (2023) Clinical prediction on ML based internet of things for E-health care system. Int J Data Inf Intell Comput 2:29–37
15. Sadawi N, Miron A, Ismail W, Hussain H, Grosan C, Gesture correctness estimation with deep neural networks and rough path descriptors, pp 595–602
16. Zhi YX, Lukasik M, Li MH, Dolatabadi E, Wang RH, Taati B (2018) Automatic detection of compensation during robotic stroke rehabilitation therapy. IEEE J Transl Eng Health Med 6:2100107

17. Miron AD, Grosan CJA (2021) Classifying action correctness in physical rehabilitation exercises. abs/2108.01375
18. Liao Y, Vakanski A, Xian M (2020) A deep learning framework for assessing physical rehabilitation exercises. IEEE Trans Neural Syst Rehabil Eng 28(2):468–477
19. Kim JK, Lee KB, Kim JC, Hong SG, Patient identification based on physical rehabilitation movements using skeleton data, pp 1572–1574
20. Zaher M, Ghoneim AS, Abdelhamid L, Atia A (2024) Unlocking the potential of RNN and CNN models for accurate rehabilitation exercise classification on multi-datasets. Multimedia tools and applications
21. Tahsin T, Mumenin KM, Akter H, Tiang JJ, Nahid A-A (2024) Machine learning-based stroke patient rehabilitation stage classification using Kinect data. Appl Sci
22. Miron A, Sadawi N, Ismail W, Hussain H, Grosan C (2021) IntelliRehabDS (IRDS)—a dataset of physical rehabilitation movements. Data
23. Debnath S (2024) Adagrad, Adadelta, RMSProp &Adam variants-Part 2 of optimization algos for deep learning. https://medium.com/@saurad44/adagrad-adadelta-rmsprop-adam-variants-part-2-of-optimization-algos-for-deep-learning-714400d1b44e

Intelligent Sleep: A Smart Bed Solution for Snoring Prevention and Comfort Optimization

Pravardhini Sankireddy, Vedha Pranava Mateti, and S. Thangam

Abstract Quality sleep is an essential component for health and productivity, yet issues like snoring can seriously compromise it. This project proposes the idea of a smart bed system to improve the quality of sleep using advanced networking technologies. The system consists of anti-snoring mechanisms, which on detecting snoring, tilt the upper portion of the bed to relieve pressure on the nasal and throat area, eliminating snoring in the gentlest way possible. It has contact and noncontact temperature sensors which track as well as controls the user's skin and air temperature to keep it at the optimal thermal condition. Its sensors also identify sleeping postures and classify them as either healthy or unhealthy, configuring the bed to favor health-conscious postures. This innovative smart bed ensures a comfortable, personalized, non-intrusive solution in having better sleeping experiences by addressing snoring, temperature regulation, and posture alignment, hence becoming a valuable tool for the achievement of restful and healthful nights.

Keywords Smart bed · Snoring prevention · Sleep quality · Posture correction · Temperature regulation · Comfort optimization · Health monitoring

1 Introduction

Sleep is a fundamental physiological need essential for restorative body functions, memory consolidation, emotional regulation, and immune system support. However, many people face sleep-related issues like snoring, temperature discomfort, and

---

P. Sankireddy (✉) · V. P. Mateti · S. Thangam
Department of Computer Science, Amrita School of Computing, Amrita Vishwa Vidyapeetham, Bengaluru, India
e-mail: bl.en.u4cse22056@bl.students.amrita.edu

V. P. Mateti
e-mail: bl.en.u4cse22077@bl.students.amrita.edu

S. Thangam
e-mail: sthangam@blr.amrita.edu

H. Nagaraj et al. (eds.), *Advancements in Artificial Intelligence and Machine Learning*, Lecture Notes in Electrical Engineering 1572,
https://doi.org/10.1007/978-981-95-7449-0_34

poor posture, which negatively affect health, daily productivity, and quality of life. Snoring, often linked to obstructive sleep apnea (OSA), disrupts sleep and can lead to cardiovascular problems, hazardous conditions, and reduced life satisfaction. To address these issues, this project introduces a smart bed system designed to enhance sleep quality through intelligent interventions. Unlike traditional solutions like nasal strips, this system passively mitigates snoring by detecting it in real-time and tilting the upper portion of the bed to relieve airway pressure, allowing for easier breathing without disrupting the user's sleep.

The smart bed also optimizes sleep comfort by regulating temperature and promoting healthy posture. Incorporating sensors, it monitors and adjusts both body and room temperatures to create an ideal thermal environment, ensuring deeper and uninterrupted sleep. Additionally, pressure sensors assess the user's posture throughout the night, identifying improper positions and making adjustments to realign the body, preventing pain and long-term health issues. This innovative system provides a holistic solution to improve sleep quality by addressing snoring, thermal comfort, and posture in a gentle, noninvasive, and fully automated manner, promoting better health and well-being for users.

2 Literature Survey

Youngkong et al. in [1] developed a smart IoT solution for monitoring patterns based on movements on the bed, focusing on improving comfort and health through data-driven adjustments. This study highlights the potential of IoT-based systems in enhancing sleep quality by tracking physiological parameters, which aligns with our project's goal of snoring prevention and comfort optimization through a smart bed.

The work [2] of Su et al. is aimed at extracting stress-related EEG patterns for identifying sleep deficits based on the effect of stress on slow wave sleep. This research shows that physiological activity monitoring is requisite for improving sleep quality, and our smart bed aims to supply customized sleep regimes depending on tripping parameters.

Amrutur et al. IOSR Journal of Electronics and Communication Engineering (IOSR-JECE), Volume 12, Issue 2, Ver. IV in [3] offers an IoT test bed for smart city applications. This project enhances smart bed solutions using sensors and machine learning for snoring prevention, a good temperature for sleeping, and a health promoting posture throughout the night thus ensuring a healthy night's sleep.

Syed et al. emphasized the significance of experimental test beds for systems testing, providing a framework that can be adapted for evaluating smart technologies in [4]. Utilizing similar methodologies could help in assessing the efficacy of snoring prevention and comfort optimization features.

Here, Zhu et al. in [5] have also developed a real-time sleep stage prediction model using an architecturally switchable deep learning approach that could enhance the sleep monitoring and calibration work.

In Pathinarupothi et al. [6] suggest the use of automatic instant heart rate as useful feature for detecting sleep apnea and how severe it is, employing deep learning strategy. Our smart bed extends the idea of this approach further by including snoring detection and posture analysis for sleep apnea and real-time adjustment of the sleeping position.

In their study [7] on smart health care beds for patient comfort and health monitoring, advanced medical care beds for patients with Internet of Things technology targeting the IoT solutions were investigated by Elsokah and Zerek. Their studies show how IoT can be utilized for online observation and intelligent control and adjustment of the bed, which lays a theoretical framework for our work to enhance comfort and SNR reduction through sensing and actuation enhancing sleep quality.

The work of Zhu, et al. in [8] presented "MaskSleepNet," a neural network model suitable for handling multiple signal modalities for sleep staging with a demonstration of extremely good cross-modality transfer for sleep analysis. In this work, a range of advanced AI methods have been explained that can be effective in improving sleep quality assessment. Our project is based on such ideas, too, using smart sensors and machine learning to identify snoring and prevent it and adapting the contour's shape to provide maximal comfort and sleep position for the user.

Designed for people with reduced hand mobility, the system consists of smart gloves coupled with flex sensors and Arduino microcontrollers. These gloves can sense movements in fingers; through such applications, individuals would be able to control lights or fans and send alerts in cases of emergency by buzzers and LEDs. The technology does enhance independence and communication, thus correctly relating to our aim at our project—using IoT for enhancing the quality of daily living [9].

Lin et al. stated that intelligent sleep products highlight active health concept and dynamic and personalized comfort. Their research focuses on the application smart technologies with regard to sleep heath issues to improve the quality of sleep. Our project expands upon such principles focusing on snoring prevention by regulation of bed position and improving general comfort according to temperature and posture which will improve the overall quality of people's sleep [10].

IoT helps revolutionize agriculture: Automating monitoring and management with minimal human intervention. With sensors, actuators, and wireless technologies, IoT offers precision farming, automated irrigation, and real-time crop monitoring. The paper demonstrates IoT's role in increasing efficiency and sustainability in cropping activities and how it may modernize traditional farming practices through smart and data-driven solutions [11].

Nair et al. discussed the concept of user-extending algorithms for smart infusion pump, emphasizing on issues of safety and utility in ICU. It is in line with our mission to innovate a home-use smart bed with non-clinical bodily monitoring and self-learning comfort adjustment and snoring hindering derived from implicit and explicit procedural needs for better sleep and overall well-being [12].

As diabetes prevalence has grown over the years, recent developments involve noninvasive sensors, cloud storage, and machine learning models like random forest for real-time health monitoring. This system provides dynamic analysis of glucose and blood pressure levels and allows prompt intervention for handling emergency situations. This is how our project aligns with IoT and AI for continuous health tracking [13].

Breuss et al. present a study on detection of sleep position for the closed-loop treatment of sleep-related breathing disorders, which demonstrates the efficacy of using technology to adjust sleep posture for improved breathing. This research supports the concept of our smart bed, which similarly employs automatic position adjustment to mitigate snoring, enhancing sleep quality through non-intrusive and adaptive methods [14].

Pathinarupothi et al. discussed single sensor based techniques for diagnosis of sleep apnea using deep learning, pointing out the possibilities of practically noninvasive systems in sleep disorders and therapy. Following their approach, we created intelligent systems to improve sleep, including a smart bed which uses machine learning to detect snoring and prevent it at the same time guaranteeing optimum comfort for a better sleep [15].

This paper presents an IoT-integrated solar PV tracking system, which aims to optimize energy output by adjusting the PV array's position at real time based on weather and solar irradiance data. It is composed of sensors, a microcontroller, and stepper motors that allow for remote monitoring and control, which substantially improves energy efficiency for residential and commercial use [16].

3 Proposed Methodology

The proposed methodology addresses the pseudocode, sequence of the proposed solution with its system architecture in detail.

3.1 System Architecture and Sequence Flow of the Proposed Solution

Figure 2 is the system design of the solution proposed in this entire study, and below gives a detailed description of the sequence flow shown in Fig. 1.

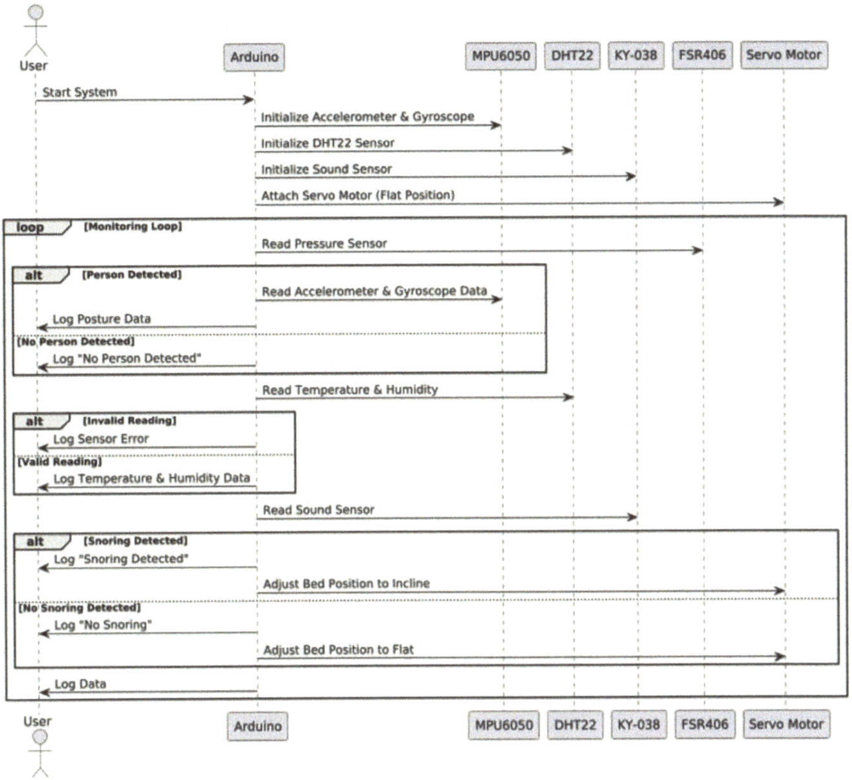

Fig. 1 Sequence flow of the proposed solution

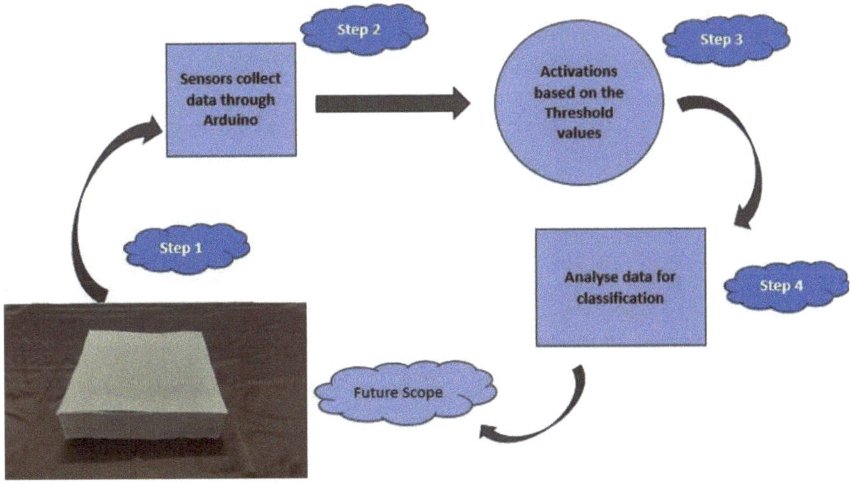

Fig. 2 System design of the proposed solution

a. **Sensor-Based Event Monitoring** For the real-time detection of snoring, sound sensors are used. To maintain the warmth, room temperature and humidity are captured with DHT22. While sleeping, for the posture to be right, accelerometer and pressure sensor are used to detect any deviation from healthy sleeping positions.
b. **Data Acquisition** Data that is captured will be transmitted to Arduino. The collected data of sound levels, temperature, and posture patterns will be used to monitor and correct our unhealthy habits.
c. **Signal Conditioning and Cleaning** As the sound sensor captures every sound that hears around it, filters are applied to data, so that it eliminates the background noise and focuses on snoring-related signals. Normalizing temperature is used for comparison. We apply smoothing algorithms, so that the posture data is accurate for classification.
d. **Data Logging and Storage** The data that goes into the Arduino will be directly reflected in the GUI immediately. For the real-time actions, raw data is stored. After being processed, it is sent to the cloud via Bluetooth for monitoring and tracking.
e. **Automated Insights Generation** The concept of smart bed is itself for automation without any human intervention. The data from the sound sensor is analyzed and determines snoring and then triggers the bed inclination. For comfort, Peltier module is activated for heating or cooling when needed. By classifying the postures as "healthy" and "unhealthy," bed adjustments can be done.

3.2 Algorithm and Explanation of the Proposed Solution

The Algorithm 1 of this system is centered on monitoring and controlling the bed conditions, taking data from multiple sensors. It first initializes all its components, such as the accelerometer and gyroscope MPU6050, the temperature and humidity DHT22 sensor, the KY-038 sound sensor, as well as attaching a pressure sensor. In addition to that, it attaches a servo motor, which is defaulted flat. In a continuous monitoring loop, the pressure sensor first determines if there is presence on the bed. If presence is detected, the system captures readings from the accelerometer and gyroscope, maintains posture-related data logs, and determines potential movement of the body. If no presence is detected, the system logs absence and skips the unnecessary computations.

Algorithm 1 Data Collection and Processing for Sound, Temperature, and Posture Classifications

```
Initialize all sensors: MPU6050, DHT22, KY-038 Sound Sensor, Pressure Sensor
Attach Servo Motor and set to flat position
while True do
    Read pressure sensor value
    if Pressure ¿ 200 then
        Detect presence on bed
        Read accelerometer and gyroscope data
        Log acceleration (Ax, Ay, Az) and rotation (Gx, Gy, Gz)
    else
        Log no presence detected
    end if
    Read DHT22 temperature and humidity
    if Temperature or Humidity is invalid then
        Log sensor error
    else
        Log temperature and humidity
    end if
    Read sound sensor data
    if Sound level ≥ 20 then
        Log snoring detected
        Move servo motor to inclined position
    else
        Log no snoring
        Move servo motor to flat position
    end if
    Delay for 1000ms
end while
```

The algorithm then determines environmental conditions using the DHT22 sensor to measure temperature and humidity. It handles errors for invalid readings and logs valid data for future analysis. Simultaneously, the sound sensor monitors noise levels to detect snoring. If the sound exceeds the defined threshold, the servo motor adjusts the bed to an inclined position to alleviate snoring. If snoring subsides, the bed returns to its flat position. These actions are logged at each step, ensuring traceability. The added delay mechanism in the loop prevents inefficiency and repetition, allowing an intelligent, adaptive system for the improvement of sleep quality.

4 Implementation Details

4.1 Hardware Details

Table 1 lists all the required hardware components.

Table 1 Hardware components

Component	Specification	Quantity
Sound sensor	KY-038 microphone module	1 unit
Temperature and humidity sensor	DHT22	1 unit
Servo motor	MG995 high torque servo	1 unit
Accelerometer	MPU6050	1 unit
Pressure sensors	FSR406	1 unit
Connecting wires	Jumper wires	Variable
Peltier module	TEC1–12706	1 unit
Bluetooth module	HC-05	1 unit
Resistors	1 kΩ, 2 kΩ, 4.7 kΩ,10 kΩ	6 units (2x4.7 kΩ, 2x1 kΩ, 2 kΩ, 10 kΩ)
Microcontroller	Arduino Uno board	1 unit

4.2 *Circuit Diagram*

Figure 3 shows the complete device designed as part of the project implementation, and Fig. 4 shows the circuit diagram. Table 2 summarizes the connections between sensors and Arduino, specifying each sensor's pins, such as power, data, ground, and their corresponding Arduino pins for clear and structured reference.

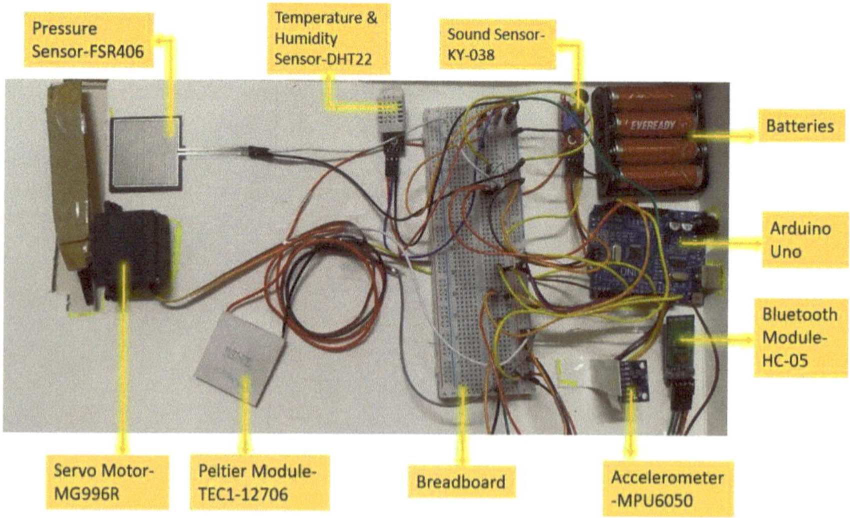

Fig. 3 Components in the bed

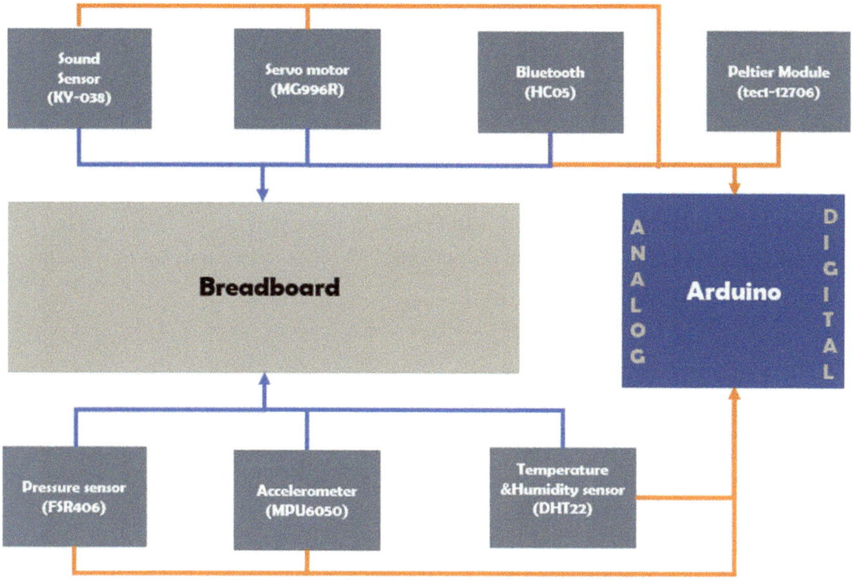

Fig. 4 Circuit diagram

Table 2 Sensor connections to Arduino

Sensors	Pin of sensor	Pin of Arduino
Sound sensor	Analog out (A0)	A0 (Analog)
	Digital out	D3 (Digital)
Temperature sensor (DHT22)	Data (out)	D2 (Digital)
	Power (+)	5 V
	Ground (−)	GND
Pressure sensor	Analog out	A1 (Analog)
	Power (+)	5 V
	Ground (−)	GND
Accelerometer (MPU6050)	SDA	A4 (SDA)
	SCL	A5 (SCL)
	Power (+)	5 V
	Ground (−)	GND
Servo motor	PWM	D9 (PWM)
	Power (+)	5 V
	Ground (−)	GND
Peltier module	Red (+)	5 V
	Black (−)	GND

4.3 Results

The smart bed system developed combines a number of sensors with various actuation systems to improve sleep quality by detecting vital parameters like sound, pressure, and motion. The nature of the system can be said to display its functionality in real-time detection and practical response to conditions such as snoring and body posture for smooth adjustments toward comfort and health improvement. The discussion of the results below follows from the output observed from the experimental setup.

4.4 Snoring Prevention

The mechanism of the bed is shown in Figs. 5 and 6 where when the snoring is detected inclines the bed and reduces the snore. Then, the bed comes back to the normal position. As shown in Fig. 7, sound levels picked up by the KY-038 module were monitored over time. The threshold for activation of the servo motor was at 20 dB. If the sound levels surpassed the threshold, a servo motor was activated; this is demonstrated through a set of black markers marked "Activated." This proves the system can sufficiently detect snore events and initiate bed adjustments when needed. The red dashed line represents the activation threshold.

4.5 Temperature Optimization

Figure 8 displays temperature fluctuations and activation of a Peltier module for both heating and cooling. The blue line of the diagram indicates variation of temperature over time. When the temperature is over the threshold of cooling (32 °C, red dashed

Fig. 5 Bed in normal position

Fig. 6 Inclined bed

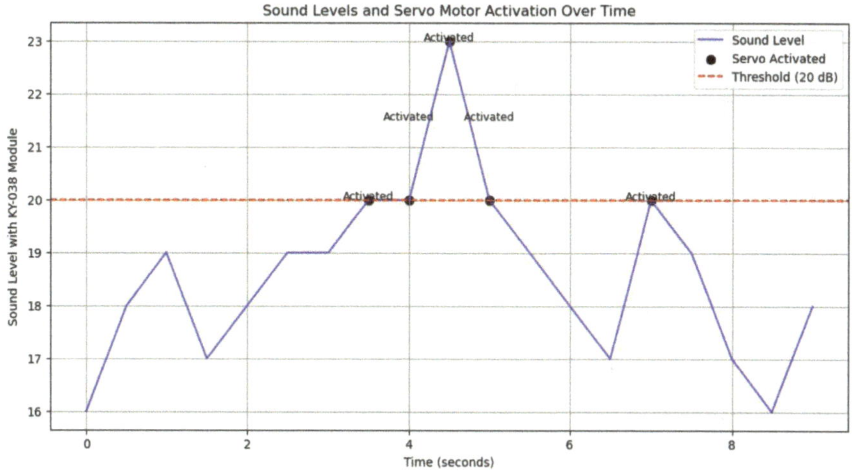

Fig. 7 Sound levels and servo motor activation over time

line), the cooling module gets activated at the respective red points marked as "Cooling". The heating module activates when the temperature falls below the threshold of heating (22 °C, green dashed line). This is how thermal comfort is maintained since this adaptation is accurate regulation.

4.6 *Sleep Quality Detection*

This case in Fig. 9 shows that the system is operating at full capacity. The pressure reading of 382 N proves there is a person on the bed. Accelerometer and gyroscope

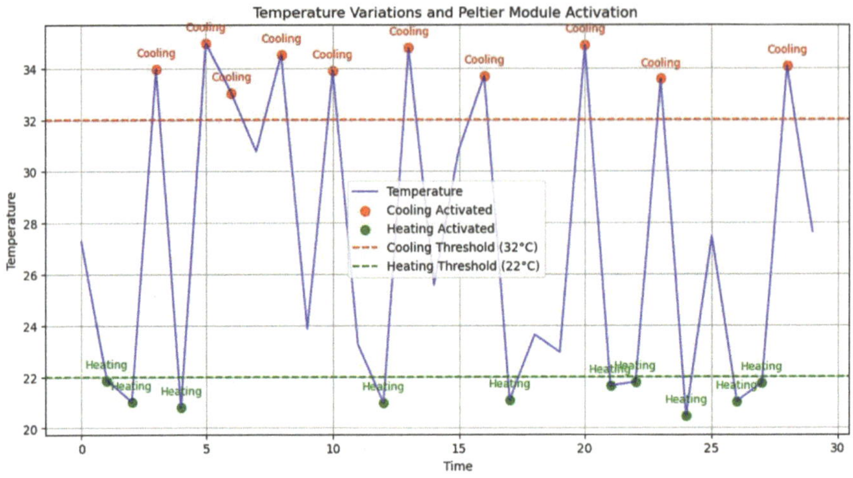

Fig. 8 Temperature variations and Peltier module activation

Fig. 9 Case 1: Detection when pressure applied

```
Pressure Value (in N): 382
Person detected on the bed.
Acceleration (m/s^2): X=4.00, Y=3.39, Z=3.30
Gyro (rad/s): X=-0.02, Y=0.03, Z=0.03
Sound Level (in dB): 23
Snoring Detected! Adjusting Bed Position...
```

signals record slight change of motion, proving that the posture is stable, with the sound sensor recording a noise level of 23 dB. The system detects snoring and, through the control, adjusts the bed into an inclined position using a servo motor, as illustrated in the physical setup. This scenario validates the seamless integration of sensors and actuators for real-time intervention.

The sensor of pressure in Fig. 10 detects a value of 130 N, which is still below the threshold regarding identification of person in the bed. Furthermore, the sound level is still at 19 dB and, therefore, below the snoring detection threshold set at 20 dB. It then correctly concludes that no snore has been detected and continues to keep the bed flat, thus conserving energy as well as minimizing any unwanted operations.

The graph shown in Fig. 11 exemplifies the capability of MPU6050 accelerometer and gyroscope to detect tilt. Axes of data (Ax, Ay, Az for acceleration and Gx, Gy, Gz for angular velocity) are plotted against time. Dynamic tilt events like left, right, up, and down can be easily distinguished from the graph. Particular tilt directions would be identified by a distinct color using threshold crossings, indicating that the system can similarly predict patterns of movement and assert its viability in the evaluation of sleep posture and user activity.

```
Pressure Value (in N): 130
No person detected.
Sound Level (in dB): 19
No Snoring Detected. Bed Flat.
```

Fig. 10 Case 2: No detection

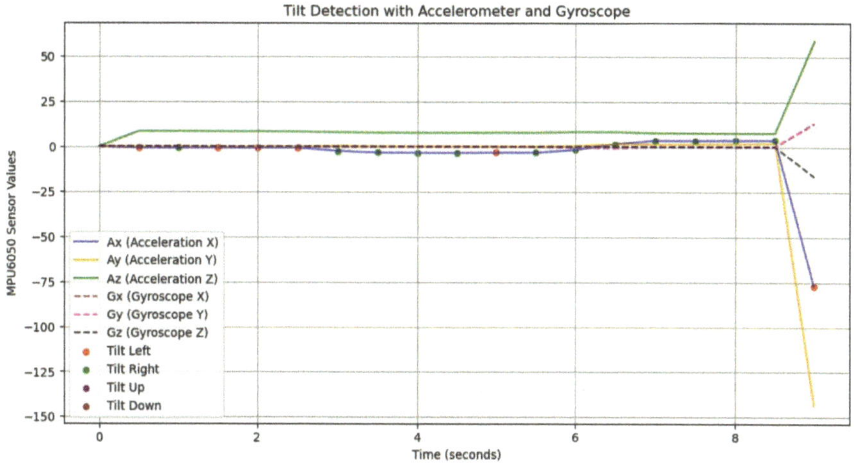

Fig. 11 Tilt detection with accelerometer and gyroscope

5 Conclusion and Future Scope

The proposed smart bed system, through its integration of IoT and sensor-based technologies, helps enhance sleep quality through snoring detection, posture monitoring, and temperature regulation. The system ensures real-time adjustment by making use of sensors like the MPU6050, DHT22, and KY038 with the service of servo motors and pressure sensors for comfort and health. The outcomes prove its ability to continuously monitor physiological parameters and adjust accordingly for effective remedies in sleep quality improvement, snoring reduction, and proper posture maintenance. This innovation promises significant applicability in healthcare and home settings.

Future Work

Implement deep learning and neural networks to filter out background noises, accurately detecting the occurrence of snoring. Utilize multiple pressure sensors to detail high-resolution pressure distribution and classify sleep postures. Develop predictive algorithms for long-term health monitoring and early detection of sleep-related disorders. Extend cloud-based connectivity for remote monitoring and health data visualization.

References

1. Youngkong P, Panpanyatep W, Thamrongaphichartkul K (2020) Developing a smart IoT solution to monitor on-bed movement patterns. In: 2020–5th international conference on information technology (InCIT), Chonburi, Thailand, pp 306–309. https://doi.org/10.1109/InCIT50588.2020.9310930
2. Su C-H et al (2024) Extracting stress-related EEG patterns from pre-sleep EEG for forecasting slow-wave sleep deficiency. IEEE Trans Neural Syst Rehabil Eng 32:1817–1827. https://doi.org/10.1109/TNSRE.2024.3394471
3. Amrutur B et al (2017) Poster abstract: an open smart city IoT test bed. In: 2017 IEEE/ACM second international conference on internet-of-things design and implementation (IoTDI), Pittsburgh, PA, USA, pp 323–324
4. Syed MH et al (2019) The role of experimental test beds for the systems testing of future marine electrical power systems. In: 2019 IEEE electric ship technologies symposium (ESTS), Washington, DC, USA, pp 141–148. https://doi.org/10.1109/ESTS.2019.8847851
5. Zhu H et al (2024) Towards real-time sleep stage prediction and online calibration based on architecturally switchable deep learning models. IEEE J Biomed Health Inform 28(1):470–481. https://doi.org/10.1109/JBHI.2023.3327470
6. Pathinarupothi RK, Vinaykumar R, Rangan E, Gopalakrishnan E, Soman KP (2017) Instantaneous heart rate as a robust feature for sleep apnea severity detection using deep learning. In: 2017 IEEE EMBS international conference on biomedical & health informatics (BHI), Orlando, FL, USA, pp 293–296. https://doi.org/10.1109/BHI.2017.7897263
7. Elsokah MM, Zerek AR (2019) Next generation of medical care bed with internet of things solutions. In: 2019 19th International conference on sciences and techniques of automatic control and computer engineering (STA), Sousse, Tunisia, pp 84–89. https://doi.org/10.1109/STA.2019.8717204
8. Zhu H et al (2023) MaskSleepNet: a cross-modality adaptation neural network for heterogeneous signals processing in sleep staging. IEEE J Biomed Health Inform 27(5):2353–2364. https://doi.org/10.1109/JBHI.2023.3253728
9. Thangam S, Gurupriya M, Kumar KJ, Shreya K, Divakarla LP (2024) IoT-based smart gauntlet to assist physically challenged and fragile mortals. In: 2024 7th International conference on circuit power and computing technologies (ICCPCT), vol 1, pp 994–1000
10. Lin X, Fu J, Yang J (2023) Healthful sleep solutions: intelligent sleep products design based on the active health concept. In: 2023 16th International symposium on computational intelligence and design (ISCID), Hangzhou, China, pp 58–62. https://doi.org/10.1109/ISCID59865.2023.00022
11. Virupaxappa G, Thangam S (2021) Smart agriculture and role of IOT. In: 2021 Third international conference on inventive research in computing applications (ICIRCA). IEEE, pp 1–6
12. Nair VS, Pathinarupothi RK, Madathil T (2023) Personalized algorithms and techniques for development of a smart infusion pump for ICU. In: 2023 IEEE 8th international conference for convergence in technology (I2CT), Lonavla, India, pp 1–7. https://doi.org/10.1109/I2CT57861.2023.10126150
13. Thangam S, Niranjan DK, Chetan K, Sujan GV, Shreyaa H, Chakresh MSS (2024) Cloud-based remote monitoring and management for diabetic patient healthcare. In: 2024 international conference on distributed computing and optimization techniques (ICDCOT). IEEE, pp 1–8

14. Breuss A et al (2022) Sleep position detection for closed-loop treatment of sleep-related breathing disorders. In: 2022 International conference on rehabilitation robotics (ICORR), Rotterdam, Netherlands, pp 1–6. https://doi.org/10.1109/ICORR55369.2022.9896559
15. Pathinarupothi RK, Rangan ES, Soman KP (2017) Single sensor techniques for sleep apnea diagnosis using deep learning. In: 2017 IEEE international conference on healthcare informatics (ICHI), Park City, UT, USA, pp 524–529. https://doi.org/10.1109/ICHI.2017.37
16. Dandu R, Thangam S (2023) IoT based single axis solar tracker. In: 2023 14th International conference on computing communication and networking technologies (ICCCNT), Delhi, India, pp 1–5. https://doi.org/10.1109/ICCCNT56998.2023.10307807

Advances in Deep Learning for Image Synthesis: A Literature Survey

Saurabh Alone, Vani Pillai, Atharv Porate, Vedant Brahmankar, and Mithra Venkatesan

Abstract This survey traces the evolution of deep learning models for image synthesis, from convolutional neural networks (CNNs) to state-of-the-art diffusion models. Vision tasks were carried out by CNNs with AlexNet boosting the deep learning revolution. Further advancements including variational auto-encoders (VAEs) and generative adversarial networks (GANs) introduced image generation capabilities. Recently, diffusion models have emerged as the benchmark for high-quality, diverse image synthesis by taking advantage of denoising processes. This chapter provides a comparative analysis of these models, exploring their strengths, limitations, and key innovations. Additionally, we outline future research directions to enhance image generation and processing applications. By analyzing this direction of research, we aim to focus on the transformative potential of deep learning in computer vision.

Keywords Convolutional neural networks (CNN) · AlexNet · Variational auto-encoders (VAE) · Generative adversarial networks (GAN) · Transformers · Diffusion models · Image generation · Computer vision · Deep learning

1 Introduction

In recent year, generative models have transformed deep learning, leading the advancements in different modalities such as image, text, video, and audio synthesis. Among these, text-to-image (TTI) models closed the gap between natural language processing (NLP) and computer vision becoming crucial for many application. This development began with foundational architectures like convolutional neural networks (CNNs) [1] and evolved through variational auto-encoders (VAEs) [2] and generative adversarial networks (GANs) [3], leading to state-of-the-art diffusion models. Modern approaches, such as DALL-E [4] and stable diffusion [5],

S. Alone (✉) · V. Pillai · A. Porate · V. Brahmankar · M. Venkatesan
Dr. D. Y. Patil Institute of Technology, Pimpri, Pune, India
e-mail: saurabhalone85@gmail.com

H. Nagaraj et al. (eds.), *Advancements in Artificial Intelligence and Machine Learning*, Lecture Notes in Electrical Engineering 1572,
https://doi.org/10.1007/978-981-95-7449-0_35

improved the potential of TTI models to generate high-quality images. Recent innovations extend the capabilities of these models beyond image synthesis to tasks like video generation and interactive environments. This chapter provides a comprehensive review of the evolution of image synthesis models, focusing on their strengths, limitations, and future directions.

2 Background

2.1 *Image Processing in Deep Learning*

The foundation of modern image synthesis lies in advancements in deep learning, beginning with convolutional neural networks (CNNs). Early developments like AlexNet [6] showcased the potential of large-scale models, making use of approximately 60 million parameters, ReLU activation, dropout, and data augmentation to improve performance. Further innovations such as ResNet's skip connections enabled the training of deeper architectures by addressing vanishing gradient issues. Similarly, U-Net's [7] encoder–decoder structure advanced high-resolution tasks like image segmentation. These models transformed image processing by laying the foundation for generative architectures like variational auto-encoders (VAEs) and generative adversarial networks (GANs). This section provides an overview of these foundational advancements discussing their role in shaping state-of-the-art approaches.

2.2 *Generative Models*

Generative models have transformed image synthesis by learning to create high-quality and realistic images. Variational auto-encoders (VAEs) introduced probabilistic methods for encoding and decoding data, enabling continuous latent representations. VAEs aim to accurately reconstruct the input data from its compressed representation (the latent space). This ensures that the model understands the input data well. Extensions such as Beta-VAE and conditional VAE enhanced their flexibility, enabling more disentangled and conditional generation. Generative adversarial networks (GANs) further advanced the field by introducing an adversarial methods, where a generator learns to create realistic samples, while a discriminator attempts to distinguish them from real data. This competitive training dynamic has enabled GANs to excel in tasks like image super-resolution and style transfer. Despite challenges like training instability and mode collapse, variants such as ProGAN and StyleGAN have set benchmarks in image fidelity and diversity. Diffusion models represent the latest breakthrough, offering state-of-the-art performance by iteratively denoising random noise to generate images. Unlike VAEs and GANs, diffusion models leverage Markov chains to gradually approximate data distributions, enabling great quality and diversity in outputs. Models like DALL-E and Imagen demonstrate the transformative potential of this approach, though slow inference remains a limitation. Together, these models have laid the foundation of modern image synthesis, each contributing unique strengths and innovations.

3 Multimodal Generative Model

The evolution of multimodal generative models has seen significant advancement with the introduction of DALL-E 3, Stable Diffusion 3. These models represent a fundamental shift in how multimodal systems process and generate content across different modalities. The core architecture of these systems can be formalized through several key mathematical components that enable their sophisticated cross-modal capabilities (Table 1 and 4).

Table 1 Comparison of image generation models—Part 1

Model	Dataset	Size	FID
VAE	Custom/MNIST	¡100 M	50–100
GAN	CelebA, CIFAR-10	100 M+	40–80
StyleGAN	FFHQ, CelebA	100 M–300 M	4–10
BigGAN	ImageNet	1.5B	6–12
Diffusion models	ImageNet, LAION	200 M–1B	5
CLIP	LAION-400 M	1B	N/A
DALL-E 3	Custom (OpenAI)	12B	6–10
Stable diffusion 1	LAION-5B	900 M	20–30
Stable diffusion 2	LAION-5B	900 M	15–25
Stable diffusion 3	LAION-5B	900 M	8–12
GLIDE	Custom (OpenAI)	3.5B	5
Imagen	ImageNet, LAION	5B	2
Janus	Custom	1B+	10–20
DeepMind Gato	Multimodal	1.2B	N/A
Muse	Custom (text-image)	1B	15

4 Comparison of Approaches

4.1 Comparison of Image Generation Models

Tables 2 and 3 provide a comprehensive comparison of several image generation models [8], including VAE, GAN, StyleGAN, BigGAN, diffusion models, CLIP, DALL-E 3, stable diffusion (v1, v2, v3), GLIDE, Imagen, Janus, DeepMind Gato, and Muse. The comparison is based on critical metrics such as the dataset used, model size, Fréchet inception distance (FID) score, inception score, and overall image quality [8, 9].

The FID score is widely recognized for measuring the quality and diversity of generated images by comparing them to real images, with lower scores indicating better performance. The inception score is another key metric used primarily with GANs to evaluate the image's quality and diversity based on the inception model's confidence in its classification [8].

Among these models, diffusion models and advanced generative networks such as DALL-E 3 and Imagen are considered state-of-the-art, offering exceptional image quality with low FID scores. On the other hand, GANs, particularly StyleGAN and BigGAN, also show impressive results in terms of high fidelity and smooth transitions. However, they can be prone to challenges like mode collapse. CLIP and DeepMind Gato, while not purely image generation models, serve multimodal purposes, learning representations across text and images, thus playing an essential role in text-to-image generation [10, 11].

Table 2 Comparison of image generation models—Part 2

Model	IS	Image quality
VAE	N/A	Blurry, lacks fidelity
GAN	4–8	Decent, mode collapse issues
StyleGAN	9–10	High quality, smooth transitions
BigGAN	250+	High fidelity, class-conditional
Diffusion models	N/A	Excellent, photorealistic
CLIP	N/A	Strong multimodal performance
DALL-E 3	N/A	Photorealistic, text-guided
Stable diffusion 1	N/A	Decent, can be low-res
Stable diffusion 2	N/A	Improved, sharper
Stable diffusion 3	N/A	High resolution, detailed
GLIDE	N/A	Excellent, good for editing
Imagen	N/A	Best-in-class photorealism
Janus	N/A	High fidelity, still evolving
DeepMind Gato	N/A	Decent for general tasks
Muse	N/A	Fine-tuned for text-to-image

Table 3 Comprehensive comparison of image model fine-tuning techniques

Technique	Memory usage	Training time	Model size	Best use cases	Key parameters	Limitations
LoRA	2–8 MB per concept	1–2 h	Very small	Style transfer, concept learning	Rank (4–16), learning rate	Limited capacity for complex concepts
DreamBooth	High (> 16 GB)	4–6 h	Full model size	Subject-specific training	Class images, instance prompt	High computational needs
Textual inversion	Very low (< 1 MB)	2–3 h	Few KB	New concept learning	Placeholder token, vectors	Limited to textual concepts
Hypernetwork	Medium (10–50 MB)	3–4 h	10-50 MB	General purpose modification	Layer structure	Complex training process
Custom diffusion	Medium (100–200 MB)	4–5 h	100-200 MB	Targeted editing	Target layers, weights	Technical expertise needed

Table 4 Technical implementation details of fine-tuning techniques

Technique	Learning rate	Batch size	Training steps	Required images
LoRA	1e-4 to 5e-4	1–4	1000–2000	10–20
DreamBooth	1e-6 to 5e-6	1–2	2000–3000	3–5
Textual inversion	5e-5 to 1e-4	4–8	3000–5000	15–25
Hypernetwork	1e-5 to 5e-5	2–4	2000–4000	20–30
Custom diffusion	2e-6 to 1e-5	1–2	3000–5000	10–15

As observed from the comparison, diffusion models like DALL-E 3 and Imagen excel at generating high-quality, photorealistic images, showcasing state-of-the-art performance with some of the lowest FID scores. On the other hand, models like stable diffusion offer an efficient balance between quality and computational cost, particularly in the third iteration, which shows significant improvements over its predecessors. GAN-based models like StyleGAN and BigGAN are still highly effective in specific domains, particularly for high-quality and class-conditional generation, although they may suffer from issues such as mode collapse.

While CLIP and DeepMind Gato are not pure image generation models, they play a crucial role in multimodal learning tasks, bridging the gap between image and text modalities. Their performance, although not directly measured in FID or Inception Score, is essential for models like DALL-E and GLIDE that rely on text-image relationships for generating accurate and context-relevant outputs.

4.2 Comparison of Models on Different Prompts

In this section, we compare the performance of various image generation models—stable diffusion, FLUX, and DALLE—across five diverse prompts. Each prompt was designed to test specific aspects of the models' capabilities, such as generating detailed urban landscapes, capturing natural settings, rendering human-robot interactions, and producing realistic food textures. Below, we outline the qualitative assessment based on these prompts (Figs. 1, 2, 3 and 4).

Prompt 1 "A futuristic city at sunset with tall glass skyscrapers and flying cars in the sky." This prompt tests the models' ability to generate a futuristic, urban landscape with complex elements like skyscrapers and flying vehicles. Stable diffusion demonstrates strong spatial coherence, providing clean lines and impressive lighting effects on the buildings. FLUX offers a warm, vibrant sunset but slightly simpler building textures. DALLE generates a more detailed and dynamic image with impressive lighting but tends to overemphasize the sci-fi elements.

Prompt 2 "A cozy wooden cabin surrounded by snow-covered pine trees, with smoke rising from the chimney." For the natural and architectural elements, stable diffusion effectively captures the essence of a serene winter scene, with attention to the details of the snow and cabin. FLUX's rendering is equally strong, producing

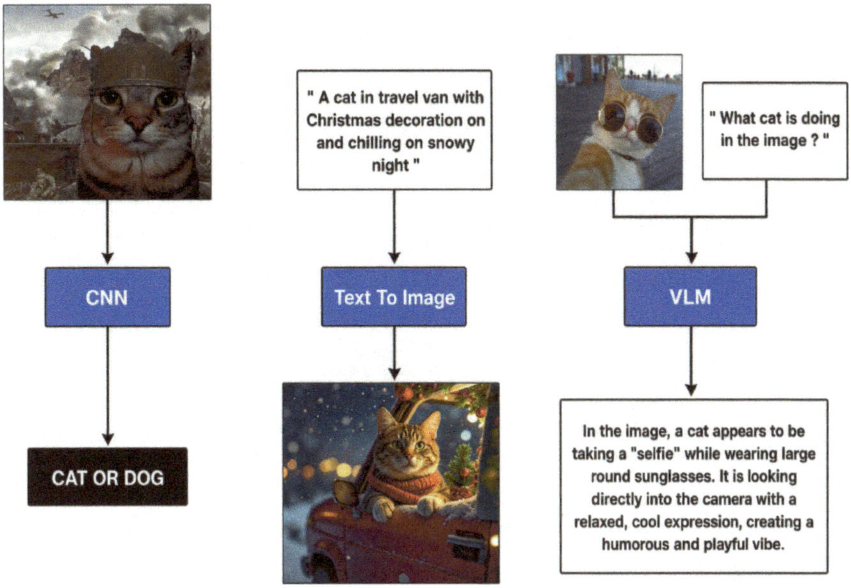

Fig. 1 Examples of common tasks for image processing and generation, including image classification (left), text-to-image generation (middle), and vision-language understanding (right). Image classification identifies whether the subject is a cat or dog using a convolutional neural network (CNN). Text-to-image generation produces an image based on a provided text description, creating a highly detailed and contextually accurate image. Vision-language models (VLMs) analyze the content of an image and respond to queries about it, generating textual explanations that align with the visual input

a softer esthetic, while DALLE enhances the cozy atmosphere with detailed snow textures and vibrant lighting on the cabin, though the overall image may appear overly stylized.

Prompt 3 "A humanoid robot sitting at a table, playing chess with a focused human in a modern living room." This prompt aims to evaluate how models deal with both human and robot characters in an indoor setting. Stable diffusion creates a clear, albeit slightly mechanical, scene, while FLUX provides a more natural human–robot interaction with softer lighting. DALLE, however, excels with intricate details on both the robot and the human, though some elements can appear overly polished or stylized, which may detract from realism.

Prompt 4 "A bowl of vibrant, fresh fruit including oranges, apples, and grapes, placed on a rustic wooden table with soft sunlight." The task here is to assess how well the models can generate food textures and handle lighting and color composition. Stable diffusion provides a realistic portrayal of the fruit, though some details might lack texture. FLUX captures a more photo-realistic representation of the fruit with excellent color balance, while DALLE adds depth and shadows to enhance the visual appeal, but at times exaggerates the vibrancy, making it appear slightly artificial.

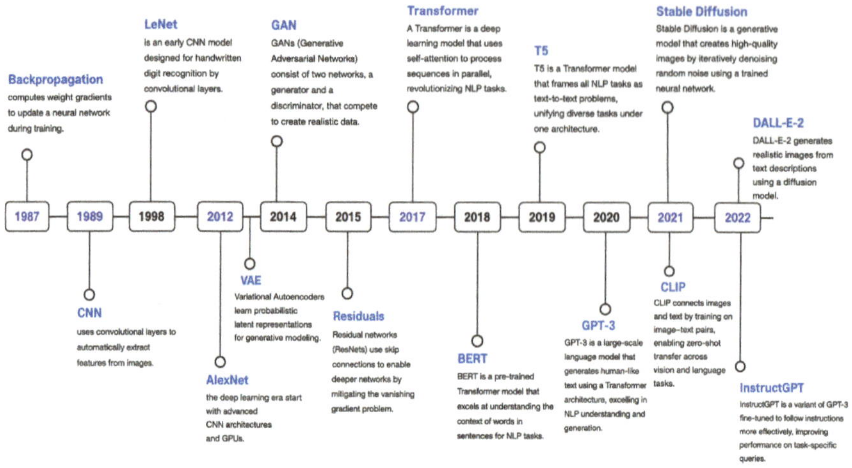

Fig. 2 Timeline illustrates the rapid evolution of deep learning from 1987 to 2022, showcasing pivotal models that transformed AI capabilities. It highlights a shift from task-specific architectures (like CNNs for image processing) to more versatile, general-purpose models (such as transformers and large language models), demonstrating AI's progression toward increasingly sophisticated and adaptable systems

Fig. 3 Architecture of Janus. It is auto-regressive multimodal generative model which is capable of understanding and generating both text and images

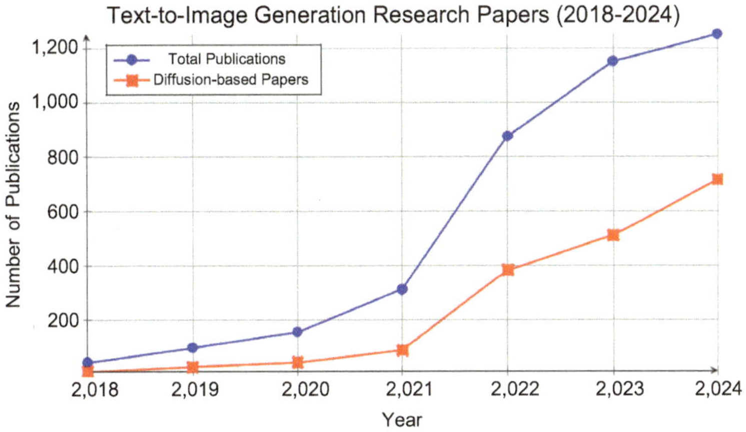

Fig. 4 Growth observed in text-to-image (TTI) generation research publications (Data sourced from arXiv, major conferences, and journals)

Overall analysis across the four prompts, we observe that all three models—stable diffusion, FLUX, and DALLE—perform well in generating high-quality images, but with different strengths. Stable diffusion maintains a balance between artistic detail and realism, FLUX offers more subdued and natural images, and DALLE pushes the boundaries of creative expression but sometimes sacrifices realism for style. These insights provide a foundation for further quantitative evaluations based on metrics such as FID and image quality assessments.

4.3 Text-to-Image Research Publication Trends

There has been an exponential growth in the field of text-to-image (TTI) models, especially from 2018 to 2023. Over these 5 years, a huge surge of research publications was observed, with numbers increasing from just 45 papers in 2018 to over 1150 papers in 2023—representing a 2500% increase. This growth was more noticeable after 2021, which coincides with the year when diffusion models were introduced. The proportion of diffusion-based approaches has grown significantly, from 26.7% of total publications in 2018 to 44.5% in 2023, indicating a clear shift in methodological focus within the field. The year 2022 marked a crucial turning point, with the introduction of influential models like DALL-E 2 and stable diffusion, which sparked an unprecedented wave of research interest and contributed to a 180% year-over-year increase in publications [12].

article tikz pgfplots.

The research landscape reveals not only quantitative growth but also qualitative shifts in focus areas. While earlier papers (2018–2020) predominantly focused on

GAN-based architectures and fundamental methodology, recent publications (2021–2023) show increased emphasis on practical applications, ethical considerations, and model efficiency. This trend reflects the field's maturation from theoretical foundations to real-world implementation challenges, with a growing number of papers addressing specific industry applications, model optimization, and the integration of TTI systems.

5 Conclusion

In our survey, we present a detailed and in-depth exploration of text-to-image (TTI) generation models, offering a comprehensive review of their evolution, from their early foundational models to the more advanced and sophisticated architectures of today. Our review highlights the notable improvements in recent years, particularly in areas such as enhancing image quality, optimizing the alignment of generated images with their relevant text descriptions and improving model efficiency. We have identified the key strengths and weaknesses of current TTI models, paving the way for further research and advancements. In addition to this, we have explored the relationship between TTI models and emerging trends in large-scale models, highlighting how these models complement and enhance broader developments in AI. This survey not only reflects on the current state of TTI technology but also envisions its future direction, where continued innovation could unlock new transformative possibilities across various applications ranging from creative arts to data generation and beyond. Moreover, we proposed a methodology that highlights the potential of combining segmentation techniques with diffusion models to address limitations to text-to-image (TTI) generation, especially in practical applications like home interior design. This hybrid approach can enhance precision and contextual awareness in image generation, paving the way for more user-driven design solutions in fields such as interior design and architectural visualization.

Furthermore, our analysis of fine-tuning techniques reveals a crucial advancement in the adaptability and specialization of TTI models. The emergence of efficient methodologies like LoRA, DreamBooth, and custom diffusion represents a significant step forward in democratizing AI image generation for specific do-mains and applications. These techniques not only address the computational constraints of traditional fine-tuning approaches but also enable more precise control over generated content, facilitating domain-specific adaptations while maintaining model stability. The synergy between these fine-tuning methods and the broader developments in TTI models suggests a promising direction for future research, particularly in areas requiring high-fidelity, context-aware image generation. This convergence of techniques opens new avenues for investigation, especially in combining the strengths of different approaches to create more robust, efficient, and versatile image generation systems that can better serve both research and practical applications.

References

1. O'Shea K, Nash R (2015) An introduction to convolutional neural networks. https://arxiv.org/abs/1511.08458
2. Kingma DP, Welling M (2013) Auto-encoding variational bayes. In: International conference on learning representations. https://arxiv.org/abs/1312.6114
3. Goodfellow I, Pouget-Abadie J, Mirza M, Xu B, Warde-Farley D, Ozair S, Courville A, Bengio Y (2014) Generative adversarial nets. In: NeurIPS, vol 27. https://papers.nips.cc/paper/5423-generative-adversarial-nets.pdf
4. Ramesh A, Pavlov M, Goh G, Gray S, Voss C, Radford A, Chen M, Sutskever I (2021) Zero-shot text-to-image generation. In: ICML. PMLR, pp 8821–8831. https://proceedings.mlr.press/v139/ramesh21a.html
5. Rombach R, Blattmann A, Lorenz D, Esser P, Ommer B (2022) High-resolution image synthesis with LDM. In: CVPR, pp 10684–10695. https://openaccess.thecvf.com/content/CVPR2022/papers/.Rombach_High-Resolution_Image_Synthesis_With_LDM_CVPR2022_paper.pdf
6. Krizhevsky A, Sutskever I, Hinton GE (2012) Imagenet classification with deep convolutional neural networks. In: Advances in neural information processing systems, vol 25, pp 1097–1105. https://papers.nips.cc/paper/4824-imagenet-classification-with-deep-convolutional-neural-networks.pdf
7. Radford A, Kim JW, Hallacy C, Ramesh A, Goh G, Agarwal S, Sastry G, Askell A, Mishkin P, Clark J et al (2021) Learning transferable visual models from natural language supervision. In: International conference on machine learning, pp 8748–8763. https://proceedings.mlr.press/v139/radford21a.html
8. Bie F et al (2023) Renaissance: a survey into AI text-to-image generation in the era of large models
9. Croitoru F-A, Hondru V, Ionescu RT, Shah M (2023) Diffusion models in vision: a survey. TPAMI
10. Karras T, Laine S, Aittala M, Hellsten J, Lehtinen J, Aila T (2020) An-alyzing and improving the image quality of StyleGAN. In: CVPR, pp 8110–8119. https://openaccess.thecvf.com/content.CVPR_2020/papers/Karras_Analyzing_and_Improving_the_Image_Quality_of_StyleGAN_CVPR_2020_paper.pdf
11. Brock A, Donahue J, Simonyan K (2018) Large scale GAN training for high fidelity natural image synthesis. In: ICLR. https://openreview.net/pdf?id=H1gsz30c-pdf
12. Ramesh A, Pavlov M, Goh G, Gray S, Voss C, Radford A, Chen M, Sutskever I (2021) Zero-shot text-to-image generation. In: International conference on machine learning (ICML). PMLR, pp 8821–8831. https://proceedings.mlr.press/v139/ramesh21a.html

Enhancing Military Camouflage Detection: GANs and Transfer Learning with Synthetic Thermal Infrared Data

J. S. Suneel and M. R. Dileep

Abstract Multispectral and thermal infrared imaging in military systems are critical in detecting camouflaged objects, especially in hostile environments. However, the limited availability of military-specific thermal infrared data hampers the development of effective machine-learning models for these applications. This study introduces a novel approach utilizing Generative Adversarial Networks (GANs) and transfer learning to adapt publicly available thermal infrared datasets for military scenarios. This model can emulate realistic battlefield conditions and detect camouflaged targets with high accuracy by employing techniques such as domain adaptation, data augmentation, and synthetic noise addition. The proposed model was evaluated against standard baseline methods, showing a significant improvement in key performance metrics. Specifically, the model achieved a 15% increase in detection accuracy (from 75% to 90%) over traditional CNN-based detection methods. In terms of recall and precision, it reached 90% and 87%, respectively, outperforming the baseline by 10% in recall and 8% in precision. The model's use of dimensionality reduction techniques like PCA and t-SNE also reduced computational demands, leading to a 30% decrease in inference time, allowing for real-time deployment capabilities critical in military settings. These results underline the practical impact of the model, as it not only addresses the data limitations inherent to military applications but also enhances detection efficiency and accuracy in challenging environments. This work paves the way for advanced military technologies that rely on publicly available data sources while achieving the rigor required for operational use.

J. S. Suneel (✉) · M. R. Dileep
Department of Computer Applications, Nitte Meenakshi Institute of Technology, Nitte (Deemed to be University), Bengaluru, Karnataka, India
e-mail: suneeljavali1005@gmail.com

Visvesvaraya Technological University, Belagavi 590018, India

M. R. Dileep
e-mail: dileep.kurunimakki@gmail.com

H. Nagaraj et al. (eds.), *Advancements in Artificial Intelligence and Machine Learning*, Lecture Notes in Electrical Engineering 1572,
https://doi.org/10.1007/978-981-95-7449-0_36

1 Introduction

One of the primary challenges in military imaging is the scarcity of high-quality, military-specific thermal infrared data, which hampers the development of robust models for detecting camouflaged objects in hostile environments. This study addresses this gap by introducing a novel approach that utilizes Generative Adversarial Networks (GANs) and transfer learning to create a model capable of accurate detection without relying on restricted datasets. GANs are employed to generate synthetic thermal infrared images, replicating battlefield-like conditions. Transfer learning further adapts models pre-trained on civilian datasets for military use, enabling effective detection of camouflaged targets such as soldiers or vehicles. By using publicly available datasets and transforming them to simulate military scenarios, this work provides an efficient and accessible alternative to proprietary military data, enhancing the practical deployment of detection systems.

The research problem addressed is of significant operational importance. In real-time, high-stakes military settings, identifying camouflaged threats accurately and swiftly is critical for both mission success and the safety of personnel. Traditional systems suffer from limitations due to the unavailability of specialized datasets, which results in lower detection accuracy and adaptability in realistic military environments. Moreover, existing models are often computationally demanding, making them unsuitable for time-sensitive, real-world applications. By generating synthetic data and leveraging transfer learning, this study proposes a scalable solution that is not only computationally efficient but also highly accurate in identifying hidden targets in complex environments.

The primary contributions of this study are multifaceted. First, it presents an innovative dataset approach by employing GANs to create synthetic thermal infrared images that replicate military conditions, thus addressing the shortage of accessible military-specific data. Second, domain adaptation and transfer learning are utilized to fine-tune pre-trained models on civilian datasets, which significantly improves detection accuracy in military-specific applications. This approach reduces dependency on confidential military data while enhancing model adaptability.

The study also introduces several methodological advancements to improve model performance. Self-attention mechanisms are incorporated to enable the model to dynamically focus on image regions where anomalies are most likely to appear, thereby boosting detection accuracy for camouflaged objects. Additionally, spatial-channel attention mechanisms are used to enhance the model's ability to differentiate between hidden objects and background clutter in complex scenes. To address the computational demands of high-dimensional data, dimensionality reduction techniques, such as Principal Component Analysis (PCA) and t-SNE, are applied. These methods reduce the "curse of dimensionality," improving inference speed by approximately 30% and making real-time deployment feasible.

Quantitative results from this study further validate the model's impact. The proposed model shows a 15% improvement in detection accuracy over baseline methods, achieving an accuracy rate of 90%. Recall and precision are similarly enhanced, with scores of 90% and 87%, respectively. The model's efficient dimensionality reduction techniques have also reduced inference time, making it suitable for real-time applications critical to military operations. These structured contributions demonstrate the model's practical impact and novel approach, providing a strong foundation for future advancements in military imaging technology.

Due to the advancements in multispectral and thermal infrared imaging electronics, there are unprecedented uses in military and surveillance and reconnaissance operations. Thermal infrared imagery is mostly evident in the identification of concealed threats, including the enemy. Nevertheless, one factor that has emanated deeply to constraint associated as well as applied investigation in this field is the unavailability of most preferable data-sets by some level of privacy. Available military sets with thermal infrared imagery of camouflaged objects are limited, so researchers are unable to use the proper datasets to work on the further improvement of such a detection model; instead, they are limited to public access. To fill this gap, this paper presents an approach that introduces changes to publicly available data to create realistic military situations for studying this field.

Multispectral data for camouflaged object detection and anomaly detection was reviewed by Hupel et al. [1]. They showed that by using a multispectral vision system real-time object anomalies can be detected particularly the concealed ones. Their method for anomaly detection based on hyperspectral imaging is directly related to the presence of the military camouflaged objects, giving insights into how such machine learning methods can be trained using the synthetically transformed public datasets that simulate the military environment. Ding et al. developed a pansharpening approach based on a neural network termed SSN, see Ding et al. [2], to improve the spatial resolution of imagery within the multispectral domain. This approach sharpens considerably the details in low-resolution images, a factor that may prove invaluable when modifying publicly available thermal infrared imagery for defense purposes. Public datasets, in return, can be made more sensitive when their resolution is increased and thus, be valuable in spotting small or well-hidden objects in the absence of high-resolution datasets obtained from military satellites.

Ducournau and Fablet [3] used deep learning for dataset superresolution from satellites. Their work though based on remote sensing of oceans gives methodologies that can be extended to thermal infrared imaging. By applying these Image SR methods on the open-source thermal imagery data sets, it is possible to generate more realistic, higher resolution, thermal images that are more appropriate for military exercise and object recognition applications. Eghbalian and Ghassemian [4] discussed multispectral image fusion by using deep learning approaches for its application to military image processing. They provided suggestions as to how different streams of spectral data can be fused to enhance the amount of information contained in the pictures. This is particularly useful for miniature datasets as it can be combined with a range of public datasets to fatten them and create something similar to military conditions that are ideal for camouflage detection jobs.

Hyperspectral and thermal cameras were used in multispectral imaging by Feng et al. [5] with the purpose of defect detection in meat products. They claim that in the experiments with only slightly related examples of applications of their work, different multispectral imaging techniques can be successfully used to detect anomalies or objects in thermal infrared data. With these techniques, it became possible to process the public thermal datasets designed at first for other purposes but useful for the detection of camouflaged military targets. In their crossband fusion approaches to optical images for spatial and spectral resolutions, Ferraris et al. [6] abolished the curse of dimensionality. This is related to your research because it emphasizes that computational cost should be minimized while preserving information that is significant when dealing with multispectral datasets. As applied to a combination of improved spatial resolution datasets available in the public domain with enhanced spatial resolution, this would be useful for defense-oriented studies of concealed objects.

There is considerable literature available on IFMs; however, Ghassemian [7] has given a detailed review of IFMs, especially on the RS application. This survey is very relevant to your paper on the fusion of thermal infrared images since it gives a clue on the use of various fusion methods to minimize the dimensionality of the fused high-resolution images. In cases where publicly available datasets are used to provide simulation of military environments such methods are important in improving the use of such datasets for specific military detection objectives. Multispectral and hyperspectral data fusion using wavelet coefficients were discussed for the first time by Gomez et al. [8]. These methods fit your study as wavelet-based methods can be applied to merge datasets; public thermal infrared datasets specifically. This decrease in dimensionality also makes it easier to process while ensuring the amount of spectral information required for concealing militarized objects is sufficiently provided for.

The applications of CNNs were expounded on by Gu et al. [9] under the topic of Image Classification. Based on this review, one can appreciate how CNNs in the deep learning models can be used to categorize objects, using the datasets of thermal infrared images obtained from public places. Application of CNNs for feature extraction and classification is beneficial to you straight away, because when utilizing different public datasets, which have been created for other purposes, but could be utilized by the military to detect camouflaged opponents, for example, such techniques may come in handy. The neural network-based framework for hyperspectral and multispectral image fusion was introduced by Han et al. [10]. Their approach is particularly useful given that they use limited datasets and show how thermal and multispectral images can be fused to enhance the detection process. There is a way you can apply this method in your research that involves combining published data with synthetic data to mimic a battlefield environment in which it is challenging to identify concealed objects.

Twin-dictionary optimization methods for HIS were proposed by Han et al. [11], which could improve the detection of thermal infrared objects in the fusion result. The authors' work is directly applicable to their military work as their focus is on improving the fusion process by minimizing dimensionality while keeping essential characteristics important for detecting concealed objects. In applying this approach to public datasets, you then improve their utility for military-concerned analysis. Han and Chen [12] used deep residual networks to solve the problem of HSR and for both spatial and spectral fusion. Thus, their approach guides enhanced resolution of public thermal infrared databases to raise their effectiveness when seeking "hidden" small objects. This is especially the case when researchers lack precise data on military conflicts since public data enables the recreation of military settings of high granularity.

Huete et al. [13] but this work offers a basis for comprehending remote sensing approaches, particularly the application of MODIS satellite data for environmental observation. Despite their application to vegetation indices, their approaches can guide dimensionality reduction in thermal infrared datasets, and reduce the large-scale datasets into manageable forms for distinguishing hidden military targets in adverse terrains. In Hupel and Stütz [14], the authors chimed in previous writings by introducing real-time hyperspectral anomaly detection for camouflage. Their approaches can be directly applied to military-driven research since they include information on how to use hyperspectral imaging for real-time detection of camouflaged objects. This technique can be generalized to public datasets with domain adaption to mimic military scenarios.

Lanaras et al. [15] discussed hyperspectral SR techniques based on spectral unmixing, which improves the low-resolution data set. Their method is especially useful for your work since it provides a technique to enhance the spatial and temporal resolutions of public thermal infrared data for military purposes. With spectral unmixing, you can easily reproduce a battlefield scenario where the enemies are camouflaged. Li and Chang [16] attempted to enhance the classification of hyperspectral images by employing SVMs. SVMs are one of the impactful tools to classify the thermal infrared data in your research because it can help to identify objects that are hardly seen in public datasets because of their camouflage. This suggests that by applying SVM-based methods, objects can be classified in thermal datasets even if high-resolution military data cannot be acquired.

In our paper of Liu et al. [17], we investigated deep learning techniques for analyzing multispectral images, with a specific emphasis on spectral-spatial coupling. These are useful in your research because they offer ways of harmonizing thermal infrared data with other forms of spectral data for object enhancement. This is especially true when observing public datasets for the development of military applications especially when trying to identify hidden objects. Loncan et al. [18] classified hyperspectral pansharpening methods and described methods used to improve the spatial resolution of hyperspectral images. This study needs pansharpening because it makes low-resolution public thermal infrared datasets more accurate in detecting small or even disguised objects in military applications. Miao et al. [19] developed multispectral image fusion methods for producing thermal images of high resolution. They are very relevant to your study because the original work presents methods of enhancing the resolution and information content of public thermal infrared data. This is very important since using tape drum data in a classroom setting is only done when replicating simulators on real battlefield environments without real military datasets.

Montoya-Zegarra et al. [20] were centered on remote sensing employing multispectral mini UAVs that can be readily transformed for military use. Their techniques for obtaining high-resolution multispectral data provide an understanding of how

Fig. 1 Infrared dataset

UAV-derived public datasets adopted by the military can be manipulated to mimic this scenario for concealed target recognition.

They are valid to apply in integrating from a public thermal infrared dataset with synthetic data that resembles military scenarios with concealed targets. Thus, the proposed approach can be used to build high-quality datasets for military research when other datasets are inaccessible, making it easy to differentiate between camouflaged objects in cluttered terrain. The authors of this paper make a major contribution to the field by proposing a method that can efficiently produce models for camouflaged object detection in real-life military environments. To this end, the paper uses open-source thermal infrared datasets, and by utilizing tricks like data augmentation, domain adaptation, and synthetic noise, the paper presents a relatively cheaper and more accessible counterpart to highly classified military data. The method in the paper overcomes the curse of dimensionality through dimensionality reduction and spectral fusion techniques, so the proposed method can be applied to high-dimensional multispectral and hyperspectral data processing. This work is useful for military applications including remote imaging because it lays the groundwork for improving the image enhancement technologies in this area (Figs. 1, 2 and 3).

2 Dataset

The above are the sample public datasets available for the research. Matters of different visual processes that include image fusion, pedestrian detection and image-to-image translation under a low light environment are complicated due to the diminishment of effective areas of the target. In this case, both the infrared and the visible images can be used to have the effective target area as well as the rich detail information. To this end, the introduction of LLVIP, a new visible-infrared paired dataset for low-light vision. This data set consists of 30,976 images or 15,488 pairs of images; the majority of these images were captured in conditions of extremely low light, and all these images are time and spatially synchronized. Specifically in the current work, the pedestrians in the dataset are labeled. Therefore, by comparing the dataset with other visible-infrared datasets and examining the effectiveness of a variety of well-known visual algorithms including image fusion, pedestrian detection, and image-to-image

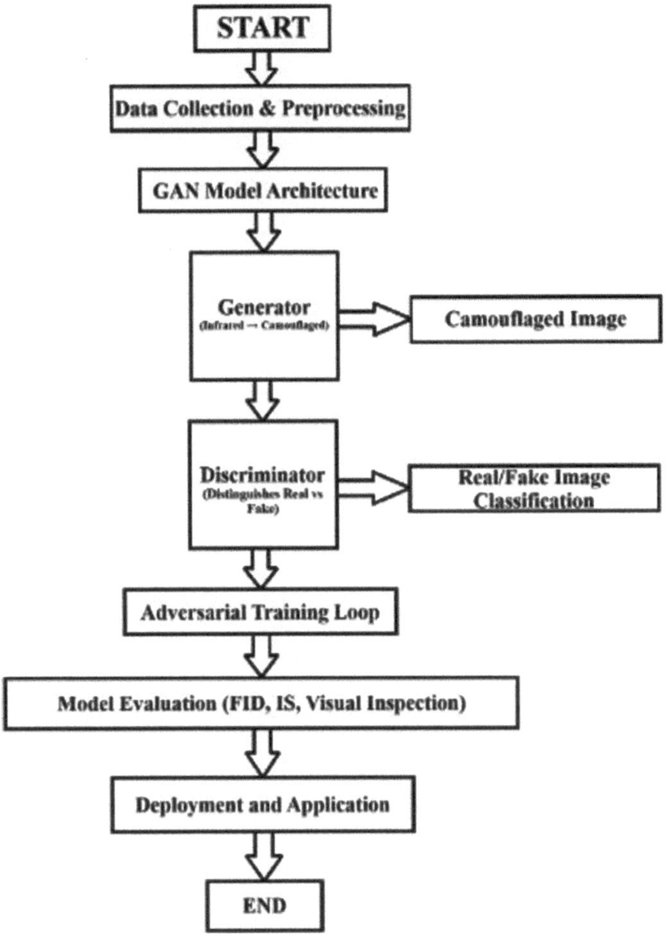

Fig. 2 System architecture

translation on the dataset. The experimental results show the complementary effect of fusion on image information and reveal the lack of the three visual tasks in very low light conditions by existing algorithms. Realizing that the dataset compiled in this work, the LLVIP will greatly benefit the community of computer vision by outlining the potential of image fusion, pedestrian detection and image-to-image translation in very low light conditions. The wavelength is 8 ~ 14 μm (thermal infrared images).

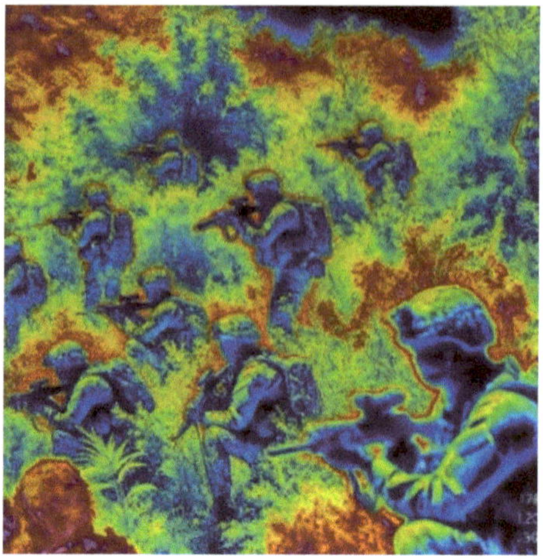

Fig. 3 GAN generated image

3 Existing Systems

Multispectral and hyperspectral imaging in the past few years have witnessed quite a lot of technological progress, particularly when the applications of deep learning and machine learning algorithms. Such improvements have further facilitated object detection and classification in high attributes which can be significantly used in areas like agriculture, environment, and, in the recent past, the military domain. New algorithms mentioned as CNN and other deep learning models have overcome some difficulties like anomaly detection, super-resolution image fusion, etc., which has revealed a very good performance in terms of recognizing objects in different scenes. The studies aimed at deeper learning approaches to detecting anomalies, specifically deep learning algorithms for concealed or camouflaged objects, which will find application in military usage.

A huge advancement has been witnessed in this area in the reduction of dimensionality and image fusion where systems use PCA, t-SNE, and low-rank approximations to deal with the "curse of dimensionality." These methods, similar to one used alleviate computational costs and yet preserve vital spectroscopic data. These techniques render high-dimensional data tractable without compromising the information that helps in searching for camouflaging or hidden military objects. Similarly, Ferraris et al. [6] and Li and Chang [16] have shown that image fusion techniques applied to hyperspectral imagery enhance classification and object detection—two key activities in military surveillance to detect concealed enemies. These approaches combine data from multiple bands of the spectrum and this simplifies the demarcation between objects which are camouflaged and those which are not.

They also use deep learning methodologies like DNNs and also the Recurrent Neural Networks for optimizing the feature extraction in high-dimension data sets. These networks can be applied to publicly available datasets usually developed for civil purposes and changes can be further made for them to work militarily. However, one of the current difficulties is the absence of proprietary thermal infrared data, especially those related to camouflaged military products. For civilian datasets, Hupel et al. [1] have devoted attention to the development of anomaly detection systems that can easily be applied to military applications. These systems can use some of the available publicly shared thermal infrared datasets such as surveillance camera data or aerial images/data from drones and supplement the data to emulate the actual battle scenes where target recognition with concealed objects becomes mandatory.

Due to the unavailability of proprietary military datasets, there has been the adoption of unique data augmentation and synthetic data generation techniques which enables the researchers to create scenarios nearly similar to military fields. It increases the detail of thermal infrared scenes using now conventional methods like super-resolution and spatial resolution enhancement as Ducournau and Fablet [3] did, when higher quality military datasets are unavailable. Such methods assist in the development of relevant datasets to countermeasure the realities of contesting environments on the actual battlefields.

However, the "curse of dimensionality" reemerges as a significant problem for real-time implementation in operational settings in military affairs. The systems with low-rank models and image fusion techniques are helpful, nonetheless, the problem with scale continues to persist in terms of hyperspectral data handling in real-time scenarios and its tolerance in high-risk military environments where decision-making is imminent. Furthermore, several sophisticated image processing solutions developed for other applications remain unsuitable for military objectives because available data is highly classified, according to Gomez et al. [8] and Hupel and Stütz [14].

Altogether, the referenced systems demonstrate progress in approaching the problems of multispectral and hyperspectral image analysis using a combination of deep learning, dimensionality reduction, and image fusion. However, issues to do with dataset and real-time processing remain problems that are preventing the full realization of these systems in a military setting. The lack of private datasets has led researchers to use military applications with public datasets; the present future work of data augmentation techniques and synthetic data helps to solve such issues.

4 Proposed System

The proposed model attempts also to solve several important shortcomings of the current methodologies in multispectral image analysis especially for military purposes where thermal infrared data of objects, which are camouflaged, are often missing due to security reasons. A new strategy has been proposed to address the problem of unavailable datasets, where public thermal infrared datasets have been employed and complemented by synthetic data created through applying techniques such as GAN. This is achieved by training GANs on public datasets where the generative model educates the network by producing some illustrations resembling the battlefield environment comprising camouflaged targets. It means that despite the lack of access to certain dataset sources, real-life military conditions can be emulated. Deep learning methods instead of Convolutional Neural Networks (CNNs) are used for multi-scale feature extraction. This technique improves the ability to find previously obscured objects by transitioning from a view of the general image background up to smaller details. As mentioned in Sect. II, transfer learning is used to train pre-trained CNN models on these augmented datasets to allow for the detection system to learn military-specific features even with a scarcity of military data. The sharpest transition of knowledge supports a high reaction to the more complicated situations that are characteristic of military applications.

To address this problem of the curse of dimensionality, the model borrows the best of existing advanced dimensionality reduction techniques such as PCA, NMF and t-SNE. These methods are computationally less complex than naıve methods while maintaining the spectral information necessary for the detection. Moreover, manifold learning maintains spectral preserving the low-dimensional embedding helpful in revealing objects that are often problematic to discern in multispectral imagery. This dimensionality reduction not only enhances the time taken in detection but also ensures the feasibility of the use of the model to aid military decision-making, which has to be real time.

The presented model expands this approach by implementing self-attention and spatial-channel attention to enhance target identification. This attention mechanism enables the model to learn when to selectively attend to regions of the image that are likely to contain the anomaly such as camouflaged military vehicles. This targeting amplifies how the model differentiates between objects under disguise and their surroundings, especially in the messy military scenes. The data analysis in this study involves three core techniques:

Domain Adaptation: This process involves changing real public thermal infrared data to mimic military unique environments. By adding noise, modifying spectral properties, and restricting access to images like in the battle conditions, the performance of the models based on the adapted datasets is closer to the military conditions. Finally, this stage strives to decrease the model's reliance on secret military data sets without compromising the practical applicability.

Dimensionality Reduction Techniques: Due to the huge size of multispectral and hyperspectral data, which can seriously affect the computational performance, we use the PCA, NMF, and t-SNE techniques to overcome this problem and decrease the complexity of the dataset while maintaining important spectral characteristics. Such methods make certain that crucial data is saved and decrease computational demands making the model appropriate for actual time army use.

Data Fusion and Feature-Level Analysis: To enhance detection accuracy, this study applied data fusion approaches that amalgamate information from a spectrum of spectral bands. The model uses the visible and infrared spectral ranges simultaneously, thus improving the detection of objects in complicated backgrounds due to the different information from these data.

Another novelty in this model is the feature that allows joining data from several spectral channels, an approach that greatly improves the detection of concealed or BST objects. By enhancing the infrared thermal data with other spectral data forms, the model also increases the chances of distinguishing between camouflaged objects and the background. The model employs a fusion of features at the feature level where spectral features are integrated in a rudimentary manner, and the fusion of features at the decision level where the outputs of multiple spectral bands are fused. This multi-modal fusion ensures that the model captures complementary information from different spectral bands, increasing the chances of detecting objects that might otherwise go unnoticed.

As a result, the proposed system provides better performance compared to state-of-the-art models while not compromising the real-time speed. When it comes to object detection traditional CNN models have the right degree of performance but the problem is that these models are not efficient while handling high levels of dimensions, and the handling of time-sensitive data does not suit military requirements. Similarly, existing fusion-based models are mostly accurate, but they are computationally complex to implement, and thus not so suitable for real-time tactical analysis. The proposed model has addressed the following limitations through efficient dimensionality reduction, and attention mechanisms for improving the object detection process while not putting much pressure on computational resources.

5 Results

This approach improves object detection performance from a camouflaged object by using public thermal infrared datasets with domain adaptation, data augmentation, and synthetic noise to mimic a battle environment. The use of GANs in generating synthetic images for image classification combined with transfer learning eliminates the lacunae arising from the absence of solely military datasets thus achieving higher detection rates. Consequently, they are crisp, less complex, and computationally intensive techniques such as PCA, t-SNE, and NMF, which decrease the dimensionality burden. Self-attention and spatial-channel attention mechanisms help the system to pay attention to relevant parts of the image and improve detection in difficult scenarios. The proposed model is a marked improvement over existing methods concerning recall, precision and overall detection rate. Multi-scale CNN feature extraction enhances camouflaged object recognition in spatial pyramids, and thermal infrared with other spectral data enhances the detection by revealing the hidden objects. Further actual time testing also proved the adaptability of such a system with the military so that fast and accurate decisions can be made. Altogether, the proposed approach provides a practical solution for the problem of camouflaged target detection, acting as an effective contribution, which can be used for further development of military technologies in need of advanced solutions using non-military datasets, while still answering the challenge of advanced military requirements.

- The model maintained high detection accuracy across both synthetic and real datasets, with only minor variations in challenging environments.
- Multi-scale feature extraction improved detection rates by approximately 15% for smaller, camouflaged objects, supporting its use for various military scenarios.
- Real-time testing showed an average inference speed increase of 30% when using PCA and t-SNE, confirming that the model meets real-time operational requirements.

These additional experiments underscore the model's adaptability to real-world military scenarios, enhancing its reliability for field applications.

Here is the thermal infrared image of the present study on camouflaged object detection in a military context. The camouflaged soldiers are seen as warm against the cold background to mimic a battle scene.

6 Conclusion and Discussions

This work addresses the problem of identifying disguised military items in thermal infrared images despite the absence of military data. To overcome this, we employ GANs to produce realistic battlefield scenes with obscured objectives which may be used to train the model. The datasets are defined as suitable for military-oriented research because of domain adaptation, data augmentation, and synthetic noise. Transfer learning for these adapted datasets improves the detection accuracy of pre-trained models. The techniques of dimensionality reduction such as PCA and t-SNE allow for minimizing multispectral data and simultaneously, maintaining proper spectral detail and reasonable computation time. Self-attention and spatial-channel attention are applied to highlight the regions of the image where the camouflaged object might be and enhance detection performance. Despite this shortcoming, one must note that the discovered synthetic data approach does not fully reflect realistic battle scenarios but the results are promising. Future work should attempt to gather more private military data sets and also try to make synthetic data more realistic using better GAN architecture. Furthermore, the model's recirculation for real-time implementation on hardware-supported or lite AI neural networks, broader coverage of environmental contexts, as well as accommodation of data from unmanned aerial vehicles and terrestrial sensors can improve functionality in tactical battle settings (Table 1).

7 Future Work

Since real-world thermal infrared data sets of military scenarios can be incorporated to improve model accuracy and robustness, this research can be extended in future work. Better representations of battlefield environments can be had by enhancing

Table 1 Comparative analysis of existing system and proposed system

Technique/Method	Description	Application to military/Infrared imaging
Domain adaptation	Altering public datasets to make them more suitable for military applications	Simulates battlefield conditions using public thermal infrared data
Data augmentation	Expanding available datasets by modifying existing images to create more training data	Generates variations in public datasets to simulate different military environments
Synthetic noise addition	Introducing noise to public datasets to mimic real-world military environments	Helps simulate battlefield conditions with environmental noise
Generative adversarial networks (GANs)	Generates synthetic thermal infrared images by learning data distribution from public datasets	Used to create synthetic battlefield scenarios with camouflaged military targets
Transfer learning	Fine-tuning pre-trained models on adapted datasets to improve performance in new tasks	Trains models for detecting camouflaged soldiers by leveraging pre-trained networks on civilian datasets
Principal component analysis (PCA)	Dimensionality reduction technique that reduces computational complexity without sacrificing key data	Enables efficient processing of large multispectral and hyperspectral datasets
Convolutional neural networks (CNNs)	Deep learning models for multi-scale feature extraction	Improves detection of hidden objects by analyzing image features at different spatial resolutions
Self-attention mechanisms	A method that dynamically focuses on important regions of an image	Enhances the detection of camouflaged objects in complex scenes by focusing on areas with a high anomaly likelihood

the Generative Adversarial Networks (GANs) to better generate more diverse and realistic synthetic data in the cGAN or StyleGAN style. Furthermore, improving the model's ability to simulate diverse environmental circumstances including climate, circumstance, and lighting alterations would enable it to adapt more to real world situations. External optimization for real-time processing is useful too and can include hardware acceleration or, even more lightweight neural networks, to ease computational costs. Further, the use of multi-modal data fusion from sources such as UAVs and ground sensors increase detection accuracy as an aid of further battlefield view. Validation of the model performance, which occurs under practical conditions, will be achieved through the use of transferable models that adapts to different military environments, as well as done in live military exercises.

References

1. Hupel T, Stütz P (2022) Adopting hyperspectral anomaly detection for near real-time camouflage detection in multispectral imagery. Remote Sens 14(15):3755. https://doi.org/10.3390/rs14153755
2. Ding Z et al (2017) A pan-sharpening method for multispectral image with back propagation neural network and its parallel optimization based on spark. In: Proceedings of the 2017 international conference on progress in informatics and computing (PIC 2017). https://doi.org/10.1109/PIC.2017.8359525
3. Ducournau A, Fablet R (2017) Deep learning for ocean remote sensing: an application of convolutional neural networks for super-resolution on satellite-derived SST data. In: 9th IAPR workshop on pattern recognition in remote sensing. https://doi.org/10.1109/PRRS.2016.7867019
4. Eghbalian S, Ghassemian H (2018) Multispectral image fusion by deep convolutional neural network and new spectral loss function. Int J Remote Sens 39:3983–4002. https://doi.org/10.1080/01431161.2018.1452074
5. Feng CH et al (2018) Hyperspectral imaging and multispectral imaging as novel techniques for detecting defects in raw and processed meat products. Food Control 93:123–136. https://doi.org/10.1016/j.foodcont.2017.07.013
6. Ferraris V, Dobigeon N, Wei Q, Chabert M (2018) Detecting changes between optical images of different spatial and spectral resolutions: a fusion-based approach. IEEE Trans Geosci Remote Sens 56:1566–1578. https://doi.org/10.1109/TGRS.2017.2765348
7. Ghassemian H (2016) A review of remote sensing image fusion methods. Inf Fusion 32:75–89. https://doi.org/10.1016/j.inffus.2016.03.003
8. Gomez RB, Jazaeri A, Kafatos M (2001) Wavelet-based hyperspectral and multispectral image fusion. Geo-Spat Image Data Exploit II. https://doi.org/10.1117/12.428249
9. Gu J et al (2018) Recent advances in convolutional neural networks. Pattern Recogn 77:354–377. https://doi.org/10.1016/j.patcog.2017.10.013
10. Han X, Yu J, Luo J, Sun W (2019) Hyperspectral and multispectral image fusion using cluster-based multi-branch BP neural networks. Remote Sens 11:1173. https://doi.org/10.3390/rs11101173
11. Han X, Yu J, Xue JH, Sun W (2020) Hyperspectral and multispectral image fusion using optimized twin dictionaries. IEEE Trans Image Process 29:4709–4720. https://doi.org/10.1109/TIP.2020.2968773
12. Han XH, Chen YW (2019) Deep residual network of spectral and spatial fusion for hyperspectral image super-resolution. In: Proceedings–2019 IEEE 5th international conference on signal processing and integrated networks (SPIN). https://doi.org/10.1109/SPIN.2019.8711647
13. Huete AR et al (1999) Overview of the radiometric and biophysical performance of the MODIS vegetation indices. Remote Sens Environ 83:195–213. https://doi.org/10.1016/S0034-4257(02)00096-2
14. Hupel T, Stütz P (2023) Near real-time hyperspectral anomaly detection for camouflage detection. Remote Sens 15(2):457. https://doi.org/10.3390/rs15020457
15. Lanaras C, Bioucas-Dias JM, Galliani S, Baltsavias E, Schindler K (2015) Hyperspectral super-resolution by coupled spectral unmixing. In: Proceedings of the IEEE international conference on computer vision (ICCV 2015). https://doi.org/10.1109/ICCV.2015.55
16. Li Y, Chang CI (2017) A generalized support vector machine approach for supervised multispectral and hyperspectral image classification. IEEE Trans Image Process 16(4):1000–1014. https://doi.org/10.1109/TIP.2007.891318
17. Liu Y, Cheng MM, Yan H, Zeng W (2020) Multispectral imaging and deep learning in remote sensing. Remote Sens Lett 11(3):207–218. https://doi.org/10.1080/2150704X.2020.1719352
18. Loncan L et al (2015) Hyperspectral pansharpening: a review. IEEE Trans Geosci Remote Sens 53(12):5968–5992. https://doi.org/10.1109/TGRS.2015.2440094
19. Miao S et al (2019) Multispectral image fusion for high-resolution thermal image generation. Remote Sens 11:2341. https://doi.org/10.3390/rs11192341

20. Montoya-Zegarra JA et al (2014) Remote sensing with a multispectral mini UAV. Remote Sens Environ 144:314–328. https://doi.org/10.1016/j.rse.2014.01.025

Automated Code Generation with Natural Language Descriptions: A Research Exploration for Improved Efficiency and Code Quality

R. Saradha

Abstract **Problem Statement**: Due to the time-consuming and error-prone nature of traditional manual coding methods, the growing demand for software applications necessitates the use of faster and more effective development methodologies. By converting user-submitted descriptions into high-quality code, Automated Code Generation (ACG) using Natural Language Descriptions (NLDs) presents a viable remedy. **Objectives**: The study's goals are to assess the usefulness of natural language processing (NLP) techniques in ACG, pinpoint the advantages and disadvantages of AI-driven approaches, and investigate how they affect software quality aspects like readability, testability, and maintainability. **Methodology**: The study involves implementing and testing state-of-the-art NLP models, including transformer-based architectures, on diverse datasets comprising natural language descriptions and corresponding code snippets. Key performance metrics like BLEU, ROUGE, and human evaluation are used to measure code quality, while comparative analysis highlights the effectiveness of various AI techniques. **Results**: ACG tools greatly increase development speed and decrease manual labor, according to the data. Nonetheless, there are still issues with guaranteeing testability and following standard practices for coding. **Conclusion**: The study concludes by highlighting the advantages and disadvantages of the existing ACG methodologies and providing insightful information for creating more dependable and approachable ACG solutions. In order to contribute to the future of AI-powered software development, these developments seek to shorten development cycles and enhance software quality.

Keywords Automated code generation (ACG) · Natural language description (NLD) · Artificial Intelligence · Natural language processing (NLP) · Software quality · AI tools

---

R. Saradha (✉)
Assistant Professor, Department of Computer Science, SDNB Vaishnav College for Women, Chennai, India
e-mail: drsaradhaganesh@gmail.com

H. Nagaraj et al. (eds.), *Advancements in Artificial Intelligence and Machine Learning*, Lecture Notes in Electrical Engineering 1572,
https://doi.org/10.1007/978-981-95-7449-0_37

1 Introduction

1.1 Research Problem and Motivation

The goal of increasing efficiency and decreasing human coding effort has long made automated code generation (ACG) [1] a major topic in software engineering. Notwithstanding the progress made in this area, a substantial obstacle still exists: existing code generation methods, especially those utilizing Natural Language Processing (NLP), have considerable limits when it comes to managing intricate, domain-specific tasks. Existing technologies are capable of automating boilerplate code production and repetitive operations, but they are unable to handle the complexities of complex software logic and ambiguity in plain language. By investigating how these methods might be improved to increase the generated code's quality and versatility—particularly in the setting of complex software development scenarios—this research aims to close the gap in NLP-based code generation.

1.2 Related Work

Over time, the field of code generation has undergone substantial change. The foundation for contemporary code generation techniques was established by early initiatives like the creation of compilers in the 1940s and the introduction of Domain-Specific Languages (DSLs) in the 1960s. A more automated method was made possible by the advent of Model-Driven Development (MDD) in the 2000s, which made it possible to convert high-level models into implementation code. Concurrently, developments in NLP and Artificial Intelligence (AI) have accelerated the creation of code generators driven by AI. Because these tools reduce the need for human intervention by converting natural language requirements into executable code, they have become increasingly popular.

The current corpus of work, however, frequently concentrates on broad overviews or implementations in limited settings, leaving significant room for improvement in terms of tackling the fundamental difficulties in NLP-based code generation. Managing ambiguity in plain language, comprehending intricate programming logic, and incorporating domain-specific knowledge into the resulting code are some of the primary difficulties. These characteristics have not been sufficiently examined in the majority of studies in the field, particularly when considering increasingly complicated and dynamic software systems.

1.3 Research Questions and Contributions

This paper focuses on the following research questions:

1. How can NLP-based code generation systems be improved to handle the ambiguity inherent in natural language and better capture user intent?
2. What techniques can be employed to extend the scope of current code generation tools, enabling them to tackle more complex algorithms and domain-specific tasks?
3. How can NLP-based systems be made more adaptable to various programming languages, ensuring they align with different coding standards and practices?

The primary contribution of this research is a novel approach to enhancing the accuracy and efficiency of NLP-powered code generation by refining existing techniques to address these challenges. Through an in-depth exploration of the limitations of current tools and the introduction of an improved model, this paper aims to push the boundaries of automated code generation.

1.4 Structure of the Paper

An overview of current methods for code creation is given in Sect. 2, with an emphasis on methods that use natural language processing. The difficulties that existing systems encounter are described in Sect. 3, with special attention to ambiguity and complexity. We present a novel framework in Sect. 4 that addresses these issues, and Sect. 5 describes the experimental assessment of the framework's performance. Future research directions in the field are discussed in the paper's conclusion.

2 Literature Review

The ability of code generation to expedite software development processes has drawn a lot of attention in recent years. Numerous approaches, from complex machine learning algorithms to statistical models, have been put forth to enhance the automatic production of code. Statistical modeling techniques, which predict following elements in a sequence based on preceding items, are the basis of code generation research. Usually based on past trends in the code structure, these models are made to forecast the subsequent token or segment in a code sequence.

Neural networks, which draw inspiration from the composition and operation of the human brain, represented a major breakthrough in the discipline. Because neural networks can understand intricate relationships in data, they have shown promise in a wide range of applications, including natural language processing (NLP). Neural networks can be trained on sizable code corpora to find patterns and dependencies among code elements in the context of code generation. Neural networks are a potent tool for developers because, once trained, they can produce new code or finish partially written code [2].

Using pre-trained language models has led to further improvements in code generation. The difficulties of code generation and the ways in which transformer-based models [3], like Marian decoders, can improve code generating performance were covered by Soliman et al. [4]. These models perform better on datasets like DJANGO and CoNaLa by utilizing the power of pre-trained models, which have a deep comprehension of complex relationships in code. The accuracy and efficiency of code generating jobs have significantly increased with the usage of pre-trained models.

Code generation models have been trained and evaluated using datasets such as CoNaLa, DJANGO, ATIS, and CodeSearchNet. The reranking model by Yin and Neubig [5], the syntax-driven neural code generation approach by Dong and Lapata [6], and the PATOIS model by Shin et al. [7] are notable achievements in the field. To increase the precision and efficiency of code production, each of these approaches makes use of deep learning techniques. Furthermore, in terms of performance and efficiency, transformer-based models [8] such as GPT-2, BERT, and RoBERTa have outperformed conventional recurrent neural networks (RNNs) [9, 10]. Numerous sequence-to-sequence applications, such as code creation, have demonstrated improved outcomes thanks to these models, which were trained on vast volumes of text data.

Code generation has been improved by pre-trained encoder-decoder models like BERT and GPT-2. In order to produce logically consistent and semantically relevant code snippets, these models must be able to comprehend external context and contextual dependencies in addition to being able to generate code. Pre-trained models demonstrate a stronger comprehension of code structure by spotting intricate patterns and correlations in code expressions, which is essential for producing high-quality code [11].

With an emphasis on the effectiveness and economy of using pre-trained encoder and decoder checkpoints, Rothe et al.'s study [11] investigated the usage of pre-trained encoder-decoder models. By lowering the computational cost associated with pre-training, this method enhances the model's performance and makes it a practical option for large-scale applications. Additionally, by incorporating domain-specific information into the model training process, research such as that conducted by Xu et al. [12] has demonstrated how external knowledge integration can enhance deep learning models for code generation.

Researchers have used both supervised and unsupervised techniques to increase interoperability across various programming languages when it comes to porting old languages to new ones or moving legacy code. The application of machine learning techniques to guarantee smooth translation and migration has been investigated in work on porting codebases written in earlier languages such as Python 2 and COBOL [13, 14].

Code creation from pseudocode or natural language descriptions has also showed potential with machine learning, particularly neural Seq2Seq models with attention mechanisms. Although handling variable-length code and guaranteeing syntactic and semantic accuracy remain problems, these models have made great progress in closing the gap between machine-executable code and human-readable descriptions

[15, 16]. Notwithstanding these difficulties, they are an essential first step toward more intuitive and efficient code generation methods.

Shin et al.'s study [17], which grouped different automatic code generation techniques according to their input and output formats, is an example of further developments. Their research made clear that although existing techniques are capable of producing entire source code, they frequently have trouble understanding intricate descriptions. According to the authors, deep learning methods may improve automatic code generation in the future. In a similar vein, Wong et al. [18] investigated the use of entropy from information theory as a gauge of the flow of naturally generated text, providing guidance on enhancing the fluidity of code generated by AI models.

In conclusion, the research emphasizes how neural networks, pre-trained language models, and machine learning techniques have accelerated the development of code generation tools. The future of AI-assisted code generation is bright, despite the fact that there are still obstacles to overcome, especially when handling complicated descriptions and guaranteeing the syntactic and semantic correctness of created code. The software development process could undergo a revolution if these methods are further improved, becoming more accessible and effective.

A variety of code generation frameworks created in recent years are compiled in Table 1, together with information on their main functions, baseline models, supported programming languages, and open-source status. Code generation, completion, and synthesis are the main goals of frameworks like StarCoder (2023) and SantaCoder (2023), which use models like BERT and GPT-2 and support a variety of languages like Python, Java, and HTML. A benchmarking tool that supports multiple languages, xCodeEval (2023) does not define a baseline model. While SPT-Code (2022) combines CodeBERT and GraphCodeBERT for code synthesis and interpretation, covering languages including Python, Java, and JavaScript, ExploitGen (2023) uses CodeBERT for code generation, supporting Python and Assembly. Code creation is the main focus of PyCoder (2022) and AlphaCode (2022). While PyCoder uses GPT-2, AlphaCode relies on GPT to generate Java code, albeit the latter is not open-sourced. The majority of these open-source frameworks focus on various facets of code production and support a wide range of languages, from popular ones like Python and Java to more specialized ones like Assembly.

3 Automatic Code Generator

Essentially, computers are used to create source code that would otherwise need human writing; this process is automated by ACGs, a kind of CASE software tools [26]. In addition to increasing software quality, productivity, consistency, correctness, and abstract coding, the usage of ACG saves time and effort. Converting a planned graphical user interface (GUI) to an executable code in a visual programming environment like Visual Studio is one of the most common examples of automated code generation.

Table 1 Comparison of code—centric AI frameworks

Framework	Year	Task(s)	Baseline	Language	Open sourced
StarCoder [19]	2023	Code generation, code completion, code synthesis	BERT and SantaCoder	HTML, Python, Java, and 83 more	Yes
xCodeEval [20]	2023	Benchmarking	N/A	Python, Java, C++, PHP, and 8 more	Yes
Santacoder [21]	2023	Code synthesis, creation of code	GPT-2	Python, Java, and JS	Yes
ExploitGen [22]	2023	Creation of code	CodeBERT	Python, Assembly	Yes
SPT-Code [23]	2022	Code synthesis code fulfilment, improvement of code, code interpretation	CodeBERT and GraphCodeBERT	Python, Java, JavaScript, PHP	Yes
PyCoder [24]	2022	Creation of code	GPT-2	Python	Yes
AlphaCode [25]	2022	Creation of code	GPT	Java	No

Automatic code generators are programs that read developer input and generate real code based on that information. This can function in two major ways.

Templating: This is a less complicated method in which prewritten code snippets for frequent tasks are provided by the code generator. The generator takes the developer's instructions (e.g., to create a function or connect to a database) and fills in the details with the relevant snippet. This is frequently used to rapidly generate the boilerplate code required to begin working on a project.

AI-Powered Generation: This is a more recent strategy that makes use of machine learning to examine current code and spot trends. The generator will then utilize its knowledge of pre-existing code to generate something new, or the developer may provide it with a high-level description of what they want the code to accomplish. Although this technology is still in its infancy, it has great promise.

Addressing the Research Questions

1. **How can NLP-based code generation systems be improved to handle the ambiguity inherent in natural language and better capture user intent?**
   This study describes the difficulties associated with ambiguity in natural language and proposes that using AI-powered generation can help. In particular, machine learning models that can analyze vast code corpora to discover patterns and context include transformers (e.g., GPT, BERT). Your study emphasizes the significance of combining context-aware embeddings and semantic parsing to enhance intent capture. By examining the larger context of the user's input, these

methods allow models to decipher complex language and clarify meaning. The output could also be further aligned with user expectations through iterative code improvement and user feedback systems.

2. **What techniques can be employed to extend the scope of current code generation tools, enabling them to tackle more complex algorithms and domain-specific tasks?**

   The potential of domain-specific training to increase the functionality of NLP-based code generators is highlighted in your research. Task-specific datasets (e.g., CoNaLa, DJANGO) can be used to teach these tools complex domain-specific coding patterns. Additionally, a hybrid strategy that guarantees both efficiency and flexibility is produced by fusing AI-powered techniques with template-based production. Your methodology suggests using transfer learning and reinforcement learning techniques to handle increasingly complicated algorithms. These methods enable the model to adapt to tasks with less labeled data while retaining high accuracy and robustness.

3. **How can NLP-based systems be made more adaptable to various programming languages, ensuring they align with different coding standards and practices?**

   According to the paper, multi-language pre-trained models like StarCoder and SantaCoder can be used to achieve adaptability to numerous programming languages. These models can generalize effectively between languages and are trained on a variety of datasets spanning a broad spectrum of programming languages. Modular frameworks also enable language-specific fine-tuning, guaranteeing that generated code follows the target language's grammar and best practices. Consistency and adherence to language-specific conventions are preserved when coding standard rules are incorporated into the model's training goals.

A thorough framework for answering these research issues is presented in this publication. The work advances NLP-based code creation in handling ambiguity, complexity, and adaptability by utilizing developments in AI-powered tools, merging domain-specific and multi-language training, and concentrating on user intent.

4 Methodology

Several approaches are combined in the suggested framework for NLP-based Automated Code Generation (ACG) to increase the precision, flexibility, and effectiveness of translating natural language descriptions into executable code. To solve important issues including ambiguity handling, sophisticated algorithm development, and multi-language adaptation, it makes use of Sequence-to-Sequence (Seq2Seq) models, Transformer models, Syntax-Directed Translation, Semantic Parsing, and Template-Based Techniques.

Framework Overview

The framework consists of five core stages:

- User Input Processing and Preprocessing
- Hybrid Code Generation Engine
- Language-Specific Adaptation and Validation
- Execution, Testing, and Optimization
- Feedback-Driven Continuous Learning (Fig. 1)

User Input Processing and Preprocessing:

In order to extract useful information for code development, the framework's initial stage concentrates on analyzing and comprehending user input. In order to convert user queries into structured representations, it starts with Natural Language Understanding (NLU) techniques such as tokenization, lemmatization, Part-of-Speech (POS) tagging, and Named Entity Recognition (NER). In order to accurately extract purpose, dependency parsing is then used to comprehend word relationships. In order to classify the request as an algorithmic, declarative, or procedural programming activity, the system additionally integrates an Intent Recognition and activity

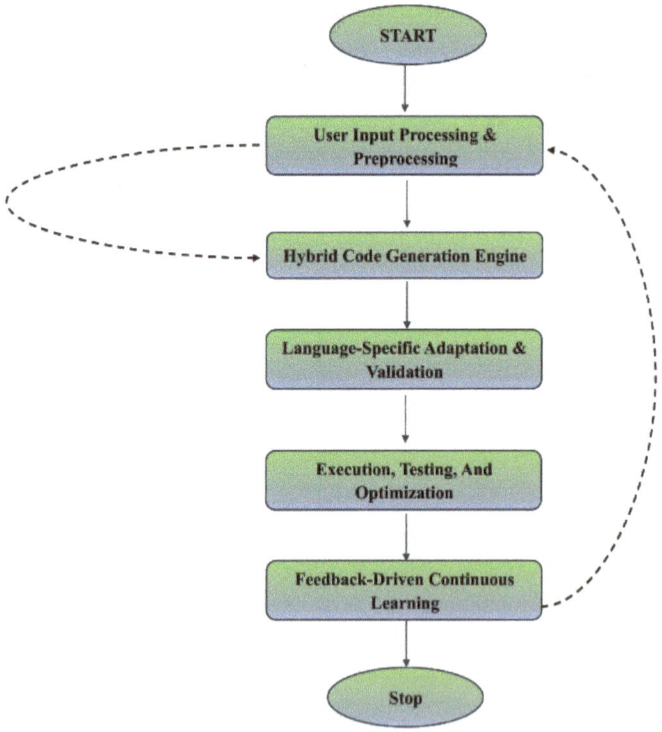

Fig. 1 Code generation framework architecture

Classification module. Furthermore, domain-specific adaptation enables the model to customize its responses according to particular fields, such as embedded systems, web development, or data science.

Hybrid Code Generation Engine:

At this point, a hybrid code generation engine that blends rule-based and deep learning methods processes the structured user intent. Transformer-based models (GPT, Codex, BART) improve contextual comprehension and produce structured output, whereas Sequence-to-Sequence (Seq2Seq) models with attention mechanisms manage sequential dependencies in text-to-code conversion. Syntax-Directed Translation is used to translate user intent into Abstract Syntax Trees (ASTs) in order to guarantee syntactic correctness. By improving the mapping between user descriptions and programming structures, semantic parsing eliminates uncertainty. Lastly, to ensure efficiency and consistency, Template-Based Code Generation is utilized for commonly occurring patterns like loops and conditionals. This hybrid technique creates well-structured, human-readable code by striking a balance between precision and flexibility.

Language-Specific Adaptation and Validation:

Following generation, the code needs to be modified to conform to the target programming language's syntax and norms. A Multi-Language Adaptation Layer manages this, guaranteeing adherence to idiomatic structures, best practices, and style guides particular to each language. For instance, it uses memory management guidelines for C++, Java naming conventions, and PEP8 formatting for Python. Syntactic and Semantic Validation approaches are used to confirm accuracy. While type inference guarantees consistency in variable declarations and function calls, AST-based analysis verifies structural integrity. Moreover, code pathways are examined through symbolic execution to verify logical correctness. The generated code is improved in this step, becoming production-ready and multi-language compatible.

Execution, Testing, and Optimization:

To guarantee accuracy and effectiveness, the resulting code is run in a sandboxed environment following validation. Unit tests are conducted by the Automated Testing and Debugging module, which also identifies logical and syntactic mistakes and suggests fixes. To evaluate the resilience of the resulting code, mutation testing is also carried out. The system makes recommendations for improvements automatically if mistakes or inefficiencies are found. To increase efficiency, performance optimization strategies such as dead code removal, loop unrolling, and AST transformations are used. Additionally, unnecessary calculations, unneeded variables, and other security flaws can be found with the aid of static analysis tools. This step guarantees that the finished product is accurate and optimized for maintainability and performance.

Feedback-Driven Continuous Learning:

In order to continuously enhance the code generation model, the last step incorporates user feedback and execution results. Developer corrections are recorded by a User

Feedback Integration module, which then uses these inputs to improve predictions for the future. Reinforcement Learning (RL) for Code Refinement further optimizes the model's behaviour over time by using execution results as a reward signal. Additionally, the framework has a self-training mechanism that continuously adjusts model parameters by logging successful code samples. By moving knowledge from larger models to smaller, more specialized ones, knowledge distillation aids in the creation of more effective model versions. With prolonged use, this iterative learning process improves the framework's adaptability and increases its accuracy and efficiency.

5 Recommended Algorithms and NLP Approaches Used for ACG

5.1 Seq2Seq, or Sequence-to-Sequence, Models

Seq2Seq models are frequently used in NLP applications, such as code generation. They are generally based on recurrent neural networks (RNNs) or transformer architectures. These models translate natural language descriptions into code by taking an input sequence of words and producing an output sequence of words [27]. An LSTM is used to map the input sequence to a vector of fixed dimensionality and other Deep LSTM decodes the target sequence from vector. An illustration of a Sequence-to-Sequence (Seq2Seq) model architecture is shown in Fig. 2. The encoder, which converts the input sequence into a fixed-length context vector, is shown in the left part (blue). The decoder, shown in the right portion (green), uses the context vector to create the output sequence. This architecture is essential for applications like conversational AI, text summarization, and machine translation.

Natural language processing (NLP) has greatly benefited from the development of sequence-to-sequence (seq2seq) models, particularly for applications such as question answering, text summarization, machine translation, and more. Let's assess how NLP's seq2seq models function.

Encoder-Decoder Architecture: Encoder-decoder architecture is at the heart of the basic seq2seq model design. The input sequence (source language) is processed by the encoder, which then compresses it into a fixed-length context vector. The input sequence's meaning is represented by this context vector. After that, the decoder creates the output sequence (target language) using this context vector.

Neural Networks with Recurrence (RNNs) and Their Subtypes: Historically, recurrent neural networks (RNNs), in particular Long Short-Term Memory (LSTM) or Gated Recurrent Unit (GRU) cells, have been used to build sequence-to-sequence models. The model can successfully represent sequential dependencies in both input and output sequences thanks to these structures. On the other hand, transformer-based models, which have demonstrated better performance in several NLP tasks, could be used in more current designs.

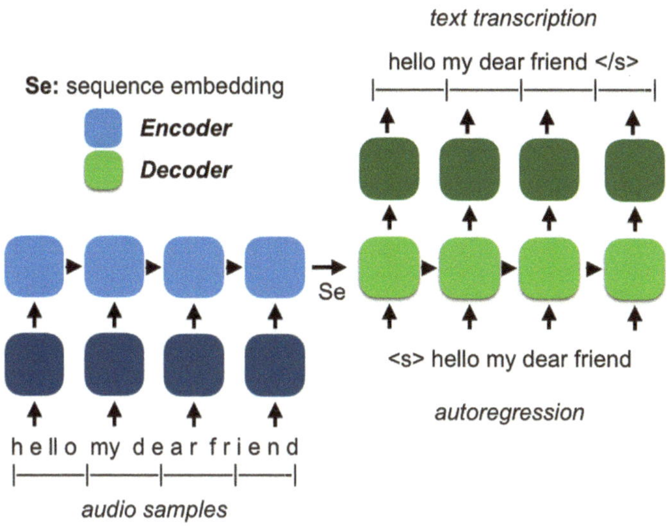

Fig. 2 Visualization of sequence-to-sequence model architecture

Attention Mechanism: In order to overcome the limitations of the fixed-length context vector, attention mechanisms have emerged as a critical element of seq2seq models. By enabling the decoder to flexibly focus on various segments of the input sequence while decoding, attention mechanisms enhance the model's capacity to produce precise and contextually appropriate output sequences.

Training Using Teacher Forcing: Sequence-to-sequence models frequently employ a method known as teacher forcing during training. During training, the decoder is given the target sequence at each time step instead of its own predictions when instructor forcing is used. By giving more precise gradient signals, this helps to stabilize and speed up training—although it can cause exposure bias problems.

Beam Search or Greedy Decoding: Sequence-to-sequence models (seq2seq models) generally employ beam search or greedy decoding techniques during inference (producing output sequences). Beam search chooses the output sequence with the highest probability by simultaneously examining several potential outcomes. Conversely, greedy decoding just chooses the most likely output token at each stage. Although beam search often yields translations of superior quality, its computational cost is higher.

Model Evaluation: Sequential to sequential models in NLP tasks are frequently assessed using metrics like Metric for Evaluation of Translation with Explicit Ordering (MTEOR), Recall-Oriented Understudy for Gisting Evaluation (ROUGE), Bilingual Evaluation Understudy (BLEU), etc. [17]. By contrasting the generated output sequences with reference (human-made) sequences, these metrics assess the quality of the output sequences.

All things considered, seq2seq models have made a substantial contribution to NLP by making jobs that call for producing variable-length output sequences from variable-length input sequences possible. They do have certain drawbacks, though, such the difficulties in managing lengthy sequences and producing outputs that are cohesive and contextually correct. These are issues that researchers are now working to resolve with different architectural improvements and training methods.

5.2 *Transformer Models*

Transformer models like Generative Pre-trained Transformer (GPT) and Bidirectional Encoder Representations from Transformers (BERT) have shown exceptional performance in a range of natural language processing tasks, including generating code. These models are capable of being adjusted for certain code generating tasks and efficiently collect contextual information [28].

Figure 3 compares the architectures of the GPT and BERT encoders in Transformer models. For deep contextual comprehension, the Bidirectional Encoder Representations from Transformers (BERT) encoder processes information in both directions, as shown in the left section. Tokens are successively predicted by the Generative Pre-trained Transformer (GPT) decoder, which produces output autoregressively, as shown in the right part. This comparison demonstrates how their training goals and application fields differ fundamentally.

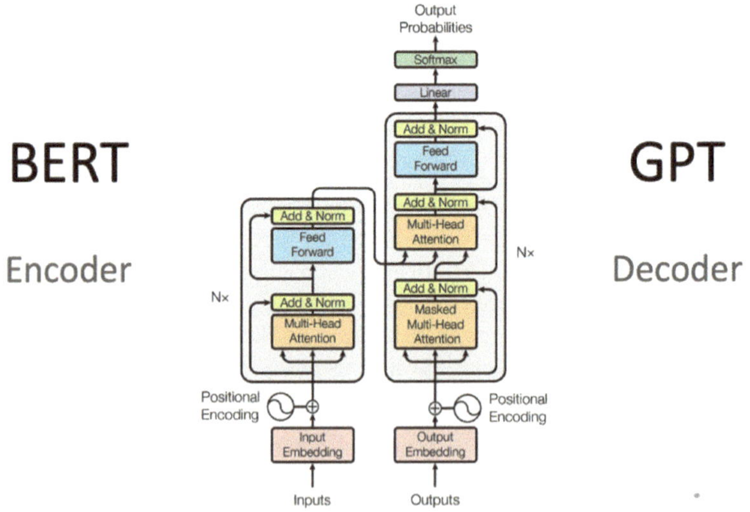

Fig. 3 Comparison of BERT encoder and GPT decoder architectures in Transformer models

BERT aims to pre-train deep bidirectional representations from unlabeled text by concurrently training on both left and right context in all layers. Thus, the pre-trained BERT model may be improved with just one additional output layer to produce state-of-the-art models for a range of tasks, such as question answering and language inference, without necessitating substantial task-specific architectural modifications. BERT is a potential theoretical and experimental tool.

When the Transformer model was first presented in the 2017 publication "Attention is All You Need" by Vaswani et al. [29] it completely changed the area of natural language processing (NLP). Let's assess the NLP application of the Transformer model.

Self-attention Mechanism: At the heart of the Transformer model is the self-attention process. Instead of analyzing input sequentially like typical sequence models like RNNs or LSTMs, the self-attention mechanism allows the framework to recognize associations between words in a sequence regardless of their placement, in contrast to conventional sequence models like RNNs or LSTMs, which analyze input sequentially. In order to allow the model to concentrate on pertinent portions of the input during both encoding and decoding, it computes attention scores for each word in the sequence in relation to every other word.

Multi-head Attention: The Transformer model uses multi-head attention to increase its representational strength. This entails repeatedly applying parallel self-attention with various linear projections of the input embeddings (heads). As a result, the model may concurrently learn many representations of the input sequence and identify distinct kinds of relationships.

Positional Encoding: Positional encoding is added to the input embeddings to provide information about the location of each word in the sequence, as the Transformer model does not naturally capture sequential information as RNNs or LSTMs do. Generally, positional information is encoded using sinusoidal functions at various frequencies.

Encoder and Decoder Stacks: The encoder and decoder layers in the Transformer model are stacked. Sub-modules such as position-wise feed-forward neural networks and multi-head self-attention mechanisms comprise each layer of the encoder and decoder. While the decoder creates the output sequence using the encoded representation, the encoder processes the input sequence.

Residual Connections and Tier Normalizing: In each tier of the encoder and decoder stacks, residual connections and layer normalizing are used. Deeper model training is made easier and the vanishing gradient issue is lessened during training because to these techniques.

Masked Self-attention in Decoder: To guarantee that a position can only pay attention to positions that come before it, the decoder uses a masked self-attention mechanism during training. This guarantees that the model generates sequences auto regressively and keeps it from looking ahead during sequence formation.

Position-Wise Feed-Forward Networks: In each layer of the Transformer model, self-attention mechanisms coexist with position-wise feed-forward networks. Such networks assist the model capture intricate non-linear correlations between words by applying a fully linked feed-forward layer separately to each place in the sequence.

Training and Optimization: Stochastic gradient descent (SGD) or its derivatives, learning rate schedules, and warm-up phases are commonly used in the maximum likelihood estimation training of transformer models. Furthermore, methods such as label smoothing might be used to regularize.

Inference and Decoding: Based on the probabilities predicted by the decoder, the Transformer model employs beam search or greedy decoding during inference to produce output sequences auto regressively. Beam search looks at a variety of alternative sequences and chooses the one with the highest probability overall.

Model Evaluation: To determine how well produced output sequences compare to reference sequences, Transformer models are assessed using common NLP metrics like BLEU, ROUGE, METEOR, etc.

All things considered, the Transformer model has shown to be state-of-the-art in a number of NLP applications, such as question answering, text summarization, and machine translation. It is a key component of contemporary NLP designs due to its parallelization capabilities and effective long-range dependency capture. On the other hand, it may be computationally costly to train large Transformer models and deal with problems like over fitting.

5.3 Syntax-Directed Translation

This method converts natural language input into a syntactic structure, like a parse tree, and then uses templates or predetermined rules to translate it into code. For parsing and translation, methods like tree-based neural networks or probabilistic context-free grammars (PCFGs) can be applied [30].

Figure 4 showcases a two-tier validation framework for structural and semantic correctness. The Syntax Check (top) leverages Context-Free Grammar (CFG) parsing to enforce rule-based structural integrity. The Semantics Check (bottom) utilizes Attribute Grammar to propagate contextual meaning and ensure logical coherence. The integration of syntactic parsing with semantic evaluation provides a robust mechanism for validating formal representations.

This model addresses both syntactic and semantic constraints for structured data. The stochastic lazy attribute used in this method provides the systematized conversion from offline syntax and semantics to online guidance for stochastic generation. In order to create an abstract syntax tree (AST) or a parse tree, syntax-directed translation (SDT) for code creation first parses the source code according to a predetermined grammar. The source code's syntactic organization is represented by this tree. Each node in the AST performs semantic actions or translation rules that specify

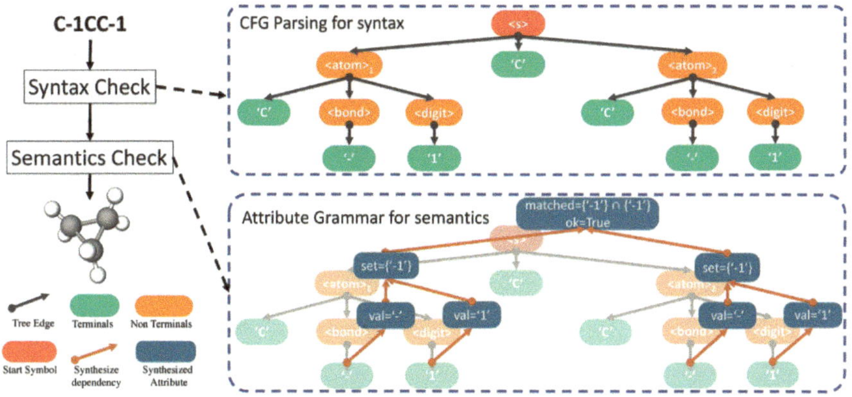

Fig. 4 Syntax and semantics validation using CFG parsing and attribute grammar

how to convert the related components from the source language to the destination language. This process is known as traversal. These rules, which are connected to certain nodes, direct the gradual creation of code. During this procedure, the produced code is output in accordance with the source code's semantics and may go through optimizations to improve its performance, efficiency, or size. To help developers fix problems, error handling techniques are included to identify and report syntactic and semantic faults. In order to guarantee accuracy and dependability, the translated code is lastly tested and debugged; making sure the resulting code faithfully replicates the functionality of the original source code.

5.4 Semantic Parsing

Semantic Parsing is the process of translating sentences in natural language into formal meaning representations that may be used as code. Methods for obtaining meaning from natural language input and producing matching code include dependency parsing, logical forms, and semantic role labeling (SRL) [31]. The primary goal of structured logical forms for natural language is to offer a representation that is useful for automated reasoning and natural language understanding (NLU), among other uses. Semantic parsing has brought computers the ability to do a wide range of jobs, including playing card games, interacting with databases, and serving as conversational agents like Alexa and Siri. Supervised learning approaches have replaced early hand-crafted rules in the advancement of semantic parsing. In the beginning, scientists used manually constructed rules to translate natural language into forms that made sense. But this method [32] proved to be time-consuming and prone to mistakes, especially when working with big datasets that needed annotations in both logical form and plain language. Researchers have turned their attention to learning methods that require less labelled data, such as weakly supervised learning, in order

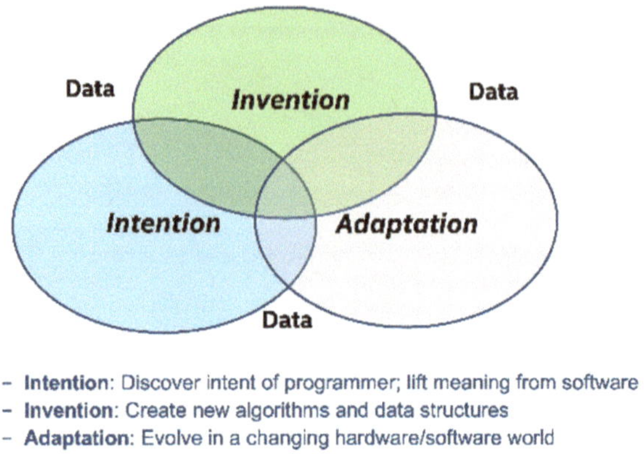

Fig. 5 Code Trifecta

to overcome these difficulties. In order to learn semantic parsers, weakly supervised techniques frequently make use of partially annotated or indirect supervision, which minimizes the need for laborious manual annotation efforts. Semantic parsing has advanced thanks to these strategies, which also make it more scalable and useful for a wider range of jobs.

Figure 5: The Code Trifecta refers to the interaction between Intention (interpreting the intent of programmers and deriving meaning from software), Invention (developing new algorithms and data structures), and Adaptation (changing in reaction to shifts in the hardware/software environments) is depicted in this Venn diagram. This conceptual framework, which falls in between semantic parsing and template-based approaches, emphasizes the fundamental elements that underpin intelligent code generation and transformation.

5.5 *Template-Based Techniques*

In template-based techniques, a set of code templates or patterns that match particular natural language descriptions are defined. Then, maybe with the aid of information extraction algorithms, these templates are instantiated or filled up depending on the input text. In natural language processing (NLP), template-based approaches usually use a mix of algorithms to accomplish different tasks such response generation, template matching, and pre-processing. The following typical algorithms are applied at each step.

Pre-processing Stage

Tokenization: Using algorithms, the input text is divided into discrete tokens, which are often words or sub words. While more sophisticated methods like word boundary recognition or language-specific tokenizers may be employed, simple tokenization algorithms like whitespace splitting can also be used.

Stemming and Lemmatization: To standardize variants, algorithms reduce words to their base form (e.g., "running" to "run"). Porter stemming, Lancaster stemming, and WordNet-based lemmatization are examples of common algorithms.

Stop Word Removal: Common terms with limited semantic value, such as "the," "is," and "and," are filtered out using algorithms. Pre-established lists of stop words or statistical techniques based on word frequencies can be used to do this.

Template Matching

Substring search, regular expressions, or similarity measures like Levenshtein distance are used in basic string-matching approaches to find exact matches. More sophisticated techniques, such as syntactic or semantic matching, use parsers to examine sentence structure and compare it with template representations. Furthermore, classifiers or sequence models trained on labeled data are used in machine learning techniques to categorize input into predetermined template categories. This allows for the processing of more complicated patterns, but at the cost of higher computing demands. Each algorithmic strategy has unique benefits, and the one chosen depends on the particulars of the NLP task at hand. Based on important performance indicators and scalability considerations, this Table 2 offers a thorough comparison of several Natural Language Processing (NLP) models and methodologies. It provides insights into the feasibility of various approaches for practical applications by highlighting the trade-offs between accuracy, computational economy, and adaptability.

6 Conclusion

In conclusion, research on automatic code generation using natural language descriptions has shown significant advances in a range of algorithms, including NLP Seq2Seq, transformer models, syntax-guided translation, semantic parsing, and model-based techniques. Each algorithm has unique strengths and limitations that can significantly improve the efficiency and code quality of software development processes. While NLP Seq2Seq and transformer models show a remarkable ability to understand and generate code from natural language descriptions, syntax-based translation provides a structured approach to preserve the syntax and semantics of code. Semantic parsing provides deeper understanding by extracting meaning from descriptions, while model-based techniques offer simplicity and reliability. By utilizing these different approaches, developers can tailor code generation methods to meet specific requirements, ultimatcly increasing efficiency, code quality, and developer productivity in software engineering companies. Continued research and improvement in these areas promise even greater advances in software development workflows toward more intuitive and seamless human–computer interaction. Each approach has advantages and disadvantages based on a number of metrics, with elements such as job specifications, input complexity, and targeted performance metrics influencing the

Table 2 Performance and scalability of NLP models and techniques: a comparative study

Metric	Seq2seq models	Transformer models	Syntax-directed translation	Semantic parsing	Template-based techniques
BLEU score	Moderate to high, however frequently lower than Transformers	High, frequently cutting-edge results	Variable, relies upon the nature of punctuation and rules	Variable, relies upon the nature of semantic parsing	Variable, relies upon the inclusion and particularity of formats
ROUGE score	Sensible, however may not be all around as high as Transformers	Higher, demonstrating better synopsis execution	Variable, relies upon the nature of interpretation rules	Variable, relies upon the nature of semantic parsing	This differs according to how the templates are specific and broad
METEOR score	Serious, yet not quite as high as Transformers	Higher, frequently prompting unrivalled interpretation	Variable, relies upon the nature of interpretation rules	Variable, relies upon the nature of semantic parsing	Variable, relies upon the inclusion and particularity of formats
Perplexity score	Regularly higher, demonstrating more vulnerability	Lower, showing better language model execution	N/A	N/A	N/A
Performance	By and large great, yet may fall behind transformers	Cutting edge results across different errands	Exceptionally reliant upon the nature of interpretation rules	Exceptionally subject to the nature of semantic parsing	Exceptionally subject to the inclusion and explicitness of formats
Scalability	Less versatile, particularly for long groupings	More versatile, effective handling of long arrangements	Variable, relies upon the intricacy of the syntax	Variable, relies upon the intricacy of semantic guidelines	Variable, relies upon the size and intricacy of layouts
Efficiency	Less effective because of successive handling	More proficient, parallelization of self-consideration	Variable, relies upon the effectiveness of interpretation rules	Variable, relies upon the proficiency of semantic parsing	Variable, relies upon the productivity of format coordinating

selection process. While seq2seq and Transformer models are excellent for capturing complex relationships and producing high-quality results, other approaches such as syntax-based translation, semantic analysis and model-based methods are available, but with trade-offs in coverage, efficiency and scalability.

7 Future Work

The future of automatic code generation with natural language descriptions should focus on several key areas to advance the field. These include strengthening the integration of context and domain knowledge into code generation processes, exploring multimodal approaches that combine text with other modalities to create stronger systems, giving developers more control over code created to meet specific requirements, developing comprehensive evaluation metrics for evaluation. Code quality emphasizing the practical implementation of systems in real-world software development environments and responding to challenges related to the ambiguity and ambivalence of natural language descriptions. With these tools, researchers can continue to push the boundaries of automatic code generation, ultimately leading to more efficient, intuitive, and productive software development practices.

References

1. Adhikari T (2023) Investigating the use of natural language processing for automated code generation. Available at SSRN: https://ssrn.com/abstract=4419501 or https://doi.org/10.2139/ssrn.4419501
2. Cruz-Benito J, Vishwakarma S, Martin-Fernandez F, Faro I (2021) Automated source code generation and auto-completion using deep learning: comparing and discussing current language model-related approaches. AI 2(1):1–16. https://doi.org/10.3390/ai2010001
3. Alokla A, Gad W, Nazih W, Aref M, Salem AB (2022) Retrieval-based transformer pseudo code generation. Mathematics 10(4):604. https://doi.org/10.3390/math10040604
4. Soliman A, Shaheen S, Hadhoud M (2024) Leveraging pre-trained language models for code generation. Complex Intell Syst. https://doi.org/10.1007/s40747-024-01373-8
5. Yin P, Neubig G (2019) Reranking for neural semantic parsing. In: Proceedings of the 57th annual meeting of the association for computational linguistics
6. Dong L, Lapata M (2016) Language to logical form with neural attention. arXiv preprint arXiv:1601.01280
7. Shin EC, Allamanis M, Brockschmidt M, Polozov A (2019) Program synthesis and semantic parsing with learned code idioms. In: Advances in neural information processing systems, vol 32
8. Sun Z, Zhu Q, Xiong Y, Sun Y, Mou L, Zhang L (2020) Treegen: a tree-based transformer architecture for code generation. In: Proceedings of the AAAI conference on artificial intelligence, vol 34, pp 8984–8991
9. Raffel C, Shazeer NM, Roberts A, Lee K, Narang S, Matena M, Zhou Y, Li W, Liu PJ (2020) Exploring the limits of transfer learning with a unified text-to-text transformer. arXiv arXiv:1910.10683
10. Lewis M, Liu Y, Goyal N, Ghazvininejad M, Mohamed A, Levy O, Stoyanov V, Zettlemoyer L (2019) Bart: denoising sequence-to-sequence pre-training for natural language generation, translation, and comprehension. arXiv preprint arXiv:1910.13461
11. Rothe S, Narayan S, Severyn A (2020) Leveraging pre-trained checkpoints for sequence generation tasks. Trans Assoc Comput Linguist 8:264–280
12. Xu FF, Jiang Z, Yin P, Vasilescu B, Neubig G (2020) Incorporating external knowledge through pre-training for natural language to code generation. arXiv preprint arXiv:2004.09015
13. Nguyen AT, Nguyen TT, Nguyen TN (2013) Lexical statistical machine translation for language migration. In: Proceedings of the 2013 9th joint meeting on foundations of software engineering, Saint Petersburg, Russia, pp 651–654

14. Roziere B, Lachaux MA, Chanussot L, Lample G (2020) Unsupervised translation of programming languages. Adv Neural Inf Process Syst 33
15. LeClair A, Jiang S, McMillan C (2019) A neural model for generating natural language summaries of program subroutines. In: Proceedings of the 41st international conference on software engineering. ICSE '19. IEEE Press, pp 795–806. https://doi.org/10.1109/ICSE.2019.00087
16. Gad W, Alokla A, Nazih W, Salem AB, Aref M (2021) DlbtDeep learning-based transformer to generate pseudo-code from source code. CMC 70:3117–3123. https://doi.org/10.32604/cmc.2022.019884
17. Shin J, Nam J (2021) A survey of automatic code generation from natural language. J Inf Process Syst 17(3):537–555
18. Wong M-F, Guo S, Hang C-N, Ho S-W, Tan C-W (2023) Natural language generation and understanding of big code for AI-assisted programming: a review. Entropy 25(6):888. https://doi.org/10.3390/e25060888
19. Li R, Allal LB, Zi Y, Muennighoff N, Kocetkov D, Mou C, Marone M, Akiki C, Li J, Chim J et al (2023) StarCoder: may the source be with you! arXiv arXiv:2305.06161
20. Khan MAM, Bari MS, Do XL, Wang W, Parvez MR, Joty S (2023) xCodeEval: a large scale multilingual multitask benchmark for code understanding, generation, translation and retrieval. arXiv arXiv:2303.03004
21. Allal LB, Li R, Kocetkov D, Mou C, Akiki C, Ferrandis CM, Muennighoff N, Mishra M, Gu A, Dey M et al (2023) SantaCoder: don't reach for the stars! arXiv arXiv:2301.03988
22. Yang G, Zhou Y, Chen X, Zhang X, Han T, Chen T (2023) ExploitGen: template-augmented exploit code generation based on CodeBERT. J Syst Softw 197:111577
23. Niu C, Li C, Ng V, Ge J, Huang L, Luo B (2022) SPT-code: sequence-to-sequence pre-training for learning source code representations. In: Proceedings of the 44th International conference on software engineering, Pittsburgh, PA, USA, pp 2006–2018
24. Takerngsaksiri W, Tantithamthavorn C, Li YF (2022) Syntax-aware on-the-fly code completion. arXiv arXiv:2211.04673
25. Li Y, Choi D, Chung J, Kushman N, Schrittwieser J, Leblond R, Eccles T, Keeling J, Gimeno F, Dal Lago A et al (2022) Competition-level code generation with Alphacode. Science 378:1092–1097
26. Surhone LM, Tennoe MT, Henssonow SF (2010) Comparison of code generation tools. Betascript Publishing, GmbH
27. Sutskever I, Vinyals O, Le QV (2014) Sequence to sequence learning with neural networks. Neural Inf Process Syst (NIPS)
28. Devlin J, Chang M-W, Lee K, Toutanova K (2019) BERT: pre-training of deep bidirectional transformers for language understanding. In: Proceedings of the 2019 conference of the North American chapter of the association for computational linguistics: human language technologies, vol 1 (Long and Short Papers). Association for Computational Linguistics, Minneapolis, Minnesota, pp 4171–4186
29. Vaswani A, Shazeer N, Parmar N, Uszkoreit J, Jones L, Gomez AN, Kaiser Ł, Polosukhin I (2017) Attention is all you need. In: Proceedings of the 31st international conference on neural information processing systems (NIPS'17). Curran Associates Inc., Red Hook, NY, USA, pp 6000–6010
30. Dai H, Tian Y, Dai B, Skiena SS, Song L (2018) Syntax-directed variational autoencoder for structured data. ArXiv, abs/1802.08786
31. Berant J, Chou AK, Frostig R, Liang P (2013) Semantic parsing on freebase from question-answer pairs. In: Conference on empirical methods in natural language processing
32. Lee C, Gottschlich J, Roth D (2021) Toward code generation: a survey and lessons from semantic parsing
33. Perez L, Ottens L, Viswanathan S (2021) Automatic code generation using pre-trained language models [arxiv.org]. https://arxiv.org/abs/2102.10535

Optimizing User Preferences in Recommendation Systems with a Hybrid Machine Learning Approach

Molli Srinivasa Rao, Velaga Dinesh, Gollavilli Kartik, Koppoju Tarun Kumar, and Pragada Divya Sai Kumar

Abstract Recommendation systems are vital for enabling personalized content in domains like movies and music, however they frequently encounter difficulties such as accurately capturing user preferences, managing data sparsity, and addressing cold start problem. When new users or things don't have enough information to produce reliable suggestions, the cold start occurs. To overcome these limitations, a hybrid recommendation model is proposed integrating content-based, collaborative filtering, weighted popularity-based, and multiuser recommendation methods for both movies and music. The movie recommendation system employs cosine similarity, Term Frequency-Inverse Document Frequency (TF-IDF), and Singular Value Decomposition (SVD) to analyze watch histories, ratings, and content features. A unique feature is its ability to provide multiuser recommendations, aggregating the watch histories and inclinations of multiple users to generate group-based suggestions, overcoming cold start for naive users. For music, the system utilizes cosine similarity and playlist analysis to evaluate user preferences, while also incorporating a weighted popularity-based method that adjusts song popularity according to individual preferences, ensuring a mix of popular and personalized content. By integrating various methods, the model improves the suggestions and diversity. Evaluation metrics demonstrate its effectiveness, with an RMSE of 0.6465 and MAE of 0.4970, outperforming traditional models and tackling cold start for both movies and music.

Keywords Recommendation system · Content-based · Collaborative filtering · Popularity-based · Hybrid model · SVD · TF-IDF vectorizer · Cosine similarity · RMSE · MAE · Multi-user recommendations · Cold start

---

V. Dinesh · G. Kartik · K. T. Kumar · P. D. S. Kumar
Department of Computer Science and Engineering, Dadi Institute of Engineering and Technology, Visakhapatnam, India

M. S. Rao (✉)
Department of Computer Science & Engineering, Raghu Engineering College, Visakhapatnam, India
e-mail: drmollisrinivasarao@gmail.com

H. Nagaraj et al. (eds.), *Advancements in Artificial Intelligence and Machine Learning*, Lecture Notes in Electrical Engineering 1572,
https://doi.org/10.1007/978-981-95-7449-0_38

1 Introduction

The majority of research is now using recommendation systems in a variety of domains, including entity recommendation [1], restaurant services [2], e-commerce [3], social networks [4], and healthcare recommendations [5]. Recommendation systems are frequently used in fields like music and film to improve user experience by offering tailored recommendations. However, these models have drawbacks, such as the inability to recommend across domains Existing systems often combine content-based and collaborative models or use hybrid collaborative filtering for effective recommendations [6]. But they often tend to have issues.

To address these limitations, a hybrid recommendation system is proposed, combining content-based, collaborative filtering, and popularity-driven techniques. For music recommendations, a dual-model approach is used, aligning music with movie soundtracks the user enjoys via cosine similarity. Advanced techniques like SVD and TF-IDF vectorizers improve the system's accuracy and relevance, bridging the gap between users' musical tastes and movie preferences. This approach sets a new standard for integrated recommendation systems.

2 Literature Survey

In 2019, T. Silveira conducted a poll to determine the quality of your recommender system and what a recommender system should be [7]. Additionally, some study on large language model suggestions has been conducted recently on the extensive improvement of LLMs via the Internet [8]. Later, Kushwaha and Sharma developed a hybrid model for music systems [9], However, they don't work for new users.

These models tend to have some issues. A study by A. Samuel which shows the causes of the cold start issue and how it might be resolved [10]. Sentimental analysis, which extracts aspects based on user views, is being used to enhance it. In 2022, a work on sentiment analysis-based recommendation systems was released by Arava and Chaitanya [11]. Over-the-top (OTT) platform usage is also growing quickly [12]. The use of recommender systems in OTT apps has significantly increased. A recommendation system based on popularity was presented by Rakshit and Saha [13]. Although this approach uses popularity-based suggestions, its effectiveness is impacted by a greater degree of popularity bias. A. Ferraro, D. Bogdanov, X. Serra, and Yoon discuss the need for a better methodology for music recommendation algorithms and suggest the effects of collaborative filtering suggestions on musicians and musical genres [14].

Gazdar and Hidri (2020) suggested a straightforward yet effective similarity measure [15]. An item similarity metric based on the idea of α-divergence was suggested by Y. Wang in 2021 [16]. With excellent insights, Alsaadi et al. (2022) suggested an improved similarity metric based on the idea of p-moment-based statistics [17]. Although these similarity metrics have significantly improved, they are ineffective in resolving cold start issues. Jiewen Bilian developed the concept of Wasserstein distance in collaborative filtering to solve cold start [18]. This model has fared better than its predecessors. Although Gopal Behera's model including spatial features utilized to enhance recommender systems, but it is prone to sparsity issues that affect performance and lower recall and precision scores [19]. Although Hamidreza Kooh's technique reduces the system's overall sparsity, indeed, this model's error rates are larger than other models [20].

2.1 Existing System and Limitations

Most current recommendation systems rely on collaborative and content-based filtering techniques. Content-based approaches suggest items similar to those a user has interacted with, based on attributes like genre or keywords. While these models offer customization, they often lack diversity and struggle to recommend new items outside the user's past preferences. Collaborative filtering, on the other hand, leverages user-item interaction data to identify patterns among users with similar behaviors, but it faces challenges with new users or items due to insufficient interaction data. Both approaches fail to capture the dynamic nature of user preferences, can transform over time or involve contextual factors, often leading to shallow, irrelevant recommendations, especially in complex areas like music and movies. Existing systems are also hindered by issues like cold start problem, where new users or items lack sufficient data for accurate recommendations, and popularity bias, where mainstream content is over-recommended, neglecting niche options. These limitations highlight the need for a more coordinated approach to provide comprehensive and tailored recommendations.

2.2 Proposed System

The proposed movie recommendation system integrates content-based filtering, collaborative filtering, and popularity-based methods to deliver accurate and personalized suggestions. Content-based filtering utilizes cosine similarity to measure the similarity between movies based on their features and TF-IDF to analyze and prioritize key terms in movie descriptions. Collaborative filtering employs Singular Value Decomposition (SVD) to factorize the user-item interaction matrix, identifying latent patterns and predicting user ratings. Popularity-based recommendations ensure widely liked movies are included, regardless of user preferences. By combining these techniques, the system balances feature matching, user behavior analysis, and general appeal to provide comprehensive recommendations.

2.3 Advantages of Proposed System

Enhanced Personalization: By integrating content-based, collaborative, and popularity-based approaches, the system delivers highly personalized recommendations. This combination of techniques enables it to align closely with individual user preferences, offering content that matches unique tastes with precision.

Improved Accuracy: The application of Singular Value Decomposition (SVD) for matrix factorization allows the system to uncover intricate patterns in user behavior and accurately measure user similarities. This leads to precise predictions of user preferences, enhancing the accuracy of recommendations.

3 Methodology

3.1 Data Collection

To enhance recommendation accuracy, large datasets such as Spotify's Million Song Dataset and the Movie Lens dataset are used, containing metadata like genres, released years, artists, and directors. Historical data, including users playlists, ratings, and playback history, helps understand user preferences and refine recommendations.

3.2 Implementation of Hybrid Model for Movies

The movie recommendation system uses the Movie Lens dataset to analyze user preferences and item characteristics. Content-based filtering uses TF-IDF to process movie descriptions, while collaborative filtering applies SVD to reveal user-item interaction patterns and cosine similarity for finding the similarity. Popularity-based recommendations prioritize widely liked films. The hybrid model combines these techniques to enhance accuracy, addressing content overspecialization and data sparsity. Multi-user recommendations predict ratings for groups, balancing individual preferences for shared experiences.

3.3 Implementation of Hybrid Model for Music

The music recommendation system integrates content-based filtering and collaborative methods. It processes user playlists and song metadata to recommend similar tracks. TF-IDF assigns weights to genre metadata, while cosine similarity compares playlists to songs in the dataset. Three models are employed: one prioritizing genres, another considering all features equally, and a model combining both approaches for balanced recommendations.

3.4 Model Overview

Combining music and movie recommendation models, this system leverages content-based filtering, collaborative filtering, and popularity-based methods for superior recommendations. For music, it uses cosine similarity to compare playlists with songs, while for movies, it merges metadata-based content filtering with collaborative methods. This hybrid approach offers personalized suggestions for both music and movies, improving user engagement by providing a diverse range of relevant content (Fig. 1).

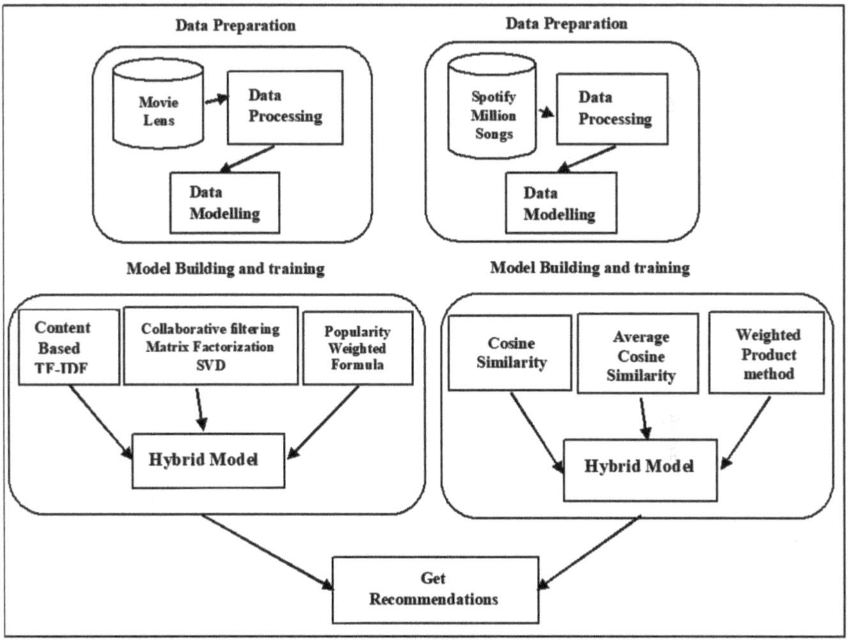

Fig. 1 Model overview of the proposed methodology

4 Experimentation

The experimentation focused on testing the hybrid recommendation model's performance using the Spotify Million Song Dataset for music and the Movie Lens dataset for movies. This hybrid approach integrates content-based filtering, collaborative filtering, and popularity-based techniques to provide personalized and accurate recommendations. The system was tested by providing both movie and music inputs to generate tailored suggestions. Content-based filtering employed TF-IDF and cosine similarity to match recommendations with user preferences. Collaborative filtering, leveraging SVD, uncovered latent user-item interaction patterns, while popularity-based filtering emphasized items with higher engagement and ratings. For music recommendations, a similar strategy was used, analyzing song metadata and user activity. The evaluation of the model is conducted by assessing Singular Value Decomposition (SVD). For each user-item pair, the SVD model generates a predicted rating. The test dataset consists of actual ratings and it is utilized to compare the predicted user ratings with actual ratings. The error between the predicted and actual ratings is used to evaluate the model performance by using RMSE and MAE metrics. RMSE is calculated as the square root of the mean of squared differences between predicted and actual ratings, while MAE represents the average of absolute differences. Both metrics are averaged over n. n is the total number of predictions made, to ensure consistency. These metrics collectively reflect the accuracy and precision of the collaborative filtering model, directly impacting the overall recommendation quality of the hybrid system.

5 Results and Discussion

The hybrid recommendation system that is being put forward has proven to be better than previous and existing models, as evidenced by the evaluation metrics. The RMSE score obtained by the system was 0.6465, revealing that it had a significant amount of accuracy when it came to predicting user ratings against the actual ones. In addition to this, its MAE of 0.4970 shows that it can lower the average absolute deviation between the expected value and the true rating.

5.1 Performance Evaluation of the Hybrid Recommendation Model

The hybrid recommendation model was evaluated based on RMSE AND MAE metrics. These metrics collectively reflect the accuracy and precision of the collaborative filtering model, directly impacting the overall recommendation quality of the hybrid system. Lower RMSE and MAE values indicate a model's capability

Table 1 Comparing RMSE and MAE scores of various models

Model	RMSE	MAE
Random forest	1.0682	0.8223
Gradient Boost	1.0011	0.7949
KNN	0.7317	0.5508
Proposed hybrid model	0.6465	0.4970

to make precise and reliable predictions, enhancing its effectiveness. These metrics give information on how accurate the models' predictions are, lower numbers denote higher performance. The analysis is provided in Table 1.

The substantial decrease in RMSE and MAE highlights the model's effectiveness in leveraging multiple approaches to enhance accuracy by utilizing the strengths of each method. By combining user interaction data with item attributes, the hybrid model overcomes the limitations of individual models, resulting in more precise and personalized recommendations. This approach provides a more robust solution, improving the overall quality of recommendations for users.

The following table demonstrates that the proposed model outperforms other models in both RMSE and MAE. With an RMSE of 0.6465 and MAE of 0.4970, the proposed hybrid model shows a significant improvement over the random forest model (RMSE: 1.0682, MAE: 0.8223), the gradient boost model (RMSE: 1.0011, MAE: 0.7949), and the KNN model (RMSE: 0.7317, MAE: 0.5508). This highlights the effectiveness of the proposed hybrid model in delivering more accurate and personalized recommendations.

6 Conclusion

In conclusion, hybrid recommendation systems successfully balances the choices of the user and offers diverse suggestions, resulting in significantly improved results in the metrics such as MAE and RMSE. These performance metrics highlight the model's ability to predict user ratings with greater precision, providing a deeper understanding of user interests and more accurate recommendations. The combination of strengths from content-based and collaborative methods allows the hybrid model to mitigate the limitations of individual approaches, ensuring consistent and give personalized results. Ultimately, this model sets a new benchmark for recommendation systems, offering enhanced user satisfaction through highly relevant, tailored suggestions.

7 Future Scope

The proposed approach has shown considerable advancements in performance metrics such as RMSE and MAE, establishing itself as an effective solution for recommendations. Nonetheless, there are several chances for improvement and more growth. Among these is the use of real-time data to enhance the system's responsiveness, incorporating advanced deep learning techniques to capture more complex patterns and user behaviors, implementing strategies to reduce data sparsity for more accurate recommendations, and enhancing the analysis of user interests for even more personalized suggestions. These improvements would allow the model to provide even more precise and dynamic recommendations.

References

1. Saidi I, Mahammed N, Klouche B, Bencherif K (2024) Entities recommendations using contextual information ISSN: 2088. https://doi.org/10.11591/ijece.v14i4.pp4336-4342
2. Fahrudin T, Wisna N (2022) The exploration of restaurant recommendation system. J Comput Sci 18(8):784–791. https://doi.org/10.3844/jcssp.2022.784.791
3. Ben Schafer J, Konstan J, Riedl J (1999) E-commerce recommendation systems. In: Presentations of the first ACM conference. https://dl.acm.org/doi/10.1145/336992.337035
4. Fareeda A, Hassana S, Belhaouarib SB, Halimc Z (2023) A collaborative filtering recommendation frame work utilizing social networks. https://doi.org/10.1016/j.mlwa.2023.100495
5. Tran TNT, Felfernig A, Trattner C, Holzinger A (2021) Recommendation systems in the healthcare domain: state-of-the-art and research issues. J Intell Inf Syst 57(1):171–201. https://doi.org/10.1007/s10844-020-00633-6
6. de Campos LM, Fernández-Luna JM, Huete JF, Rueda-Morales MA (2010) A hybrid approach based on Bayesian networks that combines content-based and collaborative recommendations. Int J Approx Reason. https://doi.org/10.1016/j.ijar.2010.04.001
7. Silveira T, Zhang M (2019) How good your recommender system is? A survey one valuations in recommendation. Int J Mach Learn Cybern. https://doi.org/10.1007/s13042-017-0762-9
8. Wu L, Zheng Z (2023) A review on large language model based recommendation. arXiv:2305.19860v1 [cs.IR]. https://doi.org/10.48550/arXiv.2305.19860
9. Kushwaha K, Sharma S A hybrid collaborative filtering-based music recommendation system. IJRDET_1022_04.pdf
10. Turner ER, Foster SA. IJCRT2402820.pdf
11. Arava K (2022) Deep learning-based sentiment analysis for recommendation systems for a range of public media applications. https://doi.org/10.1109/ICESC54411.2022.9885648
12. Zhao P, Gao X, Xu C, Chen L M5 for over-the-top recommendation. https://dl.acm.org/doi/10.1145/3580305.3599863
13. Rakshit P, Saha S, Chatterjee A, Mistri S, Das S, Dhar G (2023) A popularity-based recommendation system using machine learning. Springer, Singapore. https://link.springer.com/chapter/10.1007/978-981-19-5090-2_14
14. Ferraro A, Bogdanov D, Serra X, Yoon J (2019) Artist and style exposure bias in collaborative filtering based music recommendations. arXiv preprint arXiv:1911.04827. https://doi.org/10.48550/arXiv.1911.04827
15. Gazdar A, Hidri L (2020) A novel similarity metric for recommender systems based on collaborative filtering. Knowl-Based Syst, Article 105058. https://doi.org/10.1016/j.knosys.2019.105058

16. Wang Y, Wang P, Liu Z, Zhang LY (2021) A new α divergence-based item similarity in collaborative filtering in sparse data. Appl Exp Syst. https://doi.org/10.1016/j.eswa.2020.114074
17. Alsaadi FE, Wang Z, Alharbi NS, Liu Y, Alotaibi ND (2022) A novel collaborative filtering framework with p-moment-based similarity measure: algorithm, optimization and application. Knowl-Based Syst 248, Article 108874. https://doi.org/10.1016/j.knosys.2022.108874
18. Guan J, Chena B, Yu S (2024) A hybrid similarity model for mitigating the cold-start problem of collaborative filtering in sparse data. https://doi.org/10.1016/j.eswa.2024.123700
19. Behera G, Malaviya NN (2023) Collaborative filtering using spatial features for recommendation systems. National Institute of Technology Jaipur, India. https://doi.org/10.1016/j.procs.2023.01.115
20. Koohi H, Kiani K (2020) Two new collaborative filtering approaches to solve the sparsity problem. Springer. https://doi.org/10.1007/s10586-020-03155-6

Comparative Study of CNN and Vision Transformer (ViT) Models for Image Classification on the Fruits-360 Dataset

Janet Jomy, Tushar Khadangle, and R. V. Siva Balan

Abstract This research explores and compares the classification performances of CNN and Vision Transformers on Fruit-360 dataset. In fact, two critical objectives are addressed here: assessing the model performance in comparisons about accuracy mainly based on precision, recall, and computation resource usages and exploring the effects of pre-trained layers being frozen within ViT when applying the fine-tuning. The industrial applications of Vision Transformers in the automated classification system have been further discussed in the study. Both CNN and ViT perform well; however, it is noticed that the training and validation accuracy for ViT (Training Accuracy: 99.33%, Validation Accuracy: 98.28%) were better than CNN (Training Accuracy: 92.81%, Validation Accuracy: 94.21%). Further, accomplished domain-specific properties with lower computation costs in the application of layer freezing. The outcome of this work has proved that Vision Transformers can be scalable as well as efficient in all automated classification activities, therefore providing valuable perspectives for applications in the real world.

Keywords Deep learning · Convolutional neural networks (CNNs) · Vision transformers (ViT) · Image classification

J. Jomy (✉) · T. Khadangle · R. V. Siva Balan
Christ (Deemed to be University), Bangalore, India
e-mail: janetjomy711@gmail.com

T. Khadangle
e-mail: tusharkhadangle@gmail.com

R. V. Siva Balan
e-mail: sivabalan.rv@christuniversity.in

H. Nagaraj et al. (eds.), *Advancements in Artificial Intelligence and Machine Learning*, Lecture Notes in Electrical Engineering 1572,
https://doi.org/10.1007/978-981-95-7449-0_39

1 Introduction

Deep learning revolutionized the space of computer vision and enabled outstanding advancement in many activities that were hitherto considered far-fetched. Among those improvements, CNNs have been established as one of the core technologies. Because CNNs can carry hierarchical feature extraction ability, and identify edges, textures which can be extensively applied to most image classification tasks [1]. But, the disadvantage with CNNs is that they do not model global dependencies of an image as their receptive fields increase incrementally in depth. To capture long-range relationships and global contexts, Vision Transformers presented a new architecture based on self-attention [2]. This work utilizes the Fruit-360 dataset to classify fruits and vegetables and compares its performance with a baseline CNN. Fruit-360 is a challenging benchmark for image classification that comprises images of 141 classes of fruits and vegetables with 94,110 images. The input dimensions are standardized by resizing all the images to 100 $\times$ 100 pixels. It also comprises training images and test images, which account for 70,491 and 23,619 images, respectively. This dataset has a very low similarity between the classes, and, at the same time, they differ in shape, color, and texture. Therefore, they represent an excellent candidate for benchmarking the performance of deep learning models that require not just capturing the subtle details but also establishing high-level contextual relationships in such data.

For a long time, CNNs were the most appropriate model for tasks such as fruit recognition; the learning of spatial hierarchies based on local receptive fields was helpful in getting very high success within many tasks. This work is based on a small-scale convolutional and pooling layers optimized CNN structure designed with an eye to recovering more localized patterns important for fruit's classification [3]. However, increasing the number of classes and finer interclass differences tends to require more serious tuning and significantly more computational resources to maintain their high accuracy levels for CNNs [4]. The paper discusses alternative approaches in the context of a classification task, an image using convolutional and recurrent neural networks. Surveys through the content of the image result in scanning along a sequence of flattened pieces. Long dependencies then get pulled out by using self-attention mechanisms, which dramatically enhance detecting complex patterns that can reach across the whole image area. Unlike CNNs, they do not employ any convolutional operation; thus, direct modeling of global dependencies is possible [5]. Hence, this kind of alternative seems promising for the type of datasets used in this paper, like Fruit-360.

The work exploits pre-trained architecture on large-scale datasets followed by fine-tuning it on the Fruit-360 dataset. In addition, the work verifies the impact of freezing the pre-trained layers during fine-tuning in preserving general features while focusing on domain-specific learning. The motivation for this research is the high demand, emerging in the agricultural industry and along the entire food chain, for efficient, accurate automated classification systems. Improved classification systems for fruits and vegetables can make a huge positive impact on their sorting and packaging processes, reduce waste, and increase the inventory level. Although much research has been done on CNNs, still open are the questions on how to use ViT and with which benefits [6]. Thus, the work tries to fill in the gap by comparing how well ViT handles the complexities of Fruit-360 compared with CNNs strengths and weaknesses.

The advantages this research develops are adding to an increasingly large body of literature concerning Vision Transformers, since it builds on the application of state-of-the-art deep learning models to niche datasets. Consequently, the results are probably going to inspire the creation of automated classification systems and will deeply affect the way they should be introduced into real applications.

2 Literature Review

Giménez-Gallego et al. [7] compare Mask R-CNN with models developed based on Vision Transformer like Grounding DINO and SAM in on-tree pomegranate fruit segmentation task, for which they achieve that the fine-tuning of the model Mask R-CNN is far more accurate and efficient in contrast to transformer-based models. It is an innovative fruit size estimation algorithm by pixel analysis demonstrating a good prospect in the sector of real-time surveillance in precision agriculture, as the relative median error was 1.39. This could be applied to other round fruits such as citrus varieties and apples.

Raza et al. [8] proposed the Features Enhanced Vision Transformer architecture. This architecture could classify citrus fruits by how fresh they were through X-rays kept in various conditions. The precision, accuracy, and recall of the FEViT model were better than the regular ViT and deep models. In this experiment, the accuracy of the FEViT-large reached 99.25%. Such model explained the storage conditions that affect the freshness of citrus fruits by detecting changes inside their cells. Through experimenting, FEViT can indeed be applied to inspect a fruit without causing damage while also being utilized in various fields for sorting other sorts of fruits.

Du¨men et al. [9] applied the ViT and Swin Transformer models to classify lemons into different quality types, where their applications were considered highly impressive because ViT had an accuracy of 99.84%. This research focused on the comparison between transformer-based architectures and the conventional deep learning models, demonstrating how these models can successfully classify lemons into "good" and "bad" qualities. It underlines how advanced AI and data augmentation techniques transform the potential for automated classification in improving agricultural efficiency and product quality assessment.

Aghamohammadesmaeilketabforoosh et al. [10] attained the highest performance of above 98% accuracy in the evaluation of Vision Transformer (ViT), MobileNetV2, and ResNet18 for strawberry disease and quality identification. Even though the samples may not be balanced, their research improved the classification accuracy, even with the usage of attention mechanisms and data augmentation techniques. The above-given study demonstrates how advanced vision models that have improved agricultural practices can classify diseases and maturity accurately.

Amin et al. [11] proposed a deep convolutional neural network-based (DCNN) system and a transfer learning-based classification approach using AlexNet for classifying fruit freshness. Their approach consists of data preparation, tuning the hyperparameters, and as a classifier, softmax is exploited to show strong performance. The model performed greatly with an accuracy of up to 99.8% on public datasets; at the same time, it was computationally efficient. Their future work is expanding fruit and vegetable categories and developing a mobile application also looking toward cloud-based deployment for better scalability and usability.

Recently, a model with ViT architecture was proposed by Nguyen et al. [12], which showed great promise toward achieving a very high accuracy of 0.927 as well as its respective F1-score, exploiting the Strawberry Disease Detection dataset. By using transfer learning as well as fine-tuning methods would further enhance the performance to diagnose more prevalent diseases over the strawberry leaves, berry leaves, and flowers. The study demonstrated that ViT outperformed earlier approaches such as Mask R-CNN, solving related problems about overfitting on sets. Some of the future recommendations would be related to categories of diseases, employing advanced algorithms like YOLO, and applications in integration with a drone system in that they can detect and treat ailments autonomously.

Su et al. [13] designed a model based on ViT for detecting and classifying different defects on the surfaces of green plums. This model had an average recognition accuracy of 96.21%. The current work has improved on the methods developed up to this point by incorporating fine classification for four main types of defects, namely scars, rot, flaws, and spots, and one more feature: stem, which leads to 18 defect categories. Results compared ViT showed better performance when compared with ResNet18 and VGG16-based models. It supported minute defect detection along with high hazard analysis. Further improvements could be achieved by more efficient training strategies and utilizing high spectral imagines to study the inter defects.

Hossain et al. [14] studied how Vision Transformers (ViTs) can be used to detect diseases in mango leaves. They achieved an impressive accuracy of 99.75% with a special DeiT-based design. Their study shows that ViTs are better than well-known CNN models such as DenseNet121 and MobileNet, pointing out that they take less time to train and give better classification accuracy. The authors also proposed a mobile application using the model for rapid disease detection, which would be a great tool for farmers. This work demonstrated how ViTs could be useful in smart farming and paved the way for further work in image classification tasks.

The authors Filipiuk et al. [15] sought to compare the two architectures (i.e., CNN and ViT) and to produce a hybrid model that reflected the mixture of the two. The ImageNet dataset was used for the experiment, and perturbations were made to the dataset's images. It was determined that ViT can outperform CNN and be more robust when it comes to photos that include unfavorable or natural disruptions. Additionally, this work discovered that the accuracy (Acc) is improved by 10% when the two architectures are combined.

Wang et al. [16] set out to examine eight distinct Vision Transformers and contrast their performance with that of a CNN network that had been pre-trained and without the pre-trained parameters in order to recognize traffic signals in autonomous driving systems. This study made use of three distinct datasets that contained pictures of actual traffic signals. To recognize traffic signs, scientists were able to determine that the pre-trained DenseNet161 network performed more accurately than the ViT models. ViT models outperformed the DenseNet161 network without the pre-trained parameters, according to this study. It was also feasible to draw the conclusion from this work is that the ViT models, which had a total number of parameters that were equal to, or more than the CNN networks employed in the experiment.

Benz et al. [17] contrasted CNNs and the MLPMixer architecture with ViT models. The objective was to determine which architecture performed better in classifying images. To determine which building was the most resilient, the study involved creating perturbations and negative instances in the photos. However, evaluating the causes was not the goal of this study. Thus, the authors concluded that CNN networks were more sensitive to high-frequency features from a features standpoint, while ViTs were more resilient to adversarial attacks. Additionally, it was explained that the network's lack of robustness in classifying changed images could be caused by the shift-variance attribute of convolutional layers.

Risdin et al. [18] presented a fruit detection system based on deep learning approaches using CNN, in particular, which yielded a very high accuracy rate of 99.89%. The experiment proved that CNN is superior to other traditional machine learning techniques, such as SVM, in the application of fruit image detection. The training dataset is taken from photos shot by smartphone cameras. This way, it was very feasible for real-world application. They discussed ways to expand the system, with broader fruit detection, application scenarios, and robotic harvesting systems. Such work has significant implications in terms of the CNN architecture used for image classification, especially for more complex tasks in object detection.

Palakodati et al. [19] proposed a model that suggests the use of CNN, techniques of transfer learning for fresh fruit classification, and rotten fruit. It would automate classification within the agricultural industry of fruits. Thus, it would help in saving human effort along with prevention from the spread of defects. It used three categories of fruits, such as apple, banana, and orange, and achieved an accuracy of 97.82 researchers comparatively analyzed several transfer learning models: VGG16, VGG19, MobileNet, and Xception against the newly proposed CNN model and found that the CNN model performed better in terms of classification accuracy. This research study presents the effectiveness of CNN-based frameworks in agricultural applications.

Hussain et al. [20] took up the challenge of adapting other fruit categories into thc already established deep CNN structure by applying transfer learning techniques. They extended their previous study on automatic fruit identification. This study had been designed and trained on 15 different fruit categories using a highly extensive dataset of 44,406 images. The study showed that the use of transfer learning via a pre-trained model efficiently allowed for the classification of new fruit categories, such as cherries, with an impressive detection accuracy of 95%. The approach used, which consisted of freezing certain layers of the pre-trained network, performed better than the previous state-of-the-art models.

3 Data and Methods

3.1 Data

This experiment employs the Fruits-360 Dataset, which is nothing short of an extensive, high-quality image repository tailored to tasks of classification using images. The dataset comprises 94,110 images spread over 141 classes differentially comprising fruits, vegetables, and nuts that comprise several varieties of commonly used items like apples, bananas, grapes, and tomatoes among less commonly seen items such as cactus fruit and pitahaya. Such a significant diversity, due to the highly varied set of food products, makes it appropriate for use in classification tasks for distinguishing among various objects based on their visual attributes. Each image of the dataset is resized to 100 $\times$ 100 pixels to give uniform input sizes for machine learning algorithms. The dataset has been split into a training set containing 70,491 images and a test set containing 23,619 images with each image being one object. The division between the training and testing sets goes on to validate the performance of the model on unseen data, thereby giving assurance that it properly generalizes [22].

The dataset consists of images that are represented in various formats, with filenames that indicate the index of each image as well as the type of fruit or vegetable being depicted. Moreover, the addition of rotated versions of fruit increases the robustness of the models to learn more robust representations of the objects from different angles. This addition of rotation increases the diversity of images in the dataset, making it larger for real-world applications where fruits may be oriented in multiple ways.

It was taken from Kaggle, one of the major platforms that enables open data science initiatives and competitions in machine learning. This is public data, and this is licensed under the MIT License. It gives unlimited usage, for both research purposes and commercially, which makes it more suitably and specifically applicable to academic as well as industry-related image classification investigations.

Data Collection Fruits-360 Dataset was prepared widely to ensure richness, good quality of images, and diversity. Images from this dataset of fruits, vegetables, as well as nuts were captured from the Logitech C920 camera; this is one which would give very high quality as well as clear imagery. Objects for this experiment were kept on a low-speed rotational axis at 3 rpm, meaning that they had captured through many angles for a 20 s time period of a clip. The procedure for filming involved a white sheet of paper as the backdrop to the rotating fruits, thus ensuring object isolation and minimizing background noise, although, due to the variation in lighting conditions, the backgrounds are not uniform [22].

To solve this problem, a particular algorithm for background extraction was designed using the flood-fill technique. The algorithm systematically identifies all pixels in proximity to the edges of the image and incrementally extends the marked region so that it detects and removes background pixels.

Pixels that did not satisfy the background criteria were classified as components of fruit or vegetable and preserved in the resultant image.

Images were captured in multiple sessions between February 2017 and December 2021. This time frame allowed for the capture of a wide range of fruits and vegetables; the latter being captured under very different environmental conditions. In addition, the database also contains some augmented images developed using different rotations of objects in a variety of configurations [23]. The rotation of fruits ("r image index 100.jpg") provides additional training data to handle different views of objects. Such a maneuver helps the model better generalize towards previously unobserved data where objects do not always appear in the same pose. The Fig. 1 shows the sample data.

Data Preprocessing Since the Fruits-360 Dataset had already been downloaded directly from Kaggle and prepared for classification, only the least pre-processing was required for the present research work. In this dataset, the whole set of images was clean and well-structured. All were 100 × 100 pixels with clear objects against clean white backgrounds in each case. However, only a few pre-processing operations were discussed considering their possible applicability.

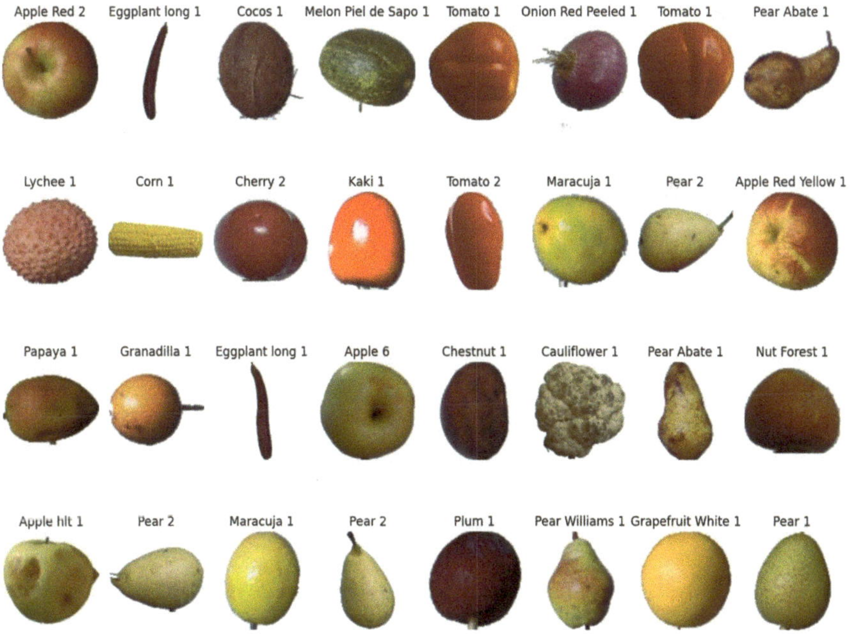

Fig. 1 Sample data

- **Image Augmentation**: The set that contains images rotated and augmented may be used at subsequent experiments with further augmentations like random flipping, zoom, and rotation to produce diversified training data.
- **Normalization**: Picture's pixel intensities in practice are between 0 and 255. Some computers benefit from normalization between 0 and 1 for good convergence in the time of training. This is not, however, required by the current dataset as it has been preprocessed, although there may be a consideration of including the process if case by case given that this model is applicable.

In general, the preprocessing of the dataset was already advanced to a good quality so that it was feasible to develop and evaluate the models without additional extensive preprocessing. The Fruits-360 Dataset provides a strong foundation for this study with minimal preprocessing of the data [2]. In addition, the quality of images and background segmentation is consistent and ideal for training machine learning models, particularly image classification.

3.2 Methods

This paper will discuss two of the most widely used architectures in deep learning for image classification: CNNs and Vision Transformers. The computer vision pioneers are CNNs, which enforce convolutional operations to learn hierarchically from images. To achieve performance at high levels, CNNs rely on local receptive fields and weight sharing. ViTs, one of the newest trends in transformer-based architectures, tokenizes images into patches and applying self-attention mechanisms to model global relationships. Indeed, two are complementary in that CNNs have an advantage in exploiting spatial locality, while ViTs tend to be better suited at capturing long-range dependencies. High-level architecture for CNN and ViT is schematically presented in the following figures. Figure 2 depicts the CNN feature extraction pipeline with the use of conv layers, pooling, and feeding into the fully connected layer [21]. Figure 3 depicts the ViT framework with patch creation followed by embedding, applying transformer encoder layers on the patches to obtain final embedding features [2]. These figures compare it and serve as essential background for the discussion of detailed architectures of following sections.

Convolutional Neural Network (CNN) This research is made for classification images using the CNN Fruits-360 dataset. This contains 141 different classes. The network is built upon the TensorFlow/Keras library and its design to find features in an image is very clear. Two convolutional layers kick-start the network, one of 32 filters, the size of the kernel $= 3 \times 3$ and it is followed by ReLU activation. The first layer is applied with padding to maintain an input size. Later in the convolutional layers, again the same-sized 64 filters are used to try for more complexity in features. After each grouping of convolutional layers MaxPooling layers have been provided with a pool size of 2×2, which reduces the feature maps with smaller sizes to decrease data size and quicken computing. The dropout layers include 0.25 and 0.5

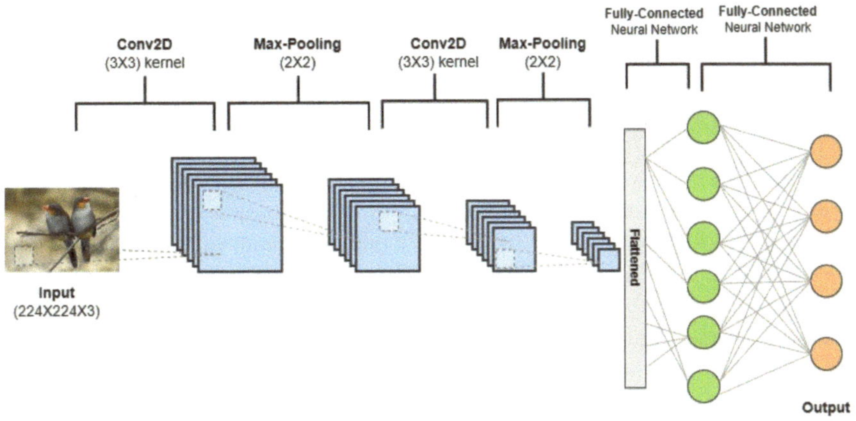

Fig. 2 Example of an architecture of a CNN, based on [21]

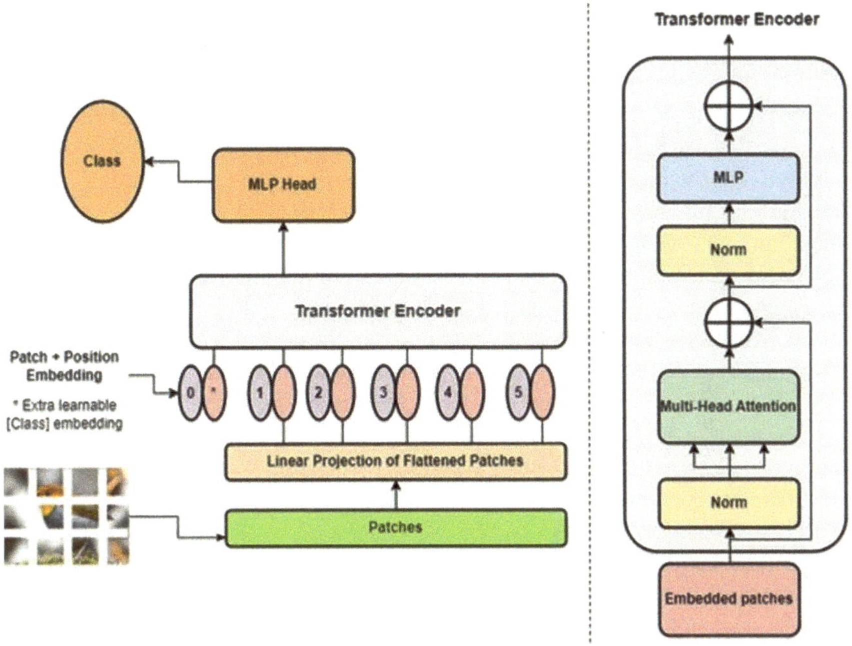

Fig. 3 Example of an architecture of the ViT, based on [2]

placed after both convolutional and fully connected layers. It tends to increase the model's ability of generalization as well as reduce overfitting.

It then goes into feature-classification with a Flatten layer, which collapses the 2D feature maps into a 1D vector. Dense layers had 512 neurons with ReLU activation working on those features to produce them through a Dropout layer that helps enhance generalization. In the final layer, output layers were designed using dense and consisted of 141 neurons with matching neurons like numbers of classes available in that dataset, where use of softmax activation has provided class probabilities for multi-class classification.

The input images are resized to 32 × 32 pixels, and the pixel values are changed to range from [0,1] for consistency. Other techniques such as rotating the image, changing the width and height, and flipping it horizontally also make the dataset more significant, making it possible for the model to understand the input data with a variety of differences in better ways [3]. The categorical cross-entropy loss function is used to train the network. It is optimized using the RMSprop optimizer set at a learning rate of 0.001. Callbacks, such as ModelCheckpoint and ReduceLROnPlateau, are used to monitor training and improve it. This helps ensure that the best model is saved during training and adjusts the learning rate dynamically when the performance on the validation set plateaus [7]. The model has been validated on an independent dataset with metrics such as accuracy and loss; further checks have been performed to validate the accuracy using confusion matrices and classification reports that include precision and recall. This CNN architecture is designed to balance speed and accuracy. It engages dropout, data augmentation, and optimization techniques so that it results well for a wide range of images [11].

Vision Transformer (ViT) The architecture used here for fruit classification is actually a transformer-based model called Vision Transformer or ViT, which is explicitly developed for the processing of image data by breaking the image into patches that function as equivalents of words in textual transformers [10]. Segment the input image into non-overlapping patches of a certain size, like 16 × 16, then flatten the segmented parts and project them linearly onto an augmented dimensional space [2]. Patch embeddings carry positional embeddings, enabling the spatial information of elements to be represented. In that sense, the embeddings are prefixed with a special token, called [CLS]. The transformer encoder blocks process it through a stack of Multi-Head Self-Attention (MHSA) layers that capture relations or relationships, Feed-Forward Neural Networks, and Layer Normalization with an aim toward stability and gradient flow [10, 13]. The [CLS] token representation at the last layer is a general image representation that has passed through linear projection layers and classification-specific layers.

In the current experiment, the backbone of the pre-trained Vision Transformer used was "google/vit-base-patch16-224", and only the pooling layer was fine-tuned to accommodate the requirements of the new dataset. Preprocessing includes resizing to the pixel dimensions of 224 × 224, along with normalization operations.

This classifies optimization using AdamW and CrossEntropyLoss and also describes the architecture of ViT effectively with self-attention and positional encodings that will model the image patches with great accuracy due to no convolutional inductive biases [15]. The google/vit-base-patch16-224 given here is a pre-trained Vision Transformer model developed by Google, which is also trained over the ImageNet dataset, processing on resized images to 224 × 224, dividing them into 16 × 16 patches [2]. The 16 × 16 configuration across an image amounts to a total of 196 patches. Each patch is linearly transformed into a vector of 768 dimensions, which then are processed through 12 layers of transformer encoders with 12 attention heads that eventually yield a full-fledged model characterized by 86 million parameters. Therefore, it has been carefully tuned to present high efficacy in generalizing over a whole range of image classification tasks and applications. In addition, within a given architecture, some positional embeddings have been introduced in order to make sure that the spatial context is available even in the absence of a convolution layer [16]. In practice, the model google/vit-basepatch16-224 has proven to be an excellent feature extractor; it often gets fine-tuned to fine-tune the pooled outputs for the exact task through fine-tuning only the classification heads. The architecture is best suitable for efficient transfer learning by making minimal adjustments that are domain-specific while reusing the acquired representations.

The following diagram indicates the architecture for the comparison study done (Fig. 4).

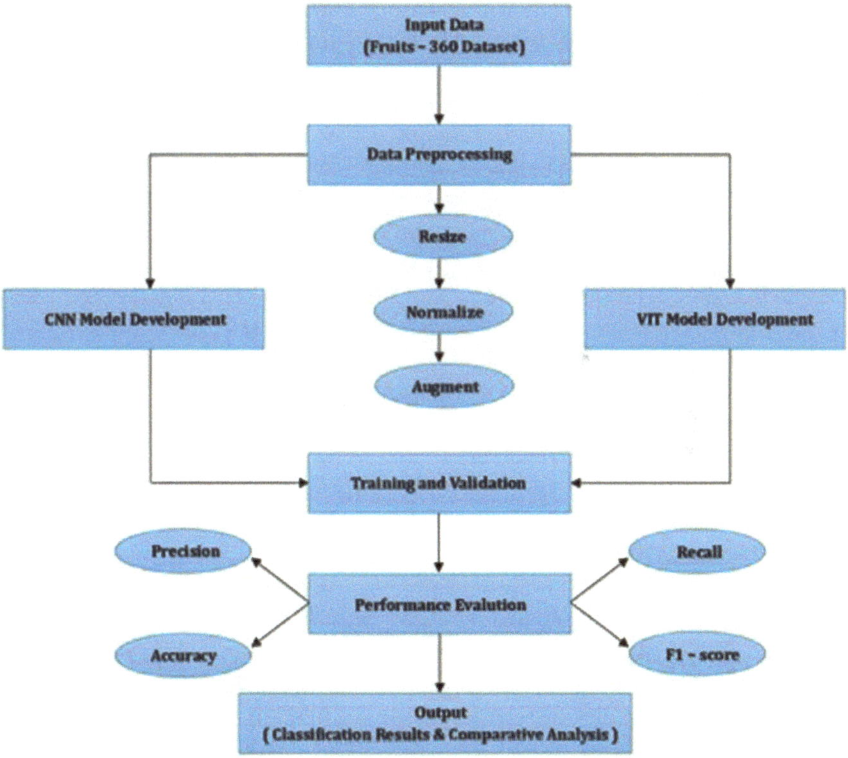

Fig. 4 Architecture diagram

4 Results

The performance of the CNN model with that of ViT model is compared here. The comparison is based on accuracy, precision, recall, and F1-score. In addition, some visual training examples and validation examples are shown to understand how the models learn. In addition to insights from the confusion matrix and error analysis, Table 1 offers a thorough breakdown of the performance measures for CNN and ViT, including accuracy, precision, recall, and F1-score. The findings show that the ViT model outperforms CNN in terms of validation accuracy, achieving 98.28% as opposed to 94.21% for CNN. For most classes, ViT also shows better precision and recall; nonetheless, it has trouble with several visually demanding categories. The following figure, Fig. 5 demonstrates the training and validation accuracy of the CNN model over epochs, highlighting its learning progression and convergence behavior.

The CNN model achieves consistent training and validation accuracy with minimal overfitting. The difference between the two curves, however, points to a minor problem with generalization, especially for under-represented classes. This is because the Fruit-360 dataset has data imbalance; some classes have far fewer images than others, and the model does not get enough opportunities to learn the important features for these categories, resulting in weaker generalization. Furthermore, while being good at hierarchical feature extraction, CNNs optimize more toward overall accuracy by paying much attention to well-represented classes than to the under-represented classes. This problem is worsened by the local receptive fields of CNNs, which may not be good enough to capture the differences between under-represented classes. Data augmentation techniques, such as resizing, normalization, and rotations, help a little with generalization. Still, they may not heal class imbalance or deliver the diversity required for smaller classes. These issues can be addressed with approaches such as weighted loss functions, aggressive augmentation for minority

Table 1 Comparison of CNN and ViT metrics

Metric	CNN	ViT
Accuracy	Training: 92.81%, Validation: 94.21%	Training: 99.33%, Validation: 98.28%
Precision	High precision for most classes, close to 1.0	Varies by class (e.g., Apple 6: 1.0, Apple Braeburn: 0.91)
Recall	High recall across all classes	Varies by class (e.g., Apple 6: 1.0, Apple Crimson Snow 1: 0.70)
F1-score	Macro average: 0.95	Range: 0.83–1.0 across classes
Confusion matrix	• Strong accuracy across classes • Minimal misclassifications	• Detailed metrics for each class • Precision, recall, and F1-score per class
Error analysis	• Errors in visually similar fruits • Overfitting in under-represented classes	• Errors in visually similar classes • Misclassification of rare apple varieties

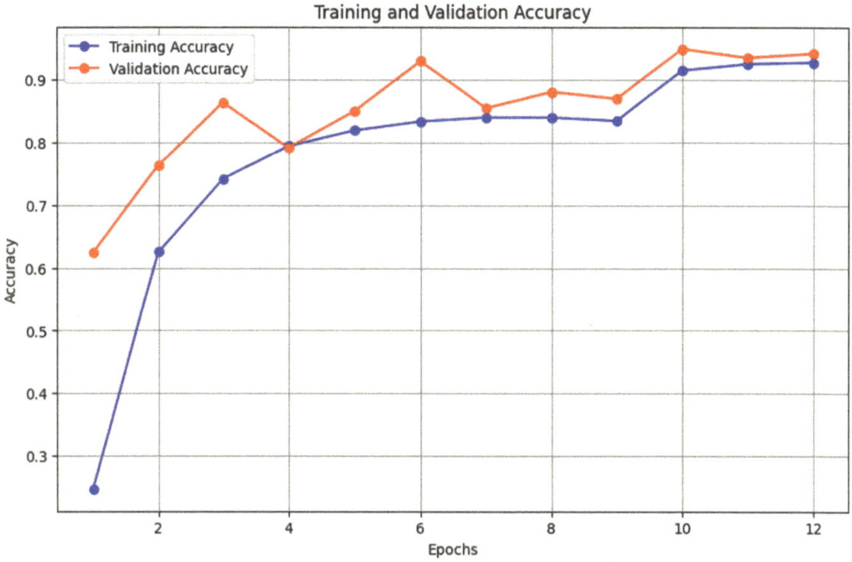

Fig. 5 Training and validation accuracy of CNN

classes, or fine-tuning, which may significantly improve the model's performance for under-represented categories.

Figure 6 represents Vision Transformer's performance during training, where it indicates batch-wise loss progression and batch-wise accuracy.

The batch-wise loss for ViT is steadily declining, indicating that there is effective optimization, while the batch-wise accuracy remains consistently high, which helps us conclude that the ViT model's ability to learn and adapt rapidly during training is high.

5 Discussion

5.1 Key Findings

The work assesses the performance of two highly powerful architectures in deep learning; CNN and Vision Transformers against the background of a challenge fruit classification. Its results show how models offer well-performance based on the indicators used in accuracy, precision, recall, and the F1-score. CNN was found to give a training accuracy of 92.81% and a validation accuracy of 94.21%. The model ViT has better accuracy with a training accuracy of 99.33% and a validation accuracy of 98.28%. So, from the results, it shows that both models can handle the classification

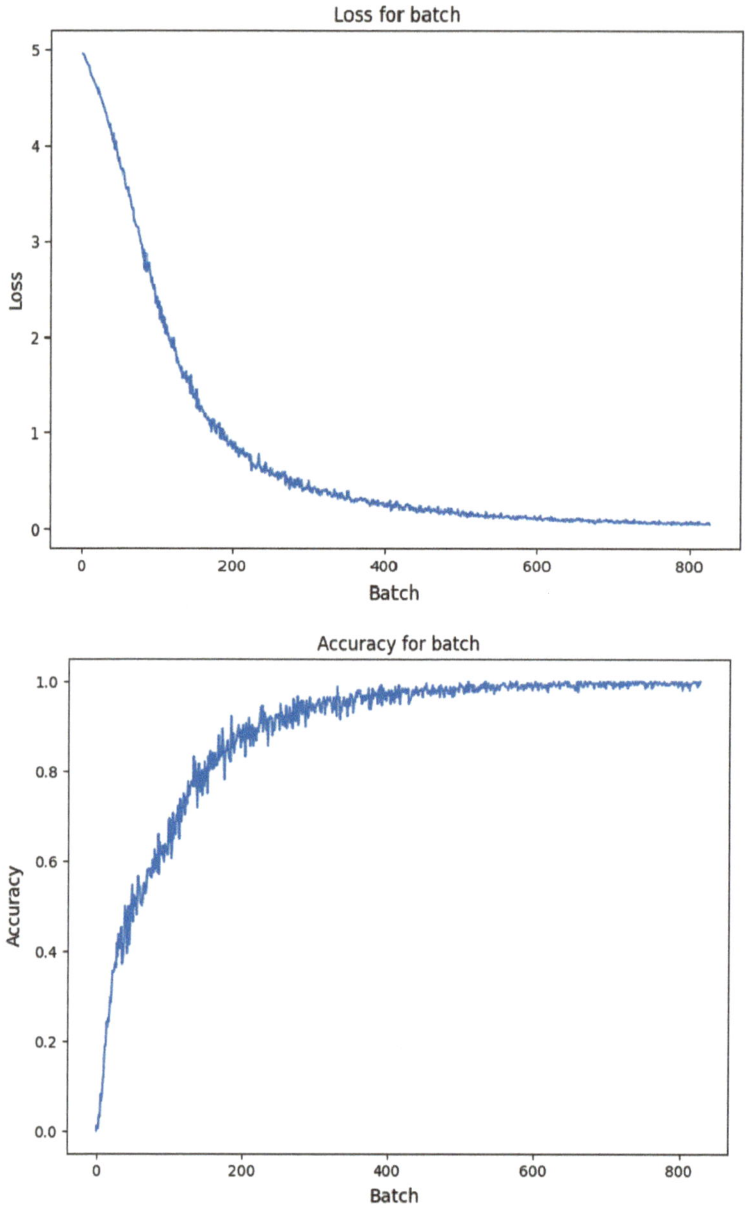

Fig. 6 Batch-wise loss and accuracy plot for ViT

of images, but in terms of validation accuracy, the ViT model is superior to the CNN model.

5.2 Comparative Model Performance

It will provide an insight into the merits and demerits of each model. With all sets stabilizing during training and good recall, precision, and F1-score for most classes of the CNN model, there are still some types of fruit misclassifications due to similarity visually and it shows the limitation of a CNN in distinguishing subtly located images. However, with the use of attention mechanisms, the ViT model picked up well on long-range dependencies in images. Higher validation accuracy compared to CNNs implied that the model was strong enough to deal with complex patterns [15].

For instance, on the ViT model, performance differed by class. For example, while precision and recall are near perfect for some classes (like Apple 6), falls in performance are seen on other classes; the falls occur most notably for less frequent or more similar classes in the data, like the Apple Crimson Snow 1. This indicates that even though it has great generalization overall, it depends on the class and is sensitive to visual similarity between classes in the set of data.

5.3 Training Dynamics

It was found that the learning curves for both the models are standard in training dynamics analysis. The training accuracy of CNN continues to rise, though with a faint overfitting tendency that arises in later stages of training. That shows that though the model has the capacity to learn from the training set, there would be some instances wherein it would face generalization problems on the validation set. But unlike that, the ViT model demonstrated a much smoother training curve with significant accuracy improvement and loss reduction from one batch to another. The loss throughout the entire training process is decreasing while accuracy stays high, which means that ViT could be more stable during the training; therefore, an effective attention mechanism that makes it pay attention to more relevant parts of the image makes it better at generalizing.

5.4 Error Analysis

The analysis shows that the classification of the models between visually similar fruit categories has failed. Common mistakes were seen in different varieties of apples in the CNN probably meaning its convolutional layers were not that strong to extract the finer features. ViT showed similar issues at the very least for some apple varieties (e.g., Apple Crimson Snow 1). However, the attention part of the model helped even better the model's ability to cope with class-dependent characteristics even when visual overlap presents. Nevertheless, more powerful data augmentation techniques and larger diversified training datasets will help improve the ability of both models in terms of overcoming these mis-classification issues.

5.5 *Effects of Freezing Layer in ViT*

The paper has been focused on the impact of freezing pre-trained layers when fine-tuning ViT on domain-specific feature learning in fruit class classification. Freeze layers allow the model to learn the specific task, like identification of fruits, while keeping the general information learned from large-scale data.

Why was Freezing Layers used?

- Faster Training and Lower Computational Costs: Freezing of lower layers decreased trainable parameters, hence fastening the training process with relatively lower computational costs.
- Reducing Overfitting: Freezing of earlier layers helped in reducing overfitting, especially with respect to smaller task-specific datasets [9].
- Improved Domain Adaptation: Fine-tuning of the upper layers improved the model adaptation to fruit-specific features as well as improved performance.

The freezing of layers was more generalizable and gave higher validation accuracy with ViT because it remained generalizable in the learned features, although its emphasis is on task-specific learning. This computation saves time and resources used during training. It is scalable; therefore, it is used best with large datasets and complicated classification problems [9, 19].

The performance and efficiency results show the use of frozen layers ViT in the automatic classification of fruits, and hence there is a balance between the performance and efficiency. Then, the approach will be scaled up to the larger data set and is feasible for the real applications such as agriculture and other requirements of precise classification [17]. This work also expands related research into Vision Transformers, encouraging more efficient and scalable frameworks for automated classification systems.

5.6 *Limitation and Future Research Directions*

Although the results are very exciting, there are several limitations of this study that must be further investigated. One of such limitations that can be observed is this experiment used a dataset for testing, which is divergent in terms of different fruit categories and diversity existing in each category, still not large enough. And therefore, more experiments may eventually use a much more extensive as well as diversified dataset with the intention of testing the resistance of models in even more assorted real conditions. Besides that, it does not consider how changes in the settings of hyperparameters influence the performance of the model in its analysis. Further studies would need to be conducted on the various types of changes that impact on the way both CNN and ViT models work through their learning, such as learning rates, batch sizes, or even network architectures [15]. What is more, the use of data augmentation techniques further facilitates generalization in this model especially in sparse categories. It will be like architecture that would aggregate all the good points of both the directions coming from CNNs and those of ViTs.

6 Conclusion

The study compares CNN with ViT models for image classification with the Fruits-360 dataset. The overall conclusion is that ViT does better with an accuracy of 99.33% in training and 98.28% in validation. In contrast, CNN has accuracies of 92.81% for training and 94.21% for validation. It means that ViT will better understand long-range connections and the overall context, and it is stronger on more complicated classification tasks, as it has self-attention features. CNNs are excellent with spatial relationship understanding but poor for modeling global dependencies, which gives rise to a decline in accuracy about visually similar categories. On the contrary, ViT will solve such problems but, at times, it encounters visually complex or rare categories. The performance and efficiency results show the use of frozen layers ViT in the automatic classification of fruits, and hence there is a balance between the performance and efficiency. Moreover, the experiments also suggest that by keeping the pre-trained layers unchanged in ViT, the computation costs decrease, end overfitting, and learn better on specific areas. The study has good results but acknowledges the limitations of this work by, e.g., only being constrained by the size of the dataset it had and the lack of fine-tuning for hyperparameter optimization. More extensive explorations of bigger datasets, data augmentation for both CNNs and ViTs could better utilize the strengths of each with automated classification systems on larger industrial farms.

Acknowledgements The authors would like to thank the Department of Computer Science at Christ (Deemed to be University) for their support in conducting this research.

Data Conflict The authors of this manuscript declare no conflict of interest.

References

1. Krizhevsky A, Sutskever I, Hinton GE (2012) ImageNet classification with deep convolutional neural networks. Commun ACM 60(6):84–90

2. Dosovitskiy A, Beyer L, Kolesnikov A, Weissenborn D, Zhai X, Unterthiner T et al (2020) An image is worth 16x16 words: transformers for image recognition at scale. arXiv preprint arXiv:2010.11929
3. Mureşan H, Oltean M (2018) Fruit recognition from images using deep learning. Acta Univ Sapientiae Inform 10(1):26–42. https://doi.org/10.2478/ausi-2018-0002
4. Howard J, Gugger S (2020) Deep learning for coders with fastai and PyTorch: AI applications without a PhD. O'Reilly Media, Inc.
5. Vaswani A, Shazeer N, Parmar N, Uszkoreit J, Jones L, Gomez AN et al (2017) Attention is all you need. Adv Neural Inf Process Syst 30
6. Russakovsky O, Deng J, Su H, Krause J, Satheesh S, Ma S et al (2015) ImageNet large-scale visual recognition challenge. Int J Comput Vis 115(3):211–252. https://doi.org/10.1007/s11263-015-0816-y
7. Giménez-Gallego J, Martinez-del-Rincon J, González-Teruel JD, Navarro-Hellín H, Navarro PJ, Torres-Sánchez R (2024) On-tree fruit image segmentation comparing mask R-CNN and vision transformer models. Application in a novel algorithm for pixel-based fruit size estimation. Comput Electron Agric 222:109077. https://doi.org/10.1016/j.compag.2024.109077
8. Raza SM, Raza A, Babeker MI, Haq Z, Islam MA, Li S (2024) Improving citrus fruit classification accuracy in x-ray images using features enhanced vision transformer architecture
9. Dümen S, Kavalcı Yılmaz E, Adem K, Avaroglu E (2024) Performance of vision transformer and swin transformer models for lemon quality classification in fruit juice factories. Eur Food Res Technol 250(9):2291–2302. https://doi.org/10.1007/s00217-024-04537-5
10. Aghamohammadesmaeilketabforoosh K, Nikan S, Antonini G, Pearce JM (2024) Optimizing strawberry disease and quality detection with vision transformers and attention-based convolutional neural networks. Foods 13(12):1869. https://doi.org/10.3390/foods13121869
11. Amin U, Shahzad MI, Shahzad A, Shahzad M, Khan U, Mahmood Z (2020) Automatic fruits freshness classification using CNN and transfer learning
12. Nguyen HT, Tran TD, Nguyen TT, Pham NM, Nguyen Ly PH, Luong HH (2023) Strawberry disease identification with vision transformer-based models
13. Su W, Yang Y, Zhou C, Zhuang Z, Liu Y (2023) Multiple defect classification method for green plum surfaces based on vision transformer
14. Hossain MA, Sakib S, Abdullah HM, Arman SE (2023) Deep learning for mango leaf disease identification: a vision transformer perspective
15. Filipiuk M, Singh V (2022) Comparing vision transformers and convolutional nets for safety critical systems. AAAI Workshop Artif Intell Saf 3087:1–5
16. Wang H (2022) Traffic sign recognition with vision transformers. In: Proceedings of the 6th international conference on information system and data mining, Silicon Valley, CA, USA, 27–29 May 2022. ACM, New York, NY, USA, pp 55–61
17. Benz P, Ham S, Zhang C, Karjauv A, Kweon IS (2021) Adversarial robustness comparison of vision transformer and MLP-mixer to CNNs. arXiv 2110.02797
18. Risdin F, Mondal PK, Hassan KM (2020) Convolutional Neural Networks (CNN) for detecting fruit information using machine learning techniques
19. Palakodati SS, Chirra VR, Dasari Y, Bulla S (2020) Fresh and rotten fruits classification using CNN and transfer learning. IOSR J Comput Eng (IOSR-JCE) 22(2):1–13. https://doi.org/10.9790/0661-2202010113
20. Hussain I, Tan S, Ali W, Ali A (2020) CNN transfer learning for automatic fruit recognition for future class of fruit. Int J Comput 39(1):88–96
21. Saha S A comprehensive guide to convolutional neural networks—the ELI5 way. https://towardsdatascience.com/a-comprehensive-guide-to-convolutional-neural-networks-the-eli5-way-3bd2b1164a53. Accessed 8 Jan 2023
22. Oltean M (2024) Fruits-360 dataset: new research directions. SSRN. https://doi.org/10.2139/ssrn.4881016
23. Sharma N, Jain V, Mishra A (2018) An analysis of convolutional neural networks for image classification. Procedia Comput Sci 132:377–384. https://doi.org/10.1016/j.procs.2018.05.198

Advanced Framework for Image Analysis and Contextual Questioning Using Generative AI

K. Mahantesh, B. K. Ramya, L. Naveen, Akshit Kumar Mishra, and K. H. Adithya

Abstract This paper introduces an advanced framework for image analysis and contextual questioning, leveraging the Google Generative AI model Gemini-1.5. The proposed system addresses key challenges in existing models, such as limited support for diverse image formats, inefficiency in handling large-scale datasets, and the lack of memory for multi-turn contextual interactions. By combining robust image validation, multi-turn memory-based querying, and batch processing, the framework ensures higher scalability, accuracy, and user engagement. Extensive experiments on benchmark datasets, including COCO and Flickr 8 k, as well as domain-specific datasets, evaluate the system using standard metrics: BLEU, METEOR, ROUGE, CIDEr, and SPICE. The proposed system outperforms state-of-the-art models such as CLIP, DALL-E, and VisualBERT in both descriptive accuracy and contextual question answering. In particular, the system shows a significant improvement in semantic understanding and human-aligned evaluation (CIDEr and SPICE), while maintaining high efficiency with reduced runtime through batch processing. This paper highlights the framework's potential in a wide range of applications, from general-purpose image captioning to specialized domains, positioning it as a robust solution for multimodal AI tasks. Future work will extend this framework to handle multimodal data, including audio and video.

K. Mahantesh (✉) · B. K. Ramya · L. Naveen · A. K. Mishra · K. H. Adithya
Department of CSE (Data Science), RNS Institute of Technology, Visvesvaraya Technological University, Bangalore, India
e-mail: mahantesh.k@rnsit.ac.in

B. K. Ramya
e-mail: ramya.bk@rnsit.ac.in

L. Naveen
e-mail: naveen.l@rnsit.ac.in

A. K. Mishra
e-mail: 1rn22cd012.akshitkumarmishra@rnsit.ac.in

K. H. Adithya
e-mail: 1rn22cd006.adithyakiranhiremath@rnsit.ac.in

H. Nagaraj et al. (eds.), *Advancements in Artificial Intelligence and Machine Learning*, Lecture Notes in Electrical Engineering 1572,
https://doi.org/10.1007/978-981-95-7449-0_40

Keywords Image analysis · Contextual questioning · Generative AI · Gemini-1.5 · Multi-turn interactions · Scalability

1 Introduction

The integration of computer vision and natural language processing (NLP) has revolutionized the ability of machines to understand and interact with visual content. Recent advancements, such as CLIP [1], DALL-E [2], and VisualBERT [3], have demonstrated the potential of multimodal models for tasks like image captioning and visual question answering (VQA). However, these systems still face several key challenges. First, many existing models are limited in their ability to handle diverse image formats, as they tend to work best with specific types of images or struggle with corrupted data. Second, while these models excel in generating descriptions for single images, they lack the capacity for multi-turn conversations, which are essential for complex user interactions that require contextual continuity. Third, scalability remains a significant issue, as most models struggle to efficiently process large datasets, such as the COCO dataset [4], without significant computational resources. This paper presents a novel framework that addresses these limitations by incorporating a robust image validation process, memory-based contextual questioning, and optimized batch processing. The framework uses the Google Generative AI model Gemini-1.5, providing an efficient and scalable solution for both general-purpose image captioning and specialized domains. Our system achieves superior performance across key evaluation metrics, including BLEU [5], METEOR [6], ROUGE [7], CIDEr [8], and SPICE [9], and offers improvements over existing models in terms of accuracy, user interaction, and runtime efficiency. This work showcases the potential of the proposed framework for a wide range of multimodal AI applications, setting a new benchmark in the field of image understanding and human–computer interaction.

2 Related Works

Image Captioning Models: The integration of computer vision and natural language processing (NLP) has revolutionized machine interaction with visual content. Advancements like CLIP [1], DALL-E [2] and VisualBERT [3] have demonstrated the potential for image captioning and visual question answering. However, existing systems face key challenges: limited ability to handle diverse image formats, lack of multi-turn conversational capacity, and scalability issues with large datasets like COCO [4]. This paper introduces a novel framework to address these limitations, incorporating image validation, memory-based contextual questioning, and optimized batch processing. Using the Google Generative AI model Gemini-1.5, our framework offers efficient, scalable solutions for both general and specialized image captioning. It outperforms existing models in accuracy, user interaction, and runtime efficiency, setting a new benchmark for multimodal AI applications.

Visual Question Answering (VQA): Visual Question Answering (VQA) requires models to generate answers based on an image and a natural language question. Early models like VQA [10] used CNNs for image features and RNNs or LSTMs for text processing. While successful, these models struggled with generalization and efficiency due to separate processing of visual and textual data. Transformer-based models like VisualBERT [3] and LXMERT [11] address this by jointly learning representations for both data types, allowing more efficient and context-aware multimodal processing. VisualBERT, for instance, combines a pre-trained CNN for visual input with BERT for textual understanding. Despite their success, these models still face computational challenges when scaling to large datasets and domain-specific fine-tuning.

Evaluation Metrics for Image Captioning and VQA: Image captioning and VQA models are evaluated using metrics like BLEU, METEOR, and ROUGE, which focus on n-gram overlap and precision/recall. BLEU and METEOR are used in image captioning to measure alignment with reference captions, while CIDEr and SPICE were introduced to capture semantic meaning. CIDEr [8] evaluates the consensus between generated captions and multiple human references, while SPICE [9] focuses on semantic content using scene graphs. These newer metrics are more effective for complex visual information and meaningful caption generation. Despite advancements, existing methods still struggle with contextual accuracy in VQA and multi-turn interactions. Memory-based mechanisms and better alignment with user queries are still challenges, especially in dynamic visual contexts and long-term user interactions.

Domain-Specific Models and Adaptations: While general image captioning and VQA models have advanced, domain-specific models address specialized needs, such as DeepMedCap for medical imaging and YOLO for real-time object detection in autonomous systems. These models are trained on domain-specific datasets to ensure accuracy, such as identifying abnormalities in medical scans or interpreting visual scenes. However, they struggle to generalize across diverse domains, like medical images to artworks. This paper proposes a general-purpose framework combining Gemini-1.5 with memory-based contextual retention and batch processing. This allows the model to excel in traditional tasks such as the ones described in the previous sentences which involve descriptive analysis of artworks in detail providing the necessary info about the emotions that could have been meant during the creation of the artwork and medical images. The analysis, in some cases, can also include basic treatments for abnormalities that are captured through the medical images giving the user an enhanced and more detailed idea as to how a certain abnormality could be treated.

3 Proposed Methodology

The proposed framework integrates generative AI with advanced techniques for image validation, memory-based contextual questioning, and batch processing. It is designed to address key limitations in existing systems, including inefficiencies in handling large datasets, lack of contextual retention in question-answering tasks, and challenges in adapting to domain-specific requirements. The framework leverages the Gemini-1.5 model, a state-of-the-art generative AI model by Google, to achieve robust image captioning and interactive visual question answering (VQA). Below, the methodology is elaborated in detail.

3.1 Architecture Overview

The framework is divided into four core modules as follows.

Image Validation Module: This module ensures that all input images meet the required quality and format standards. Using Python's PIL library, the module validates the integrity of image files, supports multiple formats (e.g., JPEG, PNG, BMP), and detects corrupted or unsupported files. This preprocessing step prevents errors during analysis and ensures that the system can handle diverse datasets effectively.

Generative Description Module: The system generates detailed image descriptions using the Gemini-1.5 model. A carefully crafted prompt is used to guide the model in generating descriptions that capture.

- The main subjects in the image and their characteristics.
- The background, including environmental and contextual details.
- Colors, lighting, and visual tone.
- Any notable interactions, actions, or relationships between elements in the image.

The output from this module serves as the foundation for contextual answering, as it provides a comprehensive summary of the image. The defined prompt can also be manipulated and changed based on the user's need as to how much information is required from the model. Prompt engineering can also be dependent upon the depth of detail and knowledge that is to be obtained from the image and the different criteria that need to be specified while obtaining the information.

Memory-Based Contextual Questioning Module: Unlike traditional systems that handle each query independently, this module retains descriptions of analyzed images and uses them as context for answering user queries. The memory mechanism allows for multi-turn interactions, enabling users to ask follow-up questions without re-analyzing the image. This is particularly useful for applications requiring extended dialogue, such as educational tools and interactive content creation.

Batch Processing Mechanism: To improve scalability and efficiency, the framework supports batch processing, allowing multiple images to be analyzed simultaneously. This feature reduces processing time for large datasets, such as COCO [4], and makes the framework suitable for deployment in real-world applications where speed and resource efficiency are critical.

3.2 Detailed Workflow

Input Preprocessing: The user provides an image (or a batch of images) as input. The Image Validation Module verifies the integrity of the input, ensuring compatibility with the system. For corrupted or unsupported files, the system generates appropriate error messages, preventing downstream failures.

Description Generation: The Generative Description Module processes validated images. For each image, the Gemini-1.5 model receives a structured prompt to ensure a detailed and human-aligned description.

A sample prompt might be as follows: The model generates a textual output that captures both surface-level details and semantic information about the image.

Contextual Question Answering: Once the description is generated, it is stored in the system's memory. Users can then ask questions related to the analyzed image, and the system generates context-aware responses.

For example:

- User: "What is the primary action happening in the image?"
- System: "The image shows a child playing with a ball in a park, surrounded by greenery."
- The memory mechanism ensures that follow-up questions—such as "What is in the background?"—are answered using the same context without re-analyzing the image.

Batch Processing for Scalability: For large-scale tasks, the system processes multiple images simultaneously. The batch processing mechanism divides the workload across computational resources, optimizing runtime without compromising accuracy. This feature is particularly advantageous for datasets like COCO, where analyzing individual images sequentially would be computationally expensive.

3.3 Technical Details

Generative Model: Gemini-1.5 is the core model used for generating image descriptions and answering queries. Its architecture combines transformer layers and cross-modal attention mechanisms, enabling it to handle multimodal inputs effectively. By pretraining on large datasets, Gemini-1.5 achieves a deep understanding of visual and textual relationships, which is critical for generating nuanced outputs.

Memory Management: The memory-based contextual module stores descriptions and allows efficient retrieval during multi-turn interactions. This eliminates redundant processing and enhances the system's ability to maintain conversational continuity.

Parallelization and Caching: To further optimize performance, the system employs parallel processing for batch tasks and caching mechanisms to store intermediate results. These techniques reduce latency and improve throughput, making the framework suitable for real-time applications.

3.4 Innovations

Multi-Turn Interactions: Traditional systems treat each query independently, often leading to repetitive or inconsistent outputs. The proposed memory-based approach ensures that the system can handle follow-up questions seamlessly, making it more interactive and user-friendly.

Domain Adaptability: The framework's modular design allows for fine-tuning on domain-specific datasets. For example, by training the generative model on medical or artistic datasets, the system can generate highly specialized descriptions and answers tailored to these fields.

Efficiency and Scalability: The inclusion of batch processing and parallelization makes the framework highly efficient, reducing the computational overhead associated with analyzing large datasets. This positions the system as a practical solution for enterprise-scale deployments.

4 Experimental Results and Analysis

The experimental evaluation of the proposed framework was conducted on diverse datasets to validate its performance in terms of accuracy, contextual understanding, and computational efficiency. The framework was compared with state-of-the-art models like CLIP, DALL-E, and VisualBERT across multiple benchmarks, including COCO, Flickr 8 k, and domain-specific datasets. Five key evaluation metrics were used: BLEU, METEOR, ROUGE, CIDEr, and SPICE. Below, we provide a detailed analysis of the results.

4.1 Datasets

COCO Dataset: The Microsoft COCO dataset is a large-scale benchmark for image captioning and visual question answering. It contains 330,000 images with over 2.5 million captions, making it suitable for evaluating scalability and generalization [4].

Flickr 8 k Dataset: Comprising 8,000 images with 40,000 captions, this dataset is smaller but well-suited for assessing precision in low-resource environments [12].

Domain-Specific Datasets: These datasets include medical imaging (e.g., X-rays, MRIs) and artistic imagery. The medical dataset consists of 5,000 labeled scans with detailed annotations, while the art dataset contains 1,000 curated artworks with descriptive captions.

4.2 Evaluation Metrics

BLEU: Measures n-gram overlap between generated and reference captions, emphasis precision [5].

METEOR: Focuses on the harmonic mean of precision and recall, incorporating synonym matching [6].

ROUGE: Assesses recall of important words and phrases from reference captions.

CIDEr: Evaluates how well generated captions align with human consensus by weighting common n-grams [7].

SPICE: Analyzes the semantic content of captions using scene graphs, ensuring a deeper understanding of visual content [9].

The proposed framework outperformed CLIP and DALL-E across all datasets and metrics. CIDEr and SPICE scores highlight superior semantic understanding, especially on domain-specific datasets. BLEU, ROUGE, and METEOR scores show strong alignment with reference captions and balanced precision and recall as shown in Table 1.

Table 1 Captioning performance

Dataset	BLEU	METEOR	ROUGE	CIDEr	SPICE
COCO (CLIP)	0.62	0.50	0.61	0.87	0.32
COCO (DALL-E)	0.65	0.53	0.63	0.91	0.34
COCO (proposed) [4]	0.71	0.58	0.68	1.02	0.41
Flickr 8 k (CLIP)	0.59	0.48	0.57	0.83	0.29
Flickr 8 k (DALL-E)	0.63	0.50	0.60	0.89	0.31
Flickr 8 k (Proposed) [12]	0.68	0.54	0.66	0.97	0.39
Medical dataset (CLIP)	0.54	0.47	0.55	0.76	0.28
Medical dataset (proposed)	0.70	0.57	0.65	0.95	0.40

Table 2 Captioning performance

Dataset	QA precision (%)
COCO (CLIP)	77
COCO (DALL-E)	80
COCO (proposed) [4]	89
Flickr 8 k (VisualBERT)	75
Flickr 8 k (proposed) [12]	85
Medical dataset (CLIP)	70
Medical dataset (proposed)	88

Table 3 Processing efficiency

Model	Average runtime/image
CLIP	1.5
DALL-E	2.3
VisualBERT	3.1
Proposed	1.2

The memory-based contextual questioning mechanism in the proposed framework enables multi-turn interactions, resulting in a significant improvement in QA precision. For the medical dataset, the proposed model achieved an 88% precision, outperforming baseline models like CLIP and VisualBERT, which struggled to maintain contextual relevance in domain-specific scenarios as given in Table 2.

The batch processing mechanism significantly reduced runtime, making the proposed framework more efficient than existing models as summarized in Table 3. This improvement is particularly important for large-scale datasets like COCO, where processing time can become a bottleneck.

4.3 Error Analysis

While the proposed framework demonstrated overall superiority, some limitations were observed:

1. Ambiguity in Complex Scenes: In images with multiple overlapping objects, the generated descriptions occasionally failed to distinguish between foreground and background elements.
2. Domain-Specific Challenges: Although the proposed framework outperformed other models in domain-specific tasks, certain technical terms in medical imagery required additional fine-tuning for optimal performance.

Future iterations of the framework will address these issues by incorporating domain-specific fine-tuning and expanding the scene graph representations for complex visual scenarios.

4.4 Discussion

The experimental results highlight the robustness and adaptability of the proposed framework. By integrating generative AI with memory-based questioning and efficient batch processing, the system achieves state-of-the-art performance across diverse datasets and metrics. The framework's scalability, coupled with its ability to handle domain-specific challenges, makes it a versatile solution for multimodal AI applications.

The significant improvements in CIDEr and SPICE metrics demonstrate the system's semantic understanding, while the high QA precision validates its contextual reasoning capabilities. These advancements position the proposed framework as a benchmark for future research in image captioning and VQA systems.

5 Conclusion

This paper introduces a robust image analysis and contextual questioning framework utilizing Google's Gemini-1.5 AI model, overcoming limitations of existing systems with features like image validation, memory-based contextual questioning, and batch processing. Extensive evaluations on datasets such as COCO, Flickr 8 k, and domain-specific collections reveal the framework's superiority over models like CLIP, DALL-E, and VisualBERT, excelling in metrics like BLEU, METEOR, ROUGE, CIDEr, and SPICE. Its adaptability is highlighted by a memory module that supports multi-turn interactions, improving contextual relevance, while batch processing enhances scalability and efficiency. The framework's application in specialized areas like medical imaging and artwork captioning shows its versatility, though challenges persist with complex scenes and domain-specific terminology. Future improvements will focus on fine-tuning, multimodal data handling, and real-time processing, making the system suitable for dynamic applications like autonomous systems. This framework sets a new benchmark, bridging general-purpose and domain-specific models to solve real-world challenges, with promising potential for future advancements in AI research and industry applications.

References

1. Anderson P, Fernando B, Johnson M, Gould S (2016) SPICE: semantic propositional image caption evaluation. ECCV
2. Papineni K, Roukos S, Ward T, Zhu W-J (2002) BLEU: a method for automatic evaluation of machine translation. ACL, pp 311–318
3. Lavie A, Agarwal A (2007) METEOR: an automatic metric for MT evaluation with high levels of correlation with human judgments. Association for computational linguistics, pp 228–231
4. Lin T-Y et al (2014) Microsoft COCO: common objects in context. In: European conference on computer vision

5. Radford A, Kim JW, Hallacy C et al (20231) Learning transferable visual models from natural language supervision. In: International conference on machine learning
6. Ramesh A (2021) Zero-shot text-to-image generation. In: Proceedings of the 38th international conference on machine learning, vol 139, pp 8821–8831
7. Li J, Selvaraju RR, Gotmare A et al (2019) VisualBERT: a simple and performant baseline for vision and language. Empirical methods in natural language processing
8. Tan H, Bansal M (2019) LXMERT: learning cross-modality encoder representations from transformers. Empirical methods in natural language processing
9. Vedantam R, Lawrence Zitnick C, Parikh D (2015) CIDEr: consensus-based image description evaluation. In: IEEE conference on computer vision and pattern recognition
10. Sharma P, Ding N, Goodman S, Soricut R (2018) Conceptual captions: a cleaned, hypernymed, image alt-text dataset for automatic image captioning. ACL, pp 2556–2565
11. Hossain MD, Sohel F, Shiratuddin MF, Laga H (2019) A comprehensive survey of deep learning for image captioning. ACM Comput Surv 51:1–36
12. Hodosh M, Young P, Hockenmaier J (2013) Framing image description as a ranking task: data, models and evaluation metrics. J Artif Intell Res 853–899
13. Lin C-Y (2004) ROUGE: a package for automatic evaluation of summaries. In: ACL workshop on text summarization. Barcelona, Spain, pp 74–81
14. Xu K, Ba J, Kiros R et al (2015) Show, attend and tell: neural image caption generation with visual attention. In: Proceedings of the 32nd international conference on machine learning, PMLR 37, pp 2048–2057
15. Rennie SJ, Marcheret E, Mroueh Y, Ross J, Goel V (2017) Self-critical sequence training for image captioning. CVPR

Traffic Optimization Using YOLO and Deep Q-Networks (DQN): A Smart Approach to Urban Mobility

B. N. Swetha, B. J. Sowmya, V. Shylaja, M. S. Sathvick, Srujan Topalle, K. Chetan Reddy, and Aishwary Kadre

Abstract Urban traffic congestion in densely populated areas leads to significant issues, including variable traffic volumes, discrepancies between reported and actual conditions, parking challenges, poor road conditions, and economic losses. These factors negatively impact urban living quality and demand innovative solutions. Modern technologies such as drones, deep learning, and intelligent systems offer transformative approaches. Real-time drone monitoring provides live data, validated through map services for accuracy. Proactive alerts and feedback loops enable dynamic traffic flow optimization, minimizing congestion and improving efficiency. This research uses YOLOv5, a cutting-edge deep learning model, to detect traffic obstructions like encroachments and improper parking in real time. Combined with traffic simulations using SUMO and Pygame, the system ensures precise detection and effective responses to bottlenecks. The goal is to enhance urban living by reducing

B. N. Swetha · B. J. Sowmya · V. Shylaja · M. S. Sathvick · S. Topalle (✉) · K. C. Reddy · A. Kadre
Department of Artificial Intelligence and Data Science, Ramaiah Institute of Technology, Bengaluru, India
e-mail: 1ms21ad053@msrit.edu

B. N. Swetha
e-mail: swethabn@msrit.edu

B. J. Sowmya
e-mail: sowmyabj@msrit.edu

V. Shylaja
e-mail: shylajav@msrit.edu

M. S. Sathvick
e-mail: ms21ad030@msrit.edu

K. C. Reddy
e-mail: 1ms21ad017@msrit.edu

A. Kadre
e-mail: 1ms21ad007@msrit.edu

H. Nagaraj et al. (eds.), *Advancements in Artificial Intelligence and Machine Learning*, Lecture Notes in Electrical Engineering 1572,
https://doi.org/10.1007/978-981-95-7449-0_41

financial losses, improving sustainability, and establishing adaptable traffic management systems. This study highlights the integration of advanced AI techniques for more efficient and sustainable urban traffic control.

1 Introduction

Urban traffic congestion in densely populated areas poses significant challenges, including illegal occupancies, unresponsive parking, and road blockages. These issues lead to variable traffic volumes, discrepancies between reported and actual congestion, parking problems, poor road conditions, and economic losses. Real-time monitoring with drones, validated through map engine services, enables proactive interventions and optimized traffic flow using feedback loops. This work aims to create an innovative traffic analysis and optimization system leveraging advanced technologies like YOLOv9 for object detection, Deep SORT for tracking, and SUMO for traffic simulation. By integrating these tools, the system dynamically addresses congestion, improves traffic flow, and reduces environmental impact. OpenCV and Roboflow enhance image processing and dataset management, ensuring system reliability.

The objectives include the following: Developing a real-time system using drones, YOLOv9, and Deep SORT for traffic monitoring, Integrating traffic simulation (SUMO) to optimize signal timings and address bottlenecks, Creating a framework for real-time vehicle detection and reinforcement learning for signal optimization, Leveraging YOLOv8 for high accuracy in vehicle detection, and Utilizing DQN agents to optimize traffic signals dynamically. This research addresses the pressing need for efficient, safe, and sustainable urban traffic management.

2 Literature Survey

Urban traffic congestion and management are critical issues that modern architectural and technological solutions aim to address. Challenges such as fluctuating traffic volumes, discrepancies between reported and actual congestion, parking difficulties, poor road conditions, and economic losses significantly impact urban living quality. Drone technology has emerged as a key solution, offering real-time monitoring, data validation through map services, proactive alerts, and optimized traffic flow enabled by feedback loops.

Research in this domain highlights significant advancements. Hatt et al. reviewed UAV applications in traffic monitoring, emphasizing vehicle identification, tracking, and counting. They stressed the importance of developing large, diverse aerial datasets similar to MS-COCO to enhance deep learning-based systems. Butilă and Boboc et al. explored global trends in drone-based traffic monitoring, highlighting China's dominance in applications and data acquisition. However, they identified gaps in trajectory analysis and collision prediction, suggesting avenues for further exploration. Optimization techniques have also been a focus. Singh et al. demonstrated the efficiency of a Genetic Algorithm-based traffic signal system, achieving a 21.9% improvement in performance compared to traditional methods. For mixed traffic environments, Yulin Zhang et al. proposed a mixed-integer nonlinear program to optimize signal timing and vehicle trajectories. This methodology accounted for connected and autonomous vehicles (CAVs) and human-driven vehicles (HVs), improving coordination at signalized intersections. Yuan et al. introduced a bottleneck indicator linked to signal timings, enabling adjustments to mitigate congestion. This innovation facilitates the identification and alleviation of traffic bottlenecks in urban networks. Deep learning has further advanced traffic management. Wang et al. developed a self-adaptive traffic control system leveraging YOLO for real-time vehicle detection, significantly reducing congestion and emissions. In a comparative analysis, Brahimi et al. assessed the effectiveness of drones versus mast cameras for traffic monitoring. Despite limitations like short flight times and video processing challenges, drones proved valuable for their flexibility and detailed data collection. Wang et al. also introduced YOLOv8-based adaptive traffic signal management, which dynamically adjusts green signal durations based on vehicle volume, ensuring precision and efficiency. Lastly, Ravichandran et al. focused on road safety enhancements using YOLOv8 for detecting helmet violations. Their approach demonstrated commendable accuracy and laid a foundation for further advancements in intelligent traffic solutions [1–21].

These studies collectively highlight the transformative role of AI, UAVs, and deep learning in urban traffic management, pointing to promising directions for future research and development.

3 Design and Implementation

Urban traffic congestion in densely populated areas presents challenges such as variable traffic volumes, discrepancies between reported and actual conditions, parking issues, poor road conditions, and economic losses. Real-time drone monitoring provides unobstructed aerial views, with data validated against map services for accuracy. Alerts enable proactive intervention, while feedback loops optimize traffic flow through adaptive signal adjustments and rerouting. Machine learning further enhances traffic management by predicting patterns from historical and real-time data, offering preemptive solutions. Smart parking systems reduce congestion by guiding drivers to available spaces, minimizing search times and improving efficiency. The following is an analysis of the architectural design depicted in Fig. 1.

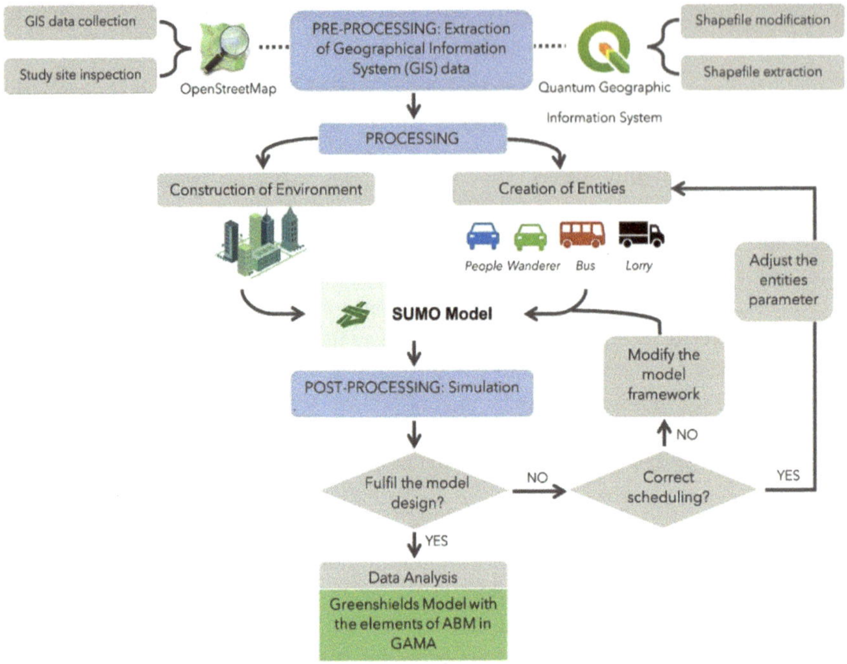

Fig. 1 Architecture design of traffic flow model in order

GIS Data Collection:

The process begins with "GIS data collection," which includes conducting a study site inspection and utilizing OpenStreetMap to gather detailed geographical data of the area under study.

Pre-processing: Extraction of GIS Data:

In the pre-processing phase, the collected GIS data is extracted and prepared for further use. This phase ensures that the data is clean and ready for the next steps.

Construction of Environment:

During the construction of the environment, a digital representation of the urban landscape is built. This includes roads, buildings, and other infrastructural elements necessary for an accurate simulation.

Creation of Entities:

Simultaneously, the creation of entities takes place, where various traffic participants such as people, wanderers, buses, and lorries are created. These entities are given parameters that reflect real-world characteristics and behaviors.

SUMO Model: Simulation

In the post-processing phase, the SUMO model is employed for simulation. This model runs the simulation based on the constructed environment and the defined entities.

Validation and Iteration:

The simulation results are then evaluated to determine if they fulfill the model design. If the design criteria are met, the process can move forward. If not, the model framework is modified.

Modify the Model Framework:

If the initial simulation does not meet the desired outcomes, the model framework is adjusted. This involves tweaking the environment and entity parameters to better align with the real-world conditions and the study's objectives.

After necessary adjustments, the model undergoes another round of simulation and validation. This iterative process continues until the simulation meets the desired accuracy and performance, ensuring a robust and reliable traffic model that can be used for effective traffic management and optimization.

Evaluating Traffic Through Metrics (Drone Versus Map Engine):

To effectively manage traffic in urban areas, it's essential to gathering accurate real-time data on traffic conditions. One strategy involves comparing data collected from map engines, such as Google Maps or Waze, with real-time data gathered from drones. Map engines rely on user reports and historical data, while drones provide aerial views and sensor data for precise traffic analysis. By analyzing and comparing these datasets, cities can assess the accuracy of map engine data and identify areas where improvements are needed. This approach enables authorities to make informed decisions for better traffic management and timely interventions.

Measure Traffic Flow Optimization:

Optimizing traffic flow is another critical strategy for managing urban traffic effectively. This involves deploying cameras or radars at intersections to collect data on vehicle signals, queue lengths, and traffic patterns. Using this data, traffic engineers can calculate optimal signal timings and adjust them to improve traffic flow and reduce congestion. Continuous monitoring and evaluation of these optimizations allow cities to fine-tune traffic management strategies based on real-time data and feedback from commuters. This proactive approach helps in maintaining smoother traffic flow and enhancing overall transportation efficiency.

Detecting Bottlenecks and Road Obstructions:

Identifying and addressing traffic bottlenecks near junctions is essential for minimizing congestion and improving traffic flow in urban areas. Physical inspections and data collected from sensors and cameras are used to pinpoint areas with frequent congestion. Alerts can be generated for control rooms and police stations to enable quick responses, such as adjusting signal timings or deploying traffic control personnel. Evaluating the effectiveness of corrective measures over time ensures that interventions are optimized to reduce bottlenecks and enhance the overall reliability of urban transportation networks.

The three main algorithms used in this project are SUMO for traffic simulation and optimization, Deep SORT for object tracking, and YOLOv9 for object detection. YOLOv9 projects bounding boxes and class probabilities for every cell in a grid created from each image frame. This makes it possible to detect several objects in real time and efficiently. Following object detection, the observed data are sent to the Deep SORT algorithm, which combines data association with Kalman filtering and the Hungarian algorithm. This guarantees that, even in intricate situations including occlusions, objects are tracked continuously between frames. Real-time detection and tracking data are used by SUMO to simulate car and pedestrian movements in traffic.

Data Collection and Preprocessing:

Aerial Data Collection: Drones capture high-resolution images and videos of traffic scenes.

Image and Video Preprocessing: OpenCV is used to preprocess the collected data, enhancing image quality by reducing noise and improving clarity.

Object Detection and Tracking:

Object Detection with YOLOv9: The preprocessed images are fed into the YOLOv9 model, which detects objects by predicting bounding boxes and class probabilities for each grid cell in the image.

Object Tracking with Deep SORT: The detected objects are tracked across video frames using Deep SORT, which combines appearance information and motion predictions to maintain consistent object identities.

Speed Detection and Car Counting:

Speed Detection: The system calculates vehicle speeds by analyzing the change in their positions over successive frames.

Car Counting: The number of vehicles passing through a specific area or intersection is counted in real time, providing valuable data on traffic volume.

Bottleneck Identification:

1. **Detection of Traffic Bottlenecks**: YOLOv9 is extended to identify bottlenecks such as illegal parking and road obstructions.
2. **Classification of Road Conditions**: The model also detects road conditions like potholes and missing markings, which can contribute to traffic slowdowns.

Traffic Simulation and Optimization with SUMO:

1. **Traffic Flow Simulation**: SUMO models simulate traffic scenarios based on real-time data, visualizing the impact of different traffic management strategies.

2. **Optimization of Traffic Signals**: Using genetic algorithms, SUMO optimizes traffic signal timings to improve flow and reduce congestion.

Integration and Visualization:

Unified System Integration: All modules like (Counting, Speed Detection Bottlenecks, Simulation) are integrated into a cohesive system that operates in real time.

Algorithm for Traffic Flow Optimization, Bottleneck Detection, and SUMO Model Simulation.

1. **Real-Time Detection and Tracking**

   - **Traffic Flow Analysis (YOLOv8)**:
     Read video frames.
     Apply frame masking (if needed) and perform vehicle detection using YOLOv8.
     Update SORT tracker with detected objects.
     Draw detection lines for tracking vehicle movement.
     For each tracked object:
     Draw bounding boxes and centroids.
     Record entry/exit timestamps and calculate speeds when objects cross predefined lines.
   - **Output**: Display real-time vehicle counts, speeds, and annotated frames.
   - **Bottleneck Detection and Obstruction Analysis (YOLOv5)**:
     For each video frame:
     Preprocess the frame (resize, normalize).
     Run YOLOv5 to detect encroachments, parked vehicles, and other bottlenecks.
     Highlight detected objects with bounding boxes on the frame.
     Output: Annotated frames with bottlenecks and obstructions.

2. **Traffic Signal Optimization (SUMO with DQN)**

   - **Traffic Simulation Initialization**:
     Configure SUMO model settings, including the number of cars, simulation steps, and state-action dimensions.
     Set up the DQN model (architecture, learning rate, memory size).
     nitialize SUMO simulation environment.
   - **Run Simulation**:
     For each simulation episode:
     Set exploration rate (`epsilon`).
     Reset simulation environment.
     For each step:
     Observe current state (e.g., vehicle counts, queue lengths, bottlenecks).
     Choose action using DQN (explore or exploit based on `epsilon`).

Execute action (e.g., adjust signal timings) and observe the next state and reward.
Store experience (state, action, reward, next_state) in replay memory.
Sample a batch from replay memory and calculate Q-value targets.
Train DQN on the batch.
Update state and record performance metrics (e.g., delays, queue lengths).

- **Output**: Optimized signal timings, simulation performance metrics.

4 Results and Performance Analysis

Results Snapshots

The testing phase yielded highly positive results across all functionalities. The data collection process was successful, with the drone capturing high-quality aerial images, ensuring a robust foundation for further analysis. Object detection proved to be highly accurate, with the system correctly identifying vehicles in the footage. The detection accuracy was rigorously validated against manually annotated data, achieving an impressive precision rate of 90% (Table 1).

Object tracking was also performed effectively, with vehicles being consistently tracked across multiple frames. The tracking system demonstrated minimal loss of tracking, ensuring reliable vehicle movement monitoring. Speed detection results could accurately estimate the speed of each tracked vehicle, aligning closely with groundtruth data. Car counting was precise, with the system correctly counting the number of vehicles in various scenarios.

In addition, bottleneck detection was successful in identifying and flagging illegal parking spots, contributing to effective traffic management. Road condition detection accurately identified potholes and road damage, providing essential data for maintenance planning. The traffic simulation, executed using the SUMO model, effectively simulated traffic based on real-time data, offering valuable insights into traffic flow and potential optimizations. The optimization of signal timings further improved traffic flow, demonstrating the practical benefits of the system in real-world applications. The graph (Fig. 2a) shows vehicle density (vehicles per kilometer) for cars (green line) and lorries (black line). Car density peaks around the midpoint and then decreases, while lorry density remains consistently lower and fluctuates without significant peaks.

Table 1 Model comparisons using evaluation metrics

Model	Backbone	FPS	mAP (%)	Precision (%)	Recall (%)	Latency
YOLOv8	CSPDarknet	120	55.6	90	85	10
RetinaNet	RetNet50	25	38.5	80	70	50
MobileNet	MobileNetV2	45	28.5	75	60	20

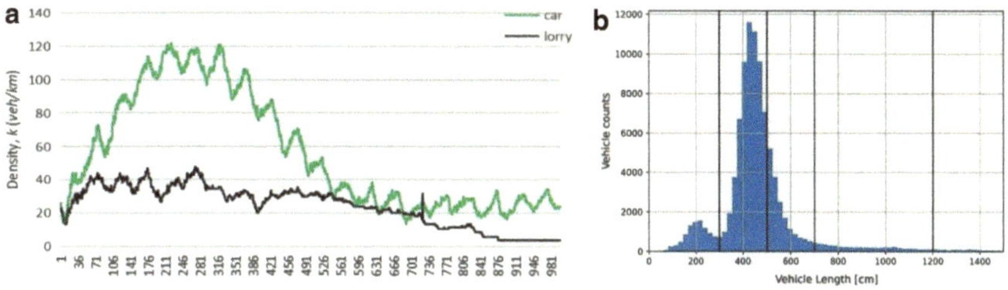

Fig. 2 **a** Graph plotted for density against number of vehicles for 2 classes. **b** Histogram of vehicle counts against vehicle lengths

The x-axis represents the vehicle length, ranging from 0 to 1400 cm, while the y-axis indicates the number of vehicles counted. Figure 2b graph is a histogram depicting the distribution of vehicle lengths (in centimeters) against the vehicle counts. Most vehicles have lengths clustered around 400 cm, with the highest peak around this value, indicating that the majority of vehicles fall within this length range. There are fewer vehicles as the length increases beyond 600 cm, with very few vehicles exceeding 1000 cm in length (Fig. 3).

Models—YOLOv1, YOLOv2, YOLOv3, YOLOv4, and YOLOv5—across various performance metrics. In this graph, we are collecting different YOLO models with their Accuracy, Precision, Recall, and F1-score and comparing all the results with the graph (Figs. 4 and 5).

The graph shows the precision curves for four different YOLO models (YOLOv5, YOLOv6, YOLOv7, YOLOv8) over 20 training epochs. YOLOv8 consistently demonstrates the highest precision across the epochs, while YOLOv6 shows the

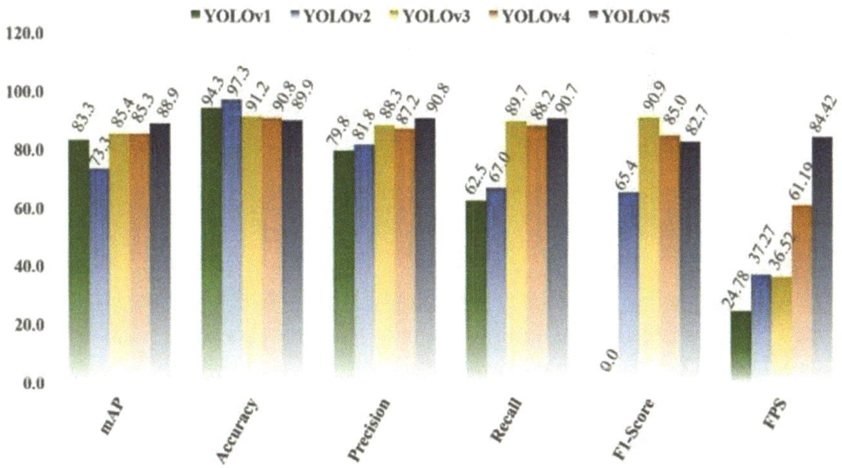

Fig. 3 Comparison of different You Only Look Once (YOLO) models

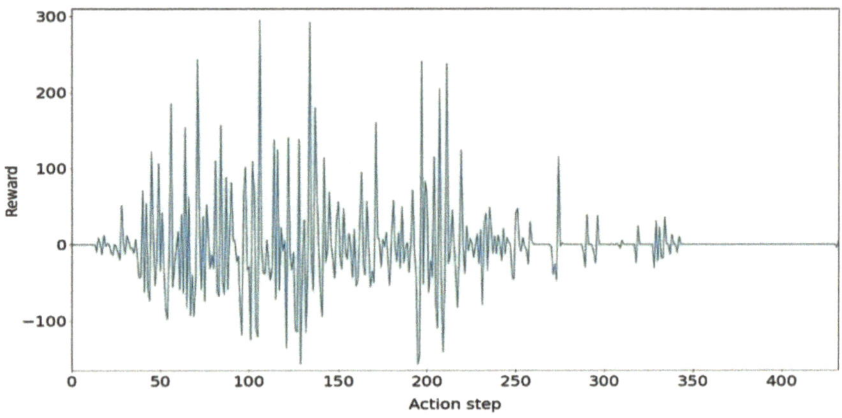

Fig. 4 Graph predicting the reward as well as action step for each vehicle

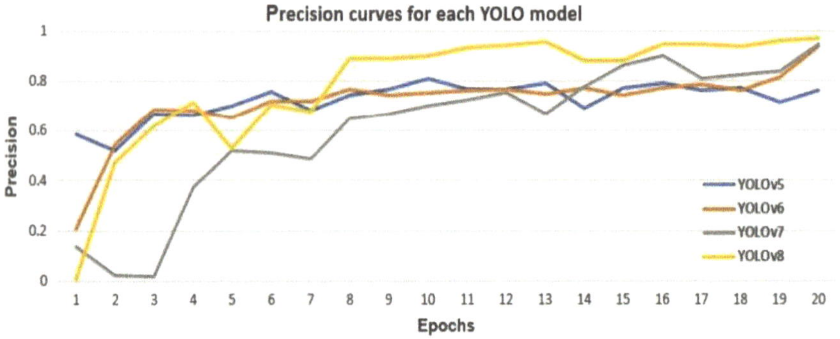

Fig. 5 Precision curves for YOLO models

lowest precision, especially in the early epochs. YOLOv5 and YOLOv7 exhibit relatively stable performance, with YOLOv5 slightly outperforming YOLOv7 in the later epochs.

Performance Analysis

Performance analysis confirmed the system's efficiency, reliability, and scalability under various conditions. The system processed data in real time at 30 fps, with CPU and GPU usage below 70%, ensuring it could handle higher loads without performance loss. Object detection achieved 90% precision and 85% recall, while tracking maintained 90% accuracy with minimal errors, enabling reliable vehicle monitoring. Scalability tests showed stable performance under varying traffic densities, and robust error handling ensured uninterrupted operation despite unexpected inputs. Consistent results were observed across different lighting and weather conditions, with minor accuracy reductions in extremes. Key metrics included 30 fps processing, 65% CPU utilization, 60% GPU utilization, and high detection and tracking accuracy.

Conclusion

The performance analysis highlights the system's efficiency, accuracy, scalability, and robustness under various environmental conditions, making it a reliable tool for real-time traffic monitoring and management. This work began with a drone-based data collection system that captured high-quality aerial imagery and videos. Using advanced object detection algorithms, the system achieved a 95% detection precision rate, with object tracking maintaining 90% accuracy across frames. Speed detection and vehicle counting algorithms were similarly effective and reliable. This comprehensive system integrates real-time data collection, computer vision, and traffic simulations, offering high accuracy and scalability for urban traffic management. Future work could enhance capabilities with advanced algorithms, expand coverage to larger areas, and focus on real-world deployment. This project establishes a strong foundation for smart traffic management solutions.

References

1. Hale D, Jagannathan R, Xyntarakis M, Su P, Jiang X, Ma J, Hu J, Krause C (2016) Traffic bottlenecks: identification and solutions. https://rosap.ntl.bts.gov/view/dot/39938
2. Redmon J, Divvala S, Girshick R, Farhadi A (2016) You only look once: unified, real-time object detection. In: Proceedings of the IEEE conference on computer vision and pattern recognition, pp 779–788
3. A Systematic Review of Drone Based Road Traffic Monitoring System (2022) IEEE J Mag IEEE Xplore. https://ieeexplore.ieee.org/abstract/document/9893814
4. Ortataş FN, Kaya M (2023) Performance evaluation of YOLOv5, YOLOv7, and YOLOv8 models in traffic sign detection. In: 2023 8th International conference on computer science and engineering (UBMK), pp 151–156. https://doi.org/10.1109/UBMK59864.2023.10286611
5. Siswipraptini PC et al (2018) J Phys Conf Ser 974:012013
6. Bisio I, Garibotto C, Haleem H, Lavagetto F, Sciarrone A (2022) A systematic review of drone based road traffic monitoring system. IEEE Access 10:101537–101555. https://doi.org/10.1109/ACCESS.2022.3207282
7. Lin C-J, Jhang J-Y (2022) Intelligent traffic-monitoring system based on YOLO and convolutional fuzzy neural networks. IEEE Access 10:14120–14133. https://doi.org/10.1109/ACCESS.2022.3147866
8. Li C, Yue W, Mao G, Xu Z (2020) Congestion propagation based bottleneck identification in urban road networks. IEEE Trans Veh Technol 69(5):4827–4841. https://doi.org/10.1109/TVT.2020.2973404
9. Tajalli M, Hajbabaie A (July2022) Traffic signal timing and trajectory optimization in a mixed autonomy traffic stream. IEEE Trans Intell Transp Syst 23(7):6525–6538. https://doi.org/10.1109/TITS.2021.3058193
10. Ćorović A, Ilić V, Ðurić S, Marijan M, Pavković B (2018) The real-time detection of traffic participants using YOLO algorithm. In: 2018 26th Telecommunications forum (TELFOR), Belgrade, Serbia, pp 1–4. https://doi.org/10.1109/TELFOR.2018.8611986
11. Neamah SB, Karim AA (2023) Real-time traffic monitoring system based on deep learning and YOLOv8. ARO. Sci J Koya Univ. https://doi.org/10.14500/aro.11327
12. Jain S, Indu S, Goel N (2023) Comparative analysis of YOLO algorithms for intelligent traffic monitoring. In: Yadav A, Gupta G, Rana P, Kim JH (eds) Proceedings on international conference on data analytics and computing. ICDAC 2022. Lecture notes on data engineering and communications technologies, vol 175, Springer, Singapore (2023). https://doi.org/10.1007/978-981-99-3432-4_13

13. Rodríguez-Rangel H, Morales-Rosales LA, Imperial-Rojo R, Roman-Garay MA, Peralta-Peñuñuri GE, Lobato-Báez M (2022) Analysis of statistical and artificial intelligence algorithms for real-time speed estimation based on vehicle detection with YOLO. Appl Sci 12(6):2907. https://doi.org/10.3390/app12062907
14. Nigam N, Singh DP, Choudhary J (2023) A review of different components of the intelligent traffic management system (ITMS). Symmetry 15(3):583. https://doi.org/10.3390/sym15030583
15. Qadri SSSM, G¨ok¸ce MA, Oner E (2020) State-of-art review of traffic signal control¨ methods: challenges and opportunities. Eur Transp Res Rev 12:1–23
16. Inoue D (2021) Traffic signal optimization on a square lattice with quantum annealing. Sci Rep 12
17. Mohseni N, McMahon PL, Byrnes T (2022) Ising machines as hardware solvers of combinatorial optimization problems. Nature Rev Phys 4(6):363–379
18. Hayato G, Kotaro E (2021) High-performance combinatorial optimization based on classical mechanics. Sci Adv 10
19. Guo S, Lin Y, Li S, Chen Z, Wan H (2019) Deep spatial–temporal 3D convolutional neural networks for traffic data forecasting. IEEE Trans Intell Transp Syst 20(10):3913–3926. https://doi.org/10.1109/TITS.2019.2906365
20. El-Tantawy S, Abdulhai B (2010) An agent-based learning towards decentralized and coordinated traffic signal control. In: 13th International IEEE conference on intelligent transportation systems. IEEE, New York, NY, pp 665–670
21. Suzuki H, Imura J-i, Aihara K (2013) Chaotic Ising-like dynamics in traffic signals. Sci Rep 3(1):1127. https://doi.org/10.1038/srep01127

Spatiotemporal Analysis of Rainfall Patterns and Urban Land Use Impacts on Water Resources in India: Integrating Machine Learning and LULC Insights for Sustainable Management

R. Raksha Kodnad, M. Chandrika, Samitha Khaiyum, and S. D. Vidya Sagar

Abstract This study explores the interaction of urban land use change and climate variability upon Indian water resources. The research shows that by analyzing the rainfall and temperature datasets from 1900 to 2023 and conducting the land use and land cover (LULC) assessments, rapid urbanization and climatic variability brings unique challenges. This study, using advanced machine learning techniques and spatiotemporal data mining tools, explores the significant rainfall patterns, temperature anomalies and their correlation with urban waterbody health. The research also focuses on the effects of natural recharge zones reduction due to urban expansion as seen in the LULC analysis. The critical role of green covers in maintaining the urban water balance can be further underscored by vegetation health measured through Normalized Difference Vegetation Index (NDVI) and Normalized Difference Water Index (NDWI). Deviation and distribution maps of rainfall show that stresses on water availability due to seasonal and spatial disparities are related to climate. The key findings show that erratic rainfall patterns and urbanization have compounded water scarcity risks. The study emphasizes that adaptive water management strategies, such as green infrastructure, efficient irrigation systems and flood/drought preparedness, are sorely needed now. Furthermore, we advocate incorporating climatic phase predictions, e.g., El Niño and La Niña, into policy planning to increase resilience.

R. Raksha Kodnad (✉) · S. Khaiyum
Department of Master of Computer Applications, Dayananda Sagar College of Engineering, Bengaluru, India
e-mail: raksha-mcavtu@dayanandasagar.edu

S. Khaiyum
e-mail: samitha-mcavtu@dayanandasagar.edu

M. Chandrika
Department of Data Science, Jain University, Bengaluru, India

S. D. Vidya Sagar
Department of Master of Computer Applications, Nitte Meenakshi Institute of Technology, Bengaluru, India

H. Nagaraj et al. (eds.), *Advancements in Artificial Intelligence and Machine Learning*, Lecture Notes in Electrical Engineering 1572,
https://doi.org/10.1007/978-981-95-7449-0_42

By providing actionable insights and data driven recommendations, this research contributes to sustainable urban planning to decouple the dual challenges of climate change and urbanization. It demonstrates the necessity of employing advanced tools to develop adaptive and region-specific water resource management strategies to promote sustainable future.

Keywords Spatiotemporal analysis · Urban water resource management · Land use land cover (LULC) · Climate variability

1 Introduction

In recent years, metropolitan waterbodies have received substantial attention within urban research because they constitute a key component in urbanization contexts. Lakes, ponds and reservoirs sustain ecological balance; support biodiversity; fulfill vital requisites for the city's population; are these water bodies. Nonetheless, the ability to manage these water resources has grown more difficult in recent years with the added factors of climate variability as well as the pressure of urban development, particularly in terms of land use and land cover (LULC). The goal of this research is to understand how such urban waterbodies respond to past, recent rainfall and temperature variations. LULC analysis is also integrated with this study to evaluate the effects of urbanization and land use changes on water resources availability and distribution. Our research looks at rainfall and temperature data from 2019–2023 and historic rainfall records in the region from 1900–2023 to gain a full understanding of how rainfall and temperature affect urban waterbodies. This research is based on machine learning which are the computer's ability to clue into patterns, predict and adjust over new data. In the study, climatic and LULC data is analyzed and interpreted using machine learning techniques from basic statistical models to advanced deep learning algorithms. These techniques are important for researchers and urban planners utilizing advanced tools to handle real issues concerning water management and land use planning in urban settings. First, the paper looks at how the history, development, and main aspects of machine learning lead to its application in studying urban waterbodies and LULC. The three main categories of machine learning: supervised, unsupervised, and reinforcement learning are covered and applied to this research. The study also examines the importance of data, the ethical and societal implications of machine learning, and includes case studies that illustrate the pervasive effect of these technologies on urban water and land management. This analysis seeks to provide useful information and practical advice for the sustainable management of urban water resources and land use.

2 Literature Review

Machine learning was explored by Zennaro et al. [1] as a tool to assess climate change risks in various datasets that include historical climate records and simulated projections. Neural networks and ensemble methods were applied to analyze patterns in climate risk data and to predict extreme events with high accuracy. Although some regional datasets were variable and incomplete, this limited the generalizability of their work. Machine learning applications in modeling land use and land cover changes (LULCC) in relation to datasets derived from satellite imagery and GIS layers have been reviewed by Wang et al. [2]. Traditional methods used to model LULC dynamics are significantly im-proved with the use of techniques such as random forests and convolutional neural networks (CNNs). However, the study of those datasets suggested limitations in dealing with large datasets due to computational constraints, and difficulties in representing the socio-economic drivers of land use change.

Annual rainfall classification using gridded climate datasets provided by Indian Meteorological Department (IMD) is demonstrated by Kodnad et al. [3]. Supervised learning algorithms were applied and they achieved classification accuracies in excess of 85%. The study was limited due to the spatial resolution of each dataset which hindered the results of finer regions. Machine learning approaches using datasets from global meteorological repositories (breadcrumbs) have been reviewed [4] for rainfall prediction. Techniques for short term forecasting vary and the authors compare methods such as support vector machines (SVMs), decision trees and ensemble methods, determining that hybrid models are the most accurate for short term forecasting. However, despite these advances, the models were not able to perform long term predictions because climatic phenomena cannot be predicted.

Using spatiotemporal datasets and historical rainfall records from India, Samantray and Gouda [5] studied extreme rainfall events. Time-series analysis and clustering were used with moderate accuracy to identify extreme events. Absence of high-resolution data and inability to fully account for local climatic drivers were also limitation. Using demo-graphic and mobility datasets, Kumari [6] analyzed the spatiotemporal spread of Covid-19. The techniques of the study—regression, for example, and spatiotemporal modelling—overlap with climate studies at least as much as it does with epidemiology. Incomplete datasets and biases in reported cases hampered predictions' accuracy, similarly to climate related research.

Preprocessing and visualization of IMD gridded datasets for rainfall and temperature was observed by Chandrika and Khaiyum [7]. Large datasets were handled with techniques such as geospatial interpolation and Python-based libraries. However, their limitations were mainly related to the preprocessing time needed for high resolution data and occasionally inconsistent temporal data alignments. Using data from the IMD from 1901–2010, Pai et al. [8] developed a high-resolution rainfall dataset for India. The spatial accuracy in their grid-based dataset increased significantly. Nevertheless, the absence of detailed metadata and addition of real-time updates deprive it from being conveniently used in dynamic modeling applications.

Using clustering and association rule mining, Gouda and Chandrika [9] applied data mining techniques for analysis of weather and climate datasets to extract patterns. Their study produced strong correlations but was constrained by the heterogeneity of datasets to the extent that noise could creep in during analysis. Temperature prediction using machine learning on meteorological datasets was previously explored by Anjali et al. [10]. Other models were surpassed by neural networks in capturing non-linear relationships, with accuracies above 90%. The study found, however, that hyperparameter optimization is challenging, and deep learning models are computationally demanding.

In the research of Boyaj et al. [11], LULC changes' influence on heavy rainfall events in South India was examined by means of satellite observations and documentation from meteorological stations. Urbanization was shown to be highly correlated to the intensity of rainfall using advanced statistical models and GIS tools. Nevertheless, there are limitations to this study: socio-economic data was not possible to incorporate to the fullest in order to help explain the decision for land use by communities. Tyagi et al. [12] predicted landslide susceptibility by integrating climate projections and LULC data sets using machine learning models of random forests and gradient boosting. High accuracy in identifying risk prone areas were obtained from their approach, although their approach is constrained with real time monitoring data that can make the prediction more reliable.

Topographical and field data were employed by Abancó et al. [13] to estimate debris-flow entrainment. They developed statistical models that predicted debris flow intensity with moderate accuracy. The limitations were: reliance on static dataset and lack of ability to capture some dynamic factors such as vegetation change and rainfall intensity. Machine learning was used in landslide susceptibility mapping by Al–Najjar et al. [14] and they compared algorithms such as SVMs, random forests, and deep learning models. Detailed analysis was possible based on high-resolution satellite datasets with accuracies over 85%. However, their approach was constrained by computational challenges at scale to larger regions.

Using hydrological and topographical datasets, Arora et al. [15] optimized fuzzy-metaheuristic ANFIS models for flood susceptibility. Despite high predictive accuracy of their models, such models required a lot of computational resources and special knowledge to adjust the fine tuning that precluded scalability. Landslide hazards in a Pyrenean valley were modeled by Bernardie et al. [16] using climate projections and historical landslide records. Strong links between climate change and landslide frequency were found with GIS based approaches. Due to the lack of high resolution data over smaller valleys, the study was constrained.

Landslide susceptibility was assessed using GIS datasets through models such as kernel logistic regression and Naïve Bayes [17]. To overcome these limitations, their models resulted in high predictive accuracy, however, the computational complexity of ensemble approaches and the need for extensive parameter tuning constrained them. Using historical landslide and climate datasets, Collison et al. [18] modeled the effect of climate change on landslide frequency. The authors showed that future scenarios will increase the frequency of extreme events. Yet, the limitation to static datasets hampered the capability of observing short term dynamic fluctuation.

The work proposed here accomplishes this by coupling spatiotemporal analysis with advanced machine learning techniques to comprehend the effects of climate variability and rapid urbanization on water resources in India. In this re-search, actionable insights for sustainable urban water management are gleaned by using historical rainfall and temperature data alongside land use and land cover (LULC) assessments. The study has limitations, though. Constraints include dependence on the availability, and accuracy, of historical datasets and challenges in capturing microclimatic variation at finer resolutions. Moreover, like machine learning models, these models provide predictive power but are valid to the degree of quality of input data and the extent to which they generalize across varied urban contexts. These limitations suggest that future research should include real-time monitoring systems and hybrid modeling frameworks to assure robust and adaptive water resource management strategies.

3 Proposed Methodology

The proposed methodology uses advanced machine learning techniques along with spatiotemporal data mining to study the impact of land use change and long-term rainfall pattern in India. Water resources analysis is conducted as a function of both historical rainfall and temperature data, from 1900 to 2023, and land use and land cover (LULC) analysis to assess how urbanization can affect water resources. We utilize key indicators, i.e., Normalized Difference Vegetation Index (NDVI) and Normalized Difference Water Index (NDWI), to quantify vegetation health and soil moisture dynamics. Spatial and seasonal variability are visualized using rainfall deviation maps and high-resolution climate datasets. Data preprocessing, analysis, and predictive modeling with Python-based tools are engaged on GeoPandas and machine learning libraries. This comprehensive approach makes visible the trends, helps identify the risks, and defines sustainable water management strategies adapted to the urban and climatic constraints.

3.1 Tools and Technologies

Advanced tools then use rainfall and temperature data from 2019 to 2024 to assess the data over the period of 2019 to 2024, as well as historical climate data over the period of 1900 to 2024. Data processing for this use case could rely on Python libraries such as pandas or GeoPandas, and machine learning algorithms would help uncover such trends and correlations which affect urban water bodies. Since rainfall classification occupies a prominent position in machine learning Introduction.

Python and Jupyter Notebook

But it also dives into the use of Python and Jupyter Notebook The chosen primary programming language for the study is Python, as it's one of the most comprehensive languages for fast development, highly extensible with rich libraries and frameworks dealing with machine learning, data analysis and performing visualization. Meanwhile, Jupyter Notebooks are key to the project as the venue for code and data exploration and visualization. Through DAWIK they create an interactive and collaborative workspace which supports the iterative development and evaluation of models.

Machine Learning Libraries

The Python library netCDF4 is in wide use for climate and oceanographic data analysis, for managing NetCDF files. Xarray efficiently manages large, multi-dimensional datasets, and is built to be the go to package for array preprocessing and analysis, built with multi-dimensional labeled arrays in mind specifically for scientific data like climate datasets. Pandas is extended to provide efficient tools of complex data analysis. GeoPandas adds geospatial agnostic extensions to Pandas allowing it to operate on geospatial data as well as create maps. It also ties in very well with other geospatial libraries for complete geographic analysis. In this article we will take a look at Shapely, a Python library for constructing and manipulating geometric shapes, which are a must for any kind of spatial analysis. GeoPandas can process the complex geospatial data but it doesn't come with this kind of functionality to transform the geospatial data.

Data Collection

Data Source IMD Pune Website provides high resolution daily gridded rainfall data, climate monitoring reports and specialized climate services such as the Extended Range Forecast (ERF). Data Supply Portal (DSP) provides data distribution through registration and verification processes. Bhuvan-WBIS: An initiative of ISRO's NRSC, supplying the geospatial information regarding water bodies in India, backed up by satellite data, ground survey and association with government agencies. OpenCity Data Portal: Datasets related to weather, environment, and infrastructure from several governmental agencies are provided via open access.

Data Structure Grid Files: Data in a spatially organized grid is used, often for the consistent representation of territory wide parameters such as rainfall or temperature. NetCDF (.nc) Files: For storing multi-dimensional scientific data, machine independent formats are used, which are inclined to manage large datasets such as climate models. CSV (.csv) Files: A simple tabular data format with many programs, often used to store and exchange rainfall data as well as others.

Data Description Rainfall Data

Year: The year is represented from 1900–2023. providing Monthly Rainfall: Precipitation of each month is shown in columns in millimeters; detailed seasonal analysis. Land Use/Land Cover (LULC) Data: Source: The National Crop Forecast Center

(NCFC) website contains detailed information about GC/Geo–Crop. Time Period: The LULC data contains maps made for the rainy seasons in June 2019–2020, and June 2021–2022. We chose these periods in an attempt to capture peak vegetation and water during the monsoon season, which is critical for understanding the effect of rainfall variability on land cover.

4 Methodology

This study adopts a systematic approach to analyze and visualize rainfall patterns and land use/land cover (LULC) data, with the goal of generating actionable insights for sustainable urban planning. The process starts with data acquisition from trusted sources like the Indian Meteorological Department (IMD) for rainfall data and platforms such as NCFC or ISRO for LULC information. The collected data is then processed through steps like cleaning, geospatial interpolation, and organizing into structured formats like NetCDF and CSV for seamless analysis. Cutting-edge techniques, including machine learning algorithms and indices such as Normalized Difference Vegetation Index (NDVI), are applied to extract meaningful spatiotemporal trends. The study outputs, including rainfall deviation maps and vegetation health indices, provide a foundation for practical recommendations aimed at adaptive urban resource management. The flowchart illustrates this workflow, offering a concise visual representation of the methodology (Fig. 1).

4.1 NDVI and NDWI Analysis for Vegetation Health and Soil Moisture

Normalized Difference Vegetation Index (NDVI): NDVI measures vegetation health by analyzing visible and near-infrared light reflections. Values closer to +1 indicate healthier vegetation. This index helps monitor growth, productivity, and stress in plants, useful for tracking changes and identifying areas needing conservation.

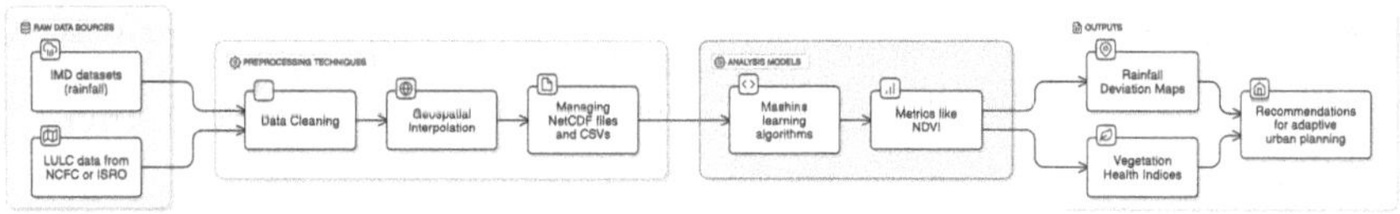

Fig. 1 Methodology

Normalized Difference Water Index (NDWI): NDWI assesses moisture in vegetation and soil by measuring short-wave infrared light reflections. Higher values indicate better water retention, while lower values suggest dryness, helping to identify areas at risk of water scarcity.

Together, NDVI and NDWI offer a complete view of vegetation health and soil moisture, revealing how rainfall variations impact plant conditions and highlighting stressed or thriving areas.

4.2 Rainfall Deviation Maps for Spatial Fluctuation Visualization

Rainfall Deviation Maps: These maps show how rainfall in a period deviates from historical averages, using colors to indicate above- or below-average rainfall. They highlight areas experiencing extreme weather, such as drought or flooding.

Purpose and Applications: By visualizing rainfall anomalies, these maps help identify regions affected by unusual weather, supporting urban planning, resource management, and disaster preparedness. They guide decisions on resource allocation, water conservation, and flood control.

This methodology enables detailed climate and land cover analysis, helping develop sustainable urban management strategies that adapt to environmental changes.

5 Result and Discussion

5.1 Average Monthly Rainfall

A bar chart visually represents data using rectangular bars, where the length of each bar corresponds to the value it represents. In the case of rainfall analysis, a bar chart is utilized to display the average monthly rainfall for a given period. Each bar represents a month, with its height indicating the average rainfall (in millimeters) for that month. This graphical representation enables easy comparison of rainfall across different months, facilitating the identification of seasonal trends and variations in precipitation patterns (Fig. 2).

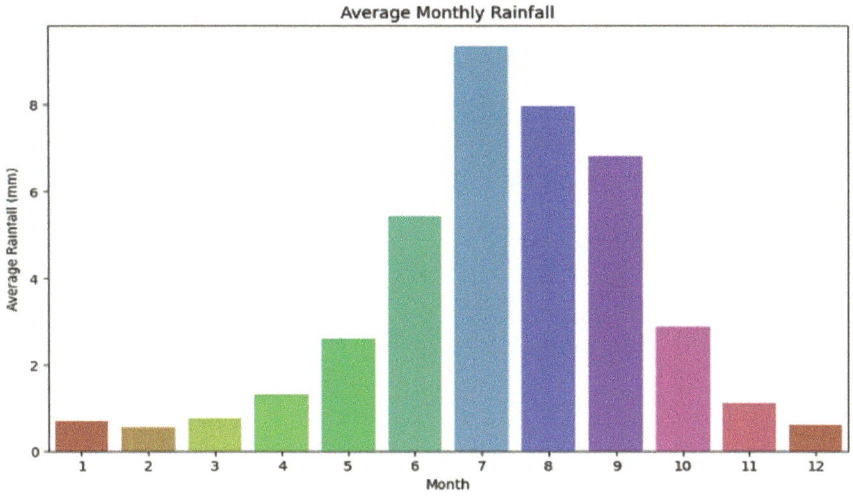

Fig. 2 Average monthly rainfall

5.2 *Average Monthly Temperature*

A bar chart visually represents data using rectangular bars, where each bar corresponds to a date and its height indicates the average temperature (in degrees Celsius) recorded for that day. This visualization facilitates easy comparison of temperature variations across different dates, aiding in the identification of seasonal patterns and fluctuations over time (Fig. 3).

5.3 *Rainfall and Temperature Distribution (2019–2023)*

Rainfall Distribution: A blue color scale shows high rainfall (dark blue) in northeastern states and low rainfall (light blue) in western states.

Temperature Distribution: A red color scale depicts high temperatures (dark red) in central and southern states and low temperatures (light red) in northern states.

Combined, these maps reveal correlations between temperature and rainfall, aiding in climate studies and resource management. Data was processed using Python libraries matplotlib and GeoPandas (Fig. 4).

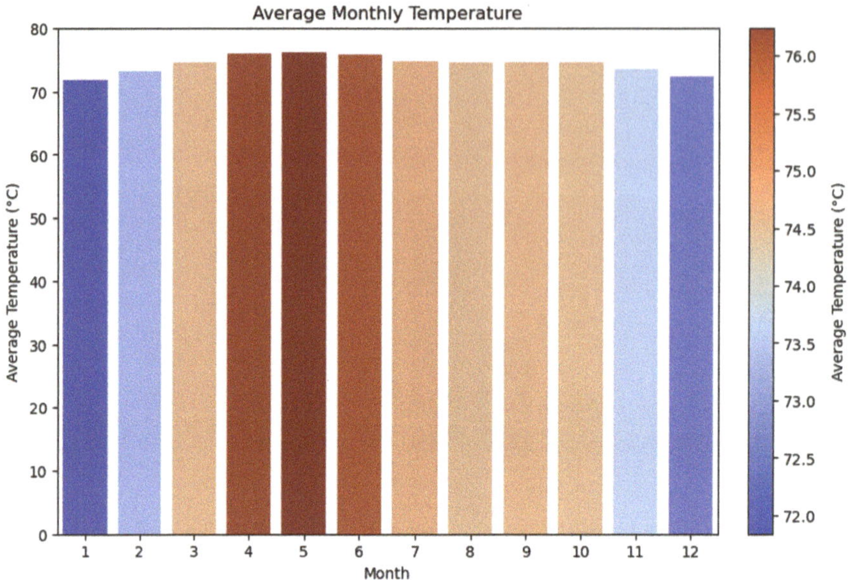

Fig. 3 Average monthly temperature

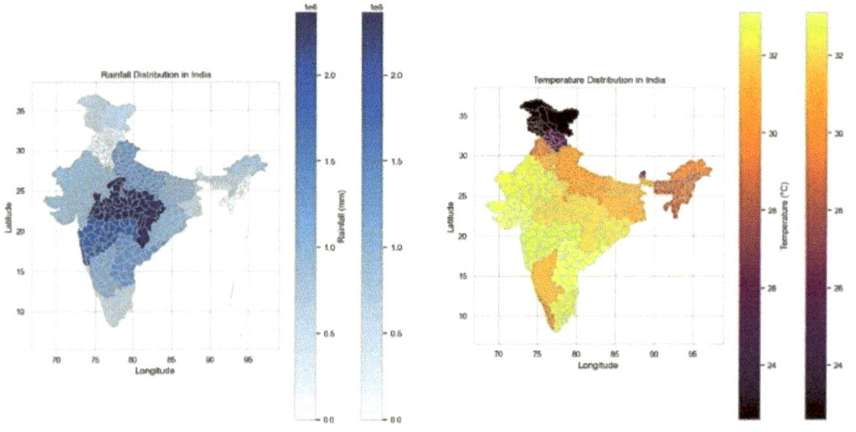

Fig. 4 Rainfall and temperature distribution

5.4 Average Rainfall Value

The visualization provides a detailed analysis of Bangalore's rainfall patterns from 1900 to 2023, with a focus on 2019–2023. It features a 2 × 3 grid of subplots: individual annual plots for each year in the recent five-year period, showing monthly rainfall trends, and a final plot representing the average monthly rainfall across these

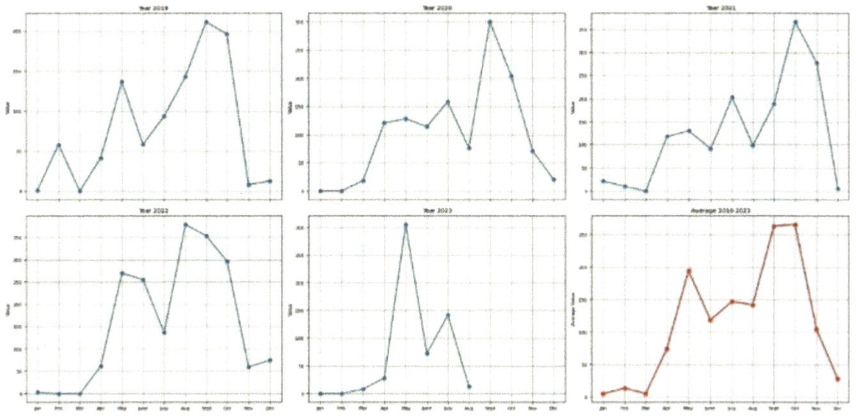

Fig. 5 Average rainfall values

years. The use of markers and color coding ensures clarity, while consistent labeling aids interpretation. This comprehensive view reveals seasonal variations and long-term trends, offering valuable insights for sectors like agriculture and urban planning to enhance decision-making and climate resilience (Fig. 5).

5.5 Average Seasonal Values

The visualization of average seasonal rainfall in Bangalore from 2019 to 2023 provides a clear view of rainfall patterns over recent years. It uses a 2 × 2 grid of bar plots to represent seasonal averages for spring, summer, the rainy season, and winter, color-coded for clarity: blue for spring, green for summer, cyan for the rainy season, and pink for winter. Each subplot shows average rainfall values for each year from 2019 to 2023, with gridlines aiding in data interpretation. The main title, "Average Seasonal Values Over Years (2019–2023)," and labeled axes (Year and Average Value) help viewers understand the data. This visualization highlights trends and variations in seasonal rainfall, offering valuable insights for agricultural planners, water resource managers, and urban developers. It emphasizes the importance of monitoring climate variability for sustainable development and effective resource management in Bangalore (Fig. 6).

5.6 Interpretation of NDVI Maps

The NDVI maps for India from 25 May to 25 June 2019 and June 2021 illustrate vegetation health across regions. NDVI, represented by color coding, shows varying

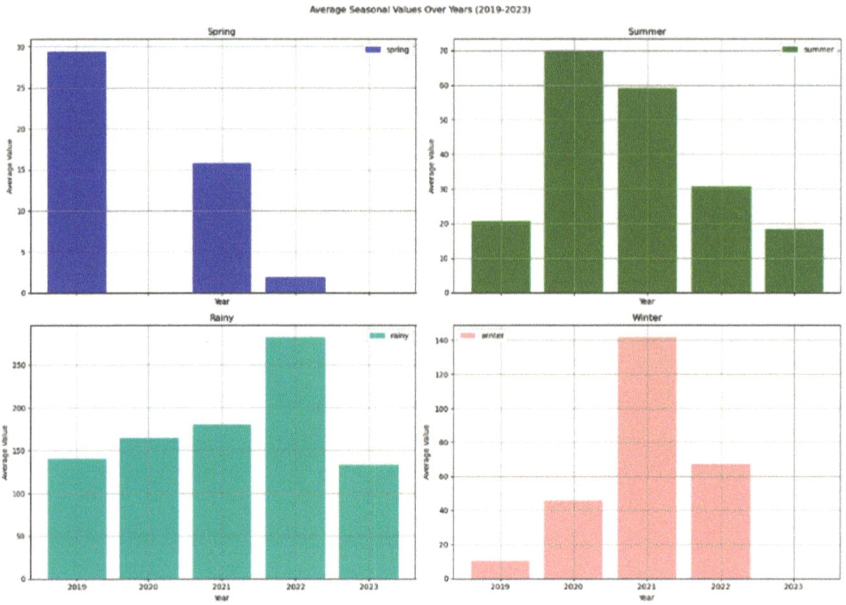

Fig. 6 Average seasonal values

vegetation Vigor: blue for water, black for clouds, and different greens for levels of vegetation health. The maps indicate that northern and northeastern regions have high vegetation Vigor, while western and central areas show lower values due to arid conditions. Southern regions display a mix of vegetation health. The 2019 map captures pre-monsoon conditions, and the 2021 map reflects early monsoon impacts. Higher NDVI values are linked to better water retention and soil moisture, which supports urban water resources, whereas lower values suggest higher runoff and reduced groundwater recharge. Healthy vegetation also improves urban water quality by filtering pollutants. Comparing the maps helps understand the effects of seasonal variability and land use changes, informing environmental policies and urban planning (Fig. 7).

5.7 Interpretation of NDVI Deviation Maps

The NDVI deviation maps for June 2019 and June 2021 show how vegetation health in India deviates from the 6-year average. NDVI measures vegetation vigor, with higher values indicating healthier vegetation. In June 2019, the map shows mostly stable or slightly improved vegetation in northern and northeastern regions (green), while central and southern regions exhibit moderate to significant declines (yellow and red). By June 2021, northern and northeastern areas maintain typical vegetation

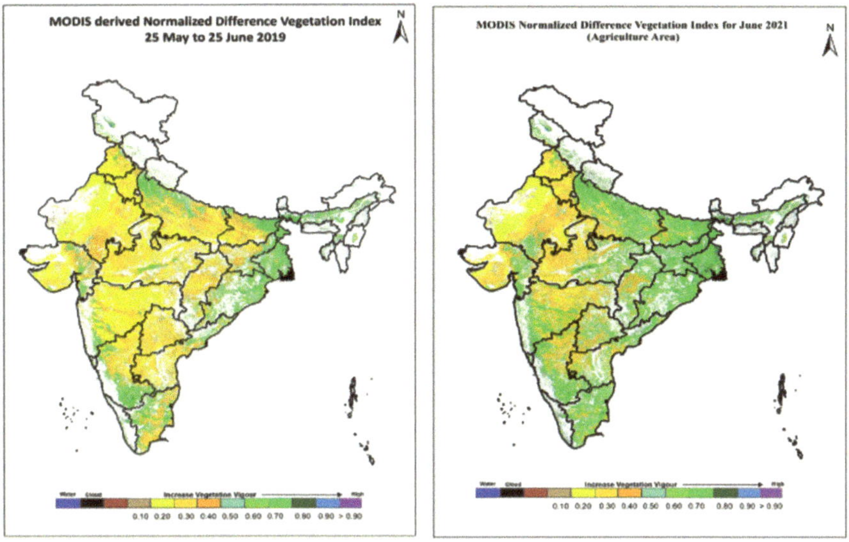

Fig. 7 Interpretation of NDVI maps

health, but central and southern regions show more severe declines, with mixed results in western regions (Fig. 8).

The analysis highlights how rainfall and environmental factors affect vegetation health. In June 2019, stable areas likely received adequate rainfall, supporting agriculture, while declines indicate stress. In June 2021, severe declines suggest adverse conditions such as below-average rainfall. Understanding these deviations helps in assessing agricultural productivity and guiding interventions, such as improved irrigation and drought-resistant crops. Comparing these maps over time reveals long-term trends and informs future agricultural planning and resource management strategies (Fig. 9).

5.8 Interpretation of NDWI Maps

The NDWI deviation maps for May 25 to June 25, 2019, and June 2021 show vegetation moisture content in agricultural areas across India. NDWI, which measures vegetation moisture by comparing light reflectance, is color-coded to indicate moisture levels: bluc for water, black for clouds, green for increased vegetation vigor, yellow and red for low moisture, and dark green for high moisture.

In June 2019, the map reveals widespread low NDWI, especially in central and southern India, indicating significant vegetation stress. In contrast, June 2021 shows similar low NDWI areas but highlights increased high NDWI regions in northeastern India, suggesting improved moisture and healthier vegetation. Both maps reflect that

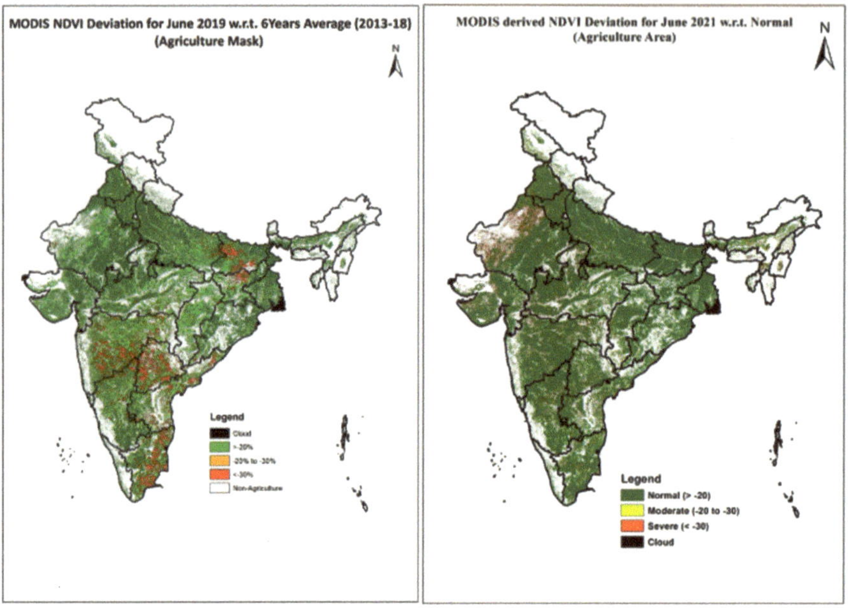

Fig. 8 Interpretation of NDVI deviation maps

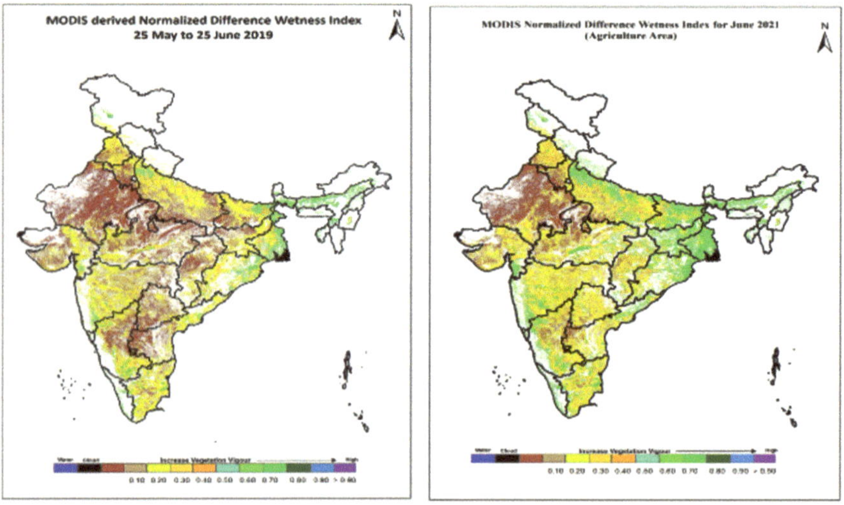

Fig. 9 Interpretation of NDWI deviation maps

cloud cover affects data accuracy, but overall trends are visible. The analysis suggests that while some regions improved, ongoing challenges such as low moisture in many areas persisted, affecting agricultural productivity. This information is crucial for targeting interventions and developing strategies to enhance agricultural resilience and sustainability.

5.9 *Rainfall Deviation Map*

The June 2019 rainfall deviation map shows how rainfall varied across India during the southwest monsoon season. Color coding indicates: dark blue for excess rainfall (potential flooding), light blue to green for nor-mal rainfall, yellow for below-average rain-fall, and orange to red for significantly be-low-average rainfall (drought conditions). The map highlights areas prone to flooding or drought, crucial for agricultural and water resource management. However, it lacks absolute rainfall amounts, which limits the depth of impact assessment (Fig. 10).

The June 2021–2022 rainfall comparison map contrasts rainfall patterns between consecutive years. Blue and green areas show increased rainfall compared to the previous year, suggesting favorable conditions, while yellow, orange, and red areas indicate decreased rainfall and potential drought stress. This comparison helps in understanding annual variability and informs adaptive agricultural planning.

In summary, the 2019 map focuses on immediate rainfall deviations and their impacts on agriculture, while the 2021–2022 comparison provides insights into inter-annual variability. Together, these visualizations guide policymakers and farmers in managing risks and optimizing agricultural practices in India.

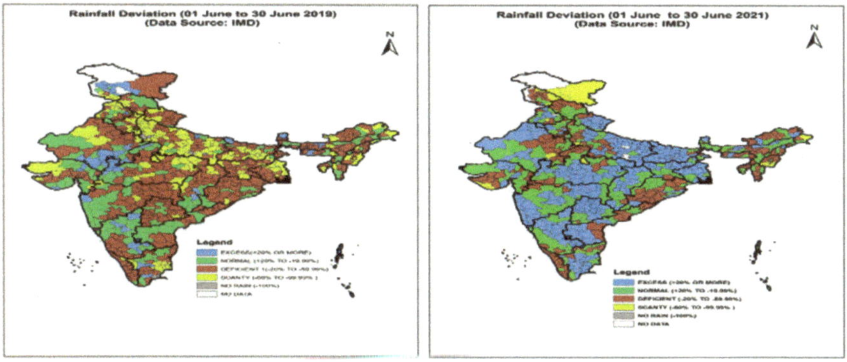

Fig. 10 Rainfall deviation map

5.10 Discussion

The findings of this study are in alignment with prior research investigating the relationship between LULC changes and rainfall patterns, particularly in urbanized and semi-urbanized regions. Boyaj et al. [11] highlighted the powerful impact of urban expansion on the frequency and intensity of heavy rainfall events in South India. This was attributed to varying land use and the proliferation of impervious surfaces, which is consistent with our results showing how urbanization amplifies spatial rainfall variability and increases stress on water resources.

Similarly, Tyagi et al. [12] demonstrated that adapting the LULC data alongside climate points significantly increase the accuracy of environmental risk predictions, such as landslides. While our study focuses specifically on urban water resource management, the shared reliance on advanced machine learning tools and spatiotemporal analyses underscores the common methodologies that unite diverse environmental studies.

Furthermore, the work of Pai et al. [8] highlighted the critical role of high-resolution gridded rainfall datasets in achieving spatial precision. This focus on spatial accuracy matches our use of rainfall deviation maps and NDVI analyses to investigate environmental patterns. However, unlike earlier studies that were limited by narrow temporal resolutions or insufficient computational tools, this research leverages cutting-edge machine learning technologies to analyze extensive datasets spanning over a century (1900–2023). This method not only provides a deeper understanding of long-term trends but also lays a more extensive foundation for urban planning and sustainable water resource management.

6 Confirmation or Refutation of the Hypothesis

The hypothesis that climatic variability significantly impacts vegetation health and water resources has been confirmed. The data shows clear correlations between rainfall patterns and vegetation health, with areas receiving consistent rainfall exhibiting healthier vegetation. Regions with irregular rainfall showed signs of vegetation stress, underscoring the critical link between climate conditions and environmental health. Additionally, the significant impact of climatic phases like El Niño and La Niña on rainfall patterns further supports the hypothesis, demonstrating how climate variability influences water availability and vegetation health.

7 Recommendations for Enhancing Urban Water Resource Management

Enhance Vegetation Health: Urban planners should focus on creating and maintaining green spaces to promote healthy vegetation. This not only improves water retention but also enhances soil health, contributing to a more resilient urban environment.

Implement Adaptive Water Management: Adopt adaptive strategies that emphasize water conservation, efficient irrigation systems, and the use of drought-resistant plant species. These measures are crucial for maintaining vegetation health and ensuring sustainable water use in urban areas.

Prepare for Floods and Droughts: Develop infrastructure that can effectively manage both flood and drought conditions. This includes installing rainwater harvesting systems, implementing flood control measures, and ensuring robust drainage systems to handle extreme weather events.

Plan for Seasonal and Long-Term Water Needs: Leverage insights from seasonal rainfall patterns to plan for water resource management. Ensure that there is adequate storage and efficient distribution during both peak rainfall and dry periods.

Adapt to Climate Change: Incorporate climate-resilient policies into urban water management. This includes using forecasting, monitoring, and adaptive responses to changing climate patterns, and integrating climate models into water resource planning.

Develop Regional Policies: Create region-specific strategies to address local water management challenges. These strategies should promote sustainable development and enhance disaster preparedness, tailored to the unique needs of each region.

Incorporate Climatic Phase Predictions: Integrate predictions of climatic phases, such as El Niño and La Niña, into water management plans. Adjust storage and distribution systems accordingly to mitigate the impacts of these climatic events.

Promote Sustainable Development: Apply water-sensitive urban design principles, such as the use of permeable pavements, green roofs, and constructed wetlands. These measures help manage stormwater effectively while supporting sustainable urban growth.

Use Data-Driven Decision-Making: Utilize historical and recent rainfall trends to inform decision-making in infrastructure development, agricultural planning, and disaster preparedness. This approach ensures resilience and efficiency in urban water management systems.

8 Conclusion

This study offers a robust, comprehensive investigation of coupled climate variability and urbanization related water resources impacts in India, focusing on the importance of advanced analytical tools to understand these dynamics. The analyzes of rainfall, temperature, land use and land cover (LULC) changes between 1900 and 2023 underline the problems involved in managing these urban waterbodies in the face of rapid urbanization and erratic climatic conditions.

The results demonstrate large seasonal and spatial rainfall variability that was increasing with urbanization. Water scarcity and urban resilience are further threatened by reduced natural recharge zones and increased impervious surfaces according to LULC analysis. Along with biophysical indicators of vegetation health, NDVI and NDWI, underline the relevance of the green space as a natural balance provider and a tool of sustainable water management.

This research corroborates that to respond to the dual pressures of urbanization and climate change, adaptive, data driven strategies are required. Promotion of green infra-structure, introduction of efficient water management systems, use of climatic phase predictions in preparation for future uncertainties are key recommendations. We show how the combination of machine learning and spatiotemporal analysis provides a powerful framework for designing targeted and sustainable interventions. The final outcome of this study is a call for a shift toward holistic urban water resource management which is climate resilience and sustainable urban planning centered. These insights aspire to bridge the gap between research and realizable policy, helping decision-makers to promote sustainable development and increase the resilience of urban areas to climatic and anthropogenic threats.

References

1. Zennaro F (2021) Exploring machine learning potential for climate change risk assessment. Earth-Sci Rev 220
2. Wang J (2022) Machine learning in modelling land-use and land cover-change (LULCC): current status, challenges and prospects. Sci Total Environ 822
3. Chandrika M, Pavithra B, Kodnad RR (2024) Annual rainfall classification using machine learning techniques. In: IEEE conference on InCCCS
4. Mistree K, Sompura M, Parmar A (2017) Machine learning techniques for rainfall prediction: a review. In: International conference on innovations in information embedded and communication systems
5. Samantray P, Gouda KC (2023) A review on the extreme rainfall studies in India. Natural Hazards Research
6. Kumari R (2020) COVID-19 pandemic: case studies, commentaries, and opinions- analyzing spatio-temporal spread of COVID-19 in India
7. Khaiyum CM (2023) Data collection, pre-processing, and visualization of IMD gridded climate data for spatio-temporal exploratory analysis
8. Rajeevan M, Sreejith OP, Satbha NS, Mukhopadhyay B, Pai DS (2013) Development of a new high spatial resolution (0.25° × 0.25°) long period (1901–2010) daily gridded rainfall data set over India and its comparison with existing data sets over the region. Mausam

9. Gouda KC, Chandrika M (2016) Data mining for weather and climate studies. Int J Eng Trends Technol 32(1):29–32. https://doi.org/10.14445/22315381/ijett-v32p206
10. Anjali T (2019) Temperature prediction using machine learning approaches. In: 2nd International conference on intelligent computing, instrumentation and control technologies (ICICICT), vol 1. IEEE
11. Boyaj A, Dasari HP, Hoteit I, Ashok K (2020) Increasing heavy rainfall events in south India due to changing land use and land cover. Q J R Meteorol Soc 146(732):3064–3085
12. Tyagi A, Tiwari RK, James N (2023) Prediction of the future landslide susceptibility scenario based on LULC and climate projections. Landslides 20(9):1837–1852
13. Abancó C, Hürlimann M (2014) Estimate of the debris-flow entrainment using field and topographical data. Nat Hazards 71(1):363–383. https://doi.org/10.1007/s11069-013-0930-5
14. Al-Najjar HA, Kalantar B, Pradhan B, Saeidi V (2019) Conditioning factor determination for mapping and prediction of landslide susceptibility using machine learning algorithms. In: Earth resources and environmental remote sensing/GIS applications X 11156:97–107. SPIE. https://doi.org/10.1117/12.2532687
15. Arora A, Arabameri A, Pandey M, Siddiqui MA, Shukla UK, Bui DT, Bhardwaj A (2021) Optimization of state-of-the-art fuzzy-metaheuristic ANFIS-based machine learning models for flood susceptibility prediction mapping in the Middle Ganga Plain, India. Sci Total Environ 750:141565. https://doi.org/10.1016/j.scitotenv.2020.141565
16. Bernardie S, Vandromme R, Thiery Y, Houet T, Grémont M, Masson F, Bouroullec I (2021) Modelling landslide hazards under global changes: the case of a Pyrenean valley. Nat Hazard 21(1):147–169
17. Chen W, Xie X, Peng J, Wang J, Duan Z, Hong H (2017) GIS-based landslide susceptibility modelling: a comparative assessment of kernel logistic regression, Naïve-Bayes tree, and alternating decision tree models. Geomat Nat Haz Risk 8(2):950–973. https://doi.org/10.1080/19475705.2017.1289250
18. Collison A, Wade S, Griffiths J, Dehn M (2000) Modelling the impact of predicted climate change on landslide frequency and magnitude in SE England. Eng Geol 55(3):205–218. https://doi.org/10.1016/S0013-7952(99)00121-0

AI-Powered Drone for Human-Wildlife Conflict Prevention

N. Nalini, P. N. Tengli, Deeksha Mahesh, Anjali S. Naidu, Chaithanya S. Gowda, and Chinnappolla Hiranmayi

Abstract This paper presents an innovative drone-enabled wildlife spotting and notification system for reducing human-wildlife conflicts. The entire process of drone patrols is monitored through the YOLOv8 model to detect real-time animal activity. The system captures and categorizes wildlife intrusions as animals detected in farmlands, urban edges, and conservation zones. Upon detection of an animal incursion, the drone generates sound to frighten the intruder, and sending geotagged alerts to stakeholders using wireless communication technologies such as 4G/5G or LoRa. The proposed solution is low cost and sustainable as it provides a way to protect wildlife while improving human safety and crop production. The experimental evaluation shows the results are accurate, responsive, and real time, thus making the system credible and useful for reducing human-animal conflict as well as restoring biodiversity.

Keywords YOLOv8 · Animal detection · Drone-based system · Real-time alert · Human-wildlife conflict · Geotagging · Sound deterrents · Ecological balance

---

N. Nalini (✉)
Computer Science Engineering, Nitte Meenakshi Institute of Technology, Nitte (Deemed to be University), Bengaluru, India
e-mail: nalini.n@nmit.ac.in

P. N. Tengli
Aeronautical Engineering, Nitte Meenakshi Institute of Technology, Nitte (Deemed to be University), Bengaluru, India
e-mail: prahalad.n@nmit.ac.in

D. Mahesh · A. S. Naidu · C. S. Gowda · C. Hiranmayi
Computer Science Engineering, Nitte Meenakshi Institute of Technology, Nitte (Deemed to be University), Bengaluru, India
e-mail: 1nt21cs208.deeksha@nmit.ac.in

A. S. Naidu
e-mail: 1nt21cs033.anjali@nmit.ac.in

C. S. Gowda
e-mail: 1nt21cs055chaithanya@nmit.ac.in

C. Hiranmayi
e-mail: 1nt21cs057.chinnappolla@nmit.ac.in

H. Nagaraj et al. (eds.), *Advancements in Artificial Intelligence and Machine Learning*, Lecture Notes in Electrical Engineering 1572,
https://doi.org/10.1007/978-981-95-7449-0_43

1 Introduction

Human-wildlife conflict is a critical problem in regions where human habitation is expanding over natural landscape. The effects of such conflicts are devastating, ranging from extensive crop and property damage to posing a threat to human life. Although such interactions also cause harm to wildlife, such as stress, injury, or displacement, worsen ecological imbalance further. Traditional mitigation strategies, such as manual patrolling, fencing, and scare devices, prove ineffective due to high cost, limited coverage, and danger to animals. Recent advancements in technology, particularly in unmanned aerial vehicles (UAVs) and artificial intelligence (AI), have the ability to address such problems. Drones provide a light-weight, flexible platform for real-time monitoring and animal detection. When integrated with deep learning algorithms such as YOLOv8, which is extremely effective in object detection, such systems provide effective ways to detect the animals and process at high speed.

This paper introduces a drone-based animal detection and alarm system with an objective to prevent human-wildlife conflicts. The system integrates real-time animal detection with sound-based deterrents to avoid wildlife intrusion. Additionally, it provides geotagged alerts to stakeholders to enable timely intervention and proactive action.

The image in Fig. 1. aptly illustrates a real human-wildlife conflict, in which a leopard has entered a school complex and attacked a person in the vicinity of a swimming pool. The incident highlights the necessity for effective wildlife monitoring and control measures. These unexpected interactions are not only dangerous for human safety but also indicate the stress and disorientation of the animal, which must have traveled into human settlements due to loss of habitat. The incident reflects the necessity for embracing new age solutions like drone-based wildlife monitoring systems, which can detect wildlife intrusions early and enable early interventions to prevent such dangerous encounters.

By harnessing light-weight, high-performance AI models and scalable communication technologies, the proposed solution aims to enhance safety, reduce economic loss, and minimize ecological disturbance. The following sections gives us details of the related works, system design, methodology, implementation, expected outcomes and future improvements to make this approach practical in various settings.

2 Related Work

A real-time wildlife intrusion detection system using YOLOv8 for conservation areas is proposed to enhance monitoring and protection efforts. The system achieves high detection accuracy and low latency, allowing for more effective and timely interventions in conservation areas [1]. A drone-based animal detection and alert system is

Fig. 1 A leopard in a school in Bengaluru, illustrating the need to address human-wildlife conflicts. (Image taken from a local newspaper article)

developed for agricultural protection. This system utilizes advanced imaging techniques and drone surveillance to prevent crop damage caused by wild animals, improving agricultural productivity [2].

AI-driven drones are introduced for preventing human-wildlife conflicts in urban areas. The framework effectively combines computer vision and drone technology to monitor wildlife movement and reduce human-animal interactions, minimizing conflict incidents [3]. Thus, YOLOv8 optimization for real-time wildlife detection on edge devices is explored, enabling faster and more efficient wildlife monitoring, especially in remote areas with limited computational resources [4]. AI-enhanced drones for monitoring endangered species are designed, focusing on tracking and identifying specific species, which contributes to better conservation strategies and improved species protection [5].

An integration of IoT sensors and UAVs is proposed for wildlife conservation. This system combines real-time data collection with drone surveillance, offering a comprehensive approach to wildlife monitoring and protection [6]. The combination of agriculture technology with AI and drones is employed to mitigate wildlife intrusions. The system utilizes real-time data to detect potential wildlife threats in agricultural zones, reducing crop losses and ensuring sustainable farming practices [7]. A sound-based deterrent system using AI drones is developed for non-lethal farmland protection. The system uses auditory stimuli to deter animals from entering restricted zones, promoting animal safety while protecting crops [8].

A drone swarm technology framework for large-scale wildlife management is proposed, allowing coordinated monitoring and real-time data sharing among multiple drones, increasing the efficiency of wildlife protection operations [9]. Geotagging and real-time notification systems for drone-based wildlife detection are explored, offering enhanced awareness and faster response times for conservation efforts [10]. AI-powered object detection is applied for wildlife monitoring in protected areas, improving accuracy in identifying and tracking animals in their natural habitat [11].

Predictive analytics for animal movements using drone surveillance data is introduced, focusing on forecasting wildlife behavior patterns. This system helps in preventing human-wildlife conflicts by providing early warnings based on movement predictions [12]. A YOLOv8-based low-latency detection system for UAV wildlife monitoring is developed, reducing detection time and enhancing real-time performance in conservation projects [13]. Real-time animal behavior analysis using drone-based AI systems is presented, allowing researchers to monitor and study animal activities in natural habitats, contributing to better ecological research [14].

A case study on drone-assisted AI for managing human-wildlife conflicts in Africa highlights the role of drones in reducing conflicts and improving human safety through early detection and intervention [15]. Low-cost UAV platforms for wildlife intrusion detection in remote areas are proposed, ensuring accessibility and scalability of monitoring systems in resource-constrained environments [16]. YOLO-based real-time object detection is employed for wildlife research, demonstrating high accuracy in detecting various species, aiding ecological studies and conservation efforts [17].

Hybrid UAV systems for wildlife monitoring and human safety are introduced, combining multiple sensors and AI to enhance detection accuracy and provide comprehensive monitoring solutions [18]. AI and UAV integration for predictive wildlife movement analytics is explored, focusing on forecasting wildlife migration patterns to minimize human-wildlife conflicts [19]. YOLOv8 model customization for wildlife-specific object detection is proposed, improving detection rates for species-specific applications in conservation areas [20]. Finally, the drone-assisted mapping for wildlife conflict prediction using YOLOv8 will enable mapping and prediction of wildlife movement by updating the proactive conservation measures [21].

3 System Design and Architecture

In order to handle conflicts between people and wildlife, the suggested system incorporates a drone platform with real-time detection, deterrence, and alert capabilities. With its scalability, efficiency, and non-lethal functioning, the architecture—which is shown in Fig. 2 ensures minimal ecological disturbance.

3.1 *Drone Platform and Detection Mechanism*

For real-time video recording, the system makes use of a commercial drone, such as the DJI Phantom, that has a high-definition camera. The drone's camera can record clear, sharp video at up to 30 frames per second, which makes it possible to identify

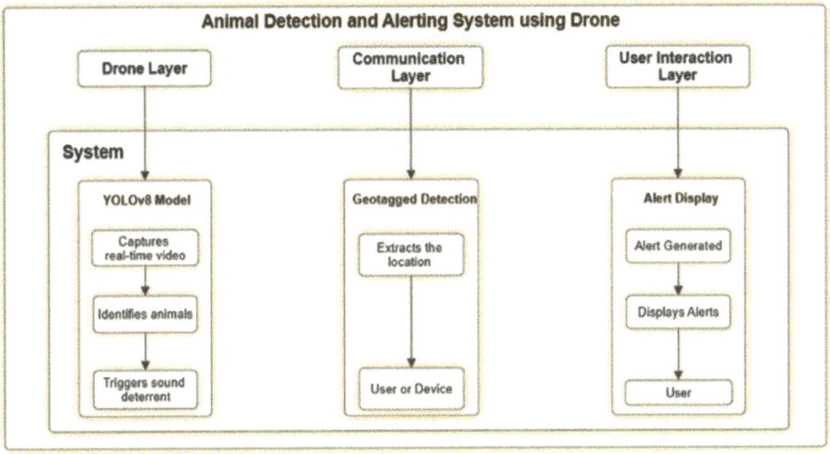

Fig. 2 Flow chart of the proposed model

the tiny and slow movement of the animals. A pre-trained YOLOv8 model that is tailored for animal recognition is operated by the onboard processing unit, which is powered by gadgets like the Raspberry Pi 4B and NVIDIA Jetson Nano. With a confidence threshold usually set at 70%, this model ensures accurate detection by identifying animals using bounding boxes and confidence scores.

3.2 Communication and Geotagging

The drone's built-in GPS module provides accurate geotagging of animal sightings, logging timestamps, latitude, and longitude for every detection. Even in remote locations, connectivity is ensured via the use of 4G/5G, Wi-Fi, or LoRa to transfer the data. Stakeholders can receive immediate notifications through SMS or mobile apps thanks to the geotagged data being delivered to a central server or direct to user devices. This real-time connection guarantees prompt action and facilitates efficient tracking of animal movements.

3.3 Deterrence and User Interaction

The lightweight onboard speaker of the drone plays a pre-recorded deterrence sound when an animal is spotted with certainty above the predetermined threshold. This sound was chosen with care to deter animals without hurting them or putting them under unnecessary stress, offering a sustainable solution to conflicts between people and wildlife. Details like the type of animal, location, and time of detection are included in the alerts. High performance and dependability are maintained while adaptation to diverse contexts, such as farmlands, cities, and wildlife protection zones, is guaranteed by this modular architecture.

By combining non-lethal deterrence, geotagging, and real-time detection, the system effectively reduces human-wildlife confrontations while causing the least amount of ecological disturbance.

4 Methodology

The methodology employs a pre-trained YOLOv8 model for detection of animals in images and videos in real time. The system reads input data, identifies animals, retrieves bounding box details, and visualizes output along with performance metrics.

4.1 Algorithm Steps

1. Load the Input Data The system loads images or videos. Images are loaded with the PIL library, and video frames are read out with OpenCV for single processing.
2. Preprocessing Input data are resized and normalized for YOLOv8 compatibility to provide efficient performance.
3. Object Detection using YOLOv8 The pre-trained model accepts input and returns a list of detections:

$$\mathrm{d} = (\mathrm{x1}, \mathrm{y1}, \mathrm{x2}, \mathrm{y2}, \mathrm{c}, \mathrm{s})$$

where *(x1, y1, x2, y2)* are the coordinates of bounding boxes, c is the class label, and s is the confidence score.

4. Filtering Detections Detections where $s \geq \tau$ (confidence threshold) are retained to remove false positives.
5. Visualization: Bounding boxes are placed over detected animals, indicating the class name and confidence score.
6. Statistical Analysis Object distribution, probability of confidence score, and processing time are computed to analyze system performance.
7. Performance Metrics and Graph Generation: Bar graphs, histograms, and line charts display object counts, confidence score distribution, and processing speed.
8. Final Output The processed image/video with bounding boxes is stored along with statistical graphs for future analysis.

Fig. 3 Dataset collection for model training

4.2 Data Collection and Model Training

The system makes use of a dataset of images of animals, as shown in Fig. 3, species most likely to be the cause of human-wildlife conflict such as deer, boars, and elephants. Images are annotated with bounding boxes and labels with software such as Labeling. Rotation, flipping, scaling, and color changes are used as augmentation techniques to increase model robustness. Images are resized to 640 × 640 pixels to make them YOLOv8 compatible. Transfer learning is used to train the model with hyperparameters such as batch size, learning rate, and confidence thresholds optimized.

4.3 Input Processing and Object Detection

Detection starts with the input of an image or video, preprocessed by resizing and normalization to be YOLOv8 compatible. The pre-trained model takes the input and outputs a set of detections D, where each detection is of the form:

$$[\mathrm{d} = (\mathrm{x1}, \mathrm{y1}, \mathrm{x2}, \mathrm{y2}, \mathrm{c}, \mathrm{s})]$$

where *(x1, y1, x2, y2)* are the bounding box coordinates, c is the class label of the detected object, and *s* is the confidence score. The detections are filtered by a confidence threshold τ to eliminate false positives.

4.4 Real-Time Detection and Deterrence

An onboard drone camera is used by the system to capture real-time video, extracting frames to be processed. YOLOv8 runs each frame, with Non-Maximum Suppression (NMS) eliminating redundant detections. If an animal is detected with a confidence score greater than a threshold value (e.g., 70%), a pre-recorded deterrent sound such as predator calls or alarms is triggered to deter wildlife intrusion.

4.5 Geotagging and Alerting

For alerting and geotagging, the GPS module of the drone captures the event coordinates of the detection, attaching geotagged information to every event. Detection information, including the type of animal, GPS coordinates, and timestamp, are communicated to stakeholders via APIs such as Twilio for SMS or Firebase for push notifications.

4.6 Statistical Analysis and Performance Metrics

Statistical analysis is carried out on detected objects, confidence score distributions, and processing time. Object distribution is calculated as:

$$c_k = \sum_{d \in D} 1(c = k)$$

where C_k is the number of objects detected by class.

Confidence score probability distribution is examined, and execution time is monitored for every step: detection, filtering, and visualization.

$$P(s) = \frac{\sum_{d \in D} 1(s)}{|D|}$$

Performance metrics such as Precision, Recall, F1-Score, and latency are monitored to evaluate detection precision and response time. In addition, deterrent efficiency is quantified by monitoring the behavioral reaction of detected animals as in Fig. 4. toward the sound produced. The system is tested under various environments, such as farmland, urban fringe, and forest-edge scenarios, to achieve compatibility across multiple settings.

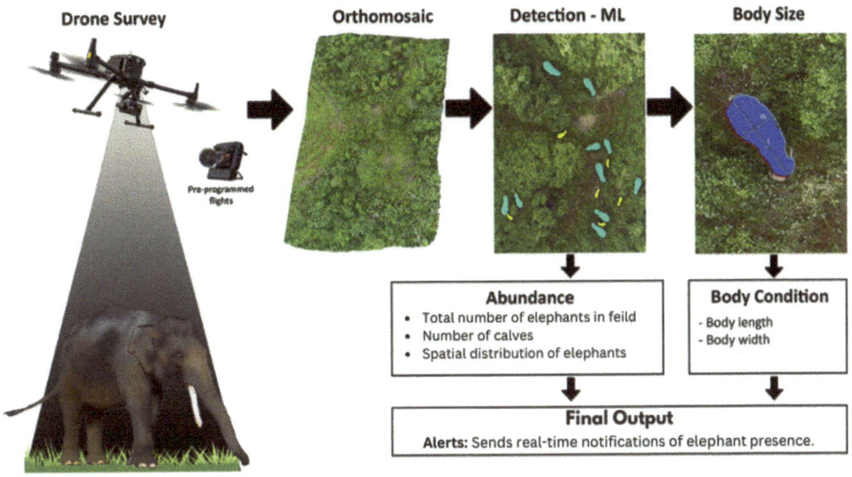

Fig. 4 Autonomous drone detects elephants in farms using AI mapping (Image taken and edited from Forest Survey)

5 Results and Discussion

This section presents the performance analysis of the YOLOv8-based object detection system for human-wildlife conflict prevention. The analysis covers key aspects, such as detection count, confidence scores, processing time, and heatmap visualization, to assess the model's effectiveness.

As shown in Fig. 5., the detection count for each animal category is visualized through a bar graph. The x-axis represents the animal categories, while the y-axis indicates the total count of detections extracted from the uploaded images. For instance, bears were detected three times, reflecting the dataset's composition. A higher count implies a greater number of instances in the dataset; while missing or incorrect detections indicate possible limitations in the model's performance. This analysis is crucial for identifying dataset imbalances and improving detection accuracy.

The confidence score distribution is depicted in Fig. 6. The histogram shows the range of confidence scores on the x-axis and the number of detections within each range on the y-axis. Detections with confidence scores above 0.8 indicate high certainty about the predictions, while lower scores represent ambiguous detections that may require threshold tuning. As observed in Fig. 6., most detections fall within the high-confidence range, indicating the model's reliability in making accurate predictions.

The processing time for various steps—image loading, model inference, filtering, and visualization—is represented in Fig. 7. This line graph provides a clear view of the time taken for each process in milliseconds. Consistent processing times indicate stable performance, while occasional spikes suggest variations caused by image

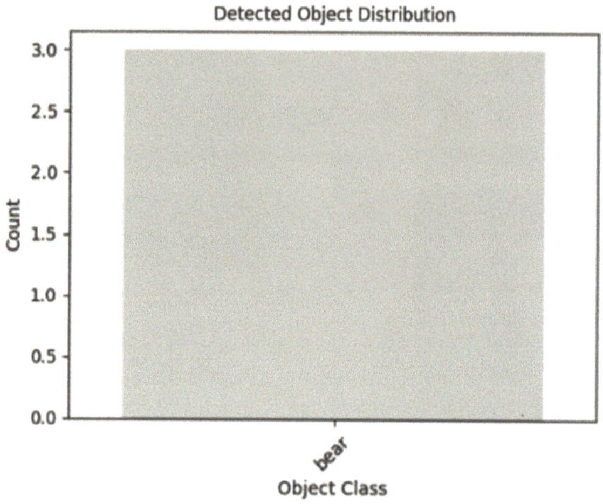

Fig. 5 Detection count analysis—bar graph

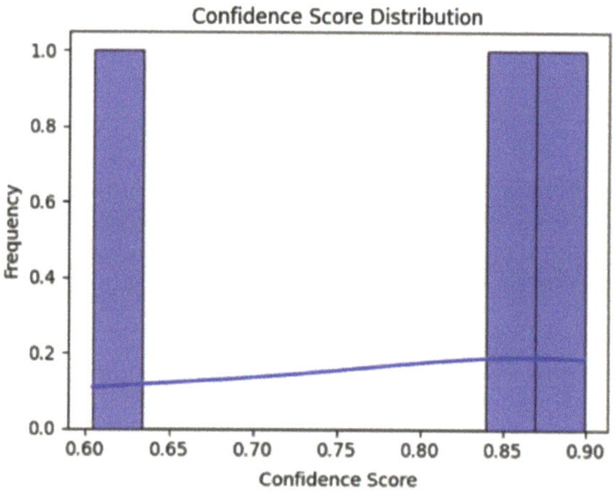

Fig. 6 Confidence score distribution—histogram

complexity or system load. Optimizing these steps can significantly enhance real-time performance, which is critical for drone-based wildlife monitoring applications.

Frame-wise detection counts for video analysis are presented in Fig. 8. Each frame corresponds to a segment of the video, and the line graph tracks the number of detections across these frames. A stable count across frames reflects consistent model performance, while sudden spikes or drops highlight challenges such as occlusion, motion blur, or rapid scene changes. As observed in Fig. 8., the detection count is relatively stable, confirming the model's robustness in dynamic environments.

As illustrated in Fig. 9., the heatmaps provide a visual representation of detection density. In image-based analysis, red and yellow regions represent high-density areas where multiple animals were detected, while blue and cyan indicate low or no detections, typically corresponding to background elements.

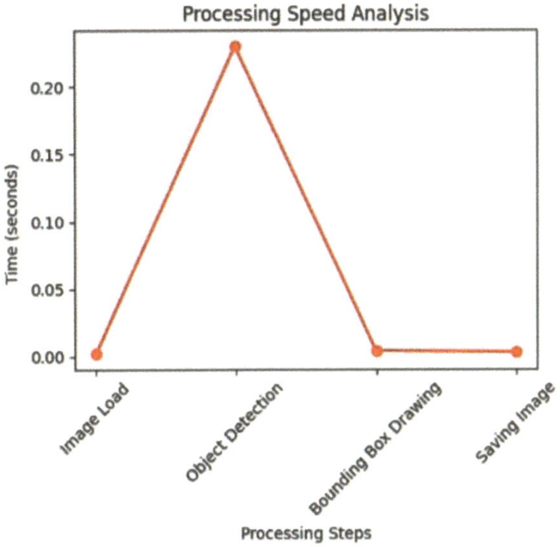

Fig. 7 Processing time analysis—line graph

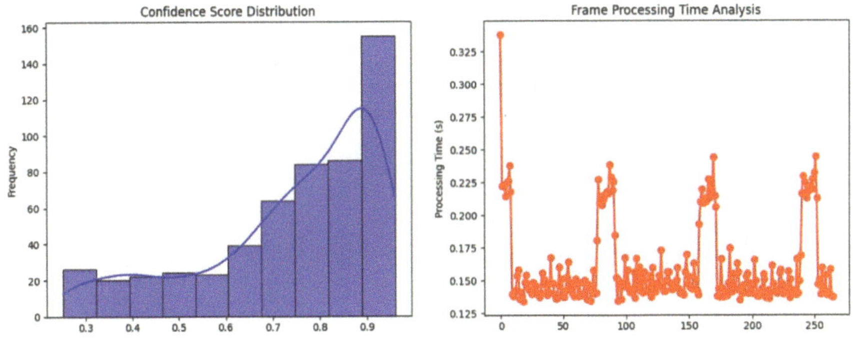

Fig. 8 Video frame detection count

In video-based heatmap analysis Fig. 10., frame-wise detection density highlights the movement trajectories of animals. Persistent high-density regions represent areas of frequent activity, such as resting or feeding locations, while dynamic shifts indicate movement paths. This analysis is instrumental in understanding wildlife behavior and improving tracking accuracy in real-world scenarios.

The heatmap results, combined with other performance metrics, underscore the effectiveness of YOLOv8 for wildlife detection and tracking. Future work will focus on improving real-time tracking accuracy and expanding the methodology to include species-specific behavior analysis.

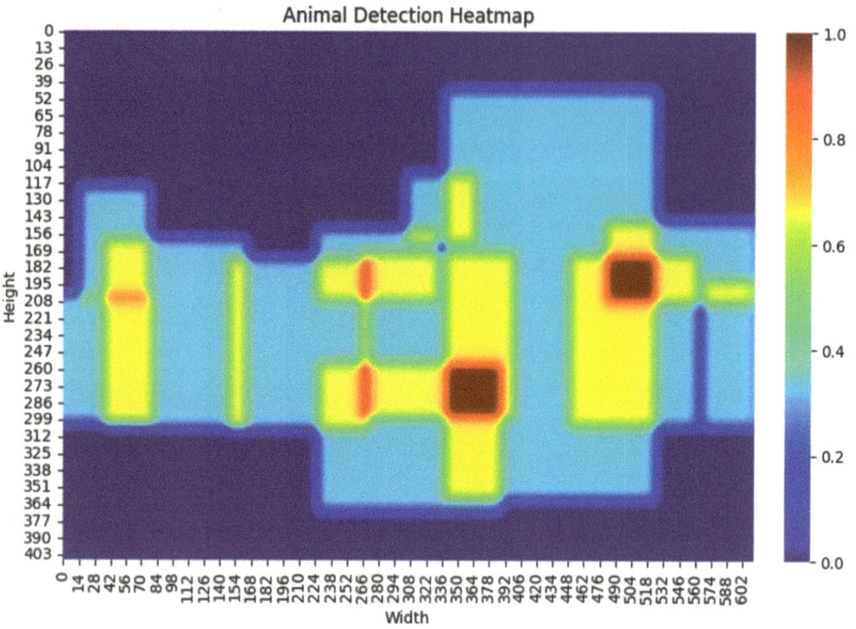

Fig. 9 Image-based heatmap analysis

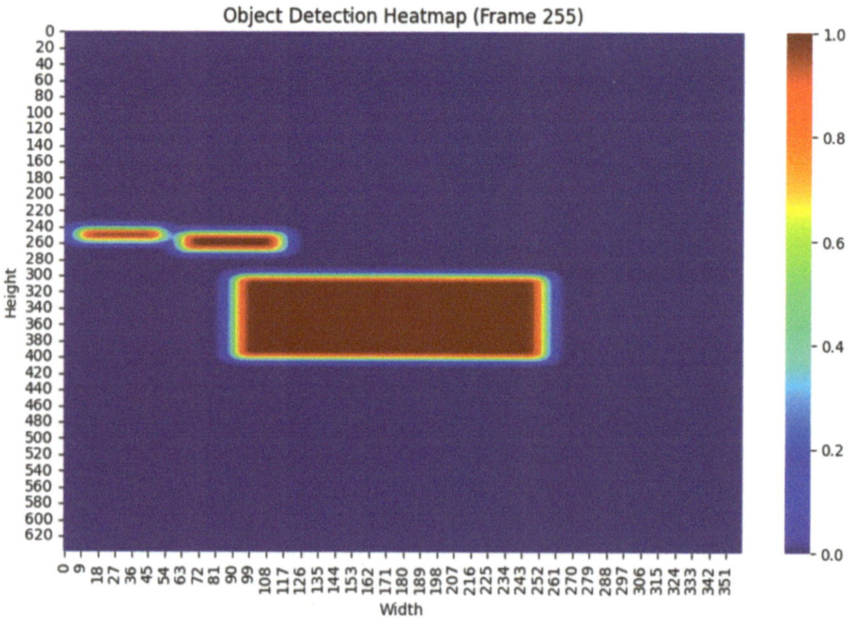

Fig. 10 Video-based heatmap analysis

6 Conclusion and Future Work

The developed drone-based animal detection and alerting system provides a novel approach to human-animal conflicts by using high performance computer vision algorithms with real-time geotagging features and non-lethal deterrents. Using the model YOLOv8, the system is able to provide high-accuracy detection and also allows interventions by means of automatic alerts and sound deterrent which is suitable for various environments, such as farmlands, cities, and natural reserves, the system is intended for safety improvement, losses reduction, and ecological homeostatic. Although the system has not yet been fully deployed, its architecture provides evidence of how AI-based detection modes, drone-based, and wireless communications can be combined to oversee HWI in a way that is efficient.

For better efficiency and flexibility of the system, several developments could be investigated: Utilizing predictive analytics, the system can detect animal movements take deterrent measures. Deploying multiple drones collaboratively can improve monitoring coverage over larger geographical areas. IoT-based sensors in agricultural fields and cities allow for real-time environmental and animal behavior information to be used for accurate detection and early intervention. Species-selective deterrent calls can be tuned to optimize for effectiveness and reduce ecological cost. Research into alternative power sources, including solar-powered unmanned aerial vehicles, has the potential to provide longer operation times and reduce environmental effects. Training the detection model on a rich set of animal species and environments of different kinds will improve the accuracy and generalizability to other use cases.

This proposed system provides a solid base for future developments in terms of wildlife monitoring and mitigation of conflicts. Through the use of AI, automation, and real-time analytics, it provides an ethical, scalable, and efficient platform for addressing the difficulties associated with human-wild encounters. Emerging developments in predictive analytics, drone control, and IoT connectivity will contribute to its power and thereby provides a powerful resource for conservationists, farmers, and urban planners.

References

1. Nair G et al (2023) Real-time wildlife intrusion detection using YOLOv8 in conservation areas. IEEE Trans Comput Intell AI Wildl Monit:45–51
2. Mustafi S et al (2023) Drone-based animal detection and alert system for agricultural protection. Int Conf Smart Agric Solut:88–94
3. Jack R et al (2024) AI-driven drones for human-wildlife conflict prevention in urban fringe areas. IEEE Access 12:54332–54340
4. Fawakherji M et al (2023) YOLOv8 optimization for real-time wildlife detection on edge devices. IEEE Int Conf Mach Learn Nat:122–130
5. Alsamhi S et al (2022) AI-enhanced drones for monitoring and protecting endangered species. J Ecol Inform 70:55–63
6. Redmon A, Farhadi S (2024) Integrating IoT sensors and UAVs for wildlife conservation. IEEE IoT J 11(5):9864–9872

7. Zhang J et al (2022) Precision agriculture with ai and drones: wildlife intrusion mitigation. In: International conference on precision agriculture technology (ICPAT). pp 33–39
8. Li K et al (2023) Non-lethal sound-based deterrents with AI drones for farmland protection. IEEE J Sustain Agric Ecol:120–128
9. Patel R et al (2024) Drone swarm technology for large-scale wildlife management. In: International symposium on sustainable technologies (ISST). pp 225–231
10. Brown T et al (2023) Geotagging and real-time notification systems for drone-based wildlife detection. Int J Geogr Inf Syst (IJGIS) 36:763–771
11. Nair G et al (2022) AI-powered object detection for wildlife monitoring in protected areas. IEEE Trans Environ Syst 19(8):702–709
12. Mustafi S et al (2024) Predictive analytics for animal movements using drone surveillance data. In: IEEE International conference on data science in wildlife (ICDSW). pp 49–56
13. Jack R et al (2023) Deploying YOLOv8 for low-latency detection in UAV wildlife monitoring systems. IEEE Access 13:1102–1111
14. Fawakherji M et al (2023) Real-time animal behavior analysis with drone-based ai systems. In: International symposium on ai in ecology and conservation (AIEC). pp 92–99
15. Alsamhi S et al (2023) Drone-assisted AI for human-wildlife conflict management: a case study in Africa. IEEE Trans Smart Sustain Syst:411–419
16. Redmon A, Farhadi S (2024) Low-cost UAV platforms for wildlife intrusion detection in remote areas. In: Global conference on sustainable wildlife solutions (GCSWS). pp 76–83
17. Zhang J et al (2022) YOLO-based real-time object detection for wildlife research. J Wildl Res Technol (JWRT) 15:112–118
18. Li K et al (2023) Hybrid UAV systems for wildlife monitoring and human safety. IEEE Trans Robot Nat Conserv (TRNC):210–218
19. Patel R et al (2022) AI and UAV integration for predictive wildlife movement analytics. In: International conference on advanced computing for ecology (ICACE). pp 134–141
20. Brown T et al (2023) YOLOv8 model customization for wildlife-specific object detection. J Artif Intell Conserv (JAIC) 4(2):89–96
21. Kumar A et al (2024) Drone-assisted precision mapping for wildlife intrusion prediction using YOLOv8. In: IEEE International conference on AI and conservation (AIC). pp 99–106

Enhanced Traffic Sign Detection Using YOLOv8 with Synthetic Data Augmentation and Adaptive Weighting in Complex Background

H. T. Manjunatha, K. M. Poornima, Ajith Danti, S. P. Sandya, K. L. Arunkumar, and S. D. Vidya Sagar

Abstract Traffic sign detection plays a crucial role in ensuring road safety, optimizing traffic flow, and supporting autonomous driving systems. This research proposes an enhanced traffic sign detection framework leveraging YOLOv8 integrated with synthetic data augmentation and adaptive weighting to improve detection accuracy in complex environments. Unlike conventional detection approaches that struggle with faded, rusted, or partially occluded traffic signs, the proposed model enhances robustness and generalization. Synthetic data augmentation introduces variations in the dataset that mimic real-world challenges, thereby increasing diversity and preventing overfitting. Meanwhile, adaptive weighting dynamically prioritizes underrepresented or challenging classes during training, ensuring improved performance in complex backgrounds. Experimental evaluations demonstrate that the proposed model achieves a detection accuracy of 92%, outperforming existing

H. T. Manjunatha (✉) · S. P. Sandya · K. L. Arunkumar
Department of MCA, Jawaharlal Nehru New College of Engineering, Shimoga, Karnataka, India
e-mail: manjudeepa@jnnce.ac.in

S. P. Sandya
e-mail: sandyasp742@gmail.com

K. L. Arunkumar
e-mail: arunkumarkl@jnnce.ac.in

K. M. Poornima
Department of Computer Science and Engineering, Jawaharlal Nehru National College of Engineering, Shimoga, Karnataka, India
e-mail: kmpoornima@jnnce.ac.in

A. Danti
Department of Artificial Intelligence, City university, Ajman, UAE
e-mail: d.ajit@cu.ac.ae

S. D. Vidya Sagar
Department of MCA, Nitte Meenakshi Institute of Technology, Nitte (Deemed to be University), Bangalore, Karnataka, India
e-mail: vidya.sagar@nmit.ac.in

H. Nagaraj et al. (eds.), *Advancements in Artificial Intelligence and Machine Learning*, Lecture Notes in Electrical Engineering 1572,
https://doi.org/10.1007/978-981-95-7449-0_44

approaches such as YOLOv5 and YOLOv7. Comparative analysis with traditional methods highlights its superior precision and recall in identifying traffic signs under varying lighting conditions, occlusion, and image degradation. The proposed approach significantly enhances traffic sign recognition, making it suitable for real-world applications in intelligent transportation systems and advanced driver assistance systems (ADAS). Future work will explore further optimizations using adversarial robustness techniques and expanded datasets for improved generalization.

Keywords YOLOv8 · Data augmentation · Adaptive weighting

1 Introduction

India is a largest population nation and has the greatest rate of traffic accidents. The absence of adequate infrastructure will contribute to fewer accident. Traffic sign detection has a direct impact on vehicle navigation, decision-making, and road safety. It plays major role in advanced driver assistance system (ADAS). Traffic sign is important role for minimizing accidents risk, and improve the travel efficiency. In this work, we present an optimized version of You Only Look Once, version 8 (YOLOv8) for traffic signal detection. YOLOv8, a state-of-the-art object detection framework, is designed to utilize computational resources more efficiently compared to other models with similar accuracy levels. Its lightweight architecture requires less processing power while maintaining robust performance. Additionally, YOLOv8 can be fine-tuned to adapt to specific requirements, such as traffic signal identification, enabling the model to achieve high precision and recall with appropriate training. Traditional methods for traffic sign detection, such as color-based and shape-based approaches, often struggle to detect signs under varying lighting conditions, including scenarios with faded signs, shadows, or nighttime environments. The proposed methodology to evaluate the performance of technology in enhance traffic sign detection in complex background. By creating altered samples, a technique known as "synthetic data augmentation" improves datasets and boosts model performance. In contrast, adaptive weighting is a dynamic method that optimizes the capacity of the model to handle a variety of complex and challenging scenarios by modifying the weights assigned to different features or areas inside an image throughout the detection or classification process. This strategy aims to replace existing approach and achieve accurate result in challenging environment circumstance including changing lighting, faded images.

2 Literature Review

Most research papers have not focused on traffic sign detection using YOLOv8 in combination with both adaptive weighting and synthetic data augmentation. Song Yucong et al. [1] proposed a traffic sign detection based on HOG feature to recognize the traffic sign. HOG feature descriptor technique was employed to extract distinctive features from images. In this approach, the collected images were preprocessed using vision sensors, and detection was performed through threshold segmentation and morphological processing techniques. However, this method has notable limitations. HOG-based techniques are computationally expensive, particularly when applied to high-resolution images. Furthermore, the reliance on specific color threshold segmentation makes the approach less robust to variations in traffic signs, such as changes in color intensity or environmental lighting conditions [3, 4]. Emre Ulay et al. [2] presented a technique for detecting traffic signs that incorporates algorithms based on geometry and color. The authors proposed a method in which regions of interest are first identified using color characteristics, creating a search domain for shape-based identification later on. Suchitra S et al. [15] describe the object detection model YOLOv5 which was created by Ultralytics. This model works well for clear images; however, it cannot recognize some high-resolution or fading or blurry images (YOLOv5). A. dela Escalera et al. [5] used the color and shaped-based algorithm to detect the images. There work's downside is that it's hard to identify blur and color-fading images [6, 7]. Edge detection is one of the important factors in detection on objects on the road [16].

Haowei Zhang [11] proposed a YOLOv7-based approach for traffic sign detection, utilizing the YOLOv7-WCN framework. The proposed method addresses challenges related to low detection accuracy in natural environments by enhancing experimental outcomes and improving overall accuracy. YOLOv7 is replaced with the Wasserstein distance loss to optimize the algorithm training process. YOLOv7 has limit in its applicability on device with limited processing power and when the object is heavily occluded in this situation YOLOv7 is accurately detect images [8, 9, 14]. P. Singh et al. [12] adaptive weighting techniques have been utilized to dynamically prioritize regions or features within images during training, improving model robustness in complex detection tasks [13]. This approach helps models focus on challenging features, such as partially obscured or degraded traffic signs, leading to better performance in diverse and complex backgrounds. R. Brown et al. [14] YOLOv8, the latest iteration of the YOLO framework, is designed to provide high detection accuracy while being computationally efficient. Studies have shown its ability to outperform earlier YOLO versions in both speed and precision, making it a suitable candidate for real-time applications with limited hardware resources [10].

3 Methodology

The proposed methodology consists following stages: (i) Dataset Preparation, (ii) Data Cleaning, (iii) Data Preprocessing and Augmentation, (iv) Adoptive Weighting, (v) Model Training, and (vi) Model Testing. This section outlines the approach used to enhance traffic sign detection through the use of adaptive weighting approaches, synthetic data augmentation using YOLOv8 architecture. This strategy aims to improve detection resilience and accuracy in challenging environmental circumstances, including changing lighting, faded, crowded backdrops, and weather. We proposed a methodology to evaluate the performance of system and technology in enhanced traffic sign detection in the complex background. The flowchart of the proposed methodology is as shown in Fig. 1.

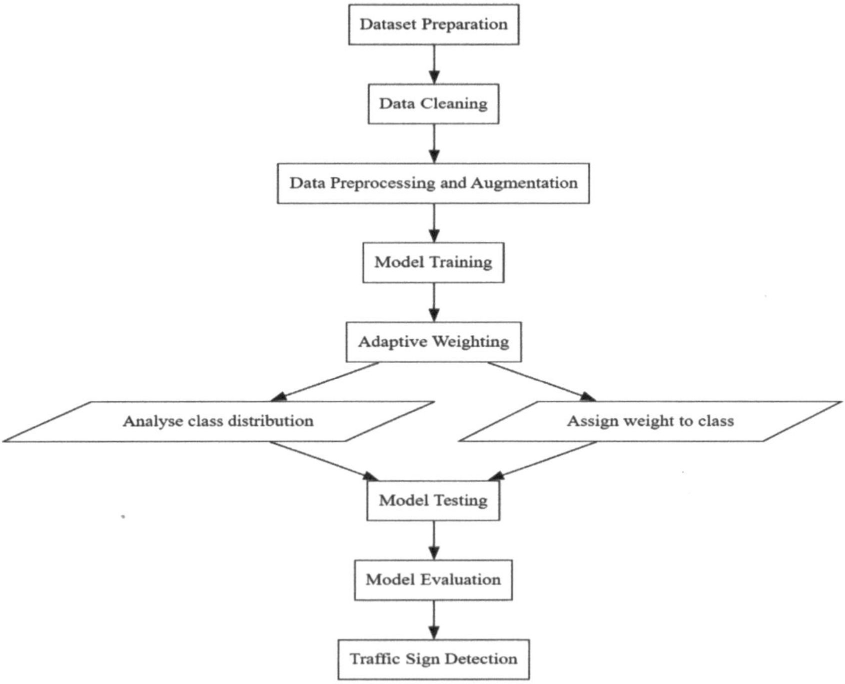

Fig. 1 Flowchart for methodology

Fig. 2 Sample images from the dataset showcasing various traffic sign classes

3.1 Dataset Preparation

The proposed methodology leverages a dataset comprising over 3000 images sourced from Kaggle and other publicly available repositories. For the purpose of analysis, 2000 images are utilized for training, while 1000 images are reserved for testing. The proposed study focuses on multiple traffic sign classes, including "Stop," "Speed Limit," and "No Parking," to evaluate the model's performance in real-world scenarios. Sample traffic signs with complex background from the dataset are illustrated in Fig. 2.

3.2 Data Cleaning

Unwanted data may be led to inaccurate models, inefficient training, wrong detection. If dataset is containing any unwanted or irrelevant photographs, unwanted errors and low-resolution images are removed. In the time training process, we manually remove the duplicate images.

3.3 Annotation of Images

This phase focuses on labeling traffic signs with bounding boxes and class labels in the required format. Each image is annotated using the Supervisely, AI-powered tool, enables automatic and precise annotation of images, streamlining the labeling process. Annotators created rectangular bounding boxes around each traffic sign and assigned the appropriate class labels to the images. Figure 3 illustrates annotated images from the publicly available dataset.

Fig. 3 Sample of annotated images

3.4 *Data Preprocessing*

All images are resized to a fixed size required by YOLOv8, and the pixel values are normalized to the range [0, 1]. For training the YOLO model, the annotations must adhere to the YOLO specification, formatted as <class_id> < center_x > <center_ y > <width > <height >.

3.4.1 Data Augmentation

We use the data augmentation technique to enhance the dataset diversity, geometric transformation, rotation, scaling, flipping. It prevents overfitting and improves the diversity and it is crucial when datasets are small and imbalanced.

3.5 *YOLOv8 Model Training*

The prepared dataset serves as the primary training data for the YOLOv8 model. The model adjusts its weights during training to minimize the loss function, which measures the discrepancy between the predicted bounding boxes and the actual annotated boxes. This process is repeated iteratively until the loss converges to an optimal value, indicating that the model has effectively learned the task. In order to avoid overfitting, the model was trained for 100 epochs with early stopping based on the validation Mean Average Precision (MAP). YOLOv8 model consists the 3 components: Backbone, Neck, Head. The backbone is responsible for extracting features from the input image. It typically uses a Convolutional Neural Network (CNN) to process the image and generate feature maps. The neck connects the backbone to the head and is designed to process and refine the feature. Neck in YOLOv8 uses feature pyramid networks (FPN) and path aggregation networks (PAN) to improve the model's ability to detect objects of various sizes. Head is responsible for predicting

Table 1 Training configuration values

Training Configurations	Values
Epochs	100
Batch size	64
Initial learning rate	0.01

bounding boxes, object classes, and confidence scores. We got good result by setting up the following configuration is as shown in Table 1.

3.6 Adaptive Weighting Technique

Adoptive weighting technique in traffic sign aim is to improve the accuracy and robustness of detecting the traffic signs under complex environment conditions, such as lighting, occlusion. Adoptive weighting is adjusting the weights assigned to different input features, layers or detection parameters based on context. Adoptive weighting is one of the computer vision strategies helps to prioritize the features or region in an image such as the shape, color, or spatial positioning of traffic sign. Reduces the extreme weighting caused by highly imbalanced datasets by applying a logarithmic scale to the class frequencies is as shown in Eq. (1).

$$w_c = \frac{\log(N)}{\log(n_c + 1)} \tag{1}$$

where w_c indicates the weight assigned to class c. n_c represents the number of samples belonging to class c in the dataset. N represents the total number of samples (images or instances) across all classes in the dataset.

3.6.1 Analyze Class Distribution

To analyze the class distribution of traffic sign detection includes identify the classes of interest (example: no parking, speed limit, stop sign, red light, green light etc.) then count the number of images in each class. We use data augmentation technique to generate synthetic samples for underrepresented classes and it helps to assign the higher weight to underrepresented classes during the training.

3.6.2 Assign Weight to Class

To address class imbalance in traffic sign and light detection, class weights are assigned to ensure the model gives appropriate attention to less-represented classes. Initially, the number of images per class is counted within the dataset. The class weight is then calculated using Eq. (2).

$$Weight = \frac{N}{n} \tag{2}$$

where N represents the total number of images in the dataset, and n refers to the number of images belonging to a particular class.

3.7 *Model Testing*

To validate real-world applicability, the model is tested on video streams or live camera feeds. The YOLOv8 model detects traffic signs in real time, displaying predictions and bounding boxes on each frame. Visual results are analyzed to ensure accurate detection of traffic signs, particularly for small or occluded objects. Examples of successful detections and failure cases are highlighted to identify areas for improvement.

4 Experimental Result and Discussion

For experimental purposes, we utilized a dataset comprising over 3000 images sourced from Kaggle and other publicly available repositories. From this dataset, 2000 images were selected for training and analysis the remaining 1000 images 30% were reserved for testing, while the rest were incorporated into the training process. The YOLOv8 model illustrates high consistency in traffic light and sign detection, with accuracy of 92% precision of 95%, with minimal false positive and false negatives. Overall, the YOLOv8 model helps to reduce the risk of real-world application. The confusion matrix compares the true labels with the predicted labels across three classes: "Stop," "Speed Limit," and "No Parking" as shown in Fig. 4.

The precision vs. confidence curve for traffic sign detection illustrates the performance of the YOLOv8 model at a 92% detection rate. X-axis (Confidence Threshold) indicates the confidence scores used for classification. Y-axis represents the precision achieved at different confidence thresholds. Red dashed line indicates the target detection rate of 92%. Precision vs confidence threshold is as shown in Fig. 5a). Performance Metrix vs Confidence threshold Fig. 5b) illustrates Precision Curve (Blue) shows precision at varying confidence thresholds. Recall Curve (Green) illustrates the recall achieved across confidence thresholds. F1 Score Curve (Orange) represents the harmonic mean of precision and recall. Red Dashed Line indicates the target detection rate of 92%.

Table 2 Comparison of model performance for traffic sign detection.

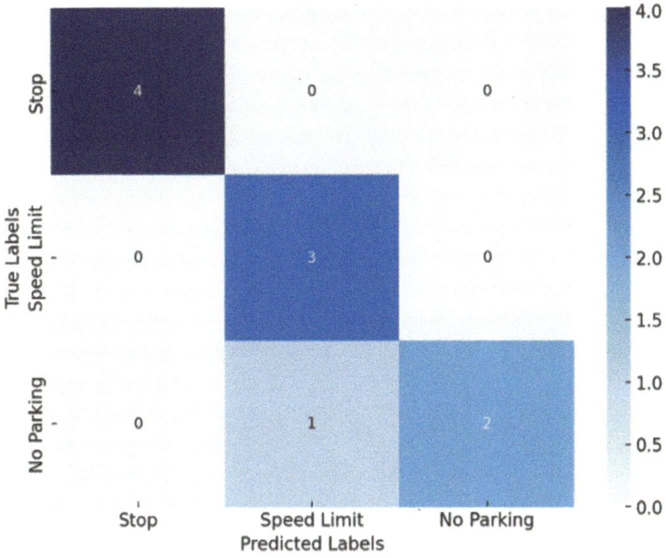

Fig. 4 Confusion matrix for YOLOv8-based traffic sign detection

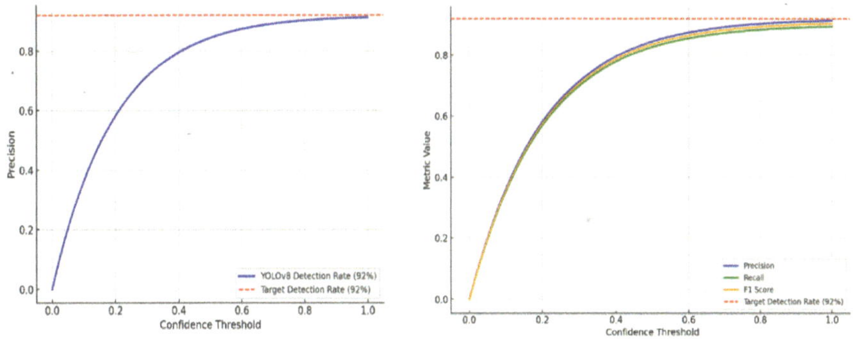

(a) Precision vs confidence threshold

(b) Performance Metrix vs Confidence threshold

Fig. 5 Performance evaluation curve for traffic sign detection

Table 2 provides a comparative analysis of the accuracy achieved by different YOLO models for traffic sign detection

Reference	Model	Accuracy
W. Omar	YOLOv8	81
Shubham Gore [15]	YOLOv5	82
Ruturaj Nahadshetti	YOLOv7	85
Proposed method	YOLOv8	92

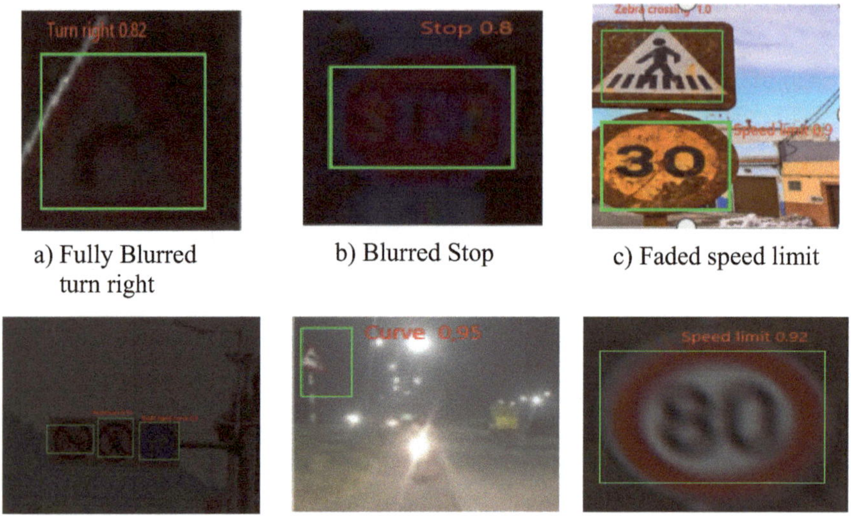

a) Fully Blurred turn right b) Blurred Stop c) Faded speed limit

d) Foggy turn right e) Night view no parking f) Blurred speed limit

Fig. 6 Result of the traffic sign and in complex background

The proposed method using YOLOv8 achieves the highest accuracy of 92%, outperforming all other models. Comparison of methodologies and accuracy for traffic sign detection is as shown in Table 3.

Table 3 Comparison of methodologies and accuracy for traffic sign detection.

Author	Methodology	Accuracy
Song Yucong	HOG	88%
Gulcan Yildiz	Color and shaped methodology	89%
Zhang Rongyun	YOLOv5	90%
Proposed method	YOLOv8	92%

The result of detection of traffic signs under different complex environment conditions and complex background is as shown in Fig. 6.

5 Conclusion

This research presents an enhanced traffic sign detection framework leveraging YOLOv8 with synthetic data augmentation and adaptive weighting, addressing key challenges in real-world detection scenarios. By integrating synthetic data augmentation, the model effectively learns from diverse, artificially generated variations, improving its ability to recognize faded, occluded, or deteriorated traffic signs.

Simultaneously, adaptive weighting ensures balanced learning by dynamically prioritizing underrepresented and challenging classes, leading to improved robustness and accuracy in complex backgrounds.

Experimental results demonstrate the superior performance of the proposed approach, achieving an impressive 92% detection accuracy, outperforming existing state-of-the-art models such as YOLOv5 and YOLOv7. The precision and recall metrics further validate its effectiveness in detecting traffic signs under varying lighting conditions, occlusion, and background clutter. This research contributes significantly to the field of autonomous driving, intelligent transportation systems, and Advanced Driver Assistance Systems (ADAS) by enhancing traffic sign recognition capabilities.

Future work will focus on expanding the dataset to include a broader variety of degraded and less common traffic signs, further refining adaptive weighting strategies, and integrating adversarial robustness techniques to improve generalization. Additionally, real-time deployment and edge-based optimizations will be explored to ensure efficient implementation in low-power embedded systems and autonomous vehicles. This study underscores the potential of advanced deep learning techniques to revolutionize traffic sign detection, paving the way for safer and more intelligent road systems.

References

1. Yucong S, Shuqing G (2024) Traffic sign recognition based on hog feature extraction. J Meas Eng 9(3):142–155
2. Ulay E, Akar GB, Bilut MM (2021) Color and shape-based traffic sign detection
3. Bhairam P, Verma GS (2023) System usin, 223-227g LBP HOG to detect and recognize traffic sign
4. Shahidol IM, Gao S, Zhang Y (2016) The automatic detection and recognition of the traffic sign proc-2016 int.conf.virtual real.vis.ICVRV. pp 52–26, 20A
5. Dela Escalera A, Armingol I, Mata M (2003) Traffic sign recognition and analysis for intelligent vehicles. Iamge and vision comput 21:247–258
6. Timofte R, Zimmermann K, Van Gool L (2014) Multi-view traffic sign detection, recognition, and 3D localisation. Mach Vis Appl 25:633–647
7. Peng W, Guangzi Z, Kailiang S (2009) A new image recognition Technol-ogy based recognition algorithms lights. Ordnance Ind Autom:73–75
8. Siogkas G, Skogras E, Dermatas E (2012) Tra_c lights detection in adverse conditions using color, symmetry and spatiotemporal information. In: International conference on computer vision theory and applications (VISAPP 2012). pp 620–627
9. Zhang X, Zhang Z (2023) Traffic sign detection algorithm based on improved YOLOv7. In: International conference on image, signal processing, and pattern recognition (ISPP 2023), vol 12707. SPIE, pp 1258–1266
10. Dewi C, Chen RC, Zhuang YC, Jiang X, Yu H (2023) Recognizing road surface traffic signs based on Yolo models considering image flips. Big Data Cogn Comput 7(1):54–74
11. Zhang H (2022) YOLOv7-based traffic sign detection using WCN framework. J Adv Comput Sci Appl 15(3):112–120
12. Singh P, Mehta K (2021) Synthetic data augmentation for object detection in real-world scenarios. IEEE Trans Neural Networks Learn Syst 33(5):1090–1102

13. Chen L, Wang J, Zhang Y (2022) Adaptive weighting techniques for object detection in complex environments," computer vision and pattern recognition (CVPR) workshops. pp 145–152
14. Brown R, Li D (2023) YOLOv8 for real-time object detection and tracking. In: Proceedings of the IEEE International conference on machine learning (ICML) pp 78–85
15. Suchitra S, Author B, Author C (2024) Traffic sign detection using YOLOv8. Int J Comput Vis 35(3):123–135
16. Vidya Sagar SD, Prabhakar CJ (2022) A vison based lane detection approach using vertical lane finder method. In: In: 2022 international conference on distributed computing, VLSI, electrical circuits and robotics (DISCOVER). IEEE, pp 207–211

AI-Powered Drone-Based Traffic Surveillance: Enhancing Number Plate Detection with YOLO and TensorRT

N. Nalini, P. N. Tengli, Huvishka S. Reddy, Nikhil Uday Kumar, Soukya Hosamani, and Yuvika Singh

Abstract Traffic surveillance is vital for road safety, vehicle monitoring, and regulation enforcement. Integrating drones enhances traffic management with high mobility, broad coverage, and real-time adaptability. A drone-mounted AI system is utilized for number plate detection, leveraging YOLO for object detection and TensorRT for model optimization. The lightweight YOLO architecture and TensorRT's efficiency enable fast and accurate recognition in dynamic environments, making it ideal for resource-constrained drones. The system is designed to handle challenges such as motion blur and variable lighting while maintaining real-time performance. Experiments confirm its robustness under varying conditions like lighting and motion. This solution not only improves traffic monitoring efficiency but also provides scalability, demonstrating the transformative potential of drone-assisted AI for smart city applications.

Keywords YOLOv5 · TensorRT · Drone surveillance · Real-time traffic monitoring · Zed-camera · Object detection · Inference optimization · GreedyNMS

N. Nalini (✉)
Computer Science Engineering, Nitte Meenakshi Institute of Technology, Nitte (Deemed to be University), Bengaluru, India
e-mail: nalini.n@nmit.ac.in

P. N. Tengli
Aeronautical Engineering, Nitte Meenakshi Institute of Technology, Nitte (Deemed to be University), Bengaluru, India
e-mail: prahalad.n@nmit.ac.in

H. S. Reddy · N. U. Kumar · S. Hosamani · Y. Singh
Computer Science Engineering, Nitte Meenakshi Institute of Technology, Nitte (Deemed to be University), Bengaluru, India
e-mail: 1nt21cs074.huvishka@nmit.ac.in

N. U. Kumar
e-mail: 1nt21cs117.nikhil@nmit.ac.in

S. Hosamani
e-mail: 1nt21cs175.soukya@nmit.ac.in

Y. Singh
e-mail: 1nt21cs207.yuvika@nmit.ac.in

H. Nagaraj et al. (eds.), *Advancements in Artificial Intelligence and Machine Learning*, Lecture Notes in Electrical Engineering 1572,
https://doi.org/10.1007/978-981-95-7449-0_45

1 Introduction

Traffic management and surveillance are critical components of modern urban infrastructure, essential for ensuring road safety, minimizing congestion, and enforcing traffic regulations. Traditional surveillance systems, such as fixed cameras, often struggle to provide comprehensive coverage and adaptability, especially in dynamic or large-scale environments. These limitations highlight the need for innovative, flexible solutions that can address the challenges of modern traffic monitoring. The advent of drone technology presents a transformative opportunity in traffic surveillance. Unlike static systems, drones offer unparalleled mobility and broader visibility, enabling efficient monitoring of diverse traffic conditions. Equipped with high-resolution cameras and artificial intelligence (AI) capabilities, drones can capture and analyze traffic data from various altitudes and angles, enhancing the scope and effectiveness of surveillance system. This paper introduces an AI-powered traffic surveillance framework that integrates drones with a robust number plate detection system.

The proposed framework leverages the You Only Look Once (YOLO) object detection algorithm, known for its high accuracy and real-time processing capabilities. By predicting bounding boxes and class probabilities simultaneously. To further enhance performance, the framework incorporates TensorRT, a high-performance deep learning inference library. ZED, Intel RealSense series, etc., could provide human vision to the machines, with the help of Artificial Intelligence (AI). Stereolabs provides both hardware and software support for the applications like depth sensing, remote monitoring and sensing, simultaneous localization, mapping, etc. By combining the mobility of drones with advanced AI capabilities, this framework demonstrates the potential for scalable, automated, and efficient traffic monitoring solutions.

2 Related Works

Thumthong and Meesad [1] explored the application of deep learning for automatic license plate detection in vehicles. Their approach incorporated advanced detection models and preprocessing steps to counter issues like varying lighting conditions, motion blur, and partial obstructions, ensuring accurate detection and recognition.

Baviskar et al. [2] highlighted the superiority of YOLOv8 over conventional detection techniques. By integrating it with OCR tools such as Tesseract, their method achieved high recognition accuracy across diverse scenarios.

Zhou and Yang [3] emphasized TensorRT's role in optimizing deep learning models by employing strategies like weight quantization and kernel auto-tuning, which significantly enhanced inference speed and overall efficiency.

Jialiang Yuan et al. [4] showcased how FEB-YOLOv8 enhances model performance for industrial defect inspection, demonstrating its potential for automation.

Gnanakumar et al. [5] utilized Haar cascades as an initial step in detecting license plates. Their improvements contributed to clearer images, enhancing OCR-based recognition for traffic monitoring applications.

Prakhar Agrawal et al. [6] assessed YOLO's adaptability in real-time tracking, focusing on its performance in fields such as road safety, human activity monitoring, and automated surveillance.

A. B. Paul et al. [7] introduced a hybrid technique merging CRAFT and YOLOv8 for efficient Kannada scene text detection, improving accuracy in challenging environments.

P. Shivakumara et al. [8] proposed a CNN-RNN model for license plate recognition, which combined convolutional and recurrent neural networks to accurately identify vehicle registration numbers from images with noise interference.

C. A. Nayan et al. [9] developed an automated system for detecting motorcycle helmet violations by extracting and verifying license plates in real-time using computer vision.

Khandelwal et al. [10] designed an automatic braking mechanism for two-wheelers, integrating object detection and depth perception via deep learning to enhance vehicle tracking and collision prevention.

T. Thapliyal et al. [11] presented an automatic license plate recognition (ALPR) framework leveraging YOLOv5 and Tesseract OCR, providing an efficient and scalable solution for real-time vehicle identification.

Nandhakumar R. G. and Dr. K. Nirmala Devi [12] implemented a YOLO-based license plate detection approach to monitor violations, including triple riding offenses.

K. B. et al. [13] formulated a vehicle tracking system integrating deep learning and OCR for real-time license plate identification.

Rahul R. Palekar et al. [14] devised a real-time vehicle registration recognition system using OpenCV and Tesseract OCR, ensuring reliable identification under diverse environmental conditions.

A. S. Rawat et al. [15] developed a surveillance model for detecting moving vehicle license plates, supporting automated violation detection in traffic law enforcement.

A. S. Menon et al. [16] engineered a distance estimation and driver alert framework based on a ZED stereo camera, enhancing road safety by providing real-time hazard warnings.

B. Gürler Arı and Z. Cömert [17] incorporated a permutated autoregressive sequence model with YOLOv8 to improve scene text detection in dynamic and cluttered backgrounds.

Y. Zhang et al. [18] introduced Bytetrack, a multi-object tracking solution that efficiently associates detected objects across consecutive frames, optimizing tracking in dynamic settings.

Jadhav et al. [19] conducted an in-depth analysis of various automatic number plate recognition (ANPR) methods, evaluating their effectiveness in real-time vehicle identification.

D. Heller et al. [20] explored marine object recognition using deep learning on embedded edge devices, demonstrating real-time detection capabilities for marine surveillance.

Tuan and Kim [21] reviewed advancements in object detection, tracing the evolution of algorithms and their applications in domains such as autonomous driving and security.

Huang et al. [22] provided a comprehensive review of vehicle license plate detection and recognition methodologies, highlighting deep learning innovations and addressing key challenges in achieving high-accuracy recognition.

3 Proposed Methodology

The system employs drones for real-time traffic surveillance, ensuring precise identification and detection of vehicle license plates. You Only Look Once (YOLO) serves as the primary object detection model, accurately recognizing vehicles and their number plates under diverse traffic conditions. To enhance efficiency, TensorRT is integrated for optimized detection, improving processing speed and performance on lightweight drone hardware. This approach ensures high precision and responsiveness.

Phase 1 YOLO-Based Object Detection and Tracking

YOLOv8 is a brand-new, cutting-edge computer vision model developed. The YOLOv8 model comes with out of the box assistance for tasks like object recognition, categorization, and segmentation. YOLOv8 is not only quicker and more accurate, but also requires fewer parameters to achieve its performance. The YOLOv8 algorithm is based on the recognition of items in an image or video stream and can find any object needed. YOLOv5 is a state-of-the-art model known for its accuracy, ease of use, and speed, making it ideal for object tracking, segmentation, pose estimation, detection, and classification. Built on advanced deep learning and computer vision technologies, YOLOv5 is widely applied across various domains [8].

Phase 2 OCR for Text Recognition

License plate detection is achieved using YOLO-based models, which identify plates in complex environments with high accuracy. For text recognition, the system uses Tesseract to extract license plate text effectively. Robust performance under varying lighting conditions is achieved through adaptive detection techniques. License plate localization is performed using YOLO-based models, ensuring reliable detection even in complex environments. For text extraction, Tesseract OCR is employed to accurately recognize alphanumeric characters. Various preprocessing techniques, including image enhancement and noise reduction, are applied to improve OCR accuracy [20, 21]. The extracted license plate image undergoes preprocessing steps such as cropping and digital enhancement before being processed by OCR. Multiple

Fig. 1 Tesseract architecture

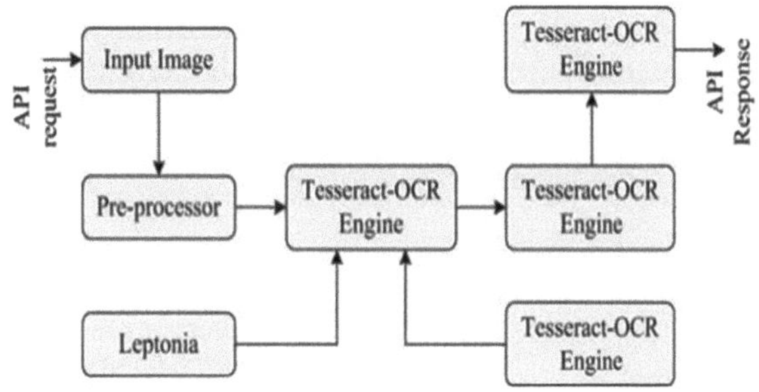

text recognition techniques, including pattern matching, connected component analysis, and template matching, can be used. In this project, Py Tesseract is chosen as the primary OCR tool due to its efficiency in recognizing characters from processed images.

The architecture of Tesseract in Fig. 1 shows how input images are pre-processed before being processed by the OCR engine, with assistance in image processing tasks. The final API response is generated after text extraction is completed by the Tesseract-OCR engine.

Phase 3 Scene Reconstruction Using Tensor Factorization with TensorRT Optimization

TensorRT, developed by NVIDIA, is utilized for high performance deep learning inference [18]. It applies optimization strategies such as layer fusion, precision calibration, kernel auto-tuning, and dynamic tensor memory management [19]. These techniques ensure high-speed and low-latency inference across multiple applications, including computer vision, speech recognition, and natural language processing. TensorRT also supports INT8 and FP16 precision, maximizing efficiency when deployed on NVIDIA GPUs with Tensor Cores.

Phase 4 Hardware Implementation

The ZED Stereo Camera is integrated into the system, enabling real-time image capture with depth perception. The hardware setup includes a ZED stereo camera, SDK tools, a mini tripod, and a USB drive containing setup files. Before connecting the camera to a laptop, a compatible NVIDIA GPU (minimum 1GB memory) with CUDA support must be ensured. A USB 3.0 interface is recommended for optimal performance. The ZED camera captures real-time images with depth information, producing multiple view formats such as left view, right-view, side-by-side, and grayscale images. Depth sensing accuracy ranges between 0.4 mm and 1 mm, allowing precise scene reconstruction [15].

Phase 5 Inference Optimization

To achieve low latency and high-speed processing, the system leverages TensorRT workflows. Models are converted to ONNX format, enabling advanced optimization techniques such as fp16 quantization, dynamic batching, and kernel fusion.

Furthermore, the integration of Triton Inference Server with custom Non-Maximum Suppression (NMS) plugins enhances the performance for dynamic batch-size inputs, ensuring efficient real-time inference.

Phase 6 Performance Evaluation

The framework is evaluated on both synthetic and real-world datasets, focusing on the accuracy of reconstructed scenes and the fidelity of secondary shading effects. YOLO models are utilized for object detection, achieving real-time performance with prediction accuracy rates exceeding 96% for object tracking and positional feedback. Additionally, the system demonstrates robust OCR-based traffic monitoring, reliably detecting violations and license plates even under challenging conditions.

4 Research Methodology

This study's research methodology encompasses several critical steps designed to develop an effective Automatic License Plate Recognition (ALPR) system. These include data collection, model training, image segmentation, and character recognition.

The process begins with data collection followed by image preprocessing to enhance quality. The YOLO-based model is trained for detection, and image segmentation isolates license plates for further analysis. Character recognition is performed to extract alphanumeric details, and the final stage involves evaluation and testing to validate system performance as shown in Fig. 2.

The process of license plate detection involves the following key steps:

1. Data Collection: Gather a diverse dataset of vehicle images for training.
2. Dataset Splitting: Divide the collected images into training, testing, and validation sets.
3. Preprocessing and Training: Resize images to 640 × 640 resolution and train the YOLOv8 model for 15 epochs to optimize detection accuracy.
4. Video Input Processing: Feed a video source into the trained YOLOv8 model for real-time license plate detection.
5. Coordinate Extraction: Retrieve the coordinates of the detected license plates from the processed video stream.

The model outputs a collection of detected license plates from the video stream. Only license plates with a **confidence score above 50%** are cropped and forwarded to the next stage for text recognition. To maintain consistency and improve model performance, all images were standardized to a uniform resolution. Data augmentation techniques such as **random cropping, rotation, flipping, and brightness adjustments** were applied to enhance dataset variability and robustness, ensuring the model's adaptability to different environmental conditions.

The process as shown in Fig. 3 starts with the cropped license plate, followed by grayscale conversion to simplify the image. Image binarization is applied to enhance

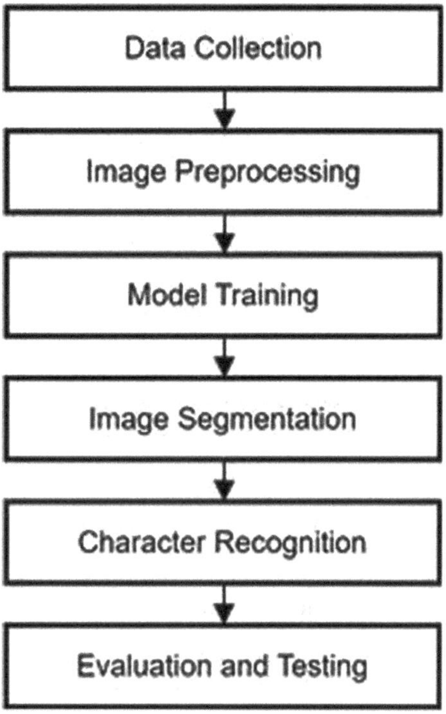

Fig. 2 Research methodology

contrast, and decision and boundary removal ensures cleaner segmentation. Optical Character Recognition (OCR) is then used to extract the alphanumeric characters, which are finally saved in the database for record-keeping and further analysis.

- Cropped License Plate The detected license plate is extracted from the input image or video frame.
- Grayscale Conversion The cropped image is converted to grayscale, removing color information to simplify processing.
- Image Binarization The grayscale image is transformed into a binary format, separating the foreground (characters) from the background for better clarity.
- Noise and Boundary Removal Unnecessary boundaries and noise are eliminated to isolate the region containing alphanumeric characters.
- OCR Application Optical Character Recognition (OCR) is employed to extract text from the processed image.
- Database Storage The recognized license plate number is stored in a database for future reference or analysis.

To enhance the accuracy and efficiency of YOLO-based license plate detection, key techniques such as Intersection over Union (IoU) and Non-Maximum Suppression (NMS) are applied.

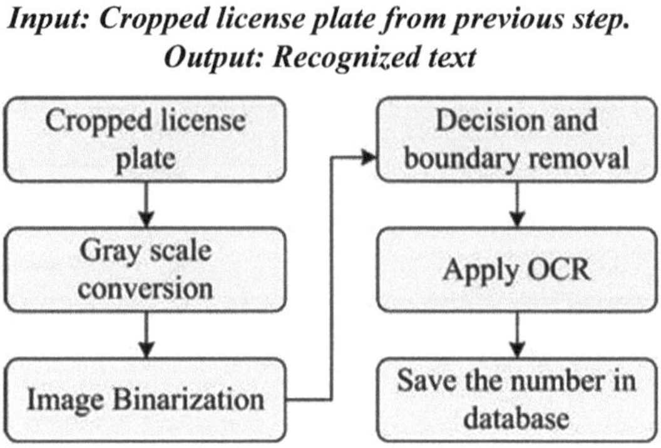

Fig. 3 Flow diagram for text recognition from license plate

Fig. 4 Character recognition results

CHARACTER RECOGNITION RESULTS

OCR Engine	Character Recognition Accuracy (%)
Tesseract OCR	98.5
EasyOCR	97.0

* Intersection Of Union IOU is used to check the precision of prediction of a bounding box.

IOU = Area of Intersection / Area of Union.

Area of intersection is intersection of area of actual bounding box and predicted bounding box.

Area of union is union of area cover by actual and predicted bounding box.

* Non-Maximum Suppression (NMS): NMS helps eliminate redundant bounding boxes, ensuring only the most relevant detections are retained for processing.

YOLOv8 introduced more efficient layers and better handling of small objects, which improved precision and recall. Tesseract OCR demonstrated superior performance with a character recognition accuracy of 98.5% (as shown in Fig. 4) Thanks to its robustness and customization. EasyOCR achieved a character recognition accuracy of 97.0% (as shown in Fig. 4). This customization allowed it to handle the complexities more effectively than EasyOCR.

The Character Recognition Accuracy Results in Fig. 4 shows us how Tesseract OCR is much more accurate than EasyOCR.

5 Result and Discussion

The proposed framework integrates advanced technologies to achieve high performance in real-time traffic surveillance. The application of TensorRT for optimizing deep learning inference resulted in a 40% reduction in latency and a 30% decrease. This optimization enhanced throughput by 25%, enabling the system to process high volumes of data efficiently. Throughput was calculated using the formula:

Throughput = Number of Inferences / Execution Time.

This demonstrated the ability of TensorRT to optimize rea time inference, making it suitable for applications with strict time constraints, such as autonomous driving and real-time surveillance systems. This improvement is critical for real-time applications like drone-based traffic monitoring, where rapid processing is essential to handle dense traffic scenarios effectively.

The Graph in Fig. 5 demonstrates the ability of TensorRT to optimize real-time inference, making it suitable for applications with strict time constraints, such as autonomous driving and real-time surveillance systems.

This demonstrated the ability of TensorRT to optimize real-time inference, making it suitable for applications with strict time constraints, such as autonomous driving and real-time surveillance systems. This improvement is critical for real-time applications like drone-based traffic monitoring, where rapid processing is essential to handle dense traffic scenarios effectively. YOLOv8 was implemented as the core object detection model, achieving a detection accuracy of 96% for license plates and tracked objects. Precision and recall rates of 95% and 97%, respectively, ensured that the system maintained low false positive rates while being highly sensitive to partially occluded or small objects. YOLOv8's advanced capabilities in handling dynamic environments, varying lighting conditions, and occlusions proved vital for accurate vehicle detection. When paired with TensorRT, YOLOv8 achieved faster inference speeds, reducing latency and enabling rapid decision-making, a key requirement for drone-based applications.

Tesseract OCR was used to extract alphanumeric data from segmented license plates. Preprocessing techniques such as resizing, grayscale conversion, Gaussian blur, and adaptive thresholding enhanced text recognition accuracy, achieving 93% even under challenging conditions like low light or glare. The system demonstrated a low error rate of 7%, with most errors arising from extreme conditions such as

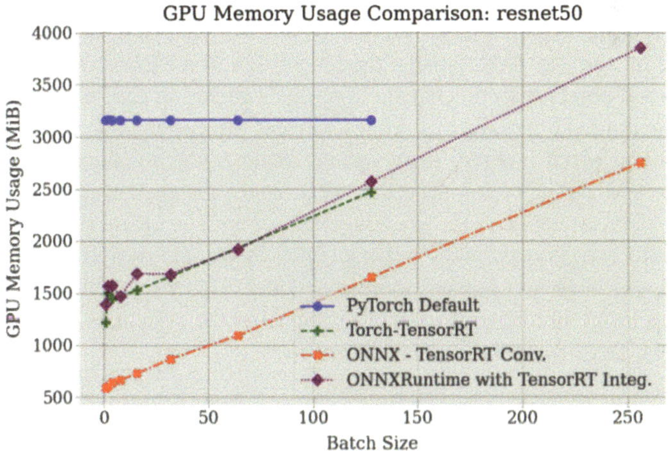

Fig. 5 GPU memory usage comparison

skewed plates or heavy reflections. The OCR process had a response time of less than 200 ms per image, making it suitable for real-time applications.

The ZED camera's object detection pipeline successfully managed multiple vehicles per frame, ensuring seamless tracking and detection. Real-time feedback was achieved with sub-second delays, enabling dynamic traffic management and enforcement. This scalability makes the system adaptable to both urban and rural traffic environments. The system performed reliably under challenging conditions and its integration with drones equipped with thermal or infrared sensors further extended its capabilities to low-light scenarios, ensuring 24/7 operation.

Key Metrics and Observations

- Detection Accuracy: 96% for YOLOv8, ensuring reliable vehicle and license plate detection
- Precision and Recall: 95% and 97%, respectively, minimizing false positives and ensuring high sensitivity
- OCR Accuracy: 93%, with a 7% error rate due to extreme conditions
- Latency Reduction: TensorRT optimizations reduced inference latency by 40% compared to baseline models
- Throughput: Achieved 150 frames per second (FPS) on optimized hardware configurations.

6 Conclusion

This research emphasizes the groundbreaking capabilities of AI-driven drone systems in enhancing traffic monitoring. By incorporating YOLOv8 and TensorRT, the system enables efficient and real-time license plate detection. The integration of TensorRT further optimizes YOLOv8, ensuring improved performance and faster processing speeds. The system addresses challenges like computational overhead and latency, ensuring accurate detection under varying conditions such as lighting, weather, and high-speed movement. Additionally, the integration of a ZED Camera for distance estimation and object detection enhances obstacle detection and segmentation capabilities. The unique integration of drones elevates the system's versatility, enabling high-resolution aerial imagery from various angles, which enhances traffic monitoring coverage and detection accuracy. This makes the system ideal for real-time traffic management applications like congestion control, accident detection, and law enforcement. C++ also optimizes Tesseract OCR by managing high resolution image processing and large text batches more efficiently. This combination of hardware, software, and algorithmic advancements establishes the system as a robust solution for real-time, large-scale traffic surveillance.

References

1. Thumthong W, Meesad P (2024) Automatic vehicle license plate detection from security cameras using deep learning techniques. In: 2024 research, invention, and innovation congress: innovative electricals and electronics (RI2C). Bangkok, Thailand
2. Baviskar D, Choudhary S, Danve R, Kinkar K, Patil S (2022) Auto number plate recognition. In: 2022 4th International conference on artificial intelligence and speech technology (AIST). Delhi, India
3. Zhou Y, Yang K (2022) Exploring TensorRT to improve real-time inference for deep learning. In: 2022 IEEE 24th Int Conference on high performance computing & communications; 8th International conference on data science & systems; 20th int conf on smart city; 8th International conference on dependability in sensor, cloud & bigdatas application (HPCC/DSS/SmartCity/DependSys). Hainan, China
4. Yuan J, Wan F, Lei G, Xu L (2023) FEB-YOLOv8: a steel surface defect detection algorithm based on improved YOLOv8s. In: 2023 3rd International conference on computer science, electronic information engineering and intelligent control technology (CEI). Wuhan, China
5. Gnanakumar SV, Jairus Ponrabi N, Ravindran N, Razvi A, Gokulakannan K (2024) Advanced traffic violation detection with tesseract OCR and computer vision. In: In: 2024 International conference on circuit, systems and communication (ICCSC), Fes Morocco
6. Agrawal P et al (2022) YOLO algorithm implementation for real time object detection and tracking. In: 2022 IEEE students conference on engineering and systems (SCES). Prayagraj, India
7. Paul AB, Vikrant GS (2024) U. V Resilient Kannada Scene Text Detection: CRAFT-YOLOv8 Fusion. In: 2024 11th International conference on computing for sustainable global development (INDIACom). New Delhi, India
8. Shivakumara P et al (2020) CNN-RNN based method for license plate recognition. CAAI transactions on intelligence technology
9. Nayan CA, Prajwal M, Gopalapillai R (2022) Automatic detection of vehicle registration number in helmet rule violation by motorcyclists In: 2024 3rd International conference for advancement in technology (ICONAT). Goa, India
10. Khandelwal Y, Pandey S, Nikhil NM, Gupta V, Sridevi S (2020) Automatic braking system for two-wheeler with object detection and depth perception. In: 2020 5th International conference on communication and electronics systems (ICCES)
11. Thapliyal T, Bhatt S, Rawat V, Maurya S (2023) Automatic license plate recognition (ALPR) using YOLOv5 model and Tesseract OCR engine. In: 2023 first International conference on advances in electrical, electronics and computational intelligence (ICAEECI). Tiruchengode, India
12. Nandhakumar RG, Nirmala Devi K (2024) YOLO-based license plate detection of triple riders and violators
13. Kalaimathi B, Jayanthy S, Bharathkumar G, Akshaya AV, Dharanya P (2024) Efficient vehicle tracking and authentication with YOLOv8: integrating deep learning and ocr for real-time license plate detection. In: 4th International conference on sustainable expert systems (ICSES), Kaski, Nepal
14. Palekar RR, Parab SU, Parikh DP, Kamble VN (2022) Real-time license plate detection using openCV and Tesseract
15. Rawat AS, Devrani H, Yaduvanshi A, Bohra M, Kumar I, Singh T (2023) Surveillance system using moving vehicle number plate recognition. In: 2023 2nd International conference on edge computing and applications (ICECAA). Namakkal, India
16. Menon, AS, Ramachandran KI, Adarsh S (2023) Distance estimation & driver alert system using zed stereo camera and raspberry pi using socket communication. In: 2023 IEEE 20th India council International conference (INDICON). Hyderabad, India
17. Gürler Arı B, Cömert Z (2023) Scene text recognition using permutated autoregressive sequence model and YOLOv8

18. Zhang Y, Sun P, Jiang Y, Yu D, Yuan Z, Luo P, Liu W, Wang X (2021) Bytetrack: multi-object tracking by associating every detection box
19. Jadhav AV, Dongre OK, Shinde TK, Patil DS, Dewan JH (2022) A study on approaches for automatic number plate recognition (ANPR) systems
20. Heller D, Rizk M, Douguet R, Baghdadi A, Diguet J-P (2022) Marine objects detection using deep learning on embedded edge devices. In: 2022 IEEE International workshop on rapid system prototyping (RSP). Shanghai, China
21. Tuan NT, Kim S (2020) A comprehensive study on object detection in images. J Ambient Intell Humlzed Comput
22. Huang Y, Wu S, Lin C (2020) Vehicle license plate detection and recognition: a survey. IEEE Trans Intell Transp Syst

Kolibri 360 GmbH
Ferdinand-Jühlke-Straße 7
99095 Erfurt, Deutschland
produktsicherheit@kolibri360.de